Epigenetics, the Environment, and Children's Health Across Lifespans

David Hollar
Editor

Epigenetics, the Environment, and Children's Health Across Lifespans

Springer

Editor
David Hollar
Pfeiffer University
Morrisville, NC, USA

ISBN 978-3-319-25323-7 ISBN 978-3-319-25325-1 (eBook)
DOI 10.1007/978-3-319-25325-1

Library of Congress Control Number: 2015958885

Springer Cham Heidelberg New York Dordrecht London

Printed on acid-free paper

Springer International Publishing AG Switzerland is part of Springer Science+Business Media (www.springer.com)

For all of our families, past and future.

Preface

Epigenetics has emerged as an important science with respect to health outcomes following several intriguing studies during the past decade. Epigenetics represents the study of gene regulatory factors that direct cell fates within the organism and that can be inherited via the germ line independent of Mendelian inheritance. It represents the next step beyond the Human Genome Project. Now that we can sequence genomes, we need to understand how genes are controlled in development. This will be a daunting task given the relationship of gene function to morphology and overall functioning; and gene–environment relationships in health, aging, and disease.

This work relates directly to our earlier Springer *Handbook of Children with Special Health Care Needs* (2012), edited by Dr. David Hollar, as well as to two unique Springer titles on epigenetics: *Epigenetics of Aging* (2010), edited by Dr. Trygve O. Tollefsbol, and *Environmental Epigenomics in Health and Disease* (2013), edited by Drs. Randy L. Jirtle and Frederick L. Tyson. The chapters in this book build upon and synergize these topics along the continuum of life from conception through aging and across generations via germ-line epigenetic inheritance. We also bring the topic in historical, biological, and health policy perspectives to public health professionals. Epigenetics has received increasing interest in public health, as evidenced by the October 2013 special issue of the *American Journal of Public Health* and conference sessions at the 141st and 142nd (i.e., 2013 and 2014) annual meetings of the American Public Health Association.

In this volume, 26 international experts and public health researchers have prepared chapters describing epigenetic mechanisms; the impact of epigenetics on prenatal, child, and lifespan development; transgenerational epigenetic inheritance; and associations between epigenetic reprogramming and morbidity/mortality. The purpose of the book is to provide health researchers, clinicians, policy developers, and consumer advocates with a resource to inform their own work and to benefit epigenetic health worldwide. The topics are by no means exhaustive, but they provide relevant research findings and applications to public health and add to the growing epigenetic research literature. The chapters are provided in no particular order of importance, but I have arranged them beginning with an overview, mechanisms, health risks, policy implications, and potential future research, although each author addresses specific topics related to their expertise.

Dr. David Hollar provides historical context, developmental biology research, general mechanisms, and applications to maternal and child health in Chap. 1. In Chap. 2, Drs. Milena Georgieva, Dessislava Staneva, and George Miloshev describe specific epigenetic molecular mechanisms and provide a fascinating model of epigenetic aging across the life span.

Dr. Marija Kundakovic follows in Chap. 3 with epigenetic mechanisms and how exposure to industrial toxins, specifically bisphenol A, impacts epigenetic reprogramming that impacts human health.

Drs. Emmy Rogakou, Vassilios Papadakis, and George Chrousos analyze the role of histones in epigenetic chromatin remodeling in Chap. 4, and they focus on the histone γH2AX as a unique biomarker of epigenetic change. In Chap. 5, Drs. Sripriya Sundararajan and Cynthia Bearer explain how environmental exposures impact epigenetic programming during prenatal development, and they provide recommendations for preterm neonatal care in the neonatal intensive care unit (NICU). Dr. Xinyin Jiang provides a thorough description (Chap. 6) of research that demonstrates the impact of maternal and child nutrition on epigenetic reprogramming and health outcomes.

Moving to child and adult lifespan environmental exposures, Dr. John Kall, Amanda Just, and Dr. Michael Aschner compare and contrast (Chap. 7) the dental research literature and make a compelling case that mercury amalgam might promote serious health problems and disease via epigenetic mechanisms. In Chap. 8, Drs. Chris Murgatroyd and Steven Bradburn describe how translational animal model research has informed our knowledge of epigenetic mechanisms in disease processes. Likewise, Dr. Ping Hu discusses how pluripotent stem cells (Chap. 9) are regulated to be static or to differentiate.

Drs. Jay Schneider and Deborah Cory-Slechta (Chap. 10) describe lifespan neurodevelopmental effects of prenatal lead, stress, and combined exposures. In Chap. 11, Jennifer S. Lewis provides a review of neurodevelopment and then relates epigenetics and physiology to post-traumatic stress disorder (PTSD).

Dr. Steven Gilbert addresses the various individual and social levels of ethical responsibility given the epigenetic effects from environmental toxins (Chap. 12). Dr. Caroline Hohensee, Tricia Varela, and Dustin Harris explore research indicating possible relationships between epigenetics and child obesity (Chap. 13), and Dr. Lisa Melvin examines (Chap. 14) similar potential relationships for child exposure to alcohol, tobacco, and other drugs (ATOD). Dr. Ankita Das explains epigenetic memory and plasticity in embryonic development (Chap. 15).

Dr. David Hollar concludes the book with Chap. 16, exploring recent theoretical work linking genomic molecular instability to cancer, aging, and diseases linked to epigenetic reprogramming. This includes discussion of targeted epigenetic reprogramming to potentially reverse negative epigenetic regulation at the gene system, cell, tissue, and organism levels. It also suggests a mathematical measure of epigenomic change using nonlinear dynamics parameters.

The authors thank their families and colleagues. I thank the authors for their expertise and teamwork in the completion of this project. I thank Springer editors Janet Kim, Khristine Queja, Bill Tucker, and Christina Tuballes for facilitating the development and publication of the book, plus Saswat Mishra and Deepthi Vasudevan for editing and proofing. I thank my family, Brooke, Paige, Virginia Dean, Beverly and Edward Merritt; Dr. William Virtue; and Drs. Vernease Miller, Barnett Parker, and my many colleagues at Pfeiffer University for their faith, interest, and support for this project.

Misenheimer, NC David Hollar
July 31, 2015

Contents

About the Authors

Michael Aschner, Ph.D. serves as the Harold and Muriel Block Chair in Molecular Pharmacology at Albert Einstein College of Medicine. He served on numerous toxicology panels (Institute of Medicine, US Environmental Protection Agency, Centers for Disease Control and Prevention) and is a member of the Neurotoxicology and Alcohol study section (NIH). Research in his lab focuses on the following topics: (1) Modulation of *C. elegans* genes (aat, skn-1, daf-16) that are homologous to mammalian regulators of MeHg uptake and cellular resistance will modify dopaminergic neurodegeneration in response to MeHg exposure. (2) Under conditions of MeHg-induced oxidative stress, Nrf2 (a master regulator of antioxidant responses) coordinates the upregulation of cytoprotective genes that combat MeHg-induced oxidative injury, and the genetic and biochemical changes that negatively impact upon Nrf2 function increase MeHg's neurotoxicity. (3) PARK2, a strong PD genetic risk factor, alters neuronal vulnerability to modifiers of cellular Mn status, particularly at the level of mitochondrial dysfunction and oxidative stress. Our studies are designed to (1) shed novel mechanistic insight into metal-induced neurodegeneration, (2) identify targets for genetic or pharmacologic modulation of neurodegenerative disorders, (3) increase knowledge of the pathway involved in oxidative stress, and (4) develop improved research models for human disease using knowledge of environmental sciences.

Cynthia F. Bearer, M.D., Ph.D. is the Mary Gray Cobey Professor of Neonatology with tenure, Chief of the Division of Neonatology, and Associate Chair for Research in the Department of Pediatrics, University of Maryland School of Medicine. Dr. Bearer is board certified in pediatrics and neonatal–perinatal medicine. She received her Ph.D. in biochemistry from Case Western Reserve University and her M.D. from the Johns Hopkins University. Dr. Bearer has published extensively on fetal and pediatric environmental health. She has been funded by NIH, CDC, and US EPA. Dr. Bearer is Chair of the Board of the Children's Environmental Health Network and is a member of the American Pediatrics Society (APS) and the Society for Pediatric Research. She currently serves as the APS representative to the Pediatric Policy Council. She has served on the Committee to Evaluate Children's

Health of the National Academy of Science and the editorial board of *Neurotoxicology* and on multiple study sections at the NIH. She is past President of the Fetal Alcohol Syndrome Study Group. She has been a member of the Scientific Advisory Board for the US EPA and the Advisory Group to the Director of the National Center for Environmental Health at the CDC. She was recently awarded a Smith College Medal. Her major research interests are the effect of the environment and environmental chemicals on the development of the central nervous system.

Steven Bradburn is a Ph.D. postgraduate student in the School of Healthcare Science at Manchester Metropolitan University. His main research interests include how inflammatory mediators can influence cognitive performance in the elderly (human and rodent models) and, additionally, how diet and exercise individually, as well as interactively, affect the molecular control of these inflammatory proteins and their relationship with neuropsychological behavior.

George P. Chrousos, M.D. is Professor and Chairman of the First Department of Pediatrics at the University of Athens, School of Medicine, Athens, Greece, and former Chief of the Pediatric and Reproductive Endocrinology Branch of the National Institute of Child Health and Human Development, National Institutes of Health, Bethesda, Maryland. Dr. Chrousos pioneered studies that elucidated the effects of stress on the organism at the behavioral, neuroendocrine, cellular, and molecular levels and made fundamental contributions to the understanding, diagnosis, and treatment of pituitary, adrenal, and stress-related pathologies, i.e., major depression, obesity/metabolic syndrome, and autoimmune/inflammatory, reproductive, and sleep disorders. He made seminal observations in the glucocorticoid signaling system and deciphered some of its key clinical implications. Dr. Chrousos is universally regarded as one of the most prominent endocrinologists. His work has been cited over 100,000 times (H-index >160), making him one of the most cited physician-scientists in both clinical medicine and biology and biochemistry and the top-cited endocrinologist in the world. He has received numerous major awards and is a member of the Academia Europaea and the Institute of Medicine, National Academy of Sciences.

Deborah A. Cory-Slechta, Ph.D. is a Professor of Environmental Medicine, Pediatrics and Public Health Sciences at the University of Rochester Medical School, whose research has focused largely on developmental neurotoxicology. This work has included the effects of developmental exposures to metals, pesticides, and air pollutants as well as combined exposures to metals and stress in animal models and human cohort studies. She previously served as Dean for Research, Chair of the Department of Environmental Medicine, and Director of the NIEHS Environmental Health Sciences Center at the University of Rochester Medical School and as Director of the Environmental and Occupational Health Sciences Institute of Rutgers University. Dr. Cory-Slechta has served on advisory panels of the NIH, the FDA, the Environmental Protection Agency, the National Academy of Sciences, the Institute of Medicine, and the Agency for Toxic Substances and

Disease Registry and on the editorial boards of the journals *Neurotoxicology*, *Toxicology*, *Toxicological Sciences*, *Toxicology and Applied Pharmacology*, and *Neurotoxicology and Teratology*.

Ankita Das, Ph.D. is a research fellow with the Carnegie Institution for Science. Her current research interests lie in studying epigenetic mechanisms, including chromatin remodeling and posttranscriptional RNA processing to understand cell lineage mechanisms in an early embryo. To address these questions, she uses the mouse model as well as human embryonic stem cell-derived neural crest stem cells. By studying the embryonic neural crest cells, she hopes to shine light on common and divergent epigenetic themes that regulate embryonic and adult stem cell plasticity. In the past, she has used zebrafish genetics to understand mechanisms of early lineage specification mechanisms.

Milena Georgieva, Ph.D. holds a Ph.D. in Molecular Biology and is an Assistant Professor at the Institute of Molecular Biology, Bulgarian Academy of Sciences. Her main interests are in the field of epigenetics, chromatin biology, and molecular genetics. Her research endeavors are aimed at elucidating the roles of linker histones in chromatin dynamics and structure maintenance during normal growth, development, and aging. Together with her colleagues, she succeeded in proving the significance of the linker histones for the maintenance of the higher-order chromatin structure organization and the importance of these upper levels of chromatin compaction for the propagation of normal cellular life and for the feasibility of cells toward stress. Furthermore, Dr. Georgieva has taken an active role in the development of new software for discrimination among different types of DNA damage on the basis of the method of Comet Assay. By the help of this software, one can easily predict the type of DNA lesions any studied genotoxin introduces in the DNA. She is currently involved in research of the epigenetic role of linker histones and chromatin remodeling in aging, as well as in detailed study of the interactions among linker histones and chromatin remodeling complexes and their significance for cellular fate. Milena is involved in popularizing science in mass media by different initiatives, some of which include writing articles in popular science journals and appearances and interviews in TV and radio programs.

Steven G. Gilbert, Ph.D., D.A.B.T. is Director and Founder of the Institute of Neurotoxicology and Neurological Disorders (INND), has a Ph.D. in toxicology, and is a Diplomat of American Board of Toxicology. He is an Affiliate Professor in the Department of Environmental and Occupational Health Sciences, University of Washington. Dr. Gilbert's research has focused on neurobehavioral effects of low-level exposure to lead and mercury on the developing nervous system. His book *A Small Dose of Toxicology: The Health Effects of Common Chemicals* was published in 2004, and the second edition is available for free as an E-book (www.asmalldoseof.org) by Healthy World Press. The book was recently translated into and published in Chinese (www.chinesesmalldose.org) and is currently being translated into German and Arabic. He started the Wiki-based web site Toxipedia (www.toxipedia.

org), which includes a suite of sites that put scientific information in the context of history, society, and culture. The last decade of his work has been driven by interest in our ethical responsibility to ensure that children have an environment in which they can reach and maintain their full potential. He believes that we have adequate knowledge, but the challenge is to use this knowledge in an ethical, responsible way that results in decisions that protect current and future generations.

Dustin Harris, B.S. is originally from Mt. Airy, NC, sometimes better known as Mayberry. Dustin graduated with honors from Queens University of Charlotte and received a Bachelor of Science in Biology. Currently, he works for HireRight in Charlotte, where he is the Supervisor of Operations and Vendor Relations, and he oversees a team responsible for maintaining relations and negotiating pricing with occupational health vendors across the world. Dustin is currently enrolled at Pfeiffer University in Charlotte, NC, and is studying toward his Masters in Health Administration and Masters in Business Administration. Dustin is an alumnus of the Phi Kappa Sigma fraternity and he lives in Charlotte, NC, with his fiancé, Jen.

Caroline Hohensee, Ph.D., M.H.A. is an Assistant Professor at Pfeiffer University in Charlotte, North Carolina. She received her Ph.D. in Health Services Research at the University of North Carolina at Charlotte (UNCC) in 2010. She is passionate about child obesity research, though she has also been involved in qualitative and quantitative research studies in several other areas of healthcare, including cancer studies and gerontology. She was born and raised in Los Angeles, California, and attended Loyola Marymount University (B.S. Biology) and the University of Southern California (Master in Health Administration and M.S. Gerontology). While living in Essen, Germany, for 5 years, she studied the language and worked in hospital management. She currently lives in Charlotte, North Carolina, with her husband and three children.

David W. Hollar, Ph.D., M.S. is an Associate Professor of Health Administration at Pfeiffer University. He received his Ph.D. in Curriculum and Teaching from the University of North Carolina at Greensboro, where he was awarded the graduate school's Outstanding Dissertation Award. He has B.S. and M.S. degrees in Biology. He successfully completed postdoctoral research in community health at the NIDRR-funded Rehabilitation Research and Training Center on Substance Abuse and Employment at Wright State University. He also has a graduate certificate in Public Health Entrepreneurship. His specialties include multivariate statistics, structural equation models, mathematical models, disability policy, and decision-making. He has numerous peer-reviewed publications on health risk factors, allostatic load, behavioral genetics, and disability policy. He edited and coauthored the *Handbook of Children with Special Health Care Needs*, published by Springer in 2012. He has served on the editorial board of the *Maternal and Child Health Journal* since 2005. He has been a member of the American Public Health Association since 2002, and he is a member of the American Association on Health and Disability. With his wife and daughter, he believes in serving God through servant leadership in the community with the Billy Graham Evangelistic Association.

Ping Hu, Ph.D. was educated in Peking University, China, and obtained her Ph.D. from State University of New York, Stony Brook, USA. She performed her research in Cold Spring Harbor Laboratory and the University of California, Berkeley, USA. Since 2011, she has been Principal Investigator in Shanghai Institute of Biochemistry and Cell Biology. She has been focused on the research on epigenetic regulation during development, especially muscle development, for a decade.

Xinyin Jiang, Ph.D., R.D. is an Assistant Professor in the Department of Health and Nutrition Sciences at Brooklyn College of the City University of New York. Her research focuses on delineating the relationship between early nutrition exposures and health. She is especially interested in the role of methyl nutrients, e.g., folate, vitamin B_{12}, choline, and betaine, in epigenetic regulation and how these nutrients mediate epigenetic changes in fetal tissues that lead to lasting functional alterations.

Amanda Just, M.S. is a freelance writer who has been researching the impact of dental amalgam mercury fillings since 2005. Her work has included reading over 1000 scientific articles about health conditions potentially related to toxic exposures, and she has represented the dental mercury issue from a consumer's perspective at meetings of the United Nations Environment Programme, the US Department of State, and the US Food and Drug Administration. She has worked with the International Academy of Oral Medicine and Toxicology since 2010 and holds a Master of Science in Education from the University of New Haven and a Bachelor of Arts in History from the College of William and Mary.

John Kall, D.M.D., F.A.G.D. received his Doctorate in Dental Medicine (D.M.D.) degree in 1977 from the University of Louisville School of Dentistry. For the next 25 years, he served as Dental Director of the Family Health Centers, Inc., a federally qualified community health center in Louisville, KY. Additionally, he founded Dental Health Center, his private practice focusing on biological, biocompatible dentistry. In 2012 upon invitation of the Philippines's Secretary of Health, he gave the presentation "Mercury-free Dentistry in Public Health" at the national conference: *Philippines—Towards Mercury-Free Dentistry*. In 2013, he coauthored "International Academy of Oral Medicine & Toxicology (IAOMT) Position Statement against dental mercury amalgam fillings for medical and dental practitioners, dental students and patients." Dr. Kall is a Fellow of the Academy of General Dentistry and a past President of the Kentucky chapter. He is a Master of the International Academy of Oral Medicine and Toxicology (IAOMT) and currently serves as the Chairman of its Board of Directors.

Marija Kundakovic, Ph.D. is an Assistant Professor in the Department of Biological Sciences at Fordham University. She is an expert in the area of behavioral and psychiatric epigenetics. She earned a Ph.D. degree in Biochemistry and Molecular Genetics from the University of Illinois in Chicago, where she did pioneering work on the epigenetics of schizophrenia. Dr. Kundakovic's postdoctoral research at Columbia University established the epigenetic mechanisms through which early-life adversity exerts sex-specific, long-term effects on brain function and behavior and revealed a candidate biomarker for the early detection of psychopathology. During her previous work at the Icahn School of Medicine at Mount Sinai, Dr. Kundakovic led a large-scale, cutting-edge epigenomic profiling of postmortem brain specimens and peripheral blood samples of control subjects, schizophrenia patients, and individuals exposed to early-life malnutrition. She has published more than 20 articles in high-impact scientific journals and 4 book chapters pertaining to the area of behavioral and psychiatric epigenetics. The main goal of Dr. Kundakovic's research is to facilitate the implementation of epigenetic findings into clinical practice and help tailor novel diagnostic, preventive, and pharmacological approaches in the area of mental health.

Jennifer S. Lewis, M.H.A., M.Ed. is a health scientist whose work focuses on health disparities, electronic health records, and facilitators/barriers to health. During the past 15 years, her work has included higher education, insurance, legal, healthcare education, and technology, and within each such area, she has served in both a training and research role. Currently, she serves in the dual role of Business Analyst and Project Manager with the University of Southern Mississippi. At Pfeiffer University with Dr. Hollar, she published research on the National Health and Nutrition Examination database to demonstrate the utility of heart age differential measures as applied to American Heart Association cardiac assessments and to show heightened risk for such differentials among persons with mobility functional limitations and disabilities. She has presented at several national conferences, including the 2014 Yale University Unite for Sight Global Health and Innovation Conference and the 2014 National Institutes of Health Minority Health and Health Disparities Conference.

Elisa Melvin, Ph.D., M.Ed. is an Assistant Professor of Health Administration at Pfeiffer University's Graduate School located in Charlotte, North Carolina. She completed her Ph.D. in Health Services Policy and Management at the University of South Carolina's Arnold School of Public Health in Columbia, South Carolina. Her background includes outcomes-oriented research in diabetes; education, motivational speaking, train-the-trainer evaluations, and educational effectiveness; and pharmaceutical and educational consulting. Her research interests include health economics and outcomes research, diabetes, and health disparities. She also advocates for literacy and economic equities for underserved communities. In addition, upon completing her doctorate, she successfully completed the University of Washington's Health Economics and Outcomes Research certificate program. Dr. Melvin currently resides in Columbia, SC, with her husband, Dr. André Melvin; they have two sons. She wants everyone to know that she gives God all of the glory for everything that has occurred in her life. Her goal is to serve others while only glorifying Him in the process. With a passion to inspire and encourage generations to live in their best, her mantra is to serve beyond one's reserve.

George Miloshev, Ph.D. is an Associate Professor of Molecular Genetics at the Institute of Molecular Biology, Bulgarian Academy of Sciences, Sofia, Bulgaria. His work is centered in the organization of DNA in chromatin. Although it has been known for a long time that chromatin in the eukaryotic nucleus is organized in the form of loops of different sizes, nothing about this level of organization has been known in detail. By developing a new method for chromatin research, Miloshev's team has been able to show that generally chromatin loop sizes in different model organisms, including yeasts, mammalian cells, etc., gravitate around certain median dimensions. These findings proved beyond doubt the great functional significance of these chromatin loops and the upper levels of chromatin organization for proper genome functioning. It was shown by the same group that in the building and maintenance of the higher-order chromatin structure and particularly of that of chromatin loops, an important role is played by the linker histone H1. Prof. Miloshev is also an active communicator of science in different types of media and is a freelance writer in popular science journals in Bulgaria. He is also a lecturer at Sofia University.

Chris Murgatroyd, Ph.D. is a Senior Lecturer in the School of Healthcare Science at Manchester Metropolitan University and researches gene–environment interactions in rodent models and human studies of early-life adversity, studying rat models of maternal separation, maternal care and aging, and childhood cohorts differing in pre- and postnatal environments.

Vassilios Papadakis, M.D., Ph.D. is a Greek native and graduate of the University of Athens. He completed residency in Pediatrics at the Columbia College of Physicians and Surgeons campus and Pediatric Hematology Oncology Fellowship at the Memorial Sloan-Kettering Cancer Center/NYH Cornell Medical Center in New York City, and then he sub-specialized in stem cell transplantation. He is Director in Pediatric Hematology–Oncology at the Agia Sofia Children's Hospital in Athens, following relocation in Greece for the past 20 years. He is board certified in the USA and holds Greek Boards in Pediatrics and Hematology. He is active in the field of leukemias, lymphomas, neuroblastoma, and histiocytoses and serves on the Board of Hellenic and International Scientific Societies, while he is Principal Investigator in international neuroblastoma studies (SIOPEN). He has authored several papers and book chapters in the field, both in Greek and English.

Emmy P. Rogakou, Ph.D. is a Senior Researcher of Epigenetics and Molecular Medicine at the First Department of Pediatrics, the University of Athens School of Medicine, Athens, Greece. Dr. Rogakou has pioneered in studies regarding the discovery of the role of the histone phosphorylation γH2AX in DNA damage response. These studies have established for the first time that epigenetics are implicated to signal transduction, apart from transcription. Today, γH2AX has been established as a specific and most sensitive epigenetic biomarker to detect DNA double-strand breaks and is in use extensively in clinical trials. Dr. Rogakou now focuses her research on novel cell processes and mechanisms where γH2AX is involved. In addition, she is developing cellular and molecular epigenetic markers to make fundamental contributions to diagnosis and treatment of cancer and other diseases. Dr. Rogakou's work has been cited over 10,000 times, making her one of the most cited among women scientists.

Jay S. Schneider, Ph.D. has over 30 years experience as a neuroscientist. He has a long-standing interest in neurotoxicology, with much of his work related either to Parkinson's disease or heavy metal neurotoxicity. His work in developmental toxicology has dealt mainly with studying effects of developmental lead exposure on behavioral and molecular biological outcomes. In particular, his lab has investigated the behavioral, molecular, and epigenetic responses of the developing brain to lead exposures in males and females during different developmental periods and how these exposures impact adult behavior, cognition, gene, and protein expression.

Dessislava Staneva, Ph.D. is an Assistant Professor of Molecular Genetics at the Institute of Molecular Biology, Bulgarian Academy of Sciences, Sofia, Bulgaria. Her work involves development of different yeasts as model systems for investigation of general cellular and molecular processes. Using the convenience of the yeast *Saccharomyces cerevisiae* as a model system, she has evidenced the physical contacts between the yeast linker histone and Arp4p, a subunit of three chromatin-modifying complexes in yeast, namely, INO80, SWR1, and NuA4, by the use of the yeast two-hybrid system. These results shed light on the way by which chromatin-remodeling complexes slide along the chromatin fiber and further proved the significance of linker histones in these dynamic processes. Moreover, she has successfully proved that the yeast *Kluyveromyces lactis* possesses the gene for the linker histone, though it has long been considered that these yeasts do not have linker histones. These results proved that the chromatin of this single-eukaryotic organism stands closer to the kingdom of all other eukaryotes and highlighted that chromatin and its organization are evolutionarily conserved.

Sripriya Sundararajan, M.D. is an Assistant Professor in the Division of Neonatology, Department of Pediatrics at the University of Maryland School of Medicine. Dr. Sundararajan is board certified in pediatrics and neonatal–perinatal medicine. She originally received her M.D. in Pediatrics from Chennai Medical College, India. She recertified in pediatrics and completed her fellowship training in neonatal–perinatal medicine from the University of Virginia. She is a member of the American Academy of Pediatrics, Section on Perinatal Pediatrics. Her major research interests include studying the signal transduction pathways involved in the pathology of lung and dermal fibrosis. She currently establishes evidence-based transfusion practices in the NICU at the University of Maryland Medical Center.

Tricia Varela, M.H.A. is a Director of Finance for Novant Health Medical Group in Charlotte, North Carolina. She is a Masters of Healthcare Administration candidate attending Pfeiffer University and has been working in Healthcare since 1998. She joined Novant Health in 2003 as the consolidated business office manager for Lakeside Family Physicians. After she earned her Bachelor of Arts in accounting and business management from Belmont Abbey College in 2009, Tricia took over as a Director of Operations in 2010 and then transitioned into her current role as Director of Finance in 2012. Tricia assumed the role of Novant Health Family Medicine Residency Program interim residency coordinator in 2013 and enjoys the opportunity to be involved with launching the inaugural residency program and working with the pioneers who will shape and educate our future family medicine providers. Tricia was born in the Florida Keys and has lived in Newfoundland, Canada, but she and her family have called North Carolina home since 2002. Tricia and her husband enjoy visiting their son in Sunset Beach, NC, attending Charlotte Hornets basketball games, and vacationing in the Appalachian Mountains.

Chapter 1
Epigenetics and Its Applications to Children's Health

David W. Hollar Jr.

1.1 Introduction

The science of epigenetics represents the study of inheritable changes in the regulation of DNA expression. Epigenetics is a normal process that is responsible for all human and other multicellular organismal development, from conception to cellular/tissue differentiation to aging and across generations via the germ-line cells. "Abnormal" gene regulatory signals are caused by cellular errors, physiological stress pathways, mobile genetic elements, environmental chemicals, and physical forces. Such signals lead to birth defects, morbidity, and diseases that emerge during one's lifetime or that are imprinted onto the genome to be transmitted via sperm or egg to future generations if the gene regulatory changes occur prior to gametogenesis (i.e., pre-adolescence for males, in utero for females). Consequently, epigenetics is of tremendous relevance to maternal and child health. Understanding epigenetic mechanisms has further applications across individual and multiple lifespans.

The epigenetic landscape was formulated by Conrad Waddington (1957). It compares a cell's development, determination, and differentiation to a ball rolling through an environment of hills and valleys. Each critical juncture in the landscape diverts the ball onto a different pathway, much as a cell bifurcates and commits to a specific tissue developmental pathway versus alternative pathways (Fig. 1.1). The further a cell develops and commits to a specific organismal role, it becomes more limited as it progresses/bifurcates down specific, generally irreversible pathways, in accordance with basic thermodynamic principles of systems converging to lowest equilibrium potentials (Ferrell 2012; Goldberg et al. 2007; Prigogine 1977, 2002; Thom 1972; Waddington 1957).

D.W. Hollar Jr., Ph.D., M.S. (✉)
Department of Health Administration, Pfeiffer University,
2880 Slater Road, Suite 100, Morrisville, NC, USA
e-mail: David.Hollar@pfeiffer.edu

D. Hollar (ed.), *Epigenetics, the Environment, and Children's Health Across Lifespans*, DOI 10.1007/978-3-319-25325-1_1

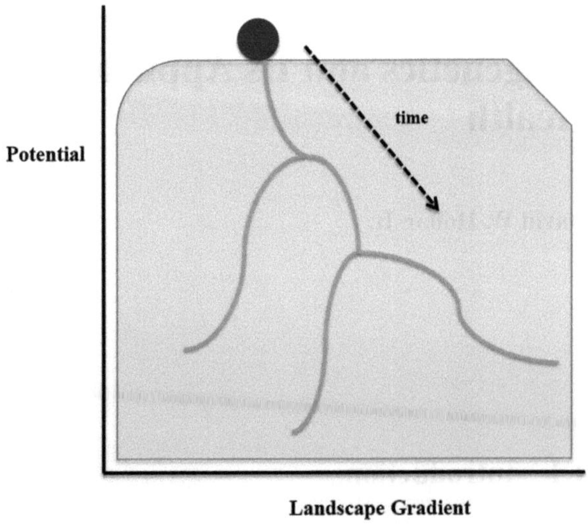

Fig. 1.1 Waddington's (1957) epigenetic landscape model

We know that cell commitments along the epigenetic landscape depend upon relatively "locked" changes in gene regulation and positional information between adjacent cells in the developing embryo, even beginning with epigenetic controls located in the maternally dominant cytoplasm of the initial zygote. The maternal epigenetic controls are growth factors and other macromolecules that establish/program the differential fates of its two daughter cells when the zygote divides and when the daughter cells and their daughter cells subsequently divide, etc., en route to generating a functionally diverse fifty-trillion-cell human body (Campbell et al. 1996; Dixon et al. 2015; Gurdon 1960, 1981; Harvey and Smith 2009; Hadorn 1965; Wolpert et al. 1998). As the early embryo grows, cells begin producing additional growth and transcriptional factors that establish further epigenetic gene regulatory pathways, thus establishing an epigenetic "memory" for each cell.

This epigenetic "memory" is remarkably stable and resilient within the dividing/differentiating cells of our bodies throughout the lifespan, even more so in the reproductive germ-line cells. Nevertheless, nuclear reprogramming of gene control regions probabilistically can lead to altered tissue phenotypes, plasias, accelerated or decelerated tissue growth, tumors and cancers, and other abnormalities (Halley-Stott and Gurdon 2013). Relaxing of epigenetic controls further contributes to aging.

In the reproductive germ line (i.e., sperm and egg), such reprogramming and error commitments to "altered programming memory" can lead to changes in gene expression among many generations of descendants in a family line. Whereas such epigenetic changes, along with genetic mutations and chromosomal rearrangements, contribute to the extensive genetic variation within and between groups of humans, the extremely complicated, subtle, and dynamic pathways of upregulated and downregulated gene expression can lead to positive, neutral, and negative health outcomes. We are concerned with epigenetic changes that produce negative health outcomes and how to counteract these changes.

1.2 Transgenerational Epigenetic Effects

The definitive study that demonstrated deleterious, transgenerational epigenetic effects was the Pembrey et al. (2006) examination of the effects of famine/food supply on 303 people as experienced by their grandparents. This retrospective study used the Avon Longitudinal Study of Parents and Children (ALSPAC) for northern Sweden. Pembrey et al. (2006) demonstrated that the grandsons of grandfathers who had experienced famine during their pre-pubertal development had significantly lower mortality risk than the grandsons of grandfathers who had experienced good nutrition during the same developmental period (i.e., gametogenesis). A similar pattern was observed for granddaughters of grandmothers during the same prepubertal period. However, the stronger converse and earlier developmental effect was significantly lower mortality for granddaughters of grandmothers who experienced good nutrition during infancy/early childhood (age 0–3 years) and in utero. There were no cross-gender, transgenerational effects (Pembrey et al. 2006).

It is important to note that the ALSPAC sample consisted of individuals born in 1890, 1905, and 1920. Consequently, these individuals and their grandparents had no exposure to pesticides and other manufactured toxins that were later invented during the mid-twentieth century. Even industrial air pollution distributions would have been limited due to the study location's proximity to the Arctic Circle. The primary, necessary and sufficient, causative agent for the variation in lower mortality risk was the same gender grandparent's exposure to famine during pre-puberty or the grandmother's (on granddaughters) good nutrition during the their own third trimester development or infancy.

Similarly, Heijmans et al. (2008) performed a comprehensive epigenetic study of individuals who were in utero during the 1944–1945 Dutch Hunger Winter toward the end of the Second World War. They found definitive proof of hypomethylation for the IGF2 gene in these individuals compared to their unexposed siblings (Heijmans et al. 2008).

Therefore, children's health epigenetically can be affected in unpredictable ways by exposure to toxic chemicals, physiological stressors such as diet/nutrition (either directly or maternally), and by these and other agents acting upon their ancestors. We argue that the pressures and pollutants generated by modern human society may be generating undesirable health effects on children. These effects can impact immediate lifespan health outcomes and disease as well as the health of future generations via the germ line. This represents a phenomenon of civilization that permeates the entire biosphere, and we all participate in these events (Carson 1962; Leopold 1949; Wilson 1984). Therefore, we seek to understand to some degree the many variables involved in epigenetic reprogramming so that society can mitigate some of the negative effects.

1.3 Public Health

Until recently, little research has examined genetic–environment interfaces with respect to public health outcomes, and almost none has addressed gene regulatory/ epigenetic effects. The National Center for Health Statistics (2012) evaluated nationally representative data measuring focal areas and objectives for health during the Healthy People 2010 decade (2001–2010). There are no genetic or epigenetic measures, although numerous Healthy People objectives examine improvements in critically important health areas such as exposures to environmental toxins that have been linked to epigenetic reprogramming.

For example, "pesticide exposures resulting in visits to a health care facility" (Objective 8-13) decreased by 69.1 % during the decade, and serum "mercury in children aged 1–5 years" (Objective 8-25e) decreased by 71.4 % (National Center 2012, p. 32). However, urine concentrations of the pesticide chloropyrifos (Objective 8-24c) increased by 0.2 μg/g creatinine during the decade, and serum DDT concentrations (Objective 8-25o) increased by 30 ng/g lipid (National Center 2012, p. 32).

Among Maternal and Child Health program objectives (Focal Area 16), the folic acid/nutrition-influenced neural tube birth defects decreased by 40.0 %, and adult female dietary intake of folic acid (Objective 16-16a) increased by 5.1 % (National Center 2012, p. 57). Nevertheless, low birth weight (i.e., less than 2500 g; Objective 16-10a) and very low birth weight infants (i.e., less than 1500 g; Objective 16-10b) increased by 0.6 % and 1.1 %, respectively. Mental retardation (16-14a) and cerebral palsy (16-14b) increased by 8.9 % and 14.5 %, respectively. Gestating mothers showed no change in alcohol use (16.17a) and binge drinking rates (16.17b) during the decade (National Center 2012, p. 57).

The U.S. Environmental Protection Agency (2013) estimated that 7 % of American children live in census tracts where hazardous air pollutants exceed the 1:10,000 person risk benchmark (p. 19). They further note (p. 51) that very few epidemiological studies have examined the human health risks from hazardous air pollutants. They noted significant reductions in the levels of organophosphate pesticide residues from 1999 to 2007–2009, with 35 % of sampled apples, 5 % of sampled carrots, 8 % of sampled grapes, and 9 % of sampled tomatoes having pesticide residues (U.S. Environmental Protection Agency 2013).

All of these results must be considered with the necessary and sufficient cause-and-effect caveat of individual genetic, epigenetic, environmental exposure, social, and physiological variation that exists between toxic contact and ultimate, latent, longitudinal health conditions. These facts introduce enormous complexity to be considered for the direct and indirect paths between suspected epigenetic agent and health outcomes.

For biomonitoring, the U.S. Environmental Protection Agency (2013) used CDC data from the biannual National Health and Nutrition Examination Survey (NHANES; approximately 10,000 household respondents for each cohort). Levels of phthalates in the urine of women age 16–49 years tended to increase during

2001–2008, and the levels were significantly higher for low socioeconomic women (p. 177). Childhood cancer increased from 153 to 161 cases per one million children during 1992–1994 to 172–175 cases per million during 2007–2009 (p. 22).

Levels of autism and Attention Deficit Hyperactivity Disorder (ADHD) in children age 5–17 years increased from 0.1 % and 6.3 %, respectively, to 1.0 % and 9.5 %, respectively, between 1997 and 2010. Most notably, with reference to Pembrey et al. (2006), the percentage of children with obesity increased from 5 to 17 % between 1980 and 2008 (U.S. Environmental Protection Agency 2013, p. 22). These results were ominously similar for the Healthy People 2010 Focus Area 19 on Nutrition and Overweight (National Center for Health Statistics 2012), where all but five of the 18 targeted objectives failed to meet final goals for the decade (pp. 62–63). Obesity increased 54.5 % in children age 6–11 years and 63.6 % in adolescents age 12–19 years (p. 62). There were decreases in grain intake and increases in saturated and total fat intake (p. 62). Iron deficiencies increased in children age 3–4 years and in nonpregnant females age 12–49 years (p. 62). If Pembrey et al. (2006) are correct, then much of the American population will experience nutrition and activity-related negative health outcomes, and their descendants might be at increased risk of premature mortality due to excessive grandparent nutrition (Fontana 2008).

1.4 Epigenetic Mechanisms

Whereas Pembrey et al. (2006) did not identify a precise epigenetic mechanism for the transgenerational lowered mortality risk from famine, they did demonstrate a relationship to genes located on the X and Y chromosomes for the sex-specific effects. Heijmans et al. (2008) did identify DNA methylation in perinatally exposed famine survivors. A strong body of molecular evidence suggests five major sources of epigenetic reprogramming at the DNA level:

1. Methylation and demethylation of consecutive cytosine-guanine nucleotides (i.e., CpG islands) by DNA methyltransferases, thus either repressing or activating gene expression (Shiraishi et al. 2002)
2. Histone methylation or acetylation, thus either activating or repressing previously inactivated or actively transcribed chromatin regions, thereby disrupting gene expression patterns, cellular commitments to specific tissue developmental states, and aberrant production or nonproduction of essential cellular structural proteins and enzymes (Dixon et al. 2015; Parle-McDermott and Ozaki 2011; Van der Muelen et al. 2014)
3. Small interfering RNAs (siRNAs), long noncoding RNAs (ncRNAs), and microribonucleic RNAs (miRNAs) that posttranscriptionally silence messenger RNA (mRNA) transcripts, thus disrupting protein synthesis and the associated cellular activities regulated by these proteins (Mercer et al. 2009; Melo and Esteller 2011, 2014; Roberts 2014; Zhong et al. 2013)

4. Endogenous retroviruses and transposable elements that make up 50 % of mammalian genomes (Löwer et al. 1996; Taruscio and Mantovani 1998; Wilkins 2010)
5. Transcriptional factors, hormones, and other gene regulatory agents, even including glucose, temperature, and light/circadian effects (Angelopoulou et al. 2012; Bauer et al. 2013; Reiter et al. 2014)

These mechanisms play normal roles in human and eukaryotic multicellular animal development. Certain genes are methylated, whereas other genes are demethylated and expressed for mRNA transcription, posttranscriptional processing, and subsequent protein translation, depending on specific cell and tissue types. Epigenetic controls exist at each information processing level (i.e., transcription, processing, translation, and posttranslation).

Given that the epigenome is malleable to environmental agents (e.g., nutrition, temperature, stressful social interactions) that can produce unknown evolutionary selection effects, the human manipulation of the environment should be a health policy topic to carefully evaluate the impact of human activities on health. Much of human technological innovation has dramatically improved the lives, health, and well-being of people across the planet, even though certain aspects of these technologies (chemical, pollution) have undesirable results. Therefore, human activities result in both positive and negative epigenetic effects, unfortunately overwhelmingly unmeasured at present. We cannot control the majority of these overwhelming interactions because they vary from person to person, tissue to tissue, and cell to cell. The best that we can do is to target specific epigenetic-related diseases and health conditions so that we can optimize health outcomes using a personalized genetic/epigenetic health approach. Such health programs should include both lifespan and transgenerational (i.e., germ line) epigenetic monitoring.

Paul Hermann Muller received the 1948 Nobel Prize in Chemistry for his 1939 discovery of dichloro-diphenyl-trichloroethane's (DDT) insecticidal properties. DDT kills insects via muscle spasms induced by the freeing of neural sodium ion channels (Muller 1946; van den Berg 2009; Tren 2010). Unquestionably, DDT has saved hundreds of millions of human lives through the eradication of malaria and other insect vector protozoan diseases (van den Berg 2009). As of 2009, only three countries (India, China, North Korea) still manufactured DDT (van den Berg 2009) following demonstration of its detrimental impact on wildlife food chains, carcinogenicity, and slow natural degradation in soils (Carson 1962; Kabasenche and Skinner2014). Ye et al. (2015) documented decreased lung functioning among adults exposed to DDT, and La Merrill et al. (2013) found pre-age 50 maternal hypertension in daughters of women exposed to DDT during the prenatal period. Kabasenche and Skinner (2014) argued several transgenerational epigenetic health effects from DDT exposure, including obesity, kidney disease, and reproductive diseases. The ethical dilemma of lives saved versus unknown lifespan and transgenerational effects must be considered, even in light of the multi-billion-dollar (U.S.) agrochemical industry (many with parallel pharmaceutical/health connections) and chemicals in almost every household, personal care, and work-related environments.

Among the factors associated with DNA epigenome modifications, Parle-McDermott and Ozaki (2011) noted that folate, methionine, and choline reduce birth defect risks by promoting overall gene methylation as well as by affecting specific gene regulation targets, including cell cycle and immune factors such as insulin-like growth factor 2 (IGF-2), tumor suppressor p16, and interleukin 10. They further identified the pro-methylation effects of the soy phytoestrogen genistein at reducing cardiovascular and cancer risks (Parle-McDermott and Ozaki 2011). Such studies highlight the gene suppression role of methylation for some, not necessarily other, health outcomes. Blazkova et al. (2009) argued that CpG methylation and transcriptional interference might partially suppress HIV in patients exhibiting HIV latency.

Christensen et al. (2009) mapped age-related and exposure-related (e.g., asbestos, smoking, alcohol) methylation patterns for over 200 individuals across non-diseased brain, blood, head/neck, cervix, placenta, lung/pleura, small intestine, kidney, and bladder tissues. They found tissue-specific changes in methylation, with an overall decrease with age, increased methylation associated with asbestos, and specific gene demethylation or methylation in association with tobacco and alcohol (Christensen et al. 2009).

The Roadmap Epigenomics Consortium et al. (2010) analyzed 111 human epigenomes, finding that histone DNA methylation across specific chromosomal domains differs across varied tissues. Five percent of methylation occurred in DNA enhancer regulatory regions, where coordinated modules of enhancers are methylated for specific tissue phenotypes. Furthermore, epigenomic variants tended to be associated with disease (Roadmap Epigenomics Consortium et al. 2010). These methylome studies promise to unravel patterns of epigenetic programming, human development, and disease.

Yu et al. (2008) demonstrated that genomic regions encoding tumor suppressor gene (TSG) antisense RNA can bind and silence regular mRNA transcripts from TSG genes, thereby preventing translation of the TSG protein that blocks cancer genes. Specifically, they found that p15 antisense RNA blocks p15, an inhibitor of leukemia in mouse ES cells. The relevance of this finding to epigenetic programming lies in the fact that approximately 70 % of the mammalian genome contains antisense genes (Yu et al. 2008).

Finally, Zamudio and Bourc'his (2010) emphasized that retrotransposable elements, including inactive endogenous retroviruses, comprise approximately 50 % of mammalian genomes. They identified 21 known repressors of transposable elements in mice and humans, including DNA methyltransferase, two histone methyltransferases, RNase H, and two RNA editing cytosine deaminases (Zamudio and Bourc'his 2010). Henke et al. (2013) showed the involvement of endogenous retroviral genes in the promotion of mouse placentogenesis, while Taruscio and Mantovani (1998) argued that endogenous retroviruses simultaneously can contribute to autoimmune and reproductive pathologies when they are modulated by toxins and steroids in the body. The clear connection between methylation and transposable element gene repression serves as the basis for Wilkins' (2010) model of cancer epigenetics as well as age-related developmental changes in health and functioning.

1.5 Human Development and Positional Information

Over the course of the first hours and days of human development, a fertilized egg, the first cell of a new individual, begins to divide and specialize, with each cell containing several billion DNA nucleotides wrapped around histone proteins on 46 chromosomes (23 from each parent) and tens of thousands of maternally derived mitochondria, each containing a much smaller but very relevant mitochondrial genome. The first cell undergoes mitosis and divides into two cells, both of which are identical to the original fertilized egg. They are identical in genetic content. However, gene-encoded regulatory proteins and other enzymatic activities stimulated by sperm entry during fertilization initiate the first epigenetic event in fertilized amphibian eggs prior to mitosis and cell division: the egg cell membrane and underlying actin-rich cortex rotates 30° (π/6 radians) toward the site of sperm entry (Wolpert et al. 1998). This movement establishes the ventral (sperm entry site) and dorsal (180° or π radians) poles of the fertilized egg, the axis along which the first cell division occurs to produce two "identical" attached daughter cells. In mammals, this dorsal–ventral axis is similarly established by concentration and movement of the inner cell mass (Wolpert et al. 1998, p. 75).

Further epigenetic changes in the first two cells lead to mitosis and perpendicular cell divisions to produce four attached cells. Additional epigenetic changes produce slight, asymmetrical division of cytoplasm as cells commit and differentiate, becoming specific tissues of the developing embryo or support structures such as the placenta. Following the 32-cell stage and up to the several hundred-cell stage, the embryo exists as a hollow ball of cells that we call the blastula, where surface cell regions already are committed to ectodermal (e.g., epidermis), endodermal (e.g., precursors of internal organs), mesodermal (e.g., muscle and nephritic/reproductive structures), and neural crest (i.e., nerve) tissues. The majority of blastular cells comprise the extra-embryonic trophectodermal tissue that forms the ectoplacental cone for implantation into the endometrium by 4 days post-conception. The tissue-committed inner cell mass of the blastula pushes dorsoventrally inward to form a solid gastrula with a gut cavity (Kiefer 2007; Wolpert et al. 1998). The initial invagination region becomes the future anus near the original dorsal pole (i.e., Nieuwkoop cytoplasmic center) of the fertilized egg, while penetration of the gut cavity cells through the gastrula produces the future mouth opening at the ventral pole (Wolpert et al. 1998).

Therefore, positional information (e.g., Nieuwkoop Center) in the cytoplasm establishes the initial cleavage plane, dorsal–ventral axis of the body plan, and subsequent gut cavity orientation many cell divisions later. Clearly, there are genetically coded regulatory proteins that direct differential gene control and expression as well as epigenetic labeling (methylation, acetylation) of control regions that direct and lock various cells into specified determination/differentiation pathways for distinctive tissue fates in the developing organism. With the gastrula stage, there is the beginning of distinctive structure and tissue specialization toward the fully functional organism.

The existence and critical role of positional information has been demonstrated by many blastular cell transplantation or ablation experiments. For example, transplantation of dorsal tissue to a ventral location leads to ventral development of the transplanted cells, and vice versa. Similarly, normal development of the blastula occurs with cell removal. However, transplantation of a few Nieuwkoop Center cells from the dorsal blastula pole to another region of the blastular sphere results in a twinned embryo having two dorsal ends and one ventral end (Wolpert et al. 1998, p. 68).

Following from the studies of embryoid bodies in teratocarcinomas (Stevens 1960, 1973), Mintz and Illmensee (1975) used genetically marked mice to demonstrate that teratocarcinoma cells transplanted into blastulae yield normal mosaic mouse offspring, thus verifying both cellular totipotency (i.e., capacity to produce any cell of the organism) and the role of positional information to differentiate the undifferentiated, transformed, and immortalized cancer cells. This work paralleled Laskey and Gurdon's (1970) research demonstrating phase shift resetting of adult amphibian *Xenopus laevis* kidney, heart, lung, testis, and skin nuclei that had been transplanted into enucleated eggs, resulting in reliable blastula and tadpole development. Such cloning experiments showing totipotency were first successful following Gurdon et al. (1958) and Gurdon's (1960, 1981) successful cloning of *Xenopus*.

Using the transparent, 2 mm-long nematode *Caenorhabditis elegans*, Deppe et al. (1978) first mapped the complete developmental fates of all cells in an individual organism, egg to adult. No other species has been fate-mapped to this extent given that *C. elegans* adults have about 1000 cells, compared to approximately 5×10^{13} cells for an adult human. Together, these early developmental studies established the totipotency of the genome; predictable cell divisions, determination, and differentiation; and the role of what ultimately has become a myriad of molecular entities that establish positional information for cell fates and positioning.

1.6 Epigenetic Regulation

For vertebrate epigenetics, research continues to unravel a staggering array of regulatory proteins and other molecules that influence gene expression during embryonic development. The extensive transforming growth factor beta (TGF-Beta) family of proteins are involved in various developmental pathways, including dorsal–ventral patterning, mesodermal development, muscle growth, blood vessel growth, the production of the vital structural matrix protein elastin, etc. Such regulation is accomplished through variant, shifting patterns during gastrulation or through organ development during the first and second trimesters of in utero human development (Hollar 2012; Kothapalli et al. 2009; Wolpert et al. 1998). The TGF-Beta genes are further regulated by other proteins (e.g., ubiquitin) that are encoded by still other genes.

Just for body patterning and embryonic tissue development, regulatory proteins also include the Xwnt family, vitellogenin (from the TGF-Beta family), brachyury, goosecoid, noggin, chordin, Xnot, Xlim, and HNF-3-Beta (Wolpert et al. 1998, p. 94). Homeobox genes such as the Hox a, b, c, and d variants (from at least 40) encode regulatory proteins that are expressed in an anterior-to-posterior dorsal gradient to establish vertebral blocks, ribs, appendage mesoderm development, and full body segmentation (Kiefer 2007; Pearson et al. 2005; Wolpert et al. 1998, p. 103).

Whereas this brief synopsis is limited to early embryogenesis, the primary emphasis is on the extensive molecular pathways that are under epigenetic gene regulatory controls at all stages of development (Hollar 2012), critically so for embryonic morphology, development, and functioning onward for every cell and tissue throughout the life of the individual. Epigenetic changes continue into postnatal development, as noted by Waterland et al.'s (2009) study of 31 methylated and 111 demethylated liver genes from embryonic development through 21 days following birth.

By the time that relatively full adult development, the end of continuous growth for most organs/tissues, is achieved during late adolescence, cells/tissues shift to replacement, as committed but undifferentiated stem cells in tissues such as dermis, bone, bone marrow, and liver replace continuously damaged cells (Eriksen et al. 1994). However, postmitotic brain, nerve, and muscle tissue cease further growth and replacement (Brunk and Terman 2002). The result is continued epigenetic changes that exponentially are recorded in replacement tissues with increased gene regulatory changes in postmitotic tissues. The methylation and acetylation patterns between young and older DNA are dramatically different. Therefore, epigenetics within an individual over time first establish body form and function, followed by postpubertal accumulation of environmentally induced epigenetic changes that are associated with aging.

Similar to the Deppe et al. (1978) fate map for *C. elegans*, Galvão et al. (2010) modeled the organization of human development to identify a global structure of cell types and differentiation steps in a network model. While it was not nearly as specific as the Deppe et al. (1978) model, the Galvão et al. (2010) map serves as a guide to future delineation of specific developmental pathways and their respective epigenetic programs.

Twin epigenetic studies have been valuable for understanding environmental exposures during adolescence and early adulthood throughout the lifespan. Frago et al. (2005) showed dramatic changes in gene expression over the lifespan: 3-year-old twin pairs have almost identical gene expression patterns on chromosomes 1, 3, 12, and 17, whereas 50-year-old twin pairs showed significant hyper- and hypomethylated regions on the same chromosomes. Furthermore, there were significant methylation differences (16.8–38.2 %) within each twin pair at age 50, especially for twins who have spent less time together (Frago et al. 2005). At the time of this book, researchers Andrew Feinberg and Christopher Mason at Johns Hopkins University and Cornell University are examining the epigenetic changes in twin American astronauts Mark (earth-bound) and Scott Kelly (low earth orbit on the

International Space Station) during a 1-year period (March 27, 2015, to March 27, 2016) (www.nasa.gov/content/twins-study).

Besides studying epigenetic changes during development, researchers also are focusing on the epigenetic reprogramming for tissue engineering. This phenomenon, transdifferentiation, was observed by Rong et al. (2003) with mouse aortic smooth muscle. They observed that excessive arterial cholesterol accumulation could transdifferentiate smooth muscle cells into macrophage-type cells. Whereas Rong et al. (2003) did not determine an exact epigenetic mechanism, the macrophage conversion has been observed with endogenous retroviral elements in other tissues (Henke et al. 2013; Taruscio and Mantovani 1998; Wilkins 2010). Takeuchi and Bruneau (2009) demonstrated that three gene regulatory proteins, Gata4, Tbx5, and Baf60c, initiated ectopic cardiac gene expression in mouse embryonic mesoderm that transdifferentiated into beating cardiomyocytes. Consequently, they identified the minimal necessary and sufficient conditions for epigenetically reprogramming cells for heart repair. Such work on epigenetic regulation of tissue repair and regeneration has the potential to treat disease and to modulate aging processes over the course of the lifespan. At the same time, the transgenerational epigenetic imprints on germ-line cells and their modification remain more elusive.

Lifespan epigenetic changes can lead to the dedifferentiation of tissues. For instance, skin is subjected to solar and environmental radiation, chemicals, heat, cold, abrasion, microorganisms, etc. As skin ages, some cells dedifferentiate to form growths and tumors, balding, atypical hair growth, etc. Such changes either require genetic changes in terms of mutation with inefficient DNA repair or gene regulatory, epigenetic changes.

Complete dedifferentiation leads to cell transformation/immortalization: cancer. Cancer cells are epigenetically characterized by hypomethylation, which leads to molecular instability and, ironically, increased methylation of gene regulatory promoter regions preceding each gene, especially genes associated with cancer (e.g., breast cancer 1 early onset BRCA1, p21, p53) (Clark 2007; Davies et al. 2011; Hayflick 2007a, b; Kelly and Jones 2010; Shen et al. 2007). As with endocrine disruptors described below, Gray et al. (2010) discussed the complexity of genetic, epigenetic, and other environmental factors that might contribute to increased breast cancer risks. Specifically, they constructed a useful, two-level conceptual model (Gray et al. 2010, p. 27) of endocrine-based cancer variables/factors that include estrogens and synthetic estrogens, radiation, circadian rhythm disruption, industrial and organic solvents, metals, dioxins, bisphenol A (BPA) in plastics, phthalates, pesticides, and food additives interacting with prenatal, genetic risk, lifestyle, exercise, nutrition, clean air, and water. Clinical trials of potential demethylating, anticancer pharmaceuticals are in the early stages, but there is a clear need for gene-specific epigenetic medications given that cancers are characterized by both hypo- and hyper-methylation (Cui and Wang 2008).

1.7 Endocrine Disruption

Skinner et al. (2011) identified major endocrine disruptors that impact ecological systems and human health. These chemicals include pesticides such as DDT (dichloro-diphenyl-trichloroethane), methoxychlor (1,1,1-trichloro-2,2-bis(4-methoxyphenyl)ethane), and parathion (*O*,*O*-diethyl-*O*-(4-nitrophenyl)phosphorothioate (Guerrero-Bosagna and Skinner 2014; Mnif et al. 2011; Skinner et al. 2011). Other chemicals include elements such as arsenic, pure heavy metals such as mercury and lead, phytoestrogens such as genistein, industrial solvents such as aromatic benzene derivatives and dioxins, fungal toxins (e.g., aflatoxin), synthetic estrogens such as diethylstilbestrol (DES), and plastic additives such as bisphenol A (BPA) (Skinner et al. 2011).

Whereas the majority of these endocrine system disruptors target estrogen and estrogen receptor pathways, thus leading to reproductive system pathologies and liver, gastrointestinal, and breast/prostate cancers, natural endocrine disruptors have positive effects. The phytochemical flavonoids and genistein (e.g., from tea and soybeans, respectively) both decrease cancer risk, albeit by opposing mechanisms. Flavonoids have been shown to increase promoter methylation/inactivation for the H-ras proto-oncogene, whereas genistein contributes to demethylation of anti-oncogene promoters for the BRCA1, GSTP1, EPHB2, and RASSF1A genes in prostate cancer cells (Guerrero-Bosagna and Skinner 2014). Negative transgenerational effects of jet fuels, dioxins, BPA, and pesticides on reproductive development, cancer, and kidney/immune disorders have been documented within a comprehensive gene methylation map (Manikkam et al. 2012).

The pervasive development and use of millions of different chemical compounds during industrialization (1850s to the present) have profoundly impacted all life on earth (Carson 1962; Wilson 1984). The long half-lives of many endocrine disruptors in soils and water make them long-term threats to food webs, even with soil turnover and global hydrogeochemical cycles. We cannot ban all chemicals because many chemicals have positive or neutral effects on human health and natural environments. Epigenetics has provided us with an understanding that we need to carefully evaluate chemicals and their myriad effects on gene control regions in different cells. This will be a daunting task given that we are only beginning to understand the complex cascades of gene regulatory cascades and their interactions. Given varied circumstances and environments, such pathways within cells will demonstrate differential upregulation and downregulation of genes.

The U.S. Food and Drug Administration (FDA) drug approval process already is overwhelming for researchers and major pharmaceutical companies, as the cost to test just one drug for effectiveness and patient safety can require at least 10 years of phased testing with clinical trials patients and costs exceeding $1 thousand million (U.S.) dollars. Even with additional oversight by the U.S. Environmental Protection Agency (EPA), the staggering array of chemicals already in the environment may require a new approach to epigenetic assessments that are both objective and practical in relation to cumulative, dosage, and synergistic effects.

The presence of endocrine disruptors and other toxins in the environment impacts complex food webs within ecosystems at a concentration rate of approximately 10 times at each trophic level of a food chain. Therefore, all individuals are exposed to increasing levels of toxins the higher they are located in the food/energy chain or web. As a prime illustration, the honeybee *Apis mellifera* pollinates approximately one-half of all agricultural plants. Mullin et al. (2010) found 98 different pesticides in the honey, wax, and pollen of sampled beehives. Whereas the effects of pesticides on honeybee health and up the food chain to humans is inconclusive, some research has focused on the billion-dollar pesticide industry, particularly the neonicotinoid insecticides (Godfray et al. 2014; Raine and Gill 2015). Again, insecticides have preserved crop yields to feed the world's eight billion people. Robust, controlled experiments are needed to identify these epigenetic and environmental effects of endocrine disruptors.

1.8 Behavioral Epigenetics

The possible role of stress in epigenetic reprogramming has emerged from a few exploratory studies led by Yehuda and Bierer (2009) and Radtke et al. (2011). Yehuda and Bierer (2009) discovered that children born to women exposed to the New York City 911 terrorist attacks exhibited significantly decreased evening salivary cortisol levels if the exposure happened during their second or third trimester of development. With supporting molecular data, Radtke et al. (2011) showed a significant increase in the methylation of glucocorticoid receptors in tissue swab samples taken from children whose mothers were exposed to intimate partner violence during the pregnancy. These findings are intriguing, and they parallel Pembrey et al.'s (2006) and Heijmans et al.'s (2008) larger nutritional transgenerational studies. However, Yehuda and Bierer's (2009) and Radtke et al.'s (2011) studies are weak given very small sample sizes, with experimental groups numbering under ten individuals in both studies. Still, these studies open the door for further research into epigenetic associations with stress-related diseases, a phenomenon documented thoroughly in humans and other species for the past half-century (Christian and Davis 1964; Selye 1950).

Most recently, the physiological effects of stress have been documented via studies of allostatic load (Hollar 2013; Hollar and Lewis 2015; Seeman et al. 2001, 2002). Seeman et al. (2001, p. 4770) described allostatic load as the "cumulative biological risk (burden)" on the body that is caused by physiological stress. Whereas allostatic load has not been explicitly linked to epigenetic effects, Yehuda and Bierer (2009), Radtke et al. (2011), Pembrey et al. (2006), and Heijmans et al. (2008) implicate allostatic load as a potential major causative agent, supporting Selye's (1950) validated adaptation syndrome for physiological stress.

Seeman et al. (2001, 2002) identified a threshold checklist of ten allostatic load measures (e.g., systolic blood pressure $\geq$148 mmHg), plus they identified individuals with fewer than three friends to have significantly higher morbidity and mortality.

Hollar (2013) found significantly increased allostatic load measures among persons with varying types of disability, especially for persons with mobility limitations. Hollar and Lewis (2015) extended these findings by using a novel, computed measure, cardiovascular age, to show significantly older hearts for persons with mobility limitations.

Both Hadrich (2011) and Gjoneska et al. (2015) have found strong correlations between autoimmune disease and epigenetic regulation of immune genes, even at early ages and in conjunction with various environmental exposures (e.g., stress, toxins). Mori (2004) demonstrated the concentration of pesticides (e.g., chlordane, DDT) in cells from human umbilical cords, arguing that the prenatal period can be highly sensitive to endocrine disruptors. Similarly, Roth and Sweatt (2011) showed epigenetic methylation in the prefrontal cortex and hippocampus of infants who were abused. Both Roth and Sweatt (2011) and Zucchi et al. (2012) argued that there may be sensitive periods during perinatal and infant/child development when stress imprinting can promote later disease in affected individuals along with increased transgenerational morbidity risks.

1.9 Implications for Maternal and Child Lifespan Health

Therefore, an already substantial but rapidly growing body of research implicates the epigenetics of gene regulation as a central necessary and sufficient component of lifespan and transgenerational morbidity (Gluckman et al. 2009; Pembrey et al. 2006; Petronis 2010). It may even play a central role in aging (Pogribny and Vanyushin 2010). Epigenetics establishes the unique diversification of organismal tissues and specializations, replicable from generation to generation, stable in many genetic regions but malleable in other regions, and cumulative over the course of individual lifespans (Johnson and Tricker 2010).

Work has commenced on DNA methylation profiling of the human genome across different tissues: the Human Epigenome Project (Eckhardt et al. 2006). This project will be far more massive than the Human Genome Project. A baseline human methylation profile will be useful for morbidity, development, and aging studies across different tissues, along with the unique epigenomic signatures that every individual experiences. Understanding epigenomic imprinting and reprogramming, both normal and aberrant, eventually will lead to disease treatments and possible modulation of longevity (Feinberg and Irizarry 2010; Morgan et al. 2005; Symonds 2010).

The ultimate implications of epigenetics is that each person is a "guardian of their genome" for future generations, as stated by Dr. Marcus Pembrey in the 2007 WGBH NOVA program, "Ghost in Your Genes" (http://www.pbs.org/wgbh/nova/transcripts/3413_genes.html). Along similar lines, Kabasenche and Skinner (2014) discussed this epigenetic consequence: the need for transgenerational environmental justice, with three tenets of (1) respect for future generation autonomy; (2) nonmaleficence to future generations; and (3) justice for future victims of current environmental toxins. The new science of epigenetics will force us to think transgenerationally in time and space to understand the vast complexities of development and inheritance. We have made tremendous advances since Jacob and Monod's (1961) first gene regulatory model, but even greater advances remain to be discovered.

Abbreviations

BPA	Bisphenol A
BRCA1	Breast cancer 1 early onset
CpG	Cytosine-guanine
DNA	Deoxyribonucleic acid
DDT	Dichloro-diphenyl-trichloroethane
DES	Diethylstilbestrol
EPHB2	Ephrin type B receptor 2
GSTP1	Glutathione S-transferase P
IGF-2	Insulin-like growth factor 2
ncRNAs	Long noncoding RNAs
mRNA	Messenger RNA
miRNAs	Microribonucleic RNAs
RASSF1A	Ras association domain containing protein 1
RNA	Ribonucleic acid
siRNAs	Small interfering RNAs
TGF-Beta	Transforming growth factor beta
FDA	U.S. Food and Drug Administration

References

Angelopoulou, R., Lavranos, G., & Manolakou, P. (2012). Sex determination strategies in 2012: Towards a common regulatory model? *Reproductive Biology and Endocrinology, 10*, 13. doi:10.1186/1477-7827-10-13.

Bauer, S. E., Wagner, S. E., Burch, J., Bayakly, R., & Vena, J. E. (2013). A case-referent study: Light at night and breast cancer risk in Georgia. *International Journal of Health Geographics, 12*, 23. doi:10.1186/1476-072X-12-23.

Blazkova, J., Trejbalova, K., Gondois-Rey, F., Halfon, P., Philibert, P., Guiguen, A., et al. (2009). CpG methylation controls reactivation of HIV from latency. *PLOS Pathogens, 5*(8), e1000554. doi:10.1371/journal.ppat.1000554.

Brunk, U. T., & Terman, A. (2002). The mitochondrial-lysosomal axis theory of aging: Accumulation of damaged mitochondria as a result of imperfect autophagocytosis. *European Journal of Biochemistry, 269*, 1996–2002.

Campbell, K. H. S., McWhir, J., Ritchie, W. A., & Wilmut, I. (1996). Sheep cloned by nuclear transfer from a cultured cell line. *Nature, 380*, 64–66.

Carson, R. (1962). *Silent spring*. Boston: Houghton Mifflin.

Christensen, B. C., Houseman, E. A., Marsit, C. J., Zheng, S., Wrensch, M. R., Wiemels, J. L., et al. (2009). Aging and environmental exposures alter tissue-specific DNA methylation dependent upon CpG island context. *PLOS Genetics, 5*(8), e1000602. doi:10.1371/journal.pgen.1000602.

Christian, J. J., & Davis, D. E. (1964). Endocrines, behavior, and population. *Science, 146*(3651), 1550–1560.

Clark, S. J. (2007). Action at a distance: Epigenetic silencing of large chromosomal regions in carcinogenesis. *Human Molecular Genetics, 16*, R88–R95.

Cui, W., & Wang, L.-H. (2008). DNA methylation: A two-edged sword as a target for anti-cancer drugs. *Bioscience Hypotheses, 1*(6), 334–335.

Davies, P. C. W., Demetrius, L., & Tuszynski, J. A. (2011). Cancer as a dynamical phase transition. *Theoretical Biology and Medical Modeling, 8*, 30. doi:10.1186/1742-4682-8-30.

Deppe, U., Schierenberg, E., Cole, T., Krieg, C., Schmitt, D., Yoder, B., et al. (1978). Cell lineages of the embryo of the nematode *Caenorhabditis elegans. Proceedings of the National Academy of Sciences of the United States of America, 75*, 376–380.

Dixon, J. R., Jung, I., Selvaraj, S., Shen, Y., Antosiewicz-Bourget, J. E., Lee, A. Y., et al. (2015). Chromatin architecture reorganization during stem cell differentiation. *Nature, 518*, 331–336.

Eckhardt, F., Lewin, J., Cortese, R., Rakyan, V. K., Attwood, J., Burger, M., et al. (2006). DNA methylation profiling of human chromosomes 6, 20, and 22. *Nature Genetics, 38*(12), 1378–1385.

Eriksen, E. F., Axelrod, D. W., & Melsen, F. (1994). *Bone histomorphometry*. New York: Raven Press.

Feinberg, A. P., & Irizarry, R. A. (2010). Stochastic epigenetic variation as a driving force of development, evolutionary adaptation, and disease. *Proceedings of the National Academy of Sciences of the United States of America, 107*(Suppl 1), 1757–1764.

Ferrell, J. E. (2012). Bistability, bifurcations, and Waddington's epigenetic landscape. *Current Biology, 22*(11), R458–R466.

Fontana, L. (2008). Caloric restriction and cardiometabolic health. *European Journal of Cardiovascular Prevention and Rehabilitation, 15*(1), 3–9.

Frago, M. F., Ballestar, E., Paz, M. F., Ropero, S., Setien, F., Ballestar, M. L., et al. (2005). Epigenetic differences arise during the lifetime of monozygotic twins. *Proceedings of the National Academy of Sciences of the United States of America, 102*(30), 10604–10609.

Galvão, V., Miranda, J. G. V., Andrade, R. F. S., Andrade, J. S., Gallos, L. K., & Makse, H. A. (2010). Modularity maps of the network of human cell differentiation. *Proceedings of the National Academy of Sciences of the United States of America, 107*(13), 5750–5755.

Gjoneska, E., Pfenning, A. R., Mathys, H., Quon, G., Kundaje, A., Tsai, L.-H., et al. (2015). Conserved epigenomic signals in mice and humans reveal immune basis of Alzheimer's disease. *Nature, 518*, 365–369.

Gluckman, P. D., Hanson, M. A., Buklijas, T., Low, F. M., & Beedle, A. S. (2009). Epigenetic mechanisms that underpin metabolic and cardiovascular diseases. *Nature Reviews Endocrinology, 5*, 401–408.

Godfray, H. C. J., Blacquiere, T., Field, L. M., Hails, R. S., Petrokofsky, G., Potts, S. G., et al. (2014). A restatement of the natural science evidence base concerning neonicotinoid insecticides and insect pollinators. *Proceedings of the Royal Society B, 281*, 20140558. doi:10.1098/rspb.2014.0558.

Goldberg, A. D., Allis, C. D., & Bernstein, E. (2007). Epigenetics: A landscape takes shape. *Cell, 128*, 635–638.

Gray, J., Nudelman, J., & Engel, C. (2010). *State of the evidence: The connection between breast cancer and the environment, from science to action* (6th ed.). San Francisco: Breast Cancer Fund. Accessed July 19, 2015, from www.breastcancerfund.org

Guerrero-Bosagna, C. M., & Skinner, M. K. (2014). Environmental epigenetics and phytoestrogen/phytochemical exposures. *Journal of Steroid Biochemistry and Molecular Biology, 139*, 270–276.

Gurdon, J. B. (1960). The developmental capacity of nuclei taken from differentiating endoderm cells of *Xenopus laevis. Journal of Embryology and Experimental Morphology, 8*, 505–526.

Gurdon, J. B. (1981). Molecular mechanisms in the control of gene expression during development. *Biochemical Society Transactions, 9*(1), 13–21.
Gurdon, J. B., Elsdale, T. R., & Fischberg, M. (1958). Sexually mature individuals of *Xenopus laevis* derived from the transplantation of single somatic nuclei. *Nature, 182*, 64–65.
Hadorn, E. (1965). Problems of determination and transdetermination. *Brookhaven Symposium on Biology, 18*, 148–161.
Hadrich, C. M. (2011). Genetic variation and epigenetic patterns in autoimmunity. *Journal of Genetic Syndromes and Gene Therapy, 2*, 1. doi:10.4172/2157-7412.10000e2.
Halley-Stott, R. P., & Gurdon, J. B. (2013). Epigenetic memory in the context of nuclear reprogramming and cancer. *Briefings in Functional Genomics, 12*(3), 164–173.
Harvey, S. A., & Smith, J. C. (2009). Visualization and quantification of morphogen gradient formation in the zebrafish. *PLOS Biology, 7*(5), e1000101. doi:10.1371/journal.pbio.1000101.
Hayflick, L. (2007a). Biological aging is no longer an unsolved problem. *Annals of the New York Academy of Sciences, 1100*, 1–13.
Hayflick, L. (2007b). Entropy explains aging, genetic determinism explains longevity, and undefined terminology explains misunderstanding both. *PLOS Genetics, 3*(12), 2351–2354.
Heijmans, B. T., Tobi, E. W., Stein, A. D., Putter, H., Blauw, G. J., Susser, E. S., et al. (2008). Persistent epigenetic differences associated with prenatal exposures to famine in humans. *Proceedings of the National Academy of Sciences of the United States of America, 105*(44), 17046–17049.
Henke, C., Reubner, M., Faschingbauer, F., Stolt, C. C., Schaefer, N., Lang, N., et al. (2013). Regulation of murine placentogenesis by the retroviral genes Syncytin-A, Syncytin-B and Peg10. *Differentiation, 85*, 150–160.
Hollar, D. (2012). Development from conception through adolescence: Physiological and psychosocial factors impacting children with special health care needs. In D. Hollar (Ed.), *Handbook of children with special health care needs* (pp. 289–306). New York: Springer.
Hollar, D. (2013). Cross-sectional patterns of allostatic load among persons with varying disabilities, NHANES: 2001–2010. *Disability and Health Journal, 6*, 177–187.
Hollar, D., & Lewis, J. (2015). Heart age differentials and general cardiovascular risk profiles for persons with varying disabilities: NHANES 2001–2010. *Disability and Health Journal, 8*, 51–60.
Jacob, F., & Monod, J. (1961). Genetic regulatory mechanisms in the synthesis of proteins. *Journal of Molecular Biology, 3*(3), 318–356.
Johnson, L. J., & Tricker, P. J. (2010). Epigenomic plasticity within populations: Its evolutionary significance and potential. *Heredity, 105*, 113–121.
Kabasenche, W. P., & Skinner, M. K. (2014). DDT, epigenetic harm, and transgenerational environmental justice. *Environmental Health, 13*, 62. doi:10.1186/1476-069X-13-62.
Kelly, T., & Jones, P. (2010). DAMD to epigenetic silence. *Proceedings of the National Academy of Sciences of the United States of America, 107*(1), 3–4.
Kiefer, J. C. (2007). Epigenetics in development. *Developmental Dynamics, 236*, 1144–1156.
Kothapalli, C. R., Taylor, P. M., Smolenski, R. T., Yacoub, M. H., & Ramamurthi, A. (2009). Transforming growth factor beta 1 and hyaluronan oligomers synergistically enhance elastin matrix regeneration by vascular smooth muscle cells. *Tissue Engineering, Part A, 15*(3), 501–511.
La Merrill, M., Cirillo, P. M., Terry, M. B., Krigbaum, N. Y., Flom, J. D., & Cohn, B. A. (2013). Prenatal exposure to the pesticide DDT and hypertension diagnosed in women before age 50: A longitudinal birth cohort study. *Environmental Health Perspectives, 121*, 594–599.
Laskey, R. A., & Gurdon, J. B. (1970). Genetic content of adult somatic cells tested by nuclear transplantation from cultured cells. *Nature, 228*, 1332–1334.
Leopold, A. (1949). *A sand county almanac*. New York: Oxford University Press.
Löwer, R., Löwer, J., & Kurth, R. (1996). The viruses in all of us: Characteristics and biological significance of human endogenous retrovirus sequences. *Proceedings of the National Academy of Sciences of the United States of America, 93*, 5177–5184.

Manikkam, M., Guerrero-Bosagna, C., Tracey, R., Haque, M. M., & Skinner, M. K. (2012). Transgenerational actions of environmental compounds on reproductive disease and identification of epigenetic biomarkers of ancestral exposures. *PLOS One, 7*(2), e31901. doi:10.1371/journal.pone.0031901.

Melo, S. A., & Esteller, M. (2011). A precursor microRNA in a cancer cell nucleus: Get me out of here! *Cell Cycle, 10*(6), 922–925.

Melo, S. A., & Esteller, M. (2014). Disruption of microRNA nuclear transport in human cancer. *Seminars in Cancer Biology, 27*, 46–51.

Mercer, T. R., Dinger, M. E., & Mattick, J. S. (2009). Long non-coding RNAs: Insights into functions. *Nature Reviews Genetics, 10*, 155–159.

Mintz, B., & Illmensee, K. (1975). Normal genetically mosaic mice produced from malignant teratocarcinoma cells. *Proceedings of the National Academy of Sciences of the United States of America, 72*(9), 3585–3589.

Mnif, W., Hassine, A. I. H., Bouaziz, A., Bartegi, A., Thomas, O., & Roig, B. (2011). Effect of endocrine disruptor pesticides: A review. *International Journal of Environmental Research and Public Health, 8*, 2265–2303.

Morgan, H. D., Santos, F., Green, K., Dean, W., & Reik, W. (2005). Epigenetic reprogramming in mammals. *Human Molecular Genetics, 14*(1), R47–R58.

Mori, C. (2004). High-risk group and high-risk life stage: Key issues in adverse effects of environmental agents on human health. *Reproductive Medicine and Biology, 3*, 51–58.

Muller, P. H. (1946). Über zusammenhänge zwischen Konstitution und insectizider Wirkung. *Helvetica Chimica Acta, 29*, 1560–1580.

Mullin, C. A., Frazier, M., Frazier, J. L., Ashcroft, S., Simonds, R., van Engelsdorp, D., et al. (2010). High levels of miticides and agrochemicals in North American apiaries: Implications for honey bee health. *PLOS One, 5*(3), e9754. doi:10.1371/journal.pone.0009754.

National Center for Health Statistics. (2012). *Healthy people 2010 final review: Progress charts*. Hyattsville, MD: U.S. Department of Health and Human Services, Centers for Disease Control and Prevention (PHS Publication No. 2012–1039).

Parle-McDermott, A., & Ozaki, M. (2011). The impact of nutrition on differential methylated regions of the genome. *Advances in Nutrition, 2*, 463–471.

Pearson, J. C., Lemons, D., & McGinnis, W. (2005). Modulating Hox gene functions during animal body patterning. *Nature Reviews Genetics, 6*, 893–904.

Pembrey, M. E., Bygren, L. O., Kaati, G., Edvinsson, S., Northstone, K., Sjostrom, M., et al. (2006). Sex-specific, male-line transgenerational responses in humans. *European Journal of Human Genetics, 14*, 159–166.

Petronis, A. (2010). Epigenetics as a unifying principle in the aetiology of complex traits and diseases. *Nature, 465*, 721–727.

Pogribny, I. P., & Vanyushin, B. F. (2010). Age-related genomic hypomethylation. In T. O. Tollefsbol (Ed.), *Epigenetics of aging* (pp. 11–28). New York: Springer.

Prigogine, I. (1977). *Time, structure, and fluctuations*. Stockholm: The Nobel Foundation.

Prigogine, I. (2002). Dynamical roots of time symmetry breaking. *Philosophical Transactions of the Royal Society of London A, 360*, 299–301.

Radtke, K. M., Ruf, M., Gunter, H. M., Dohrmann, K., Schauer, M., Meyer, A., et al. (2011). Transgenerational impact of intimate partner violence on methylation in the promoter of the glucocorticoid receptor. *Translational Psychiatry, 1*, e21. doi:10.1038/tp.2011.21.

Raine, N. E., & Gill, R. J. (2015). Tasteless pesticides affect bees in the field. *Nature, 521*, 38–40.

Reiter, R. J., Tan, D. X., Korkmaz, A., & Rosales-Corral, S. A. (2014). Melatonin and stable circadian rhythms optimize maternal, placental and fetal physiology. *Human Reproduction Update, 20*(2), 293–307.

Roadmap Epigenomics Consortium, Kundaje, A., Meuleman, W., Ernst, J., Bilenky, M., Yen, A., et al. (2010). Integrative analysis of 111 reference human epigenomes. *Nature, 518*, 317–330.

Roberts, T. C. (2014). The microRNA biology of the mammalian nucleus. *Molecular Therapy – Nucleic Acids, 3*(8), e188. doi:10.1038/mtna.2014.40.

Rong, J. X., Shapiro, M., Trogan, E., & Fisher, E. A. (2003). Transdifferentiation of mouse aortic smooth muscle cells to a macrophage-like state after cholesterol loading. *Proceedings of the National Academy of Sciences of the United States of America, 100*(23), 13531–13536.

Roth, T. L., & Sweatt, J. D. (2011). Annual research review: Epigenetic mechanisms and environmental shaping of the brain during sensitive periods of development. *Journal of Child Psychology and Psychiatry, 52*(4), 398–408.

Seeman, T. E., McEwen, B. S., Rowe, J. W., & Singer, B. H. (2001). Allostatic load as a marker of cumulative biological risk: MacArthur studies of successful aging. *Proceedings of the National Academy of Sciences of the United States of America, 98*(8), 4770–4775.

Seeman, T. E., Singer, B. H., Ryff, C. D., Love, G. D., & Levy-Storms, L. (2002). Social relationships, gender, and allostatic load across two age cohorts. *Psychosomatic Medicine, 64*, 395–406.

Selye, H. (1950). Stress and the general adaptation syndrome. *The British Medical Journal, 1*(4667), 1382–1392.

Shen, L., Catalano, P. J., Benson, A. B., O'Dwyer, P., Hamilton, S. R., & Issa, J. P. J. (2007). Association between DNA methylation and shortened survival in patients with advanced colorectal cancer treated with 5-fluorouracil-based chemotherapy. *Clinical Cancer Research, 13*(20), 6093–6098.

Shiraishi, M., Oates, A. J., & Sekiya, T. (2002). An overview of the analysis of DNA methylation in mammalian genomes. *Biological Chemistry, 383*(6), 893–906.

Skinner, M. K., Manikkam, M., & Guerrero-Bosagna, C. (2011). Epigenetic transgenerational actions of endocrine disruptors. *Reproductive Toxicology, 31*(3), 337–343.

Stevens, L. C. (1960). Embryonic potency of embryoid bodies derived from a transplantable testicular teratoma of the mouse. *Developmental Biology, 2*, 285–297.

Stevens, L. C. (1973). A developmental genetic approach to the study of teratocarcinogenesis. *BioScience, 23*(3), 169–172.

Symonds, M. E. (2010). Epigenomics – grand challenge: Much more than the developmental origins of adult health and disease. *Frontiers in Genetics, 1*, 1. doi:10.3389/fgene.2010.00001.

Takeuchi, J. K., & Bruneau, B. G. (2009). Directed transdifferentiation of mouse mesoderm to heart tissue by defined factors. *Nature, 459*, 708–712.

Taruscio, D., & Mantovani, A. (1998). Human endogenous retroviral sequences: Possible roles in reproductive physiopathology. *Biology of Reproduction, 59*, 713–724.

Thom, R. (1972). *Structural stability and morphogenesis*. New York: W. A. Benjamin.

Tren, R. (2010). DDT and malaria prevention. *Environmental Health Perspectives, 118*(1), A14–A15.

U.S. Environmental Protection Agency. (2013). *America's children and the environment* (3rd ed.). Washington, DC: Author (Publication No. EPA 240-R-13-001).

van den Berg, H. (2009). Global status of DDT and its alternatives for use in vector control to prevent disease. *Environmental Health Perspectives, 117*, 1656–1663.

Van der Muelen, J., Speleman, F., & Van Vlierberghe, P. (2014). The H3K27me3 demethylase UTX in normal development and disease. *Epigenetics, 9*(5), 658–668.

Waddington, C. H. (1957). *The strategy of the genes: A discussion of some aspects of theoretical biology*. London: Allen & Unwin.

Waterland, R. A., Kellermayer, R., Rached, M.-T., Taterian, N., Gomes, M. V., Zhang, J., et al. (2009). Epigenomic profiling indicates a role for DNA methylation in early postnatal liver development. *Human Molecular Genetics, 18*(16), 3026–3028.

Wilkins, A. S. (2010). The enemy within: An epigenetic role of retrotransposons in cancer initiation. *Bioessays, 32*, 856–865.

Wilson, E. O. (1984). *Biophilia: The human bond with other species*. Cambridge, MA: Harvard University Press.

Wolpert, L., Beddington, R., Brockes, J., Jessell, T., Laurence, P., & Meyerowitz, E. (1998). *Principles of development*. New York: Oxford University Press.

Ye, M., Beach, J., Martin, J. W., & Senthilselvan, A. (2015). Association between lung function in adults and plasma DDT and DDE levels: Results from the Canadian Health Measures Survey. *Environmental Health Perspectives, 123*(5), 422–427.

Yehuda, R., & Bierer, L. (2009). The relevance of epigenetics to PTSD: Implications for the DSM-IV. *Journal of Trauma and Stress, 22*(5), 427–434.

Yu, W., Gins, D., Onyango, P., Muldoon-Jacobs, K., Karp, J., Feinberg, A. P., et al. (2008). Epigenetic silencing of tumour suppressor gene p15 by its antisense RNA. *Nature, 451*, 202–206.

Zamudio, N., & Bourc'his, D. (2010). Transposable elements in the mammalian germline: A comfortable niche or a deadly trap? *Heredity, 105*, 92–104.

Zhong, S.-H., Liu, J.-Z., Jin, H., Lin, L., Li, Q., Chen, Y., et al. (2013). Warm temperatures induce transgenerational epigenetic release of RNA silencing by inhibiting siRNA biogenesis in Arabidopsis. *Proceedings of the National Academy of Sciences of the United States of America, 110*(22), 9171–9176.

Zucchi, F. C. R., Yao, Y., & Metz, G. A. (2012). The secret language of destiny: Stress imprinting and transgenerational origins of disease. *Frontiers in Genetics, 3*, 96. doi:10.3389/fgene.2012.00096.

Chapter 2
Epigenetic Significance of Chromatin Organization During Cellular Aging and Organismal Lifespan

Milena Georgieva, Dessislava Staneva, and George Miloshev

2.1 Entering the Puzzling World of Aging

Aging has long been considered as a major socioeconomic problem. It leads to consecutive worsening of the overall organismal health, which in turn results in steady increase of public expenses in medical and social fields (Upadhya et al. 2015; Rolf 2015). The mere fact that the overall life expectancy is steadily increasing and that almost all human populations are characterized by a growing number of elderly people turns this problem into a huge burden (Guerrero et al. 2015; McKenzie et al. 2015). Hence, it is of utmost importance for all governments to assure healthy aging of its elderly people which sounds as an impossible task without full comprehension of the general mechanisms of aging. The last has been a challenge for scientists in the field of gerontology for many years, even for decades. And this is not surprising at all. Aging is a complex biological process which leads to loss of functional reserve of multiple organ systems and to increased susceptibility to stress and age-associated maladies like cancer, neurodegenerative diseases, diabetes, etc. It is also associated with increased incidence of genetic mutations and prevalence of chronic diseases which finally lead to functional dependence and an inadequate life (Campisi 2000; Miquel 2014). Determined by a combination of genetic and environmental factors, the process of getting old is highly individualized and yet poorly understood. Extensive studies are under way with the main emphasis being set on elucidation of the intimate molecular mechanisms of the aging process.

Recently, some researchers in the field of aging (López-Otín et al. 2013) have been able to shortlist the nine most common hallmarks of the aging process and denoted them as major. These nine major hallmarks include genomic instability,

M. Georgieva, PhD (✉) • D. Staneva, PhD • G. Miloshev, PhD (✉)
Laboratory of Yeast Molecular Genetics, Institute of Molecular Biology "Roumen Tsanev", Bulgarian Academy of Sciences, 1113 Sofia, Bulgaria
e-mail: milenaki@chromatinepigenetics.com; gmlab@chromatinepigenetics.com

D. Hollar (ed.), *Epigenetics, the Environment, and Children's Health Across Lifespans*, DOI 10.1007/978-3-319-25325-1_2

telomere attrition, epigenetic alterations, loss of proteostasis, deregulated nutrient sensing, mitochondrial dysfunction, cellular senescence, stem cell exhaustion, and altered intercellular communication. Deep understanding of these common aging features could be a major step toward development and application of specific pharmaceutical approaches, the aim of which would be to ameliorate, slow, and successfully manipulate the process of getting old. Of course, scientists in the field are far from getting involved in the search for the elixir for eternal life, but do endeavor to spend lots of efforts in order to get a clearer picture of the most intimate mechanisms of aging. The task could be very hard because these well classified hallmarks of aging seem interdependent making studies in the field even more complicated. For example, DNA damage could lead to genomic instability but could be a consequence of it. Many factors, some extrinsic like UV irradiation, chemicals, and genotoxins, and others, intrinsic like ROS (reactive oxygen species) production, account for DNA damage and are one of the main culprits for the aging of the cells and the organism (Garinis et al. 2008). The picture gets even more complicated since the metabolic stress that all cells undergo could be the primary cause for increased ROS production. Apparently following this line we enter in the vicious circle of trying to understand who is who and what is the cause and what the consequence of aging. No doubt, therefore, that the complexity of aging is considered as "one of the major stumbling blocks of all biological investigations of aging" (Martin 2005).

Bearing in mind that all organisms have DNA and that the proper functioning of the genetic material is crucial for the cellular and organismal homeostasis, it is easy to predict that the genome itself could be the foundation on which aging acts. No doubt that if we scrutinize the way by which the genome ages and moreover, its dynamics during aging, we shall be able to understand the general mechanisms of this complicated process.

2.2 Chromatin as a Modulator of Cellular Fate

2.2.1 The Consecutive Steps of Chromatin Compaction in the Eukaryotic Nucleus

Each of the 60 trillion cells that make up the human body (with some minor exceptions like the human erythrocytes) contains 2 m of genomic DNA in its nucleus. This is quite a huge amount of genetic material that has to be packed and organized in the confined space of the cell nucleus (Fig. 2.1). The organization of DNA and, importantly, its proper functioning recently turned into one of the most interesting puzzles in biology. Two important issues have to be solved by the cells. The first is about the way the genomic DNA is packaged to fit the limited space in the cell nucleus which is only 10 μm in diameter, and the second question is about the way DNA works in

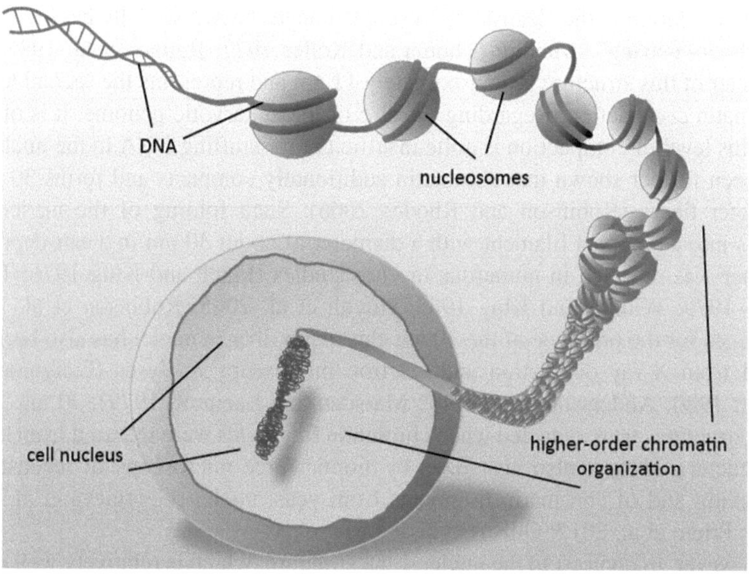

Fig. 2.1 A drawing of chromatin compaction in the eukaryotic nucleus. DNA in the eukaryotic nucleus is organized together with histone proteins in chromatin. Nucleosomes are the main building blocks of chromatin which further compact together with linker histones in higher-order chromatin structures thus preserving the genome against stress but also maintaining its functionality

this very compacted state. As a matter of fact almost all articles regarding chromatin, epigenetics, and DNA start with one and the same sentence, and it goes like that: "DNA is not naked in the cells but is organized together with histone proteins in chromatin." Indeed chromatin is the platform where DNA meets histones—the last are considered to be at the heart of DNA compaction, its regulation, and dynamics (Hayes et al. 1990; Littau et al. 1965; Wolffe 1998). Together with it they build the nucleosomes. Each nucleosome consists of two molecules of each of the four core histones (H2A, H2B, H3, and H4), around which 147 bp of DNA are wrapped together. The nucleosome represents the basic repeat unit of chromatin (Richmond et al. 1984; Luger et al. 1997). It is widely accepted that this is the first level of chromatin compaction. By a physical point of view, the nucleosome represents an obstacle to all processes that require access to the molecule of DNA.

Therefore, chromatin has to be modified, reorganized, and de-compacted in order DNA to be transcribed, replicated, and repaired. This process of chromatin reorganization is done by specialized enzyme complexes, called chromatin modifying and chromatin remodeling complexes, which can chemically modify histones and can slide nucleosomes along the molecule of DNA. Nucleosomes are not only basic units of chromatin compaction but also exert important regulatory functions and manipulate the activity of the genome. They form nucleosome arrays, which in relation to the electron microscopic images have poetically been called the "beads-on-a-string" structure (Thoma and Koller 1977; Rattner et al. 1982). The diameter of this structure is approximately 11 nm and represents the second level of chromatin compaction. Regarding the size of the eukaryotic genome, it is obvious that this level of compaction is quite insufficient for stuffing DNA in the nucleus. It has been further shown that chromatin additionally compacts and forms 30 nm in diameter fibers (Robinson and Rhodes 2006). Such folding of the nucleosome arrays into a compact filament with a diameter of about 30 nm in a salt-dependent manner was reported in numerous in vitro studies (Finch and Klug 1976; Thoma et al. 1979; Widom and Klug 1985; Huynh et al. 2005; Robinson et al. 2006). Evidence for the presence of the 30 nm chromatin fiber in nuclei has also been provided from X-ray diffraction and electron microscopy analyses (Langmore and Schutt 1980; Andersson et al. 1982; Marsden and Laemmli 1979). 30 nm "fiber-like" structures were detected when chromatin fragments were released from human cell nuclei and were also visualized by atomic force microscopy of reconstituted chromatin and of chromatin fragments from yeast nuclei (Georgieva et al. 2012, 2015; Prieto et al. 2012; Gilbert et al. 2004).

However, in contrast to the nucleosome structure, which is relatively well understood, the detailed organization of these 30-nm diameter rod-like structures is considered as quite controversial. Recently, their existence in vivo has been questioned thus opening ardent discussions of how exactly chromatin is structured above the 11-nm chromatin filaments (Fussner et al. 2012; Maeshima et al. 2014). The dispute probably has been induced by the different methodologies exploited for studying of this particular level of chromatin organization. It is true that some of the most meaningful results on this structure were obtained by in vitro experiments (Robinson et al. 2008; Robinson and Rhodes 2006; Routh et al. 2008). Though, for more than four decades, lots of data were obtained by both in vivo and in vitro experiments favoring the existence of the 30 nm chromatin fibers. Interestingly, the obtained results have proposed the existence of at least two different rod-like fibers in the nuclei of eukaryotes: the first with a diameter of 30 nm (the so-called "one-start model") and the second with a diameter of 24–26 nm (the "two-start model"), both, as noted by Grigoryev and coworkers (2009) able to appear consecutively in a single fiber. Importantly, in a critical review by Ghirlando and Felsenfeld (2013), it was stated that the 30 nm fiber is not an indicator of the transcriptional status of chromatin but rather represents a structurally more compacted state of organization.

2.2.2 *The Elusive Higher-Order Chromatin Structures*

Chromatin organization above the 10 and 30 nm in diameter structures is believed to employ important regulatory roles on the genetic information. Treatment of isolated nuclei or chromosomes with high salt solutions that remove histones resulted in the formation of a "halo" of supercoiled DNA loops surrounding a denser core (Cook and Brazell 1975; Earnshaw and Laemmli 1983). It has been proven that these chromatin loops were anchored to an insoluble proteinaceous nuclear matrix (or karyoskeleton) by DNA sequences termed as matrix attachment regions (MARs) and scaffold attachment regions (SARs), respectively.

Another aspect of the higher-order structure of chromatin is manifested by its spatial, 3D organization and motion. In the cell nucleus, each chromosome occupies a limited volume, chromosome territory, and seems relatively static being physically anchored in this distinct and limited nuclear space (Figs. 2.1 and 2.2). However, the findings from live microscopy and other experiments that enable tracking of chromatin dynamics in vivo have revealed large-scale chromatin motions and have proved the functional significance of these motions for the cells and the organism (Gasser 2002; Chuang et al. 2006). Three principle types of chromatin motion were recognized and were accepted as general. Interestingly, the three appeared quite different in time scale, range, and frequency of the movements as well as in energy dependence (Soutoglou and Misteli 2007). Moreover, it has been shown that these

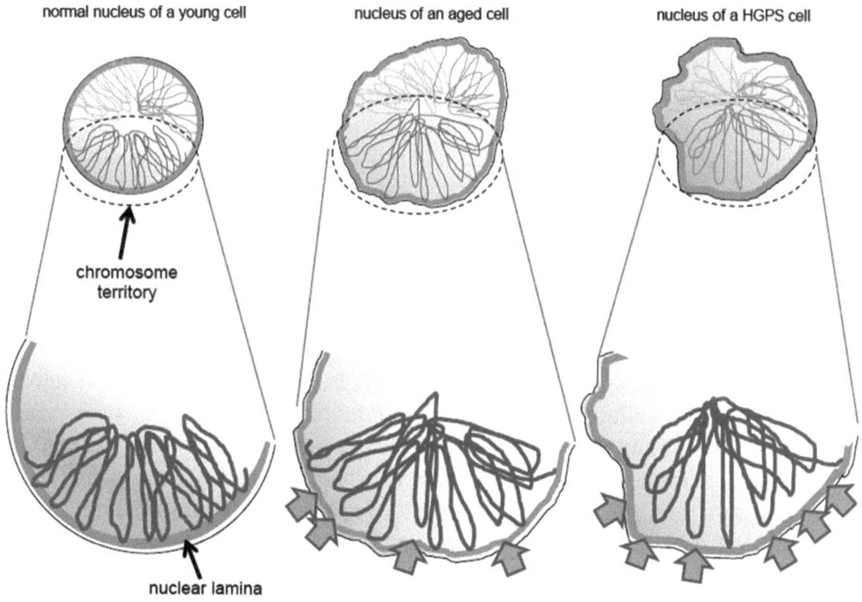

Fig. 2.2 A schematic representation of higher-order chromatin loop organization in three cellular conditions: a normal young cell, an aging cell, and a pathologically changed cell from a patient with Hutchinson–Gilford Progeria syndrome (HGPS). Normally, interphase chromosomes are making numerous contacts with the nuclear lamina. Due to changes in chromatin organization and perturbations in lamina organization during the processes of cellular aging and in premature aging syndromes, many of these contacts between the chromosomes and the lamina are lost (shown with *arrows* in the figure). This leads to formation of longer chromatin loops, reorganization of chromosome territories, and loss of heterochromatin in the nucleus. In cells from the premature aging syndrome, HGPS, the nuclear lamina is severely dilapidated and therefore most of the chromosome contacts with it are disrupted. The diminished contacts between the lamina and the chromosomes in aged cells and in HGPS cells probably alter gene expression in a similar fashion and therefore aging and progeria syndromes represent similar clinical outcomes

chromatin spatial rearrangements influence cellular fate and aging of the organisms (Chandra et al. 2015).

Global chromatin reorganization and changes in nuclear architecture with functional aftereffects in the regulation of gene expression are associated with differentiation, development and aging (Solovei et al. 2009; Oberdoerffer and Sinclair 2007). According to the conventional view of the three-dimensional (3D) genome organization within the nucleus, heterochromatin preferentially localizes at the nuclear periphery, whereas euchromatin preferentially resides in the nuclear interior. A remarkable exception of this conventional nuclear architecture is the inverted chromatin organization in photoreceptor cells of the nocturnal mammalian eye. Using FISH, some authors were able to characterize in detail the higher-order chromatin dynamics during terminal differentiation of rod cells of mouse retinas (Solovei et al. 2009). Interestingly, at birth rod nuclei have a conventional architecture, but postnatal differentiation of rod cells is accompanied with global spatial movement of higher-order chromatin structures and with the establishment of the so-called "inverted" nuclear organization.

Finally, in terminally differentiated mouse rods heterochromatin concentrated in the nuclear center, whereas euchromatin as well as nascent transcripts and splicing machinery lined the nuclear border (Solovei et al. 2009). The functional effect of this inverted pattern of nuclear architecture is to provide additional sensitivity of nocturnal mammal eye in the poor light environments. These results unambiguously prove that chromatin dynamics exert functional significance during organismal lifespan.

2.3 Chromatin: Static and Dynamic at the Same Time

Cells are persistently exposed to intrinsic as well as extrinsic environmental signals that could destine cellular fate toward proliferation, differentiation, quiescence, senescence, or malignancy. All these processes are accompanied and governed by dynamic reorganization of chromatin (Shin et al. 2011; Apostolou and Hochedlinger 2013; Pegoraro and Misteli 2009; Adams 2007; Misteli 2010; Burton and Torres-Padilla 2014). We have tried to demonstrate some of the best known chromatin large-scale reorganization during normal development, aging, and premature aging syndromes in Fig. 2.2. In order to regulate all biological processes related to the molecule of DNA, chromatin has to hold a flexible, versatile structure that can respond to internal and external signals during cellular lifespan.

To answer these challenges, chromatin exhibits a highly dynamic equilibrium between an open conformation (e.g., 11-nm beads-on-a-string) up to much more compacted, regarded even as a closed chromatin state (the so-called higher-order structures). This dynamics assures both accurate expression of the genetic information as well as promises genomic stability. For example, in the time course of the cell cycle, from mitosis through interphase up to the next mitosis, chromatin undergoes global structural reorganization—from fully condensed mitotic chromosomes through less condensed, more diffused interphase chromatin conformation and then again to compacted chromosome entities. In order to meet the requirements of

DNA-related processes, chromatin is not expected to be steadily and uniformly compacted across the genome in vivo, but to be dynamically modulated by local differences in chromatin components including numerous architectural and regulatory proteins as well as complex combinations (cross-talks) among diverse set of epigenetic modifications. No doubt that the regulation of chromatin dynamics comes through a variety of mechanisms including DNA methylation (Karymov et al. 2001; Ma et al. 2005), posttranslational modifications (PTMs) of histones (Ito 2007; Gelato and Fischle 2008), histone variants (Happel and Doenecke 2009; Clausell et al. 2009), recruitment of chromatin remodeling complexes (Corona et al. 2007; Clapier and Cairns 2009), and noncoding RNAs (Li 2014; Dhanasekaran et al. 2013). All these epigenetic regulators act in concert to exert the complex mechanisms that modulate chromatin conformation which in turn regulates gene expression in development, differentiation, aging, and disease. For example, these epigenetic marks as well as the level of chromatin condensation are distinct between normally proliferating, cancerous, and senescent cells (Oh et al. 2013; Rodriguez-Paredes and Esteller 2011).

Due to the development of new techniques, chromatin dynamics has been under extensive studies recently. It is reasonable to believe that the versatile elaboration of experimental techniques such as FISH (fluorescent in situ hybridization), 3C (chromosome conformation capture), and its high-throughput modifications (4C, 5C, Hi-C, ChIA-PET, etc.) (Dekker et al. 2002; Sanyal et al. 2011), together with AFM (atomic force microscopy), ChCA (chromatin comet assay) (Georgieva et al. 2008, 2012), and the very recently developed high resolution live microscopy, and the gathering of valuable information about higher-order chromatin structure and its dynamics will exponentially increase in the future.

2.4 Chromatin Structure and Dynamics as Major Players in Aging

Chromatin has been accepted as a major player in aging as it regulates the plasticity of the genome and governs its susceptibility to intrinsic and extrinsic DNA damaging signals. It also modulates the contacts of DNA with all factors that participate in replication, transcription, and repair. The proper contacts among DNA and all these factors guarantee their accurate functioning, contribute to the preservation of the genome against genotoxic stress, and prevent the accumulation of mutations which could be detrimental for the cells (Downs and Cote 2005; Cho et al. 2004; Seeber et al. 2014; Rodrigues et al. 2014).

As we have already discussed, the nature of chromatin is very contradictory. At one hand, it is regarded as a static unit that guards the genome and protects the genetic material against DNA damage. On the other hand, chromatin appears quite dynamic. It "breaths in" and "out" with the surrounding environment and thus allows the cells to adapt by changing the expression of certain genes. In some cases, this dynamic property appears to counter the aging process and to extend organismal lifespan (Longo and Kennedy 2006; Gotta et al. 1997; Kennedy et al. 1997). However, the stochastic component of chromatin structure might also contribute to the breakdown of nuclear, cellular, and tissue functions and consequently to lead to aging and age-associated diseases (Lazarus et al. 2013; Sedivy et al. 2008). Generally, aging involves major changes in chromatin structure and organization (Zane et al. 2014; Pegoraro and Misteli 2009; Muñoz-Najar and Sedivy 2011). It has long been believed that during aging, nuclei show loss of heterochromatin (Ishimi et al. 1987). Global changes in chromatin organization have been found to be associated with aging including the formation of senescence-associated heterochromatin foci (SAHF) in euchromatic DNA and a gradual loss of perinuclear heterochromatin (Oberdoerffer and Sinclair 2007). Deterioration of chromatin organization triggers in turn transcriptional alterations, accumulation of DNA damage, and increased genomic instability (Pegoraro and Misteli 2009; Feser and Tyler 2011; O'Sullivan and Karlseder 2012).

Though recent studies demonstrate that the situation is much more complicated as heterochromatin can be either reduced or increased during senescence (Sedivy et al. 2008; Pegoraro and Misteli 2009). It has been speculated that these SAHF silence proliferation-promoting genes and thus lead to senescence-associated phenotypes (Corpet and Stucki 2014). Remarkably, although SAHF appear to result from the condensation of almost entire chromosomes, DNA sequences that are typically contained in constitutive heterochromatin, such as pericentromeres and telomeres, actually appear to be excluded from the bulk of the condensed chromosomes (Zhang et al. 2009a; Narita et al. 2003). This was further proven by other authors who have shown that the shortened telomeres in mice lacking telomerase have reduced heterochromatin compared to telomeres from normal cells (Benetti et al. 2007a). The molecular mechanism of SAHF formation remains poorly understood although histone chaperones Asf1a and HIRA have been reported to play a role in their formation (Zhang et al. 2005). Other authors have even suggested that SAHF are a novel type of chromatin condensation involving alterations in the way by which DNA interacts with its binding proteins (Funayama et al. 2006).

They have suggested that the induction of a specific senescence phenotype is accompanied by loss of linker histone H1 from chromatin and accumulation of HMGA2 at its place. This boldly highlights the significance of chromatin structural proteins such as H1 and HMGA2 for the ongoing process of chromatin structure remodeling during aging.

It can be inferred by this that the eu- and heterochromatic states of chromatin are quite dynamic during organismal lifespan, and it is quite easy to envision that this dynamics is basically directed by epigenetic mechanisms.

2.5 Epigenetic Changes in Chromatin During Aging

2.5.1 DNA Methylation: The Aging Clock of the Organism

Epigenetics plays crucial role during development. It shapes the way the genome works in response to all kinds of changes in the surrounding environment (Wolffe and Matzke 1999; Wolffe 1998). Recent twin studies and studies with long-lived families have shown that only 20–30 % of the variations in the aging among different individuals are determined by genetic factors. Interestingly, these are relevant mainly for survival at advanced age. The other 70–80 % of variations have been shown to be associated with stochastic events, i.e., certainly to some nongenetic factors (Herskind et al. 1996). Essential for development, epigenetics inevitably becomes misregulated and misdirected during disease and aging, and importantly, these changes are more or less conserved among all species (Sedivy et al. 2008; Wood and Helfand 2013; Zane et al. 2014).

A major part of the epigenetic research has been focused on the study of modifications of DNA and histone proteins, as these are the mechanisms that generally influence chromatin structure and organization. DNA methylation, which is the first epigenetic phenomenon that has been linked to aging and disease, is also the best-studied DNA modification (Riggs 1975; Iyer et al. 2011; Fraga et al. 2005a).

Recently, some interesting dynamics of DNA methylation during aging have been explored (Zane et al. 2014; Muñoz-Najar and Sedivy 2011; Lazarus et al. 2013). It has been demonstrated that there is a consistent reduction in global DNA methylation during aging (Singhal et al. 1987; Wilson et al. 1987) while at certain places, namely the CpG islands of gene promoters, DNA methylation is increased with age (Kim et al. 2005). In general, global DNA demethylation in aging occurs mainly at repetitive DNA elements and in genomic regions with facultative heterochromatin which leads to overall deheterochromatinization of the genome (Sedivy et al. 2008). In contrast, hypermethylation of CGIs (CpG islands) is found mainly at promoters of stem cells' genes and also at promoters of tumor suppressor genes (Issa et al. 2001; Waki et al. 2003). Both epigenetic phenomena explain two of the hallmarks of aging (López-Otín et al. 2013): the observed exhaustion of stem cells during aging and the higher incidence of cancer with the advancement of age. The first is due to increased methylation at the promoters of stem cells genes, especially those responsible for their pluripotency, while the second results from the aberrant hypermethylation of promoters of tumor suppressor genes. Interestingly, it has been observed that there is an increase in 5-methylcytosines within the ribosomal DNA (rDNA) clusters in livers of old rats, which could explain the decrease in ribosomal RNA (rRNA) levels that also occurs during aging (Oakes et al. 2003). And though DNA methylation patterns at first sight seem quite contradictory, there is logic for exploiting it as an aging marker.

Not surprisingly, the level of DNA methylation has been proposed as a biological clock for aging. For example, some authors (Mitteldorf 2015; Hannum et al. 2013) offer the level of DNA methylation in stem cell niches in an aging organism as a marker, even as a biological clock for assessing the age of the organism. In a fascinating work by Horvath (Horvath 2013), a multi-tissue predictor of age that allows estimation of DNA methylation age of most human tissues has been made freely available to the public and is gaining lots of popularity (Marioni et al. 2015; Horvath et al. 2015).

Advanced age is always associated with increased risk for cancer. On the other hand, epigenetic alterations are hallmarks of both aging and cancer. Therefore, it is normal to assume that aging-associated DNA methylation changes actively contribute to cancer susceptibility and growth (Fraga et al. 2007). No doubt that both processes (aging and cancer) share almost the same DNA methylation marks. Table 2.1

Table 2.1 Changes in DNA methylation pattern and in DNMT levels in aging, HGPS, and cancer

	Function	Aging/ senescence	HGPS	Cancer
DNA methylation				
DNA hypo methylation	Activation	Increase: in repeat-rich regions (Fraga et al. 2007; Zane et al. 2014; Ehrlich 2009; Wilson et al. 1987; Fuke et al. 2004)	Increase: in hypermethylated CpG islands (Heyn et al. 2013)	Increase: in repeat-rich regions (Zane et al. 2014; Fraga et al. 2007; Ehrlich 2009)
DNA hyper methylation	Silencing	Increase: at promoters CGIs (Zane et al. 2014; Fraga et al. 2007; Ahuja et al. 1998; Issa et al. 1994)	Increase: at unmethylated CpG sites (Heyn et al. 2013)	Increase: at promoters CGIs (Zane et al. 2014; Fraga et al. 2007; Ehrlich 2009)
DNMT				
DNMT1	Maintenance of DNA methylation	Decrease (Casillas et al. 2003; Lopatina et al. 2002)	ND	Increase (Hermann et al. 2004; Robertson et al. 1999)
DNMT3a	De novo DNA methylation	Decrease (Casillas et al. 2003; Lopatina et al. 2002)	ND	Increase (Hermann et al. 2004; Robertson et al. 1999)
DNMT3b	De novo DNA methylation	Increase (Casillas et al. 2003; Lopatina et al. 2002)	ND	Increase (Hermann et al. 2004; Robertson et al. 1999)

ND No data available

summarizes the plethora of data concerning the role of DNA methylation in aging, cancer, and premature aging syndromes. What catches the attention immediately is the fact that most DNA patterns are shared among different cases. For example, global DNA hypomethylation of repeat-rich regions is observed in numerous cancers (Ehrlich 2009). DNA hypomethylation of repeat elements has been proposed to contribute to cellular transformation by promoting chromosomal rearrangements and elevating mutational rate. Apart from hypomethylation of repeat elements, loss of methylation from genic regions also causes aberrant gene expression and promotes tumorigenesis (Zane et al. 2014). In addition, promoters of genes that are targets of hypermethylation during aging overlap with DNA hypermethylated genes in many cancers (McGarvey et al. 2008). For example, aberrant hypermethylation at promoters of *CDKN2A*, *LOX*, *RUNX3*, and *TIG1*, which act as tumor suppressor genes, has been detected during aging (Yanagawa et al. 2007; So et al. 2006a, b) and was further accepted as a strong evidence that there is age-dependent onset of different types of cancer.

2.5.2 The Epigenetic Make-Up of Histone Proteins Serves as a Strong Regulator of Genome Activity

The epigenetic characteristics of young and old cells are not limited only to DNA methylation. Histones (core histones H2A, H2B, H3 H4, and the linker histones, H1) can undergo diverse and reversible posttranslational modifications that occur predominantly on their unstructured "tail" domains. Currently, more than ten different types of histone modifications are known to affect chromatin structure and function. The most prominent histone modifications include acetylation, methylation, ubiquitylation, ADP-ribosylation, phosphorylation, sumoylation, and the list is ever growing (Bannister and Kouzarides 2011; Zhang et al. 2009a).

These covalent marks can be written and erased by site-specific enzymes such as histone acetyltransferases (HATs) and deacetylases (HDACs), histone methyltransferases (HMTs) and demethylases (KDM) (Cuthbert et al. 2004; Shi et al. 2004; Wang et al. 2004), ubiquitin ligases and deubiquitinases, and many others specific for a particular modification. The known histone modifications and some of their functional consequences are very well discussed in (Bannister and Kouzarides 2011). The best known of these PTM's are shortlisted in Table 2.2, where their role in aging, cancer, and the premature aging syndromes are very well discussed. Posttranslational modifications not only alter the interaction of histone proteins with DNA but also influence inter-nucleosomal interactions and thus affect the overall chromatin structure. Moreover, these modifications could also recruit specific chromatin-associated proteins, like HP1 (heterochromatin protein 1), which recognizes and binds to H3K9me or the polycomb-group proteins (PcG), which recognize and bind to H3K27me or other complexes necessary for remodeling and maintenance of the global chromatin higher-order organization. These are also important for the establishment of distinct, probably specific, chromatin structures in defined genomic regions. Thus, local series of less compact, a more active open chromatin regions and a more condensed, closed chromatin patches were observed along the chromatin fiber (Dekker 2008; Filion et al. 2010; Gilbert et al. 2004). There are many studies showing the significance of PTM's of histones for the aging process. Moreover, many of them have been accepted as markers of aging, cancer, and HGPS.

Table 2.2 Alterations in the histone posttranslational modifications during aging, in premature aging syndromes, and cancer

Chromatin feature	Function	Aging/senescence	HGPS	Cancer
Histone PTMs				
Histone acetylation	Active chromatin	Decrease (Ryan and Cristofalo 1972; Adams 2007)	Increase (O'Sullivan and Karlseder 2012)	Increase (Dhanasekaran et al. 2012)
H2Aub	Silent chromatin	Decrease (Lazarus et al. 2013)		
H2Bub	Active chromatin	Decrease (Lazarus et al. 2013)		
H3K4me1,2	Active chromatin	Decrease (O'Sullivan and Karlseder 2012)		Decrease in: lung (Barlesi et al. 2007), adenocarcinomas (Seligson et al. 2009), prostate (Seligson et al. 2009; Ellinger et al. 2010b; Bianco-Miotto et al. 2010), breast (Elsheikh et al. 2009), pancreas (Manuyakorn et al. 2010)
H3K4me3	Active chromatin	Decrease (Cheung et al. 2010; Kuzumaki et al. 2010; Adams 2007)		Decrease: breast (Dhanasekaran et al. 2012; Elsheikh et al. 2009), promoter CGIs (Zane et al. 2014)
		Increase (Zane et al. 2014)		Increase: liver (He et al. 2012), kidney/renal (Ellinger et al. 2010a), glioblastoma (Nagarajan et al. 2014)

H3K9me1	Heterochromatin	Increase (O'Sullivan et al. 2010)	Decrease (Lazarus et al. 2013; Columbaro et al. 2005)	Decrease: kidney/renal (Rogenhofer et al. 2012)
H3K9me2,3	Repressed chromatin	Decrease (Gravina and Vijg 2010; O'Sullivan et al. 2010; Zane et al. 2014)	Decrease (Gravina and Vijg 2010; O'Sullivan and Karlseder 2012; Scaffidi and Misteli 2006; Columbaro et al. 2005; Shumaker et al. 2006)	Decrease: lung (Song et al. 2012; Seligson et al. 2009), prostate (Seligson et al. 2009), breast and colorectal (Leszinski et al. 2012), leukemia (Muller-Tidow et al. 2010), pancreatic adenocarcinoma (Manuyakorn et al. 2010)
		Increase: in SAHF and OIS (oncogene-induced senescence) (O'Sullivan and Karlseder 2012; Zane et al. 2014)		Increase: breast (Leszinski et al. 2012), myeloma (Deligezer et al. 2011), stomach (Park et al. 2008), in promoters CGIs (O'Sullivan and Karlseder 2012; Zane et al. 2014)
H3K9ac	Active chromatin	Decrease (Kawakami et al. 2009)		Decrease: lung (Song et al. 2012), breast (Elsheikh et al. 2009)
		Increase (O'Sullivan and Karlseder 2012; O'Sullivan et al. 2010; Zane et al. 2014)	Increase (Scaffidi and Misteli 2006)	Increase: lung (Barlesi et al. 2007)

(continued)

Table 2.2 (continued)

Chromatin feature	Function	Aging/senescence	HGPS	Cancer
H3K18ac	Transcriptional activation			Decrease: lung (Seligson et al. 2009), prostate (Seligson et al. 2009; Bianco-Miotto et al. 2010), breast (Elsheikh et al. 2009), pancreas (Manuyakorn et al. 2010)
H3K27me1				Decrease: kidney (Rogenhofer et al. 2012)
H3K27me3	Repressive, silent chromatin	Decrease (O'Sullivan and Karlseder 2012; Zane et al. 2014)	Decrease (O'Sullivan and Karlseder 2012; Zane et al. 2014; Shumaker et al. 2006)	Decrease: metastatic prostate (Ellinger et al. 2012), kidney (Leszinski et al. 2012; Rogenhofer et al. 2012), pediatric high-grade gliomas (Bender et al. 2013)
				Increase: localized prostate (Ellinger et al. 2012), gastric-gene specific (Zhang et al. 2009b), liver (He et al. 2012), prostate, breast, ovarian, pancreatic and esophageal (Dhanasekaran et al. 2012; Fullgrabe et al. 2011), promoter CGIs (Zane et al. 2014)
H3K56ac	Chromatin assembly and DNA repair	Decrease (O'Sullivan and Karlseder 2012; O'Sullivan et al. 2010; Dang et al. 2009)		Increase: many cancers (Sharman 2010; Dhanasekaran et al. 2012)

H3S10ph	Condensed chromatin	Decrease (O'Sullivan et al. 2010)		
		Increase (Kawakami et al. 2009)		
H4K5ac	Chromatin replication	Decrease (O'Sullivan and Karlseder 2012)		
		Increase (O'Sullivan et al. 2010)		Increase: lung (Van Den Broeck et al. 2008)
H4K8ac	Active chromatin	Increase (Huang et al. 2007)		Increase: lung (Van Den Broeck et al. 2008)
H4K12ac	Active chromatin	Decrease (O'Sullivan and Karlseder 2012)		Decrease: lung, prostate, colorectal cancers (Dhanasekaran et al. 2012; Van Den Broeck et al. 2008), breast (Elsheikh et al. 2009)
		Increase (Huang et al. 2007)		
H4K16ac	Open chromatin	Decrease (O'Sullivan and Karlseder 2012; O'Sullivan et al. 2010; Zane et al. 2014; Dang et al. 2009)	Decrease (O'Sullivan et al. 2010)	Decrease: lung (Van Den Broeck et al. 2008; Song et al. 2012), breast (Elsheikh et al. 2009; Fraga et al. 2005b), various cancers (Fraga et al. 2005b)
		Increase: in silent loci		
H4K20me1	Active chromatin	Increase (O'Sullivan et al. 2010)		Decrease: prostate (Behbahani et al. 2012)

(continued)

Table 2.2 (continued)

Chromatin feature	Function	Aging/senescence	HGPS	Cancer
H4K20me2	Silent chromatin	Increase (O'Sullivan et al. 2010)		Decrease: prostate (Behbahani et al. 2012)
H4K20me3	Constitutive heterochromatin, silent chromatin	Decrease: in aged cycling fibroblasts		Decrease: various cancers (Dhanasekaran et al. 2012; Fraga et al. 2005b), lung (Van Den Broeck et al. 2008), breast (Elsheikh et al. 2009; Leszinski et al. 2012), bladder and colorectal (Leszinski et al. 2012)
		Increase: global and in SAHF (Lazarus et al. 2013; Zane et al. 2014)	Increase (Shumaker et al. 2006)	Increase: breast (Leszinski et al. 2012), myeloma (Deligezer et al. 2011)
H4R3me2	Active chromatin			Decrease: breast (Elsheikh et al. 2009)
H3/H4-ac		Decrease: in SAHF (Lazarus et al. 2013)		Decrease: prostate (Ellinger et al. 2010b)

2.5.2.1 *Saccharomyces cerevisiae*: The Golden Model for Chromatin and Aging Research

Early work on epigenetics, chromatin, and aging has been extensively done on *Saccharomyces cerevisiae*. In yeast, aging can be studied using two main approaches: replicative lifespan (RLS) and chronological lifespan (CLS). RLS is defined as the number of buds produced before cell death. In practice, the replicative lifespan is measured by counting the number of divisions achieved by a cell whose buds are removed one by one by microdissection.

Yeast replicative aging is comparable to aging phenomena observed in asymmetrically dividing cells of higher eukaryotes such as stem cells (Henderson and Hughes 2014; Lindstrom et al. 2011; Lindstrom and Gottschling 2009). Alternatively, the yeast chronological aging is akin to the aging of nondividing cells such as neurons (Longo and Fabrizio 2012; Longo and Kennedy 2006). CLS is defined as the time a cell survives in a nondividing state, with survival being measured as cell wall integrity or as ability to form a colony. Aging is then characterized based on the cells distribution in regard to their chronological lifespans, obtained by measuring survival with time in a stationary phase culture. Both models of yeast aging are accepted as golden models helping to understand aging in higher eukaryotes.

In these cells, inactivation of the histone deacetylase, Sir2, has led to shortened replicative lifespan (Kennedy et al. 1997). Conversely, activation of Sir2 extended lifespan. This phenomenon could be well apprehended if the logic behind the molecular mechanisms is explained in detail. The anti-aging effect of Sir2 in yeast was, at least in part, due to the translocation of a Sir2-containing protein complex from telomeres to ribosomal DNA (rDNA) repeats. These repeats are prone to recombination and form extrachromosomal rDNA circles (ERCs), which by complicated mechanisms shorten yeast lifespan. At the rDNA repeats, Sir2-mediated histone deacetylation, and consequent heterochromatinization by the Sir2-associated proteins (Sir3 and Sir4), prevents recombination and formation of ERCs, thereby extending lifespan (Kaeberlein et al. 2007; Kennedy et al. 1997). Notably, orthologs of Sir2 have been shown to possess strong anti-aging functions in many other species, including nematodes, flies, and human (Herskovits and Guarente 2014; Guarente 2013). In particular, transgenic overexpression of mammalian *SIRT1* (the closest homolog to invertebrate Sir2) improves aspects of health during aging but does not increase longevity in mice (Herranz and Serrano 2010). The mechanisms of the beneficial effects of SIRT1 are complex and interconnected, including improved genomic stability that prevents the genome against fast aging.

Another epigenetic mark that turned out to be firmly connected to aging is the acetylation of histones. Histone acetylation and deacetylation seem quite important epigenetic modifications throughout the whole organismal lifespan. For example, increased global levels of H4K16ac were detected in old cells. Higher levels of H4K16ac have been demonstrated also at specific regions of the genome of old cells, for example, at the X core and X elements within the telomeric regions (Dang et al. 2009). A correlation between the increased levels of H4K16ac with age and a decreased silencing of reporter genes inserted at these telomere proximal DNA elements was also reported in this chapter (Dang et al. 2009). It has been discussed that the increased levels of H4K16ac in old cells lead to a more open chromatin structure. The N-terminal tail of H4 is a crucial player in the formation of a fully compacted 30 nm chromatin fiber (Dorigo et al. 2003; Robinson et al. 2008). Therefore, it is not surprising that increased H4K16ac is accepted determinative for the more open state of chromatin and for the genome instability observed in aged cells. Interestingly, this specific epigenetic modification has been linked to the linker histone H1, one of the factors responsible for genome stability (Downs and Cote 2005; Downs et al. 2003; Georgieva et al. 2012). It was shown recently that loss of the tumour suppressor protein phosphatase PTEN leads to dissociation of histone H1 from chromatin, elevation of histone H4 acetylation at lysine 16, and decompaction of chromatin (Chen et al. 2014b).

2.5.2.2 Linker Histones: Multiple Roles in Aging and Development

Linker histones are important structural proteins of chromatin. They bind to DNA at its nucleosome entry/exit sites and are involved in the formation and maintenance of higher-order chromatin structure. Studies on H1 mobility in different cell types exposed to various treatments have suggested that modulation of H1-chromatin interactions are one of the earliest events leading to changes in the structure of the chromatin fiber (Catez et al. 2006). Represented by 11 subtypes in higher eukaryotes which are cell and tissue-specific, it is not surprising that linker histones have long been a challenge for scientists. Importantly, it is proven that these histones are essential for development. The reduction of H1 content with 50 % led to severe embryonic defects and finally to death in mice (Fan et al. 2005). Moreover, recently it has been shown that the levels of H1, especially of H1c (one of the subtypes of H1 family), are crucial for the process of terminal differentiation of retina rod cells which requires global reorganization of chromatin and concentration of heterochromatin in the nuclear center (Popova et al. 2013). As already has been discussed in the previous sections, the linker histone H1 is linked to aging. Its loss during senescence is associated with the formation of SAHF (Funayama et al. 2006). As was mentioned, studies on linker histones in higher eukaryotes are impeded by the fact that higher eukaryotes have many subtypes of H1 and some of them are replaceable, i.e., they can be exchanged by other subtypes of linker histones. We have summarized the best known changes in linker and core histones together with other structural chromatin proteins in Table 2.3 during aging, cancer, and HGPS. What is easily seen and very well expressed is the link among these changes in the three discussed states of the organism: the process of getting old, the process of malignization, and the detrimental premature aging features of HGPS. Knowing these changes is a prerequisite for all steps toward successful managing of them. For this, we need model organisms that could offer the most relevant platform for these studies.

Because of its single-copy gene coding for H1, which furthermore is nonessential, the yeast *S. cerevisiae* is a wonderful model for studies of linker histones. *S. cerevisiae* cells offer vivid opportunities for studying the role of linker histones in higher-order chromatin organization in vivo and, moreover, the role of this structure

Table 2.3 Changes in the level of histones and nonhistone chromatin proteins in aged and cancer cells compared to normal cells

Chromatin proteins	Function	Aging/senescence	HGPS	Cancer
H2A, H2B, H3, H4 protein level	Chromatin organization	Decrease (O'Sullivan and Karlseder 2012; O'Sullivan et al. 2010; Zane et al. 2014)		
Histone H1	Higher-order chromatin organization	Decrease in SAHF (Funayama et al. 2006)		Decrease: H1.0 in gliomas (Gabrovsky et al. 2013) H1.0, H1.1, H1.4, H1x in ovarian (Medrzycki et al. 2012)
				Increase: H1.3 in ovarian (Medrzycki et al. 2012)
H2A.1	Chromatin assembly	Decrease (Rogakou and Sekeri-Pataryas 1999)		Increase: liver (Khare et al. 2011)
H2A.2	Chromatin assembly	Increase (Rogakou and Sekeri-Pataryas 1999)		Decrease: liver (Khare et al. 2011)
macro-H2A	Heterochromatin	Increase: in SAHF (Lazarus et al. 2013; O'Sullivan and Karlseder 2012)		Decrease: lung (Sporn et al. 2009), cervical (Novikov et al. 2011)
γ-H2AX(H2AX S139ph)	DNA damage repair	Increase: in SAHF (Lazarus et al. 2013; Gravina and Vijg 2010)	Increase (Gravina and Vijg 2010; Scaffidi and Misteli 2006)	Increase: in several cancers (Sedelnikova and Bonner 2006)
H2A.Z	Active and silent chromatin			Increase: sporadic colorectal, metastatic breast carcinomas (Dalvai and Bystricky 2010)

(continued)

Table 2.3 (continued)

Chromatin proteins	Function	Aging/senescence	HGPS	Cancer
H3.1, H3.2	Replication-coupled chromatin assembly	Decrease (Rogakou and Sekeri-Pataryas 1999)		
H3.3	Transcription-coupled chromatin assembly	Increase (Rogakou and Sekeri-Pataryas 1999)		
CENP-A	Centromere-specific H3 variant	Not changed (Pegoraro et al. 2009)	Decrease (Pegoraro et al. 2009)	Increase
Sirt proteins	Histone deacetylation	Decrease: Sirt1 (O'Sullivan and Karlseder 2012) Sirt6 (Zane et al. 2014)		Decrease: Sirt1 in mutant BRCA1 tumor cells (Wang et al. 2008) Sirt2 in gliomas (Sharman 2010; Inoue et al. 2007) Sirt6 in pancreatic, colorectal, hepatocellular cancers (Yuan et al. 2013)
		Increase: Sirt1 (O'Sullivan et al. 2010) Sirt2 (O'Sullivan et al. 2010)		Increase: Sirt1 in colorectal (Chen et al. 2014a), breast, ovarian, and prostate cancers (Zhang et al. 2009c) Sirt3 in breast cancer (Sharman 2010; Ashraf et al. 2006) Sirt6 in prostate cancer (Liu et al. 2013) Sirt7 in breast cancer (Sharman 2010; Ashraf et al. 2006)

HP1	Heterochromatin	Decrease—global	Decrease (Gravina and Vijg 2010; O'Sullivan and Karlseder 2012; Scaffidi and Misteli 2006; Shumaker et al. 2006)	
		Increase: in SAHF (Lazarus et al. 2013; O'Sullivan and Karlseder 2012; Adams 2007)		Increase: in several cancer cell lines (Lazarus et al. 2013)
HMGA		Increase: in SAHF (Funayama et al. 2006)		
EZH2 (HMT)	Histone methylation	Decrease (Lazarus et al. 2013)	Decrease (Shumaker et al. 2006)	Increase: in glioblastoma multiforme, melanoma, lymphoma, breast, renal, ovarian (Dhanasekaran et al. 2012), prostate (Ellinger et al. 2012) and pancreas (Manuyakorn et al. 2010) cancers

in aging. Recently, we have demonstrated that the yeast linker histone, Hho1p, is involved in the regulation of the aging processes. Mutant yeast cells lacking the gene for the linker histone are viable but inherit highly disordered higher-order chromatin structures which result in perturbed chronological lifespan (Uzunova et al. 2013).

2.5.3 *The Elusive Higher-Order Chromatin Structures Dynamics during Aging*

During the past decade, scientists have started to see some of the principles of chromatin folding dynamics and their implication in different cellular processes. The first discovery was that individual chromosomes occupy discrete domains in the interphase nucleus—named chromosome territories (Tark-Dame et al. 2011; Cremer and Cremer 2006). This finding raised a lot of discussions in the field of chromatin biology and induced further investigations which led to the observation that the mammalian interphase chromosomes are made up of a large number of structural loops, each of which is on average ~1 Mb, that correspond to DNA replication units (Ryba et al. 2010). Generally, chromatin loops are accepted as an important aspect of chromatin organization. They represent the so-called higher-order chromatin structures and demonstrate high dynamics during aging and development. They bring together distant regulatory elements that control gene expression, such as promoters and enhancers (Kadauke and Blobel 2009), and thus control cellular destiny. For instance, enhancers, insulators, and locus control regions can lie at distances from several kb up to one Mb or more away from the genes they regulate and may be located in *cis* (upstream or downstream) or in *trans* (on a different chromosome) (Woodcock 2006).

Thus, chromatin looping is crucial for transcriptional regulation: activation and repression, coordination of initiation and termination, and boundary function. Loop formation is most likely to be a prerequisite for rather than a consequence of transcriptional activation as it occurs prior to gene activation (Palstra 2009). Furthermore, chromatin segments containing active genes loop out from their chromosome territories to reach out a transcription factory for their coordinated expression (Fraser 2006). Figure 2.2 brilliantly illustrates the dynamics in chromatin loop organization during normal development, aging, and some premature aging syndromes. The notion that chromatin loops are important for the overall genome organization is also supported by recent studies in yeast showing that the linker histone is important for the maintenance of bulk chromatin loop organization (Georgieva et al. 2012).

The overall organization of chromatin in loops has been assessed by the method of Chromatin Comet Assay (ChCA) which allows estimation of chromatin loop organization at the level of a single cell (Georgieva et al. 2008, 2012). It has been shown that the general loop size of wild type nuclei is in the range of 0.3 Mb, while when the gene for the linker histone has been deleted, the global chromatin decompaction appeared and the increase in the size of chromatin loops reached 0.4 Mb.

This proves that the yeast linker histone, Hho1p, is involved in the global chromatin compaction, loop formation, and probably in the organization of loop attachment sites. Interestingly, this compaction proved to be also important for the normal aging of these cells as further experiments in following the chronological lifespan of the mutants have demonstrated that the absence of the linker histone leads to features resembling premature aging phenotypes (Uzunova et al. 2013), thus strengthening the idea that chromatin higher-order organization is a prerequisite for cellular normality.

In conclusion, chromatin loops can exert different but very important functions in the nucleus. At one hand, they could bring distantly located sequence elements into spatial proximity, thus allowing communication between these sites and proper transcription or repair. It could also induce the opposite, spatial segregation of certain genomic regions from each other ensuring their independent functions. On the other hand, bulk chromatin loop organization has an impact on cellular morphology, which links certain chromatin structural units with proper execution of the main cellular processes. The observed participation of chromatin loop organization in the aging control is an intriguing phenomenon which needs additional investigations in order to reveal the exact molecular mechanisms involved in it. The last statement highlights the necessity of further studies on the dynamics of folded chromatin conformations beyond the mere detection of long-range interactions.

2.6 Premature Aging Syndromes and Age-Associated Diseases in the Context of Chromatin

As we have already discussed in the previous sections, many recent works suggest that aging is driven by epigenetic changes and that epigenetic perturbations can lead to different progeroid syndromes. We believe that collectively these works propose that understanding and manipulating the epigenome are a good perspective for improving age-related pathologies and for extending healthy lifespan.

2.6.1 Chromatin Structure, Aging, and the Premature Aging Phenotype of HGPS

Accumulation of comprehensive experimental data endorses the concept of the causal role of the disturbed chromatin structure and function in the process of aging (Pegoraro and Misteli 2009; Misteli 2010; Feser and Tyler 2011; O'Sullivan and Karlseder 2012).

A unique opportunity to better understand the link between chromatin structure and aging offers the premature aging syndrome, called Hutchison–Gilford progeria syndrome (HGPS). HGPS is a rare, sporadic genetic disease, which occurs in approximately 1 in 4 million live births (Pollex and Hegele 2004; Gordon et al.

2014). The average life expectancy of HGPS patients is 12–15 years. During this reduced lifespan they progressively accumulate features that resemble those of an aged individual like severe growth retardation, lipodystrophy, skeletal dysplasia, joint contractures, alopecia, skin and nail defects, progressive cardiovascular disease, heart attacks, and strokes (Merideth et al. 2008).

HGPS is caused by an autosomal dominant mutation in the LMNA gene coding for two proteins, lamin A/C (expressed by alternative splicing), major structural components of the nuclear lamina. Nuclear lamina has crucial role in a plethora of cellular events, in particular it maintains nuclear shape and provides sites for peripheral heterochromatin binding (Goldman et al. 2002). Being key structural proteins of the nucleus, lamins play important roles in epigenetics, chromatin organization, DNA replication, regulation of gene expression, and DNA repair, cell proliferation, and differentiation as well as in viral infections (Dechat et al. 2008; Goldman et al. 2002). At present, over 100 mutations in LMNA gene has been reported causing more than 15 distinct diseases designated as laminopathies (Hegele 2005).

The point mutation C1824T in exon 11 of the LMNA gene leads to the activation of a cryptic splice donor site and to production of an internally truncated form of lamin A, which lacks 50 amino acid residues near the C-terminus. This dominant negative isoform of lamin A is referred to as progerin (De Sandre-Giovannoli et al. 2003; Eriksson et al. 2003). In HGPS cells, expression of progerin interferes with lamin A that in turn abrogates nuclear lamina functionality and brings numerous phenotypes, including epigenetic alterations, disorganization of chromatin, abnormal nuclear morphology (Scaffidi and Misteli 2006; Gordon et al. 2014; Kudlow et al. 2007; Dechat et al. 2008; Schreiber and Kennedy 2013). Some of these epigenetic alterations are shortlisted in Tables 2.1, 2.2, and 2.3, where one can find interesting data about DNA methylation, PTMs of histones, and the level of expression of histone proteins in aging, cancer, and HGPS.

Interestingly, most of these epigenetic characteristics are shared among these organismal states and propose good models for studying of age-associated alterations in chromatin and epigenetics. In addition, increased levels of unrepaired DNA damage, genome instability and telomere shortening, altered gene expression, and defective stem cell homeostasis were detected during HGPS development (Scaffidi and Misteli 2006; Kudlow et al. 2007; Dechat et al. 2008; Schreiber and Kennedy 2013). Cells from aged individuals and HGPS patients display intriguing similarities with respect to changes in the epigenome and chromatin organization though in progeroid cells these chromatin changes seem to be more pronounced (Sedivy et al. 2008; Dechat et al. 2008). These include loss of heterochromatin structure, alterations in the patterns of heterochromatin-associated histone PTMs (elevated histone acetylation; decrease of H3K9me3 and H3K27me3; and upregulation of H4K20me3), histone variants (increase of phospho-H2AX and HGPS-specific loss of the centromeric protein CENP-A), and the level of key architectural chromatin proteins (e.g., reduction of heterochromatin protein HP1, chromatin chaperons RBBP4/7, the histone mehtyltransferase EZH2), as well as deregulation of chromatin remodeling complexes (NuRD, the nucleosome remodeling and histone deacetylase complex) (Goldman et al. 2004; Scaffidi and Misteli 2006; Shumaker et al. 2006; Pegoraro and Misteli 2009; Pegoraro et al. 2009).

Generally, in the eukaryotic cell nucleus, heterochromatin is highly organized and may be visualized by electron microscopy as nucleoplasmic heterochromatic foci or peripheral heterochromatin, associated with regions of the lamina (Dechat et al. 2008). Both heterochromatin structures are absent in skin fibroblasts from HGPS patients, pointing out the general loss of heterochromatin in HGPS nuclei (Goldman et al. 2004; Columbaro et al. 2005; Dechat et al. 2008). To highlight the molecular mechanisms involved in the disruption of heterochromatin structures, several research groups examined the level of epigenetic marks—histone PTMs associated with either facultative (H3K27me3) or constitutive heterochromatin (H3K9me3 and H4K20me3) (Goldman et al. 2004; Scaffidi and Misteli 2006; Shumaker et al. 2006; Pegoraro and Misteli 2009).

In addition to changes in histone PTMs, alterations in the amount of some histone variants and key chromatin proteins have been detected in HGPS cells (Tables 2.2 and 2.3). These include increase in γ-H2AX (Scaffidi and Misteli 2006) and loss of the centromeric protein CENP-A (Pegoraro et al. 2009) in HGPS. In contrast, no decrease in CENP-A was detected in physiologically aged cells. Several other nuclear proteins were down-regulated in HGPS including all isoforms of the heterochromatin protein HP1, methyltransferase EZH2, histone deacetylase HDAC1, LAP2 group of lamin A-associated proteins, histone chaperons RBBP4 and RBBP7 (Pegoraro et al. 2009; Scaffidi and Misteli 2006; Shumaker et al. 2006). The reduction of key heterochromatin proteins correlates with the loss of peripheral heterochromatin and abrogation of the normal nuclear architecture.

In human somatic female cells, the inactivated X chromosome can be identified as a heterochromatin domain usually associated with the nuclear lamina. Silencing of inactivated X is provided by H3K27me3 and Xist RNA (Kohlmaier et al. 2004). EZH2 (enhancer of zeste homolog) methyltransferase is a subunit of the Polycomb Repressive Complex 2 (PRC2) that has an essential role in the epigenetic maintenance of repressive chromatin states through trimethylation of H3K27 (Margueron et al. 2009). In contrast, in fibroblast cells from a female HGPS patient, the H3K27me3 mark is lost on the inactive X chromosome with a concomitant down-regulation of EZH2 which leads to some decondensation of this chromosome (Shumaker et al. 2006). Importantly, the alterations in chromatin regulation precede the nuclear shape changes (Shumaker et al. 2006) pointing out the pivotal role of chromatin in pathological premature aging.

Furthermore, genome-wide analysis of H3K27me3 distribution revealed patches of decreased H3K27me3 in HGPS nuclei. In particular, gene-poor regions of HGPS fibroblasts experience a reduction of H3K27me3 compared with the control (McCord et al. 2013). As in previous reports, loss of H3K27me3 in HGPS was associated with downregulation of EZH2 since the mRNA level of EZH2 was significantly reduced in four HGPS fibroblast cell lines compared to normal control samples (McCord et al. 2013; Shumaker et al. 2006; Margueron et al. 2009). Genome-wide analyses revealed a correlation between alterations in patterns of H3K27me3 deposition, genome organization, and DNA-lamin A/C interactions in HGPS skin fibroblasts (McCord et al. 2013). Based on this, authors proposed a model in which accumulation of progerin in the nuclear lamina leads to reduction in the repressive histone mark H3K27me3 in heterochromatin (most possibly through the downregulation of EZH2) and disrupts interactions between heterochromatin and nuclear lamina. These changes may then result in transcriptional misregulation and global loss of spatial chromatin compartmentalization (McCord et al. 2013). In addition, downregulation of the constitutive heterochromatin mark H3K9me3 and loss of pericentric heterochromatin were detected in HGPS cells (Scaffidi and Misteli 2006; Shumaker et al. 2006; Pegoraro and Misteli 2009). The loss of heterochromatinization of pericentric regions induces transcription of pericentromeric satellite III repeats (Shumaker et al. 2006).

In contrast to the reduction in H3K9 and H3K27 trimethylation state, an increase of H4K20me3 was detected in progeroid cells (Shumaker et al. 2006), and this may relate to the telomere shortening observed in these cells. Several lines of evidence argue that the disruption of heterochromatin structure in progeria syndrome can be attributed to deregulation of the Nucleosome Remodeling and Deacetylation (NuRD) complex (Pegoraro et al. 2009). NuRD is a ubiquitous ATP-ase-dependent chromatin remodeling complex that has been shown to associate with pericentromeric heterochromatin and to maintain transcriptional repression at specific promoters (Bowen et al. 2004; Helbling Chadwick et al. 2009). NuRD is a multicomponent complex, containing the histone deacetylases HDAC1 and HDAC2, histone-binding proteins RBBP4 and RBBP7, metastasis-associated protein MTA3, the ATPases CHD3 and CHD4 (chromodomain-helicase-DNA-binding protein) as subunits (Zhang et al. 1999). Besides NuRD, histone chaperone proteins RBBP4 and RBBP7 are shared components of Polycomb PRC2 complex involved in the establishment of heterochromatin (Kuzmichev et al. 2002), and RBBP4 is also a subunit of the CAF-1 complex, which assembles chromatin upon DNA replication and DNA damage repair (Verreault et al. 1996).

RBBP4/7 interact with the amino acid residues 562–664 fragment of lamin A in vivo and neither protein binds to progerin, which lacks residues 607–657 of mature lamin A (Pegoraro et al. 2009) demonstrating that RBBP4/7 mediate physical tethering of NuRD to nuclear lamina. Therefore, the accumulation of progerin in HGPS cells abolishes this interaction that consequently affects the proper chromatin remodeling which in turn triggers deregulation of cellular functions. Further, protein levels of several NuRD components including MTA3, RBBP4, and RBBP7 as well as the HDAC1 amount and activity are reduced in HGPS cells and normal cells expressing exogenous progerin (Pegoraro et al. 2009). The observation that shRNA-mediated silencing of HDAC1, MTA3, CHD3, or CHD4 in normal cells was sufficient to recapitulate age-dependent chromatin defects, e.g., loss of H3K9me3 and HP1γ heterochromatin foci, heterochromatin deregulation, and increased DNA damage, indicates the importance of NuRD in aging-associated chromatin defects (Pegoraro et al. 2009). Moreover, the loss of RBBP4, RBBP7, and HDAC1 subunits of NuRD is not limited to premature aging, but reduced levels of these proteins are also a feature of normally aged cells (Pegoraro et al. 2009).

These findings show that the NuRD complex is deregulated in premature as well as normally aged cells and outline the molecular bases of aging-associated chromatin disorganization. Telomere attrition is one out of the numerous hallmarks of aging in mammals (Cao et al. 2011; López-Otín et al. 2013). Accelerated telomere shortening has also been observed in fibroblasts from HGPS individuals (Huang et al. 2008; Decker et al. 2009). Notably, ectopic expression of progerin in normal fibroblasts recapitulates the faster telomere shortening (Decker et al. 2009). Normally, telomeres and subtelomeric regions are enriched in the H3K9me3 and H4K20me3, epigenetic marks that are characteristic of constitutive heterochromatin (Blasco 2007), and the decrease in H4K20me3 has been associated with telomere elongation (Benetti et al. 2007b). Changes in these histone modifications, namely decrease in H3K9me3 and the increase in H4K20me3 (Shumaker et al. 2006) could account for the enhanced telomere attrition in HGPS cells.

The increased level of persistent DNA damage is another typical feature of physiological and premature aging as already discussed. Cells from HGPS patients and from old individuals displayed focal accumulation of phosphorylated histone H2AX (γ-H2AX). The presence of foci containing γ-H2AX is indicative for unrepaired DNA damage which in turn compromises genomic stability (Liu et al. 2005). However, HGPS individuals do not show higher predisposition to neoplasia that is in contrast to the aged individuals and those affected by other premature syndromes in which the risk of cancer development is significantly high (Hennekam 2006; Fraga et al. 2007).

In addition, similar to cells from HGPS patients and aged individuals, the percentage of cells containing multiple prominent phospho-H2AX foci increases in cells depleted of RBBP4 and RBBP7 (Pegoraro et al. 2009) that implicates chromatin remodeling complexes in the induction of chromatin defects associated with premature aging. Therefore, it is worth noting that the occurrence of the observed structural chromatin defects came about prior to DNA damage (Pegoraro et al. 2009). Revealing which of the age-related changes to chromatin causing aging will benefit development of therapeutic strategies for HGPS and other age-related and accelerated aging diseases.

2.6.2 Age-Associated Diseases: A Challenge for the Biology of Aging

Alzheimer's disease (AD) is a neurodegenerative disorder, which results from not only genetic but also from environmental factors. Epigenetic mechanisms, such as DNA methylation, chromatin remodeling and miRNAs, which may induce alterations in gene expression, are thought to be involved in Alzheimer's disease. The most characteristic feature of AD is cognitive impairment that together with neuropathological changes, extracellular amyloid plaques, intracellular neurofibrillary tangles and loss of synapses and neurons are similar in the early-onset and late-onset forms of the disease. Both forms the early- and late-onset of AD have genetic components. Mutations in three genes identified in the early-onset familial form, i.e., amyloid precursor protein (APP), presenilin-1, and presenilin-2, established the central role of amyloid in the disease. However, it should be noted that mutations in these genes are present in only 13 % of patients with the early-onset form of the disease. The technological advances, such as large-scale genome-wide association studies, have led to identification of more than ten risk genes for the late-onset form of AD (Bettens et al. 2013).

Early onset familial AD accounts for less than 1 % of all AD cases (Lambert and Amouyel 2007). Early-onset Alzheimer's appears to be linked to a genetic defect on chromosome 14, to which late-onset Alzheimer's is not linked (Campion et al. 1999). The genes implicated in these forms of the disease are the gene encoding for APP, located on chromosome 21q21, the gene encoding for presenilin 1 (PSEN1), located on chromosome 14q24.3, and that encoding for presenilin 2 (PSEN2), on chromosome 1q31–q42 (Ertekin-Taner 2007). APP mutations account for less than 0.1 % of AD patients. Mutations in APP are located near the cleavage sites of the protein resulting in increased production of the Amyloid beta peptide (Aβ). The average age of disease onset for this mutation is between 40s and 50s. The majority of AD cases have complex etiology due to both environmental and genetic factors, which alone do not seem sufficient for causing the disease. Several works are demonstrating direct epigenetic modulation of the disease.

The methylation status of 12 specific genes that have been implicated in AD pathology has been reported to exhibit significant "epigenetic drift." It was observed in some of the CpG sites within the DNA-methyltransferase 1 (DNMT1) promoter (Mastroeni et al. 2011). Several studies are demonstrating a correlation between dietary factors, the epigenome, and AD pathology. Nutritional deficit could lead to hyperhomocysteinemia due to alteration in the homocysteine/*S*-adenosylmethionine (Hcy/SAM) (Obeid and Herrmann 2006). Several reports have demonstrated alterations in histone proteins in AD. Phosphorylation of histone H3, a key step in the activation of the mitotic machinery, is increased to a hyperphosphorylated state in AD hippocampal neurons. A nonnuclear form of histone H1 appears to be upregulated in astrocytes and neurons in brain regions that are rich in AD pathology. H1 preferentially binds Aβ-42, as well as Aβ-like structures of numerous proteins (Duce et al. 2006). In addition, the H1 molecule has been shown to be a major target for poly-ADP ribosylation in areas of AD brain.

Aging is universally considered to be one of the most striking risk factors for AD. Why aging should be a risk factor for AD (and other age-related disorders), however, is not well understood, particularly at a mechanistic level. Progressive age-related decline in total methylcytosine has been reported in various organisms. It has been speculated that progressive, age-related, genome-wide hypomethylation may be due to parallel DNMT1 deficits. Age-dependent increase in *S*-adenosylhomocysteine (SAH) relative to SAM might also play a role (Mastroeni et al. 2011). Age-dependent hypomethylation of a number of specific genes related to AD has been reported. For example, methylation of cytosines in the APP promoter, particularly GC-rich elements from approximately −270 to −182 bp, is significantly lower in autopsy cases of 70 years old individuals compared to cases with much younger individuals (Mastroeni et al. 2011). These age-related modifications on DNA methylation alter APP expression and consequently can affect the progressive Aβ deposition with aging in the brain (Tohgi et al. 1999).

Parkinson's disease (PD) is another neurodegenerative disorder associated with aging, affecting the central nervous system and by so effecting the motor functions of the individual producing the so-called Parkinsonian gait characterized by small shuffling steps, dyskinesia presented in the form of hyperkinesia or hypokinesia, and in extreme cases akinesia (Morris et al. 1998). The pathophysiology of the disease is presented as progressive loss of dopaminergic neurons in the *substantia nigra* (SN) *pars compacta* and the frequent histological findings of Lewy bodies composed from the aggregation of alpha-synuclein proteins (Weintraub et al. 2008).

The majority of PD onsets are sporadic which accounts for about 90–95 % of the cases, but the causative factors of the disease are far from concluded. There are many models that link genetics, environmental pollution, as well possible contributions of epigenetics to PD. In 5–10 % of the cases of the disease are genetically provoked meaning that the onset of the disease has a genetic factor. Several genes and chromosomal loci, linked to the development of familial forms of Parkinson disease, named PARK1-16, are associated with autosomal dominant, recessive, and X-linked forms of the disease (Hardy et al. 2009). SNCA gene encodes for alpha-synuclein, a protein abundant in brain tissue and when mutated or elevated has the propensity to form the building blocks for the formation of Lewy bodies in PD. Three point mutations and multiplications in the SNCA gene are known to be linked to autosomal dominant PD (Klein and Schlossmacher 2007).

Another gene that is considered to be a factor for both sporadic and familial forms of the disease is LRRK2. The LRRK2 gene encodes for the large-280 kDa leucine-rich repeat kinase multifunctional protein also called dardarin. The most observed mutations in both sporadic and dominant variants of PD are the increase of catalytic activity of dardarin which causes neuronal toxicity and dopaminergic neuron degeneration (Abdullah et al. 2014). Some authors postulate that change in gene transcription exerts a modulating effect on the central nervous system (CNS) and pathophysiology of neurodegenerative diseases. Epigenetic linkage in neurodegenerative diseases is yet inconclusive and more data are needed; however, there is evidence showing the role of DNA methylation in PD and the role of chromatin remodeling in the persisting effects of dopamine on brain function. Strong evidence indicates that DNA methylation in PD is based on the deregulation of homocysteine (Hcy). Generally, the levels of homocysteine (Hcy) are variable among individuals.

This is the result of genetic or environmental factors that influence the levels of dietary folate thus having a major impact on homocysteine (Hcy) levels (Giles et al. 1995). In the metabolism of methylation reactions, folate is converted into L-methylfolate (L-MTHF) by the enzyme dihydrofolate (DHF) reductase in order to be absorbed by the intestine. The next step is the association of L-MTHF with Hcy and methionine synthetase producing methionine. Methionine is converted to *S*-adenosylmethionine (SAM) and through the joint action of methyltransferases the product *S*-adenosylhomocysteine (SAH) is formed from which subsequently Hcy is generated. The generated Hcy reenters the methylation cycle and serves for generating methyl groups required for methylation reactions on DNA CpG islands or onto the lysine and arginine residues of histone proteins. Increased concentrations of plasma total Hcy have been reported in patients with PD; this effects the SAM/SAH ratio by increasing SAH and decreasing SAM, leading to an overall decrease in methylation potential (Blandini et al. 2001). Better cognitive functions were associated with increased SAM/SAH ratio in patients with PD. This suggests a possible role of methylation in neurodegenerative disorders like PD (Obeid et al. 2009). There are several studies that submit evidence proving the central role of histone modification in neurotoxicity and thus the role of epigenetics in it.

2.7 Chromatin Regulation of Viral Infection

2.7.1 *Differences in Susceptibility Between Species*

Whether a pathogen can cause disease in a host is dependent not only on the virulence of the pathogen but also on the genetic background, the health status, and the age of the host. The differences in susceptibility may be related to a number of factors among which the age can have an overall effect on disease resistance, with the very young and the very old being more susceptible to infection by a wide variety of pathogens. Stress in the form of extreme exertion, shock, a change in environment, climate change, nervousness, or muscle fatigue can have a negative impact on health. Each of these conditions is thought to increase the release of cortisol from the adrenal cortex, causing a suppression of the inflammatory response, thereby facilitating infection. Many viruses introduce DNA into the host-cell nucleus, where they must either embrace or confront chromatin factors as a support or obstacle to completion of its life cycle. Compared to the eukaryotic cell, viruses have compact and rapidly evolving genomes.

Chromatin dynamics and epigenetic modifications play major roles in viral and host chromosome biology (Tempera and Lieberman 2014). In some cases, viruses may use novel or viral-specific epigenetic modifying activities, which may reflect variant pathways that distinguish their behavior from the bulk of the cellular chromosome. The role of chromatin in virus biology depends largely on the lifestyle of the virus, but for all viruses that transverse the nucleus, interactions with chromatin are unavoidable. A number of recent discoveries in virology underscore the importance of chromatin dynamics in the regulation of essential viral processes, including entry, gene expression, and replication (Lieberman 2006).

Cellular chromatin forms a dynamic structure that maintains the stability and accessibility of the host DNA genome. Viruses that enter and persist in the nucleus must, therefore, contend with the forces that drive chromatin formation and regulate chromatin structure (Lieberman 2006; Robertson 2005). The complex and dynamic properties of nuclear chromatin present a formidable challenge to viral gene expression and genome propagation. Recently, a wealth of information has been uncovered regarding the structure, function, and regulation of chromatin during complex cellular processes such as differentiation, recombination, aging, and carcinogenesis. Recent studies have also revealed that chromatin has a major role in the life cycle of many viruses, and that viruses have coevolved with numerous strategies for modulating chromatin-related processes (Lieberman 2006). The genomes of many DNA viruses persist for considerable lengths of time in the host-cell nucleus. The small DNA tumor viruses such as simian virus 40 (SV40) and polyoma virus are assembled into nucleosomal minichromosomes during the DNA replication process. Because these viruses use cellular replication enzymes, it is thought that the viral minichromosomes are assembled through a mechanism that is indistinguishable from nucleosome assembly on replicating cellular chromosomes (Lieberman 2006).

Most viruses package their nucleic acid genomes at very high molecular density with specialized viral packaging proteins to form a capsid particle. Adenovirus genomes are packaged as linear double stranded DNA with viral core proteins that form a chromatin-like structure. The viral DNA is covalently linked with the viral encoded terminal binding protein (TP) and core proteins VII, V, and X (Haruki et al. 2006). Core protein VII has sequence similarity to histone H3 and the basic sperm-specific protein. Transcription of the newly infecting genome utilizes a template that decondenses during entry into the nucleus. The cellular activator, referred to as the Template Activating Factor (TAF), is required for the replication competence of the viral DNA (Haruki et al. 2006).

Herpes Simplex Virus (HSV), like adenovirus, enters cells as a linear double stranded DNA molecule. Several virion-associated proteins are essential for productive infection. The HSV VP22 protein binds to TAF1B, similar to that of Ad VII protein (van Leeuwen et al. 2003). VP16 is another HSV-encoded virion protein that plays a very active role in transcription activation of the viral immediate early genes. VP16 recruits histone modifying and nucleosome remodeling enzymes which indicate that nucleosomes are assembling onto the viral genome during viral entry into the nucleus. The precise details of nucleosome assembly during the early stages of herpes virus entry have not been completely elucidated (Lieberman 2008).

Retrovirus and lentivirus genomes also enter the nucleus as double stranded DNA, but this process appears to differ significantly from that of the constitutively double stranded nuclear DNA viruses, like adeno- and herpes viruses. One major difference is that the double stranded DNA genome is synthesized by virion-associated reverse transcriptase in the cytoplasm. The newly synthesized cytoplasmic DNA forms a pre-integration complex (PIC) that consists of viral and cellular proteins (Suzuki and Craigie 2007). Entry into the nucleus is cell cycle dependent and utilizes the cellular nuclear import machinery, often through a nuclear localization signal on one or more of the PIC components. Once in the nucleus, PIC components like Lens Epithelium-Derived Growth Factor, (LEDGF) have been implicating in anchoring to chromatin. HIV, for example, was found to integrate at chromatin sites enriched in euchromatic histone modifications, including H3 K4 methylation and histone H3 and H4 acetylation (Wang et al. 2007). LEDGF appears to be a decisive and essential factor in mediating the chromatin attachment of the PIC and viral integration (Llano et al. 2006).

The role of chromatin in the organization of latent viruses may help to provide insight into genome stability in general. Viruses, although smaller than cellular chromosomes, confront many of the same challenges as the cellular chromosome. How viruses manage these tasks in the absence of canonical centromeres and telomeres, and sometimes lacking dedicated origins of DNA replication, will be important to define in the near future (Lieberman 2008).

Viruses may trigger in the infected target cells a pathway of programmed cell death (apoptosis), which is characterized by distinct morphological changes such as the formation of condensed fragments of nuclear chromatin. Apoptosis is an energy-dependent process that can be triggered from the activation of calcium-dependent endonucleases whose function is to cleave genomic DNA. The phenomenon of apoptosis seems to play an important role in the control of virus infection since several viruses have developed ways to counteract the apoptosis. Herpes virus, Epstein–Barr virus, adenoviruses, and African swine fever virus have all been reported to code a protein that functions as an inhibitor of apoptosis (Ackermann et al. 2000).

2.7.2 Age-Dependent Organism Susceptibility to Viral Infections

Deterioration of the immune system function is common in advanced aging. Older people experience enhanced susceptibility to infections. This susceptibility is due to many factors but also due to decline in the immune response. The immune system is the adaptive defense system in vertebrates, which evolved to protect them from invading pathogenic microorganisms, viruses, and cancer. It has the ability to generate a variety of immuno-competent cells and molecules that specifically recognize and eliminate pathogens.

Older organisms may succumb to viral infection due to exaggerated immune response as it is proven that in old mouse models there is an elevated immune response as aged mice produce higher serum levels of inflammatory mediator IL-17 than their younger counterparts upon herpes virus infection. Major contributor for the elevated IL-17 are the NKT cells in the aged mice. Aging induces increase in RORyT transcription factor within NKT cell promoter of IL-17. During herpes virus infection, these aged cells produce exaggerated levels of IL-17 leading to worse outcomes in aged host. This challenges the more accepted notion that reduced immunity with aging is the dominant reason for the aged hosts to be more susceptible to viral infections. In unison is the notion that aged, but not young, mice succumb to systemic herpes viral infection due to exaggerated inflammation (Goldstein 2010).

Advanced age also increases the susceptibility of the organism to influenza infection. The mechanism underlying the impaired immune response to infection is not well understood. There are data showing that the advanced age affects dendritic cells' (DC) functions. It has also been shown that the monocyte-derived DC cells from the aged organism are impaired in their capacity to secrete IFN-1 and INF-3, both playing a role in organismal defense against viral respiratory infections. Some authors advocate that the reduction of IFN-1 and IFN-3 is a result of age-associated modifications in the chromatin structure, namely the increase in H3K4me3 and H3K9me3 at the promoters of IFN-1 and IFN-3 (Prakash et al. 2013). This was accompanied by decreased association of these promoters with activator histone marks like H3K4me3 in aged DCs after activation with influenza, thus highlighting the significance of chromatin organization for the age-dependent susceptibility of organisms to viral infections.

2.8 Conclusion

Knowing the intricate mechanisms of aging, age-associated diseases, premature aging syndromes, and the dependence of organismal susceptibility to viruses on aging is a challenge for scientists. It could open new horizons in the field of pharmacogenomics and moreover in the field of yet not-developed pharmacoepigenomics. The last is going to be on the top of the iceberg called "personalized medicine" as it offers brilliant opportunities for scientists to know, control, and heal all age-associated epigenetic changes in the organisms.

Acknowledgements The authors are thankful to Djulia Milcheva and Mathew Serkedjiev for the fruitful discussions and comments on the section regarding age-associated diseases. The authors highly acknowledge Toni Efremov for all technical assistance during the preparation of the current book chapter.

Abbreviations

4C	Circularized Chromosome Conformation Capture
5C	Carbon-Copy Chromosome Conformation Capture
AFM	Atomic force microscope
bp	Base pairs
CGIs	CpG islands
ChCA	Chromatin Comet Assay
ChIA-PET	Chromatin Interaction Analysis with Paired-End Tag Sequencing
DNA	Deoxyribonucleic acid
DNMT	DNA methyltransferase
FISH	Fluorescence In Situ Hybridization
H1	Linker histone
HGPS	Hutchinson-Gilford progeria syndrome
Hi-C	Genome-wide chromosome conformation capture
HMGA	High Mobility Group protein A
HMT	Histone Methyltransferase
HP1	Heterochromatin Protein 1
Kb	Kilobases
Mb	Megabases
Nm	Nanometers
ROS	Reactive Oxygen Species
SAHF	Senescence-Associated Heterochromatin Foci

References

Abdullah, R., Basak, I., Patil, K. S., Alves, G., Larsen, J. P., & Moller, S. G. (2014). Parkinson's disease and age: The obvious but largely unexplored link. *Experimental Gerontology*. doi:10.1016/j.exger.2014.09.014.

Ackermann, H. W., Tremblay, M., & Berthiaume, L. (2000). *Viral pathogenesis in diagrams*. Boca Raton, FL: CRC Press.

Adams, P. D. (2007). Remodeling of chromatin structure in senescent cells and its potential impact on tumor suppression and aging. *Gene, 397*(1–2), 84–93. doi:10.1016/j.gene.2007.04.020.

Ahuja, N., Li, Q., Mohan, A. L., Baylin, S. B., & Issa, J. P. (1998). Aging and DNA methylation in colorectal mucosa and cancer. *Cancer Research, 58*(23), 5489–5494.

Andersson, K., Mahr, R., Bjorkroth, B., & Daneholt, B. (1982). Rapid reformation of the thick chromosome fiber upon completion of RNA synthesis at the Balbiani ring genes in Chironomus tentans. *Chromosoma, 87*(1), 33–48.

Apostolou, E., & Hochedlinger, K. (2013). Chromatin dynamics during cellular reprogramming. *Nature, 502*(7472), 462–471. doi:10.1038/nature12749.

Ashraf, N., Zino, S., Macintyre, A., Kingsmore, D., Payne, A. P., George, W. D., et al. (2006). Altered sirtuin expression is associated with node-positive breast cancer. *British Journal of Cancer, 95*(8), 1056–1061. doi:10.1038/sj.bjc.6603384.

Bannister, A. J., & Kouzarides, T. (2011). Regulation of chromatin by histone modifications. *Cell Research, 21*(3), 381–395. doi:10.1038/cr.2011.22.

Barlesi, F., Giaccone, G., Gallegos-Ruiz, M. I., Loundou, A., Span, S. W., Lefesvre, P., et al. (2007). Global histone modifications predict prognosis of resected non small-cell lung cancer. *Journal of Clinical Oncology, 25*(28), 4358–4364. doi:10.1200/JCO.2007.11.2599.

Behbahani, T. E., Kahl, P., von der Gathen, J., Heukamp, L. C., Baumann, C., Gutgemann, I., et al. (2012). Alterations of global histone H4K20 methylation during prostate carcinogenesis. *BMC Urology, 12*, 5. doi:10.1186/1471-2490-12-5.

Bender, S., Tang, Y., Lindroth, A. M., Hovestadt, V., Jones, D. T., Kool, M., et al. (2013). Reduced H3K27me3 and DNA hypomethylation are major drivers of gene expression in K27M mutant pediatric high-grade gliomas. *Cancer Cell, 24*(5), 660–672. doi:10.1016/j.ccr.2013.10.006.

Benetti, R., Garcia-Cao, M., & Blasco, M. A. (2007a). Telomere length regulates the epigenetic status of mammalian telomeres and subtelomeres. *Nature Genetics, 39*(2), 243–250. doi:10.1038/ng1952.

Benetti, R., Gonzalo, S., Jaco, I., Schotta, G., Klatt, P., Jenuwein, T., et al. (2007b). Suv4-20h deficiency results in telomere elongation and derepression of telomere recombination. *The Journal of Cell Biology, 178*(6), 925–936. doi:10.1083/jcb.200703081.

Bettens, K., Sleegers, K., & Van Broeckhoven, C. (2013). Genetic insights in Alzheimer's disease. *The Lancet Neurology, 12*(1), 92–104. doi:10.1016/S1474-4422(12)70259-4.

Bianco-Miotto, T., Chiam, K., Buchanan, G., Jindal, S., Day, T. K., Thomas, M., et al. (2010). Global levels of specific histone modifications and an epigenetic gene signature predict prostate cancer progression and development. *Cancer Epidemiology, Biomarkers & Prevention, 19*(10), 2611–2622. doi:10.1158/1055-9965.EPI-10-0555.

Blandini, F., Fancellu, R., Martignoni, E., Mangiagalli, A., Pacchetti, C., Samuele, A., et al. (2001). Plasma homocysteine and l-dopa metabolism in patients with Parkinson disease. *Clinical Chemistry, 47*(6), 1102–1104.

Blasco, M. A. (2007). The epigenetic regulation of mammalian telomeres. *Nature Reviews Genetics, 8*(4), 299–309. doi:10.1038/nrg2047.

Bowen, N. J., Fujita, N., Kajita, M., & Wade, P. A. (2004). Mi-2/NuRD: Multiple complexes for many purposes. *Biochimica et Biophysica Acta, 1677*(1-3), 52–57. doi:10.1016/j.bbaexp.2003.10.010.

Burton, A., & Torres-Padilla, M. E. (2014). Chromatin dynamics in the regulation of cell fate allocation during early embryogenesis. *Nature Reviews Molecular Cell Biology, 15*(11), 723–734. doi:10.1038/nrm3885.

Campion, D., Dumanchin, C., Hannequin, D., Dubois, B., Belliard, S., Puel, M., et al. (1999). Early-onset autosomal dominant Alzheimer disease: Prevalence, genetic heterogeneity, and mutation spectrum. *American Journal of Human Genetics, 65*(3), 664–670. doi:10.1086/302553.
Campisi, J. (2000). Cancer, aging and cellular senescence. *In Vivo, 14*(1), 183–188.
Cao, K., Blair, C. D., Faddah, D. A., Kieckhaefer, J. E., Olive, M., Erdos, M. R., et al. (2011). Progerin and telomere dysfunction collaborate to trigger cellular senescence in normal human fibroblasts. *The Journal of Clinical Investigation, 121*(7), 2833–2844. doi:10.1172/JCI43578.
Casillas, M. A., Jr., Lopatina, N., Andrews, L. G., & Tollefsbol, T. O. (2003). Transcriptional control of the DNA methyltransferases is altered in aging and neoplastically-transformed human fibroblasts. *Molecular and Cellular Biochemistry, 252*(1–2), 33–43.
Catez, F., Ueda, T., & Bustin, M. (2006). Determinants of histone H1 mobility and chromatin binding in living cells. *Nature Structural & Molecular Biology, 13*(4), 305–310. doi:10.1038/nsmb1077.
Chandra, T., Ewels, P. A., Schoenfelder, S., Furlan-Magaril, M., Wingett, S. W., Kirschner, K., et al. (2015). Global reorganization of the nuclear landscape in senescent cells. *Cell Reports, 10*(4), 471–484. doi:10.1016/j.celrep.2014.12.055.
Chen, X., Sun, K., Jiao, S., Cai, N., Zhao, X., Zou, H., et al. (2014a). High levels of SIRT1 expression enhance tumorigenesis and associate with a poor prognosis of colorectal carcinoma patients. *Scientific Reports, 4*, 7481. doi:10.1038/srep07481.
Chen, Z. H., Zhu, M., Yang, J., Liang, H., He, J., He, S., et al. (2014b). PTEN interacts with histone H1 and controls chromatin condensation. *Cell Reports, 8*(6), 2003–2014. doi:10.1016/j.celrep.2014.08.008.
Cheung, I., Shulha, H. P., Jiang, Y., Matevossian, A., Wang, J., Weng, Z., et al. (2010). Developmental regulation and individual differences of neuronal H3K4me3 epigenomes in the prefrontal cortex. *Proceedings of the National Academy of Sciences of the United States of America, 107*(19), 8824–8829. doi:10.1073/pnas.1001702107.
Cho, K. S., Elizondo, L. I., & Boerkoel, C. F. (2004). Advances in chromatin remodeling and human disease. *Current Opinion in Genetics & Development, 14*(3), 308–315. doi:10.1016/j.gde.2004.04.015.
Chuang, C. H., Carpenter, A. E., Fuchsova, B., Johnson, T., de Lanerolle, P., & Belmont, A. S. (2006). Long-range directional movement of an interphase chromosome site. *Current Biology, 16*(8), 825–831. doi:10.1016/j.cub.2006.03.059.
Clapier, C. R., & Cairns, B. R. (2009). The biology of chromatin remodeling complexes. *Annual Review of Biochemistry, 78*, 273–304. doi:10.1146/annurev.biochem.77.062706.153223.
Clausell, J., Happel, N., Hale, T. K., Doenecke, D., & Beato, M. (2009). Histone H1 subtypes differentially modulate chromatin condensation without preventing ATP-dependent remodeling by SWI/SNF or NURF. *PLoS One, 4*(10), e0007243. doi:10.1371/journal.pone.0007243.
Columbaro, M., Capanni, C., Mattioli, E., Novelli, G., Parnaik, V. K., Squarzoni, S., et al. (2005). Rescue of heterochromatin organization in Hutchinson-Gilford progeria by drug treatment. *Cellular and Molecular Life Sciences, 62*(22), 2669–2678. doi:10.1007/s00018-005-5318-6.
Cook, P. R., & Brazell, I. A. (1975). Supercoils in human DNA. *Journal of Cell Science, 19*(2), 261–279.
Corona, D. F., Siriaco, G., Armstrong, J. A., Snarskaya, N., McClymont, S. A., Scott, M. P., et al. (2007). ISWI regulates higher-order chromatin structure and histone H1 assembly in vivo. *PLoS Biology, 5*(9), e232. doi:10.1371/journal.pbio.0050232.
Corpet, A., & Stucki, M. (2014). Chromatin maintenance and dynamics in senescence: A spotlight on SAHF formation and the epigenome of senescent cells. *Chromosoma, 123*(5), 423–436. doi:10.1007/s00412-014-0469-6.
Cremer, T., & Cremer, C. (2006). Rise, fall and resurrection of chromosome territories: a historical perspective. Part II. Fall and resurrection of chromosome territories during the 1950s to 1980s. Part III. Chromosome territories and the functional nuclear architecture: Experiments and models from the 1990s to the present. *European Journal of Histochemistry, 50*(4), 223–272.

Cuthbert, G. L., Daujat, S., Snowden, A. W., Erdjument-Bromage, H., Hagiwara, T., Yamada, M., et al. (2004). Histone deimination antagonizes arginine methylation. *Cell, 118*(5), 545–553. doi:10.1016/j.cell.2004.08.020.

Dalvai, M., & Bystricky, K. (2010). The role of histone modifications and variants in regulating gene expression in breast cancer. *Journal of Mammary Gland Biology and Neoplasia, 15*(1), 19–33. doi:10.1007/s10911-010-9167-z.

Dang, W., Steffen, K. K., Perry, R., Dorsey, J. A., Johnson, F. B., Shilatifard, A., et al. (2009). Histone H4 lysine 16 acetylation regulates cellular lifespan. *Nature, 459*(7248), 802–807. doi:10.1038/nature08085.

De Sandre-Giovannoli, A., Bernard, R., Cau, P., Navarro, C., Amiel, J., Boccaccio, I., et al. (2003). Lamin a truncation in Hutchinson-Gilford progeria. *Science, 300*(5628), 2055. doi:10.1126/science.1084125.

Dechat, T., Pfleghaar, K., Sengupta, K., Shimi, T., Shumaker, D. K., Solimando, L., et al. (2008). Nuclear lamins: Major factors in the structural organization and function of the nucleus and chromatin. *Genes & Development, 22*(7), 832–853. doi:10.1101/gad.1652708.

Decker, M. L., Chavez, E., Vulto, I., & Lansdorp, P. M. (2009). Telomere length in Hutchinson-Gilford progeria syndrome. *Mechanisms of Ageing and Development, 130*(6), 377–383. doi:10.1016/j.mad.2009.03.001.

Dekker, J. (2008). Mapping in vivo chromatin interactions in yeast suggests an extended chromatin fiber with regional variation in compaction. *The Journal of Biological Chemistry, 283*(50), 34532–34540. doi:10.1074/jbc.M806479200.

Dekker, J., Rippe, K., Dekker, M., & Kleckner, N. (2002). Capturing chromosome conformation. *Science, 295*(5558), 1306–1311. doi:10.1126/science.1067799.

Deligezer, U., Akisik, E. Z., Akisik, E. E., Kovancilar, M., Bugra, D., Erten, N., Holdenrieder, S., & Dalay, N. (2011). H3K9me3/H4K20me3 ratio in circulating nucleosomes as potential biomarker for colorectal cancer. In P. B. Gahan (Ed.), *Circulating nucleic acids in plasma and serum* (pp. 97–103). Dordrecht: Springer.

Dhanasekaran, K., Arif, M., & Kundu, T. K. (2012). Cancer: An epigenetic landscape. In T. K. Kundu (Ed.), *Epigenetics: Development and disease* (Subcellular Biochemistry, Vol. 61). Netherlands: Springer.

Dhanasekaran, K., Kumari, S., & Kanduri, C. (2013). Noncoding RNAs in chromatin organization and transcription regulation: An epigenetic view. *Sub-Cellular Biochemistry, 61*, 343–372. doi:10.1007/978-94-007-4525-4_15.

Dorigo, B., Schalch, T., Bystricky, K., & Richmond, T. J. (2003). Chromatin fiber folding: Requirement for the histone H4 N-terminal tail. *Journal of Molecular Biology, 327*(1), 85–96.

Downs, J. A., & Cote, J. (2005). Dynamics of chromatin during the repair of DNA double-strand breaks. *Cell Cycle, 4*(10), 1373–1376. doi:10.4161/cc.4.10.2108.

Downs, J. A., Kosmidou, E., Morgan, A., & Jackson, S. P. (2003). Suppression of homologous recombination by the Saccharomyces cerevisiae linker histone. *Molecular Cell, 11*(6), 1685–1692.

Duce, J. A., Smith, D. P., Blake, R. E., Crouch, P. J., Li, Q. X., Masters, C. L., et al. (2006). Linker histone H1 binds to disease associated amyloid-like fibrils. *Journal of Molecular Biology, 361*(3), 493–505. doi:10.1016/j.jmb.2006.06.038.

Earnshaw, W. C., & Laemmli, U. K. (1983). Architecture of metaphase chromosomes and chromosome scaffolds. *The Journal of Cell Biology, 96*(1), 84–93.

Ehrlich, M. (2009). DNA hypomethylation in cancer cells. *Epigenomics, 1*(2), 239–259. doi:10.2217/epi.09.33.

Ellinger, J., Kahl, P., Mertens, C., Rogenhofer, S., Hauser, S., Hartmann, W., et al. (2010a). Prognostic relevance of global histone H3 lysine 4 (H3K4) methylation in renal cell carcinoma. *International Journal of Cancer, 127*(10), 2360–2366. doi:10.1002/ijc.25250.

Ellinger, J., Kahl, P., von der Gathen, J., Heukamp, L. C., Gutgemann, I., Walter, B., et al. (2012). Global histone H3K27 methylation levels are different in localized and metastatic prostate cancer. *Cancer Investigation, 30*(2), 92–97. doi:10.3109/07357907.2011.636117.

Ellinger, J., Kahl, P., von der Gathen, J., Rogenhofer, S., Heukamp, L. C., Gutgemann, I., et al. (2010b). Global levels of histone modifications predict prostate cancer recurrence. *The Prostate, 70*(1), 61–69. doi:10.1002/pros.21038.

Elsheikh, S. E., Green, A. R., Rakha, E. A., Powe, D. G., Ahmed, R. A., Collins, H. M., et al. (2009). Global histone modifications in breast cancer correlate with tumor phenotypes, prognostic factors, and patient outcome. *Cancer Research, 69*(9), 3802–3809. doi:10.1158/0008-5472.CAN-08-3907.

Eriksson, M., Brown, W. T., Gordon, L. B., Glynn, M. W., Singer, J., Scott, L., et al. (2003). Recurrent de novo point mutations in lamin A cause Hutchinson-Gilford progeria syndrome. *Nature, 423*(6937), 293–298. doi:10.1038/nature01629.

Ertekin-Taner, N. (2007). Genetics of Alzheimer's disease: A centennial review. *Neurologic Clinics, 25*(3), 611–667. doi:10.1016/j.ncl.2007.03.009. v.

Fan, Y., Nikitina, T., Zhao, J., Fleury, T. J., Bhattacharyya, R., Bouhassira, E. E., et al. (2005). Histone H1 depletion in mammals alters global chromatin structure but causes specific changes in gene regulation. *Cell, 123*(7), 1199–1212. doi:10.1016/j.cell.2005.10.028.

Feser, J., & Tyler, J. (2011). Chromatin structure as a mediator of aging. *FEBS Letters, 585*(13), 2041–2048. doi:10.1016/j.febslet.2010.11.016.

Filion, G. J., van Bemmel, J. G., Braunschweig, U., Talhout, W., Kind, J., Ward, L. D., et al. (2010). Systematic protein location mapping reveals five principal chromatin types in Drosophila cells. *Cell, 143*(2), 212–224. doi:10.1016/j.cell.2010.09.009.

Finch, J. T., & Klug, A. (1976). Solenoidal model for superstructure in chromatin. *Proceedings of the National Academy of Sciences of the United States of America, 73*(6), 1897–1901.

Fraga, M. F., Agrelo, R., & Esteller, M. (2007). Cross-talk between aging and cancer: The epigenetic language. *Annals of the New York Academy of Sciences, 1100*, 60–74. doi:10.1196/annals.1395.005.

Fraga, M. F., Ballestar, E., Paz, M. F., Ropero, S., Setien, F., Ballestar, M. L., et al. (2005a). Epigenetic differences arise during the lifetime of monozygotic twins. *Proceedings of the National Academy of Sciences of the United States of America, 102*(30), 10604–10609. doi:10.1073/pnas.0500398102.

Fraga, M. F., Ballestar, E., Villar-Garea, A., Boix-Chornet, M., Espada, J., Schotta, G., et al. (2005b). Loss of acetylation at Lys16 and trimethylation at Lys20 of histone H4 is a common hallmark of human cancer. *Nature Genetics, 37*(4), 391–400. doi:10.1038/ng1531.

Fraser, P. (2006). Transcriptional control thrown for a loop. *Current Opinion in Genetics & Development, 16*(5), 490–495. doi:10.1016/j.gde.2006.08.002.

Fuke, C., Shimabukuro, M., Petronis, A., Sugimoto, J., Oda, T., Miura, K., et al. (2004). Age related changes in 5-methylcytosine content in human peripheral leukocytes and placentas: An HPLC-based study. *Annals of Human Genetics, 68*(Pt 3), 196–204. doi:10.1046/j.1529-8817.2004.00081.x.

Fullgrabe, J., Kavanagh, E., & Joseph, B. (2011). Histone onco-modifications. *Oncogene, 30*(31), 3391–3403. doi:10.1038/onc.2011.121.

Funayama, R., Saito, M., Tanobe, H., & Ishikawa, F. (2006). Loss of linker histone H1 in cellular senescence. *The Journal of Cell Biology, 175*(6), 869–880. doi:10.1083/jcb.200604005.

Fussner, E., Strauss, M., Djuric, U., Li, R., Ahmed, K., Hart, M., et al. (2012). Open and closed domains in the mouse genome are configured as 10-nm chromatin fibres. *EMBO Reports, 13*(11), 992–996. doi:10.1038/embor.2012.139.

Gabrovsky, N., Georgieva, M., Laleva, M., Uzunov, K., & Miloshev, G. (2013). Histone H1.0–a potential molecular marker with prognostic value for patients with malignant gliomas. *Acta Neurochirurgica, 155*(8), 1437–1442. doi:10.1007/s00701-013-1802-1.

Garinis, G. A., van der Horst, G. T., Vijg, J., & Hoeijmakers, J. H. (2008). DNA damage and ageing: New-age ideas for an age-old problem. *Nature Cell Biology, 10*(11), 1241–1247. doi:10.1038/ncb1108-1241.

Gasser, S. M. (2002). Visualizing chromatin dynamics in interphase nuclei. *Science, 296*(5572), 1412–1416. doi:10.1126/science.1067703.

Gelato, K. A., & Fischle, W. (2008). Role of histone modifications in defining chromatin structure and function. *Biological Chemistry, 389*(4), 353–363. doi:10.1515/BC.2008.048.

Georgieva, M., Harata, M., & Miloshev, G. (2008). The nuclear actin-related protein Act3p/Arp4 influences yeast cell shape and bulk chromatin organization. *Journal of Cellular Biochemistry, 104*(1), 59–67. doi:10.1002/jcb.21600.

Georgieva, M., Roguev, A., Balashev, K., Zlatanova, J., & Miloshev, G. (2012). Hho1p, the linker histone of Saccharomyces cerevisiae, is important for the proper chromatin organization in vivo. *Biochimica et Biophysica Acta, 1819*(5), 366–374. doi:10.1016/j.bbagrm.2011.12.003.

Georgieva, M., Staneva, D., Uzunova, K., Efremov, T., Balashev, K., Harata, M., et al. (2015). The linker histone in Saccharomyces cerevisiae interacts with actin-related protein 4 and both regulate chromatin structure and cellular morphology. *The International Journal of Biochemistry & Cell Biology, 59*, 182–192. doi:10.1016/j.biocel.2014.12.006.

Ghirlando R, Felsenfeld G (2013) Chromatin structure outside and inside the nucleus. *Biopolymers 99*: 225–232.

Gilbert, N., Boyle, S., Fiegler, H., Woodfine, K., Carter, N. P., & Bickmore, W. A. (2004). Chromatin architecture of the human genome: Gene-rich domains are enriched in open chromatin fibers. *Cell, 118*(5), 555–566. doi:10.1016/j.cell.2004.08.011.

Giles, W. H., Kittner, S. J., Anda, R. F., Croft, J. B., & Casper, M. L. (1995). Serum folate and risk for ischemic stroke. First National Health and Nutrition Examination Survey epidemiologic follow-up study. *Stroke, 26*(7), 1166–1170.

Goldman, R. D., Gruenbaum, Y., Moir, R. D., Shumaker, D. K., & Spann, T. P. (2002). Nuclear lamins: Building blocks of nuclear architecture. *Genes & Development, 16*(5), 533–547. doi:10.1101/gad.960502.

Goldman, R. D., Shumaker, D. K., Erdos, M. R., Eriksson, M., Goldman, A. E., Gordon, L. B., et al. (2004). Accumulation of mutant lamin A causes progressive changes in nuclear architecture in Hutchinson-Gilford progeria syndrome. *Proceedings of the National Academy of Sciences of the United States of America, 101*(24), 8963–8968. doi:10.1073/pnas.0402943101.

Goldstein, D. R. (2010). Aging, imbalanced inflammation and viral infection. *Virulence, 1*(4), 295–298. doi:10.4161/viru.1.4.12009.

Gordon, L. B., Rothman, F. G., Lopez-Otin, C., & Misteli, T. (2014). Progeria: A paradigm for translational medicine. *Cell, 156*(3), 400–407. doi:10.1016/j.cell.2013.12.028.

Gotta, M., Strahl-Bolsinger, S., Renauld, H., Laroche, T., Kennedy, B. K., Grunstein, M., et al. (1997). Localization of Sir2p: The nucleolus as a compartment for silent information regulators. *The EMBO Journal, 16*(11), 3243–3255. doi:10.1093/emboj/16.11.3243.

Gravina, S., & Vijg, J. (2010). Epigenetic factors in aging and longevity. *Pflügers Archiv - European Journal of Physiology, 459*(2), 247–258. doi:10.1007/s00424-009-0730-7.

Grigoryev, S. A., Arya, G., Correll, S., Woodcock, C. L., & Schlick, T. (2009). Evidence for heteromorphic chromatin fibers from analysis of nucleosome interactions. *Proceedings of the National Academy of Sciences of the United States of America, 106*(32), 13317–13322. doi:10.1073/pnas.0903280106.

Guarente, L. (2013). Introduction: Sirtuins in aging and diseases. *Methods in Molecular Biology, 1077*, 3–10. doi:10.1007/978-1-62703-637-5_1.

Guerrero, N., Mendes de Leon, C. F., Evans, D. A., & Jacobs, E. A. (2015). Determinants of trust in health care in an older population. *Journal of the American Geriatrics Society*. doi:10.1111/jgs.13316.

Hannum, G., Guinney, J., Zhao, L., Zhang, L., Hughes, G., Sadda, S., et al. (2013). Genome-wide methylation profiles reveal quantitative views of human aging rates. *Molecular Cell, 49*(2), 359–367. doi:10.1016/j.molcel.2012.10.016.

Happel, N., & Doenecke, D. (2009). Histone H1 and its isoforms: Contribution to chromatin structure and function. *Gene, 431*(1-2), 1–12. doi:10.1016/j.gene.2008.11.003.

Hardy, J., Lewis, P., Revesz, T., Lees, A., & Paisan-Ruiz, C. (2009). The genetics of Parkinson's syndromes: A critical review. *Current Opinion in Genetics & Development, 19*(3), 254–265. doi:10.1016/j.gde.2009.03.008.

Haruki, H., Okuwaki, M., Miyagishi, M., Taira, K., & Nagata, K. (2006). Involvement of template-activating factor I/SET in transcription of adenovirus early genes as a positive-acting factor. *Journal of Virology, 80*(2), 794–801. doi:10.1128/JVI.80.2.794-801.2006.

Hayes, J. J., Tullius, T. D., & Wolffe, A. P. (1990). The structure of DNA in a nucleosome. *Proceedings of the National Academy of Sciences of the United States of America, 87*(19), 7405–7409.

He, C., Xu, J., Zhang, J., Xie, D., Ye, H., Xiao, Z., et al. (2012). High expression of trimethylated histone H3 lysine 4 is associated with poor prognosis in hepatocellular carcinoma. *Human Pathology, 43*(9), 1425–1435. doi:10.1016/j.humpath.2011.11.003.

Hegele, R. (2005). LMNA mutation position predicts organ system involvement in laminopathies. *Clinical Genetics, 68*(1), 31–34. doi:10.1111/j.1399-0004.2005.00447.x.

Helbling Chadwick, L., Chadwick, B. P., Jaye, D. L., & Wade, P. A. (2009). The Mi-2/NuRD complex associates with pericentromeric heterochromatin during S phase in rapidly proliferating lymphoid cells. *Chromosoma, 118*(4), 445–457. doi:10.1007/s00412-009-0207-7.

Henderson, K. A., & Hughes, A. L. (2014). Mother-daughter asymmetry of pH underlies aging and rejuvenation in yeast. *Elife, 3*, e03504. doi:10.7554/eLife.03504.

Hennekam, R. C. (2006). Hutchinson-Gilford progeria syndrome: Review of the phenotype. *American Journal of Medical Genetics Part A, 140*(23), 2603–2624. doi:10.1002/ajmg.a.31346.

Hermann, A., Gowher, H., & Jeltsch, A. (2004). Biochemistry and biology of mammalian DNA methyltransferases. *Cellular and Molecular Life Sciences, 61*(19–20), 2571–2587. doi:10.1007/s00018-004-4201-1.

Herranz, D., & Serrano, M. (2010). SIRT1: Recent lessons from mouse models. *Nature Reviews Cancer, 10*(12), 819–823. doi:10.1038/nrc2962.

Herskind, A. M., McGue, M., Holm, N. V., Sorensen, T. I., Harvald, B., & Vaupel, J. W. (1996). The heritability of human longevity: A population-based study of 2872 Danish twin pairs born 1870-1900. *Human Genetics, 97*(3), 319–323.

Herskovits, A. Z., & Guarente, L. (2014). SIRT1 in neurodevelopment and brain senescence. *Neuron, 81*(3), 471–483. doi:10.1016/j.neuron.2014.01.028.

Heyn, H., Moran, S., & Esteller, M. (2013). Aberrant DNA methylation profiles in the premature aging disorders Hutchinson-Gilford Progeria and Werner syndrome. *Epigenetics, 8*(1), 28–33. doi:10.4161/epi.23366.

Horvath, S. (2013). DNA methylation age of human tissues and cell types. *Genome Biology, 14*(10), R115. doi:10.1186/gb-2013-14-10-r115.

Horvath, S., Garagnani, P., Bacalini, M. G., Pirazzini, C., Salvioli, S., Gentilini, D., et al. (2015). Accelerated epigenetic aging in Down syndrome. *Aging Cell*. doi:10.1111/acel.12325.

Huang, S., Risques, R. A., Martin, G. M., Rabinovitch, P. S., & Oshima, J. (2008). Accelerated telomere shortening and replicative senescence in human fibroblasts overexpressing mutant and wild-type lamin A. *Experimental Cell Research, 314*(1), 82–91. doi:10.1016/j.yexcr.2007.08.004.

Huang, J. C., Yan, L. Y., Lei, Z. L., Miao, Y. L., Shi, L. H., Yang, J. W., et al. (2007). Changes in histone acetylation during postovulatory aging of mouse oocyte. *Biology of Reproduction, 77*(4), 666–670. doi:10.1095/biolreprod.107.062703.

Huynh, V. A., Robinson, P. J., & Rhodes, D. (2005). A method for the in vitro reconstitution of a defined "30 nm" chromatin fibre containing stoichiometric amounts of the linker histone. *Journal of Molecular Biology, 345*(5), 957–968. doi:10.1016/j.jmb.2004.10.075.

Inoue, T., Hiratsuka, M., Osaki, M., & Oshimura, M. (2007). The molecular biology of mammalian SIRT proteins: SIRT2 in cell cycle regulation. *Cell Cycle, 6*(9), 1011–1018.

Ishimi, Y., Kojima, M., Takeuchi, F., Miyamoto, T., Yamada, M., & Hanaoka, F. (1987). Changes in chromatin structure during aging of human skin fibroblasts. *Experimental Cell Research, 169*(2), 458–467.

Issa, J. P., Ahuja, N., Toyota, M., Bronner, M. P., & Brentnall, T. A. (2001). Accelerated age-related CpG island methylation in ulcerative colitis. *Cancer Research, 61*(9), 3573–3577.

Issa, J. P., Ottaviano, Y. L., Celano, P., Hamilton, S. R., Davidson, N. E., & Baylin, S. B. (1994). Methylation of the oestrogen receptor CpG island links ageing and neoplasia in human colon. *Nature Genetics, 7*(4), 536–540. doi:10.1038/ng0894-536.

Ito, T. (2007). Role of histone modification in chromatin dynamics. *Journal of Biochemistry, 141*(5), 609–614. doi:10.1093/jb/mvm091.

Iyer, L. M., Abhiman, S., & Aravind, L. (2011). Natural history of eukaryotic DNA methylation systems. *Progress in Molecular Biology and Translational Science, 101*, 25–104. doi:10.1016/b978-0-12-387685-0.00002-0.

Kadauke, S., & Blobel, G. A. (2009). Chromatin loops in gene regulation. *Biochimica et Biophysica Acta, 1789*(1), 17–25. doi:10.1016/j.bbagrm.2008.07.002.

Kaeberlein, M., Burtner, C. R., & Kennedy, B. K. (2007). Recent developments in yeast aging. *PLoS Genetics, 3*(5), e84. doi:10.1371/journal.pgen.0030084.

Karymov, M. A., Tomschik, M., Leuba, S. H., Caiafa, P., & Zlatanova, J. (2001). DNA methylation-dependent chromatin fiber compaction in vivo and in vitro: Requirement for linker histone. *FASEB Journal, 15*(14), 2631–2641. doi:10.1096/fj.01-0345com.

Kawakami, K., Nakamura, A., Ishigami, A., Goto, S., & Takahashi, R. (2009). Age-related difference of site-specific histone modifications in rat liver. *Biogerontology, 10*(4), 415–421. doi:10.1007/s10522-008-9176-0.

Kennedy, B. K., Gotta, M., Sinclair, D. A., Mills, K., McNabb, D. S., Murthy, M., et al. (1997). Redistribution of silencing proteins from telomeres to the nucleolus is associated with extension of life span in S. cerevisiae. *Cell, 89*(3), 381–391.

Khare, S. P., Sharma, A., Deodhar, K. K., & Gupta, S. (2011). Overexpression of histone variant H2A.1 and cellular transformation are related in N-nitrosodiethylamine-induced sequential hepatocarcinogenesis. *Experimental Biology and Medicine, 236*(1), 30–35. doi:10.1258/ebm.2010.010140.

Kim, J. Y., Tavare, S., & Shibata, D. (2005). Counting human somatic cell replications: Methylation mirrors endometrial stem cell divisions. *Proceedings of the National Academy of Sciences of the United States of America, 102*(49), 17739–17744. doi:10.1073/pnas.0503976102.

Klein, C., & Schlossmacher, M. G. (2007). Parkinson disease, 10 years after its genetic revolution: Multiple clues to a complex disorder. *Neurology, 69*(22), 2093–2104. doi:10.1212/01.wnl.0000271880.27321.a7.

Kohlmaier, A., Savarese, F., Lachner, M., Martens, J., Jenuwein, T., & Wutz, A. (2004). A chromosomal memory triggered by Xist regulates histone methylation in X inactivation. *PLoS Biology, 2*(7), E171. doi:10.1371/journal.pbio.0020171.

Kudlow, B. A., Kennedy, B. K., & Monnat, R. J., Jr. (2007). Werner and Hutchinson-Gilford progeria syndromes: Mechanistic basis of human progeroid diseases. *Nature Reviews Molecular Cell Biology, 8*(5), 394–404. doi:10.1038/nrm2161.

Kuzmichev, A., Nishioka, K., Erdjument-Bromage, H., Tempst, P., & Reinberg, D. (2002). Histone methyltransferase activity associated with a human multiprotein complex containing the Enhancer of Zeste protein. *Genes & Development, 16*(22), 2893–2905. doi:10.1101/gad.1035902.

Kuzumaki, N., Ikegami, D., Tamura, R., Sasaki, T., Niikura, K., Narita, M., et al. (2010). Hippocampal epigenetic modification at the doublecortin gene is involved in the impairment of neurogenesis with aging. *Synapse, 64*(8), 611–616. doi:10.1002/syn.20768.

Lambert, J. C., & Amouyel, P. (2007). Genetic heterogeneity of Alzheimer's disease: Complexity and advances. *Psychoneuroendocrinology, 32*(Suppl 1), S62–70. doi:10.1016/j.psyneuen.2007.05.015.

Langmore, J. P., & Schutt, C. (1980). The higher order structure of chicken erythrocyte chromosomes in vivo. *Nature, 288*(5791), 620–622.

Lazarus, A., Banerjee, K. K., & Kolthur-Seetharam, U. (2013). Small changes, big effects: Chromatin goes aging. *Sub-Cellular Biochemistry, 61*, 151–176. doi:10.1007/978-94-007-4525-4_8.

Leszinski, G., Gezer, U., Siegele, B., Stoetzer, O., & Holdenrieder, S. (2012). Relevance of histone marks H3K9me3 and H4K20me3 in cancer. *Anticancer Research, 32*(5), 2199–2205.

Li, L. C. (2014). Chromatin remodeling by the small RNA machinery in mammalian cells. *Epigenetics, 9*(1), 45–52. doi:10.4161/epi.26830.

Lieberman, P. M. (2006). Chromatin regulation of virus infection. *Trends in Microbiology, 14*(3), 132–140. doi:10.1016/j.tim.2006.01.001.

Lieberman, P. M. (2008). Chromatin organization and virus gene expression. *Journal of Cellular Physiology, 216*(2), 295–302. doi:10.1002/jcp.21421.

Lindstrom, D. L., & Gottschling, D. E. (2009). The mother enrichment program: A genetic system for facile replicative life span analysis in Saccharomyces cerevisiae. *Genetics, 183*(2), 413–422. doi:10.1534/genetics.109.106229.

Lindstrom, D. L., Leverich, C. K., Henderson, K. A., & Gottschling, D. E. (2011). Replicative age induces mitotic recombination in the ribosomal RNA gene cluster of Saccharomyces cerevisiae. *PLoS Genetics, 7*(3), e1002015. doi:10.1371/journal.pgen.1002015.

Littau, V. C., Burdick, C. J., Allfrey, V. G., & Mirsky, S. A. (1965). The role of histones in the maintenance of chromatin structure. *Proceedings of the National Academy of Sciences of the United States of America, 54*(4), 1204–1212.

Liu, B., Wang, J., Chan, K. M., Tjia, W. M., Deng, W., Guan, X., et al. (2005). Genomic instability in laminopathy-based premature aging. *Nature Medicine, 11*(7), 780–785. doi:10.1038/nm1266.

Liu, Y., Xie, Q. R., Wang, B., Shao, J., Zhang, T., Liu, T., et al. (2013). Inhibition of SIRT6 in prostate cancer reduces cell viability and increases sensitivity to chemotherapeutics. *Protein & Cell*. doi:10.1007/s13238-013-3054-5.

Llano, M., Saenz, D. T., Meehan, A., Wongthida, P., Peretz, M., Walker, W. H., et al. (2006). An essential role for LEDGF/p75 in HIV integration. *Science, 314*(5798), 461–464. doi:10.1126/science.1132319.

Longo, V. D., & Fabrizio, P. (2012). Chronological aging in Saccharomyces cerevisiae. *Sub-Cellular Biochemistry, 57*, 101–121. doi:10.1007/978-94-007-2561-4_5.

Longo, V. D., & Kennedy, B. K. (2006). Sirtuins in aging and age-related disease. *Cell, 126*(2), 257–268. doi:10.1016/j.cell.2006.07.002.

Lopatina, N., Haskell, J. F., Andrews, L. G., Poole, J. C., Saldanha, S., & Tollefsbol, T. (2002). Differential maintenance and de novo methylating activity by three DNA methyltransferases in aging and immortalized fibroblasts. *Journal of Cellular Biochemistry, 84*(2), 324–334.

López-Otín, C., Blasco, M. A., Partridge, L., Serrano, M., & Kroemer, G. (2013). The hallmarks of aging. *Cell, 153*(6), 1194–1217. doi:10.1016/j.cell.2013.05.039.

Luger, K., Mäder, A. W., Richmond, R. K., Sargent, D. F., & Richmond, T. J. (1997). Crystal structure of the nucleosome core particle at 2.8 Å resolution. *Nature, 389*(6648), 251–260. doi:10.1038/38444.

Ma, Y., Jacobs, S. B., Jackson-Grusby, L., Mastrangelo, M. A., Torres-Betancourt, J. A., Jaenisch, R., et al. (2005). DNA CpG hypomethylation induces heterochromatin reorganization involving the histone variant macroH2A. *Journal of Cell Science, 118*(Pt 8), 1607–1616. doi:10.1242/jcs.02291.

Maeshima, K., Imai, R., Tamura, S., & Nozaki, T. (2014). Chromatin as dynamic 10-nm fibers. *Chromosoma, 123*(3), 225–237. doi:10.1007/s00412-014-0460-2.

Manuyakorn, A., Paulus, R., Farrell, J., Dawson, N. A., Tze, S., Cheung-Lau, G., et al. (2010). Cellular histone modification patterns predict prognosis and treatment response in resectable pancreatic adenocarcinoma: Results from RTOG 9704. *Journal of Clinical Oncology, 28*(8), 1358–1365. doi:10.1200/JCO.2009.24.5639.

Margueron, R., Justin, N., Ohno, K., Sharpe, M. L., Son, J., Drury, W. J., 3rd, et al. (2009). Role of the polycomb protein EED in the propagation of repressive histone marks. *Nature, 461*(7265), 762–767. doi:10.1038/nature08398.

Marioni, R. E., Shah, S., McRae, A. F., Chen, B. H., Colicino, E., Harris, S. E., et al. (2015). DNA methylation age of blood predicts all-cause mortality in later life. *Genome Biology, 16*(1), 25. doi:10.1186/s13059-015-0584-6.

Marsden, M. P., & Laemmli, U. K. (1979). Metaphase chromosome structure: Evidence for a radial loop model. *Cell, 17*(4), 849–858. doi:10.1016/0092-8674(79)90325-8.

Martin, G. M. (2005). Epigenetic drift in aging identical twins. *Proceedings of the National Academy of Sciences of the United States of America, 102*(30), 10413–10414. doi:10.1073/pnas.0504743102.

Mastroeni, D., Grover, A., Delvaux, E., Whiteside, C., Coleman, P. D., & Rogers, J. (2011). Epigenetic mechanisms in Alzheimer's disease. *Neurobiology of Aging, 32*(7), 1161–1180. doi:10.1016/j.neurobiolaging.2010.08.017.

McCord, R. P., Nazario-Toole, A., Zhang, H., Chines, P. S., Zhan, Y., Erdos, M. R., et al. (2013). Correlated alterations in genome organization, histone methylation, and DNA-lamin A/C interactions in Hutchinson-Gilford progeria syndrome. *Genome Research, 23*(2), 260–269. doi:10.1101/gr.138032.112.

McGarvey, K. M., Van Neste, L., Cope, L., Ohm, J. E., Herman, J. G., Van Criekinge, W., et al. (2008). Defining a chromatin pattern that characterizes DNA-hypermethylated genes in colon cancer cells. *Cancer Research, 68*(14), 5753–5759. doi:10.1158/0008-5472.can-08-0700.

McKenzie, S. J., McLaughlin, D., Clark, J., & Doi, S. A. (2015). The burden of non-adherence to cardiovascular medications among the aging population in Australia: A meta-analysis. *Drugs & Aging*. doi:10.1007/s40266-015-0245-1.

Medrzycki, M., Zhang, Y., McDonald, J. F., & Fan, Y. (2012). Profiling of linker histone variants in ovarian cancer. *Frontiers in Bioscience, 17*, 396–406.

Merideth, M. A., Gordon, L. B., Clauss, S., Sachdev, V., Smith, A. C., Perry, M. B., et al. (2008). Phenotype and course of Hutchinson-Gilford progeria syndrome. *The New England Journal of Medicine, 358*(6), 592–604. doi:10.1056/NEJMoa0706898.

Miquel, P. A. (2014). Aging as alteration. *Interdisciplinary Topics in Gerontology, 39*, 187–197. doi:10.1159/000358906.

Misteli, T. (2010). Higher-order genome organization in human disease. *Cold Spring Harbor Perspectives in Biology, 2*(8), a000794. doi:10.1101/cshperspect.a000794.

Mitteldorf, J. (2015). How does the body know how old it is? Introducing the epigenetic clock hypothesis. *Interdisciplinary Topics in Gerontology, 40*, 49–62. doi:10.1159/000364929.

Morris, M., Iansek, R., Matyas, T., & Summers, J. (1998). Abnormalities in the stride length-cadence relation in Parkinsonian gait. *Movement Disorders, 13*(1), 61–69. doi:10.1002/mds.870130115.

Muller-Tidow, C., Klein, H. U., Hascher, A., Isken, F., Tickenbrock, L., Thoennissen, N., et al. (2010). Profiling of histone H3 lysine 9 trimethylation levels predicts transcription factor activity and survival in acute myeloid leukemia. *Blood, 116*(18), 3564–3571. doi:10.1182/blood-2009-09-240978.

Muñoz-Najar, U., & Sedivy, J. M. (2011). Epigenetic control of aging. *Antioxidants and Redox Signaling, 14*(2), 241–259. doi:10.1089/ars.2010.3250.

Nagarajan, R. P., Zhang, B., Bell, R. J., Johnson, B. E., Olshen, A. B., Sundaram, V., et al. (2014). Recurrent epimutations activate gene body promoters in primary glioblastoma. *Genome Research, 24*(5), 761–774. doi:10.1101/gr.164707.113.

Narita, M., Nunez, S., Heard, E., Narita, M., Lin, A. W., Hearn, S. A., et al. (2003). Rb-mediated heterochromatin formation and silencing of E2F target genes during cellular senescence. *Cell, 113*(6), 703–716.

Novikov, L., Park, J. W., Chen, H., Klerman, H., Jalloh, A. S., & Gamble, M. J. (2011). QKI-mediated alternative splicing of the histone variant MacroH2A1 regulates cancer cell proliferation. *Molecular and Cellular Biology, 31*(20), 4244–4255. doi:10.1128/MCB.05244-11.

O'Sullivan, R. J., & Karlseder, J. (2012). The great unravelling: Chromatin as a modulator of the aging process. *Trends in Biochemical Sciences, 37*(11), 466–476. doi:10.1016/j.tibs.2012.08.001.

O'Sullivan, R. J., Kubicek, S., Schreiber, S. L., & Karlseder, J. (2010). Reduced histone biosynthesis and chromatin changes arising from a damage signal at telomeres. *Nature Structural & Molecular Biology, 17*(10), 1218–1225. doi:10.1038/nsmb.1897.

Oakes, C. C., Smiraglia, D. J., Plass, C., Trasler, J. M., & Robaire, B. (2003). Aging results in hypermethylation of ribosomal DNA in sperm and liver of male rats. *Proceedings of the*

National Academy of Sciences of the United States of America, 100(4), 1775–1780. doi:10.1073/pnas.0437971100.

Obeid, R., & Herrmann, W. (2006). Mechanisms of homocysteine neurotoxicity in neurodegenerative diseases with special reference to dementia. *FEBS Letters, 580*(13), 2994–3005. doi:10.1016/j.febslet.2006.04.088.

Obeid, R., Schadt, A., Dillmann, U., Kostopoulos, P., Fassbender, K., & Herrmann, W. (2009). Methylation status and neurodegenerative markers in Parkinson disease. *Clinical Chemistry, 55*(10), 1852–1860. doi:10.1373/clinchem.2009.125021.

Oberdoerffer, P., & Sinclair, D. A. (2007). The role of nuclear architecture in genomic instability and ageing. *Nature Reviews Molecular Cell Biology, 8*(9), 692–702. doi:10.1038/nrm2238.

Oh, J. H., Gertych, A., & Tajbakhsh, J. (2013). Nuclear DNA methylation and chromatin condensation phenotypes are distinct between normally proliferating/aging, rapidly growing/immortal, and senescent cells. *Oncotarget, 4*(3), 474–493.

Palstra, R. J. (2009). Close encounters of the 3C kind: Long-range chromatin interactions and transcriptional regulation. *Briefings in Functional Genomics & Proteomics, 8*(4), 297–309. doi:10.1093/bfgp/elp016.

Park, Y. S., Jin, M. Y., Kim, Y. J., Yook, J. H., Kim, B. S., & Jang, S. J. (2008). The global histone modification pattern correlates with cancer recurrence and overall survival in gastric adenocarcinoma. *Annals of Surgical Oncology, 15*(7), 1968–1976. doi:10.1245/s10434-008-9927-9.

Pegoraro, G., Kubben, N., Wickert, U., Gohler, H., Hoffmann, K., & Misteli, T. (2009). Ageing-related chromatin defects through loss of the NURD complex. *Nature Cell Biology, 11*(10), 1261–1267. doi:10.1038/ncb1971.

Pegoraro, G., & Misteli, T. (2009). The central role of chromatin maintenance in aging. *Aging, 1*(12), 1017–1022.

Pollex, R. L., & Hegele, R. A. (2004). Hutchinson-Gilford progeria syndrome. *Clinical Genetics, 66*(5), 375–381. doi:10.1111/j.1399-0004.2004.00315.x.

Popova, E. Y., Grigoryev, S. A., Fan, Y., Skoultchi, A. I., Zhang, S. S., & Barnstable, C. J. (2013). Developmentally regulated linker histone H1c promotes heterochromatin condensation and mediates structural integrity of rod photoreceptors in mouse retina. *The Journal of Biological Chemistry, 288*(24), 17895–17907. doi:10.1074/jbc.M113.452144.

Prakash, S., Agrawal, S., Cao, J. N., Gupta, S., & Agrawal, A. (2013). Impaired secretion of interferons by dendritic cells from aged subjects to influenza: Role of histone modifications. *Age, 35*(5), 1785–1797. doi:10.1007/s11357-012-9477-8.

Prieto, E., Hizume, K., Kobori, T., Yoshimura, S. H., & Takeyasu, K. (2012). Core histone charge and linker histone H1 effects on the chromatin structure of Schizosaccharomyces pombe. *Bioscience, Biotechnology, and Biochemistry, 76*(12), 2261–2266. doi:10.1271/bbb.120548.

Rattner, J. B., Saunders, C., Davie, J. R., & Hamkalo, B. A. (1982). Ultrastructural organization of yeast chromatin. *Journal of Cell Biology, 93*(1), 217–222.

Richmond, T. J., Finch, J. T., Rushton, B., Rhodes, D., & Klug, A. (1984). Structure of the nucleosome core particle at 7 resolution. *Nature, 311*(5986), 532–537. doi:10.1038/311532a0.

Riggs, A. D. (1975). X inactivation, differentiation, and DNA methylation. *Cytogenetics and Cell Genetics, 14*(1), 9–25.

Robertson, K. D. (2005). DNA methylation and human disease. *Nature Reviews Genetics, 6*(8), 597–610. doi:10.1038/nrg1655.

Robertson, K. D., Uzvolgyi, E., Liang, G., Talmadge, C., Sumegi, J., Gonzales, F. A., et al. (1999). The human DNA methyltransferases (DNMTs) 1, 3a and 3b: Coordinate mRNA expression in normal tissues and overexpression in tumors. *Nucleic Acids Research, 27*(11), 2291–2298.

Robinson, P. J., An, W., Routh, A., Martino, F., Chapman, L., Roeder, R. G., et al. (2008). 30 nm chromatin fibre decompaction requires both H4-K16 acetylation and linker histone eviction. *Journal of Molecular Biology, 381*(4), 816–825. doi:10.1016/j.jmb.2008.04.050.

Robinson, P. J., Fairall, L., Huynh, V. A., & Rhodes, D. (2006). EM measurements define the dimensions of the "30-nm" chromatin fiber: Evidence for a compact, interdigitated structure. *Proceedings of the National Academy of Sciences of the United States of America, 103*(17), 6506–6511. doi:10.1073/pnas.0601212103.

Robinson, P. J., & Rhodes, D. (2006). Structure of the '30 nm' chromatin fibre: A key role for the linker histone. *Current Opinion in Structural Biology, 16*(3), 336–343. doi:10.1016/j.sbi.2006.05.007.

Rodrigues, H. F., Souza, T. A., Ghiraldini, F. G., Mello, M. L., & Moraes, A. S. (2014). Increased age is associated with epigenetic and structural changes in chromatin from neuronal nuclei. *Journal of Cellular Biochemistry, 115*(4), 659–665. doi:10.1002/jcb.24705.

Rodriguez-Paredes, M., & Esteller, M. (2011). Cancer epigenetics reaches mainstream oncology. *Nature Medicine, 17*(3), 330–339. doi:10.1038/nm.2305.

Rogakou, E. P., & Sekeri-Pataryas, K. E. (1999). Histone variants of H2A and H3 families are regulated during in vitro aging in the same manner as during differentiation. *Experimental Gerontology, 34*(6), 741–754.

Rogenhofer, S., Kahl, P., Mertens, C., Hauser, S., Hartmann, W., Buttner, R., et al. (2012). Global histone H3 lysine 27 (H3K27) methylation levels and their prognostic relevance in renal cell carcinoma. *BJU International, 109*(3), 459–465. doi:10.1111/j.1464-410X.2011.10278.x.

Rolf, M. K. (2015). Addressing the needs of our aging population through managed long-term care. *Home Healthcare Now, 33*(3), 177–178. doi:10.1097/nhh.0000000000000201.

Routh, A., Sandin, S., & Rhodes, D. (2008). Nucleosome repeat length and linker histone stoichiometry determine chromatin fiber structure. *Proceedings of the National Academy of Sciences of the United States of America, 105*(26), 8872–8877. doi:10.1073/pnas.0802336105.

Ryan, J. M., & Cristofalo, V. J. (1972). Histone acetylation during aging of human cells in culture. *Biochemical and Biophysical Research Communications, 48*(4), 735–742.

Ryba, T., Hiratani, I., Lu, J., Itoh, M., Kulik, M., Zhang, J., et al. (2010). Evolutionarily conserved replication timing profiles predict long-range chromatin interactions and distinguish closely related cell types. *Genome Research, 20*(6), 761–770. doi:10.1101/gr.099655.109.

Sanyal, A., Bau, D., Marti-Renom, M. A., & Dekker, J. (2011). Chromatin globules: A common motif of higher order chromosome structure? *Current Opinion in Cell Biology, 23*(3), 325–331. doi:10.1016/j.ceb.2011.03.009.

Scaffidi, P., & Misteli, T. (2006). Lamin A-dependent nuclear defects in human aging. *Science, 312*(5776), 1059–1063. doi:10.1126/science.1127168.

Schreiber, K. H., & Kennedy, B. K. (2013). When lamins go bad: Nuclear structure and disease. *Cell, 152*(6), 1365–1375. doi:10.1016/j.cell.2013.02.015.

Sedelnikova, O. A., & Bonner, W. M. (2006). GammaH2AX in cancer cells: A potential biomarker for cancer diagnostics, prediction and recurrence. *Cell Cycle, 5*(24), 2909–2913.

Sedivy, J. M., Banumathy, G., & Adams, P. D. (2008). Aging by epigenetics - A consequence of chromatin damage? *Experimental Cell Research, 314*(9), 1909–1917. doi:10.1016/j.yexcr.2008.02.023.

Seeber, A., Dion, V., & Gasser, S. M. (2014). Remodelers move chromatin in response to DNA damage. *Cell Cycle, 13*(6), 877–878. doi:10.4161/cc.28200.

Seligson, D. B., Horvath, S., McBrian, M. A., Mah, V., Yu, H., Tze, S., et al. (2009). Global levels of histone modifications predict prognosis in different cancers. *The American Journal of Pathology, 174*(5), 1619–1628. doi:10.2353/ajpath.2009.080874.

Sharman, E. H. (2010). Sirtuins and mammalian aging. In S. Bondy & K. Maiese (Eds.), *Aging and age-related disorders oxidative stress in applied basic research and clinical practice*. New York: Humana.

Shi, Y., Lan, F., Matson, C., Mulligan, P., Whetstine, J. R., Cole, P. A., et al. (2004). Histone demethylation mediated by the nuclear amine oxidase homolog LSD1. *Cell, 119*(7), 941–953. doi:10.1016/j.cell.2004.12.012.

Shin, D. M., Kucia, M., & Ratajczak, M. Z. (2011). Nuclear and chromatin reorganization during cell senescence and aging - a mini-review. *Gerontology, 57*(1), 76–84. doi:10.1159/000281882.

Shumaker, D. K., Dechat, T., Kohlmaier, A., Adam, S. A., Bozovsky, M. R., Erdos, M. R., et al. (2006). Mutant nuclear lamin A leads to progressive alterations of epigenetic control in premature aging. *Proceedings of the National Academy of Sciences of the United States of America, 103*(23), 8703–8708. doi:10.1073/pnas.0602569103.

Singhal, R. P., Mays-Hoopes, L. L., & Eichhorn, G. L. (1987). DNA methylation in aging of mice. *Mechanisms of Ageing and Development, 41*(3), 199–210.

So, K., Tamura, G., Honda, T., Homma, N., Endoh, M., Togawa, N., et al. (2006a). Quantitative assessment of RUNX3 methylation in neoplastic and non-neoplastic gastric epithelia using a DNA microarray. *Pathology International, 56*(10), 571–575. doi:10.1111/j.1440-1827.2006.02010.x.
So, K., Tamura, G., Honda, T., Homma, N., Waki, T., Togawa, N., et al. (2006b). Multiple tumor suppressor genes are increasingly methylated with age in non-neoplastic gastric epithelia. *Cancer Science, 97*(11), 1155–1158. doi:10.1111/j.1349-7006.2006.00302.x.
Solovei, I., Kreysing, M., Lanctot, C., Kosem, S., Peichl, L., Cremer, T., et al. (2009). Nuclear architecture of rod photoreceptor cells adapts to vision in mammalian evolution. *Cell, 137*(2), 356–368. doi:10.1016/j.cell.2009.01.052.
Song, J. S., Kim, Y. S., Kim, D. K., Park, S. I., & Jang, S. J. (2012). Global histone modification pattern associated with recurrence and disease-free survival in non-small cell lung cancer patients. *Pathology International, 62*(3), 182–190. doi:10.1111/j.1440-1827.2011.02776.x.
Soutoglou, E., & Misteli, T. (2007). Mobility and immobility of chromatin in transcription and genome stability. *Current Opinion in Genetics & Development, 17*(5), 435–442. doi:10.1016/j.gde.2007.08.004.
Sporn, J. C., Kustatscher, G., Hothorn, T., Collado, M., Serrano, M., Muley, T., et al. (2009). Histone macroH2A isoforms predict the risk of lung cancer recurrence. *Oncogene, 28*(38), 3423–3428. doi:10.1038/onc.2009.26.
Suzuki, Y., & Craigie, R. (2007). The road to chromatin - nuclear entry of retroviruses. *Nature Reviews Microbiology, 5*(3), 187–196. doi:10.1038/nrmicro1579.
Tark-Dame, M., van Driel, R., & Heermann, D. W. (2011). Chromatin folding – from biology to polymer models and back. *Journal of Cell Science, 124*(Pt 6), 839–845. doi:10.1242/jcs.077628.
Tempera, I., & Lieberman, P. M. (2014). Epigenetic regulation of EBV persistence and oncogenesis. *Seminars in Cancer Biology, 26*, 22–29. doi:10.1016/j.semcancer.2014.01.003.
Thoma, F., & Koller, T. (1977). Influence of histone H1 on chromatin structure. *Cell, 12*(1), 101–107.
Thoma, F., Koller, T., & Klug, A. (1979). Involvement of histone H1 in the organization of the nucleosome and of the salt-dependent superstructures of chromatin. *The Journal of Cell Biology, 83*(2 Pt 1), 403–427.
Tohgi, H., Utsugisawa, K., Nagane, Y., Yoshimura, M., Genda, Y., & Ukitsu, M. (1999). Reduction with age in methylcytosine in the promoter region -224 approximately -101 of the amyloid precursor protein gene in autopsy human cortex. *Brain Research. Molecular Brain Research, 70*(2), 288–292.
Upadhya, B., Taffet, G. E., Cheng, C. P., & Kitzman, D. W. (2015). Heart failure with preserved ejection fraction in the elderly: Scope of the problem. *Journal of Molecular and Cellular Cardiology*. doi:10.1016/j.yjmcc.2015.02.025.
Uzunova, K., Georgieva, M., & Miloshev, G. (2013). Saccharomyces cerevisiae linker histone-Hho1p maintains chromatin loop organization during ageing. *Oxidative Medicine and Cellular Longevity, 2013*, 437146. doi:10.1155/2013/437146.
Van Den Broeck, A., Brambilla, E., Moro-Sibilot, D., Lantuejoul, S., Brambilla, C., Eymin, B., et al. (2008). Loss of histone H4K20 trimethylation occurs in preneoplasia and influences prognosis of non-small cell lung cancer. *Clinical Cancer Research, 14*(22), 7237–7245. doi:10.1158/1078-0432.CCR-08-0869.
van Leeuwen, H., Okuwaki, M., Hong, R., Chakravarti, D., Nagata, K., & O'Hare, P. (2003). Herpes simplex virus type 1 tegument protein VP22 interacts with TAF-I proteins and inhibits nucleosome assembly but not regulation of histone acetylation by INHAT. *The Journal of General Virology, 84*(Pt 9), 2501–2510.
Verreault, A., Kaufman, P. D., Kobayashi, R., & Stillman, B. (1996). Nucleosome assembly by a complex of CAF-1 and acetylated histones H3/H4. *Cell, 87*(1), 95–104.
Waki, T., Tamura, G., Sato, M., & Motoyama, T. (2003). Age-related methylation of tumor suppressor and tumor-related genes: An analysis of autopsy samples. *Oncogene, 22*(26), 4128–4133. doi:10.1038/sj.onc.1206651.

Wang, G. P., Ciuffi, A., Leipzig, J., Berry, C. C., & Bushman, F. D. (2007). HIV integration site selection: Analysis by massively parallel pyrosequencing reveals association with epigenetic modifications. *Genome Research, 17*(8), 1186–1194. doi:10.1101/gr.6286907.

Wang, Y., Wysocka, J., Sayegh, J., Lee, Y. H., Perlin, J. R., Leonelli, L., et al. (2004). Human PAD4 regulates histone arginine methylation levels via demethylimination. *Science, 306*(5694), 279–283. doi:10.1126/science.1101400.

Wang, R. H., Zheng, Y., Kim, H. S., Xu, X., Cao, L., Luhasen, T., et al. (2008). Interplay among BRCA1, SIRT1, and Survivin during BRCA1-associated tumorigenesis. *Molecular Cell, 32*(1), 11–20. doi:10.1016/j.molcel.2008.09.011.

Weintraub, D., Comella, C. L., & Horn, S. (2008). Parkinson's disease–Part 1: Pathophysiology, symptoms, burden, diagnosis, and assessment. *The American Journal of Managed Care, 14*(2 Suppl), S40–48.

Widom, J., & Klug, A. (1985). Structure of the 300A chromatin filament: X-ray diffraction from oriented samples. *Cell, 43*(1), 207–213.

Wilson, V. L., Smith, R. A., Ma, S., & Cutler, R. G. (1987). Genomic 5-methyldeoxycytidine decreases with age. *The Journal of Biological Chemistry, 262*(21), 9948–9951.

Wolffe, A. P. (1998). Epigenetics. Introduction. *Novartis Foundation Symposium, 214*, 1–5.

Wolffe, A. P., & Matzke, M. A. (1999). Epigenetics: Regulation through repression. *Science, 286*(5439), 481–486.

Wood, J. G., & Helfand, S. L. (2013). Chromatin structure and transposable elements in organismal aging. *Frontiers in Genetics, 4*, 274. doi:10.3389/fgene.2013.00274.

Woodcock, C. L. (2006). Chromatin architecture. *Current Opinion in Structural Biology, 16*(2), 213–220. doi:10.1016/j.sbi.2006.02.005.

Yanagawa, N., Tamura, G., Oizumi, H., Kanauchi, N., Endoh, M., Sadahiro, M., et al. (2007). Promoter hypermethylation of RASSF1A and RUNX3 genes as an independent prognostic prediction marker in surgically resected non-small cell lung cancers. *Lung Cancer, 58*(1), 131–138. doi:10.1016/j.lungcan.2007.05.011.

Yuan, H., Su, L., & Chen, W. Y. (2013). The emerging and diverse roles of sirtuins in cancer: A clinical perspective. *OncoTargets and Therapy, 6*, 1399–1416. doi:10.2147/OTT.S37750.

Zane, L., Sharma, V., & Misteli, T. (2014). Common features of chromatin in aging and cancer: Cause or coincidence? *Trends in Cell Biology*. doi:10.1016/j.tcb.2014.07.001.

Zhang, K., Chen, Y., Zhang, Z., & Zhao, Y. (2009a). Identification and verification of lysine propionylation and butyrylation in yeast core histones using PTMap software. *Journal of Proteome Research, 8*(2), 900–906. doi:10.1021/pr8005155.

Zhang, Y., Ng, H. H., Erdjument-Bromage, H., Tempst, P., Bird, A., & Reinberg, D. (1999). Analysis of the NuRD subunits reveals a histone deacetylase core complex and a connection with DNA methylation. *Genes & Development, 13*(15), 1924–1935.

Zhang, R., Poustovoitov, M. V., Ye, X., Santos, H. A., Chen, W., Daganzo, S. M., et al. (2005). Formation of MacroH2A-containing senescence-associated heterochromatin foci and senescence driven by ASF1a and HIRA. *Developmental Cell, 8*(1), 19–30. doi:10.1016/j.devcel.2004.10.019.

Zhang, Y., Zhang, M., Dong, H., Yong, S., Li, X., Olashaw, N., et al. (2009b). Deacetylation of cortactin by SIRT1 promotes cell migration. *Oncogene, 28*(3), 445–460. doi:10.1038/onc.2008.388.

Zhang, L., Zhong, K., Dai, Y., & Zhou, H. (2009c). Genome-wide analysis of histone H3 lysine 27 trimethylation by ChIP-chip in gastric cancer patients. *Journal of Gastroenterology, 44*(4), 305–312. doi:10.1007/s00535-009-0027-9.

Chapter 3
In Utero Bisphenol A Exposure and Epigenetic Programming of Neurobehavioral Outcomes

Marija Kundakovic

3.1 Bisphenol A Exposure

3.1.1 Introduction

Bisphenol A (BPA) is a ubiquitous environmental toxicant present in many consumer products. BPA is well known as an estrogenic endocrine disruptor and early studies of BPA were focused on its toxic effects related to reproductive function and disruption of sexual dimorphism. Although sex specificity is still a common theme in BPA-associated effects, BPA has also been linked to changes in nonreproductive neurobehavioral outcomes as well as to cognitive dysfunction. This chapter will focus on in utero BPA exposure and its lasting effects on the brain function and behavior in the prenatally exposed offspring and subsequent generations. And, in particular, I will review evidence implying that BPA may exert its enduring effects through epigenetic mechanisms and will discuss possible implications of these findings.

3.1.2 BPA: Exposure in Humans

BPA is one of the world's highest-production volume chemicals, mainly used in the production of polycarbonate plastics and epoxy resins (Vandenberg et al. 2010). Polycarbonate plastics are widely used for food and drink packaging containers whereas resins are utilized as protective coatings on metal products including bottle tops and food cans. BPA is also found in dental materials (Chapin et al. 2008), plastic-containing medical devices (Huygh et al. 2015), as well as in cash register receipts (Biedermann et al. 2010). The widespread use of BPA-containing products

M. Kundakovic, Ph.D. (✉)
Department of Biological Sciences, Fordham University, Bronx, NY 10458, USA
e-mail: mkundakovic@fordham.edu

D. Hollar (ed.), *Epigenetics, the Environment, and Children's Health Across Lifespans*, DOI 10.1007/978-3-319-25325-1_3

results in nearly ubiquitous human exposure worldwide (Vandenberg et al. 2010). As an example, recent studies have shown that more than 90 % of the US population has detectable levels of BPA in urine samples (Calafat et al. 2005, 2008). Human exposure mainly occurs through diet as polymers containing BPA can be hydrolyzed, leading to BPA leaching into food and drink containers (Welshons et al. 2006).

When compared, the concentration of urine BPA is higher in children and adolescents than in adults, and, on average, females have higher urinary BPA concentrations than males (Calafat et al. 2008). Most importantly for the current discussion, BPA has been detected in the urine and serum of pregnant women as well as in placental tissue, amniotic fluid, umbilical cord blood, and the urine from newborn infants (Schonfelder et al. 2002; Vandenberg et al. 2010). At mid-gestation, levels of BPA were reported to be several fold higher in amniotic fluid than in maternal serum (Ikezuki et al. 2002). These data suggest that BPA is able to cross the placenta and may accumulate in the embryo/fetal compartment after repeated maternal exposure, most likely due to the inability of the fetus to efficiently metabolize BPA (Taylor et al. 2008).

In many animals, BPA can disrupt normal development and induce adverse effects in multiple tissues, and this has raised concerns that ubiquitous BPA exposure is negatively affecting the health of entire ecosystems and human populations. Canada was the first to officially declare BPA a "toxic substance" (*Canada Gazette*, October 13, 2010, Vol. 144, No. 21). In 2010, the U.S. Food and Drug Administration (FDA) released a statement that there is "some concern about the potential effects of BPA on the brain, behavior, and prostate gland in fetuses, infants, and young children" (www.fda.gov; January, 2010). Later on, the FDA officially banned the use of BPA-based polycarbonate resins in baby bottles and sippy cups (July, 2012) as well as BPA-based epoxy resins as coatings in packaging for infant formula (July, 2013). However, the agency still maintains that BPA currently used in food containers and packaging is safe (www.fda.gov). In response to early public concerns, the European Food Safety Authority (EFSA) considerably reduced the "tolerable daily intake" (TDI) levels for BPA, which are the levels considered safe for humans. However, most recently, the EFSA announced that "BPA poses no health risk to consumers of any age group (including unborn children, infants and adolescents) at current exposure levels" (www.efsa.europa.eu; January, 2015). While the potential toxicity of BPA in humans remains controversial, it is important to summarize available scientific evidence that invites further examination and reevaluations of the BPA regulatory policies.

3.1.3 Modeling BPA Exposure in Animals: Low-Dose Studies and Non-monotonous Effects

Numerous animal studies have shown that BPA acts as a toxicant for developing tissues. Prenatal exposure of rodent fetuses to BPA results in various adverse effects, including altered development of the male and female reproductive tract and the mammary gland, increased prostate gland volume, disruption of sexual differentiation of the brain, accelerated growth and puberty, higher incidence of breast and prostate cancer, increased body weight, altered reproductive function and sexual behavior, immune dysregulation, and deficits in many nonreproductive behaviors and cognitive function (Chapin et al. 2008; Clayton et al. 2010; Kundakovic and Champagne 2011; Richter et al. 2007).

These findings from animal studies suggest that BPA may act as a toxicant for developing human tissues as well and have stimulated debate regarding the impact of human exposure to this compound. However, it has been questioned whether animal studies really model exposure to BPA in humans, and three major issues have dominated the discussions: the dose and the nature of exposure and the differences in the metabolism between rodents and humans. As for the dose, the early studies of BPA toxicity employed very high doses of this compound, and some of those studies served for the calculation of BPA reference dose in the USA, which is 50 μg/kg/day. The U.S. Environmental Protection Agency (EPA) established the reference dose for BPA based on the lowest-observed-adverse-effects level (LOAEL) derived from high-dose toxicological studies (50 mg/kg/day) and by applying a 1000-fold safety factor (vom Saal and Welshons 2006). However, it is important to note that numerous more recent animal studies have shown significant BPA-induced effects at the doses that are significantly below the human reference dose (50 μg/kg/day), which is still considered safe for humans. Other issues that have been raised are regarding differences in metabolism between humans and rodents as well as in the nature of exposure (Vandenberg et al. 2009). In this regard, recent pharmacokinetic scaling experiments have estimated that exposures up to 400 μg/kg/day in animals produce blood concentrations of the unconjugated, bioactive form of BPA in the range of human blood concentrations (Vandenberg et al. 2012, 2007) and therefore may be considered environmentally relevant.

Another important BPA characteristic, which is shared with other endocrine disruptors, is that BPA effects often follow non-monotonous dose–response curves. This means that, for certain outcomes, lower doses (which may be more environmentally relevant) can induce more profound effects than higher BPA doses (which are regularly used in toxicological studies for risk assessment). This has been shown by a number of in vitro and in vivo animal studies and supported by evidence from epidemiological studies (Vandenberg et al. 2012). For instance, Fig. 3.1 shows experimental data from the mouse offspring that was prenatally exposed to three different doses of BPA (2, 20, and 200 μg/kg/day) (Kundakovic et al. 2013a). These data indicate that environmentally relevant doses of BPA may induce very different effects on gene expression and behavior dependent on the level of exposure and support the view that appropriate risk assessment of BPA neurotoxicity should involve multiple low-dose exposures (Vandenberg et al. 2012).

In the following section, I will first summarize low-dose animal studies that indicate effects of BPA on brain development and behavior in the offspring of BPA-treated mothers. I will then review available human evidence on association between prenatal maternal BPA exposure and behavioral outcomes in children. Although comparisons of animal and human findings are not straightforward, available evidence points out that animal studies are informative for human health risk assessments.

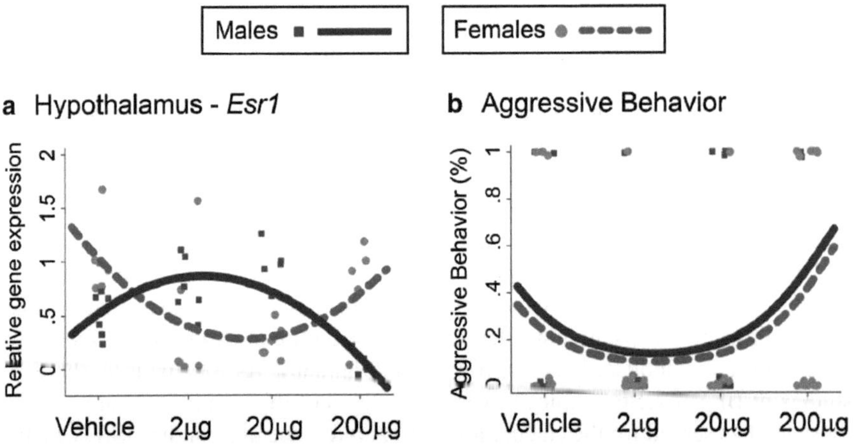

Fig. 3.1 Non-monotonous effects of in utero BPA exposure: The effects of maternal exposure to vehicle and 2, 20, and 200 μg/kg/day BPA on (**a**) estrogen receptor alpha (*Esr1*) gene expression in the juvenile hypothalamus and (**b**) adult aggressive behavior in the Balb/c mouse offspring. Figure 3.1 (**a**) and (**b**) are reproduced from Kundakovic et al. (2013a), Sex-specific epigenetic disruption and behavioral changes following low-dose in utero bisphenol A exposure, *PNAS*, 110(24), 9956–9961, Fig. 2d, p. 9957 and Fig. 5f, p. 9959, respectively; used with permission from *Proceedings of the National Academy of Sciences, USA*

3.2 BPA: Prenatal Programming of Neurobehavioral Outcomes

3.2.1 Evidence from Animal Models

So far, numerous low-dose rodent studies have provided evidence that maternal BPA exposure disrupts offspring neurodevelopment resulting in lasting structural and behavioral changes (for a more detailed review regarding the dose, route, and timing of BPA exposure in rodent studies, please see Kundakovic and Champagne 2011). The following domains have been reported to be changed in the offspring following in utero BPA exposure: neuronal differentiation, morphology, and connectivity; sexual differentiation of the brain; social, sexual, and maternal behavior; anxiety-like behavior; and learning and memory, among others.

Studies in mice by Nakamura et al. (2006) showed that a low-dose maternal BPA exposure affects neocortical development in offspring by accelerating neuronal differentiation and migration at mid-gestation. Those changes likely contribute to changes observed in later life in the adult brain, which include abnormal neuronal positioning and aberrant connectivity between the cortex and thalamus (Nakamura et al. 2007). Prenatal BPA exposure was also shown to impact neuronal morphology in the developing hippocampal CA1 region, resulting in reduction of spine density, which persists well into adulthood (Kimura et al. 2015). It is likely that observed structural changes translate into changes in behavior and learning/memory in later life.

One of the most notable findings in the studies of BPA effects on brain development concerns the *loss or reversal of sexual dimorphism*. It is well known that sexual differentiation of the brain results in sex differences at the level of both brain structure and behavior, and BPA has been shown to decrease or completely abolish many of these differences. For instance, prenatal BPA exposure has been shown to diminish sex differences in the structure of several brain regions, including locus coeruleus (Kubo et al. 2003), anteroventral periventricular preoptic area (APVP) (Rubin et al. 2006), and the bed nucleus of the stria terminalis (Funabashi et al. 2004). It is remarkable that the study of BPA effects on APVP (Rubin et al. 2006) included the doses in the nanogram (ng) range (25 and 250 ng/kg/day) applied using osmotic pump, which better mimics human episodic exposure, and still showed a remarkable loss of sexual dimorphism in offspring, not only at the level of brain structure but those changes were also accompanied with behavioral alterations (Rubin et al. 2006). Other studies have shown that BPA does not affect all sexually dimorphic brain regions, suggesting that its effects are cell type- and brain region-specific (Funabashi et al. 2004). In addition, BPA can induce sex-specific effects in certain brain regions that are not sexually dimorphic (Tando et al. 2007).

BPA also affects *social behavior*, with some aspects of social behavior being affected by BPA in a sex-specific way, whereas other domains are affected similarly in both sexes (Kundakovic et al. 2013b; Wolstenholme et al. 2011, 2012). For instance, maternal BPA exposure during pregnancy or lactation resulted in impaired sexual performance in males and increased sexual motivation in female offspring (Farabollini et al. 2002). BPA was also shown to abolish sex differences in play behavior. Early-life BPA treatment induced defeminization in social interactions (decreased play with males) and masculinization of female play behavior (increased play with females) in juvenile female offspring (Dessi-Fulgheri et al. 2002; Porrini et al. 2005). We have shown a decrease in male play behavior (chasing) with increasing exposure to prenatal BPA (Kundakovic et al. 2013a). Interestingly, Kawai et al. (Kawai et al. 2003) showed that prenatal BPA exposure increases aggression in adult male mice. However, we have demonstrated that BPA effect on aggression is dose-dependent and follows a U-shaped curve with lower concentrations decreasing aggression and higher concentrations increasing aggressive behavior in both sexes (Kundakovic et al. 2013a). BPA also induced loss or reversal in sexual dimorphism regarding sniffing behavior in both juvenile and adult mice (Kundakovic et al. 2013a). It is also interesting to note that in utero BPA exposure impaired later-life maternal behavior in female offspring (Palanza et al. 2002).

Developmental BPA exposure has been shown to increase *anxiety-related behavior* in both rats and mice. We and others have shown that BPA abolishes sex differences in open-field behavior in mice, increasing the anxiety-like behavior in female offspring (Kundakovic et al. 2013a). Using another test (elevated plus maze) and strain of mice (C57BL/6J), Gioiosa et al. showed a male-specific increase in anxiety-like behavior (Gioiosa et al. 2007). In rats, female offspring developmentally exposed to BPA was also shown to spend significantly less time exploring novel environment, which also suggests anxiogenic effect (Adriani et al. 2003).

Finally, several studies have shown that prenatal BPA exposure affects *learning and memory* in rodent models. In rats, BPA treatments during gestation and lactation resulted in impaired learning of passive and active avoidance tasks as well as impaired recognition and spatial memory at adult age (Goncalves et al. 2010; Negishi et al. 2004). In mice, the combination of prenatal and postnatal BPA exposures led to significant deficits in working and recognition memory (Tian et al. 2010), passive avoidance memory (Miyagawa et al. 2007; Xu et al. 2010), and spatial memory (Jasarevic et al. 2013; Xu et al. 2010) during the juvenile period (Xu et al. 2010) or in adulthood (Jasarevic et al. 2013; Miyagawa et al. 2007; Tian et al. 2010; Xu et al. 2010). The memory impairments were associated with reduced expression of *N*-methyl-D-aspartic acid (NMDA) receptors in the cortex and the hippocampus (Tian et al. 2010; Xu et al. 2010) as well as reduced levels of estrogen receptor beta (ERβ) in the hippocampus (Xu et al. 2010), suggesting a neurobiological substrate of BPA-induced memory and learning deficits.

3.2.2 Human Studies

As compared to animal studies, human evidence on BPA neurotoxicity is much more limited. However, there are now six studies coming from four separate US birth cohorts suggesting that high maternal BPA exposure during pregnancy may contribute to behavioral problems in children (see Table 3.1 for the summary of findings).

Braun et al. examined the samples from the Health Outcomes and Measures of the Environment Study, a prospective birth cohort from the greater Cincinnati, Ohio, area with the majority of participants being Caucasians and coming from medium-income households (Braun et al. 2009, 2011). In the first report, higher maternal urinary BPA concentrations (around 16 weeks of gestation) were associated with increased externalizing behaviors (hyperactivity, aggression) in 2-year-old children, particularly among females (Braun et al. 2009). In their follow-up study, Braun et al. showed that the sex-specific effect of BPA exposure may persist further into childhood (Braun et al. 2011). In particular, gestational BPA exposure was associated with higher scores for measures of anxiety, depression, and hyperactivity as well as with poorer emotional regulation at age 3, especially among girls. Interestingly, the latter study did not find association between childhood BPA exposure and behavior, emphasizing the importance of gestational exposure for BPA effect on neurobehavioral outcomes, at least in early childhood (Braun et al. 2011).

In the Columbia Center for Children's Environmental Health (CCCEH) cohort, Perera et al. also found sex-specific effects of maternal BPA exposure on child behavior, although boys were more affected than girls in their study (Perera et al. 2012). This study included African-American and Dominican women and their children, and BPA levels were measured in single spot urine samples collected from the mothers during the third trimester of pregnancy (34 weeks on average). In their first study, the authors reported child behavior between 3 and 5 years of age (Perera et al. 2012). Among boys, higher prenatal BPA exposure was associated with disturbed emotional regulation and increased aggressive behavior. On the contrary, in

Table 3.1 Studies of neurodevelopmental effects of in utero BPA exposure in humans

Cohort	Examined samples	Child age	Behavioral outcome	References
Health Outcomes and Measures of the Environment Study; Cincinnati, Ohio ($N=249$)	Maternal (three spot) urine samples: 16 and 26 weeks of gestation and at birth	2 years	Increased externalizing behaviors and global behavioral scores, particularly in girls	Braun et al. (2009)
Health Outcomes and Measures of the Environment Study; Cincinnati, Ohio ($N=244$)	Maternal (three spot) urine samples: 16 and 26 weeks of gestation and at birth	3 years	Higher scores for measures of anxiety, depression, and hyperactivity; poorer emotional regulation, particularly in girls	Braun et al. (2011)
Columbia Center for Children's Environmental Health Cohort (CCCEH) ($N=198$)	Maternal (one spot) urine samples: 34 weeks of pregnancy	3–5 years	Disturbed emotional regulation and increased aggressive behavior in boys; lower scores for anxiety/ depression and reduced aggressive behavior in girls	Perera et al. (2012)
Center for the Health Assessment and of Mothers and Children of Salinas (CHAMACOS), California ($N=292$)	Maternal (two spot) urine samples: 14 and 27 weeks of pregnancy	7 and 9 years	Increased internalizing scores in boys at age 7	Harley et al. (2013))
Study for Future Families II (SFFII) ($N=153$)	Maternal (one spot) urine samples: 27 weeks of pregnancy	6–10 years	Increased internalizing and externalizing behaviors, depressed behavior, somatic problems, and oppositional/ defiant traits in boys	Evans et al. (2014)
Columbia Center for Children's Environmental Health Cohort (CCCEH) ($N=250$)	Maternal (one spot) urine samples: 34 weeks of pregnancy	7–9 years	Increased internalizing and externalizing composite scores in boys; decreased internalizing scores in girls	Roen et al. (2015)

girls higher maternal BPA levels were associated with lower scores for anxiety/ depression and reduced aggressive behavior. Similar to the study of Braun et al., childhood BPA exposure did not predict those behavioral outcomes. In their follow-up study, the authors examined children at 7–9 years of age (Roen et al. 2015). A high prenatal BPA concentration was associated with increased internalizing and externalizing behaviors in boys and with the decreased internalizing behaviors in girls. However, at this age, high postnatal BPA concentration was shown to be associated with increased internalizing and externalizing behaviors in girls.

Harley et al. examined another birth cohort: Center for the Health Assessment of Mothers and Children of Salinas (CHAMACOS) in California, with participants being primarily of Hispanic origin (Harley et al. 2013). In this study, prenatal urinary BPA concentrations were associated with increased internalizing problems, including anxiety and depression, in boys at age 7; there were no associated behavioral problems in girls. Interestingly, they also showed associations between childhood BPA levels and behavioral changes in both boys and girls at age 7, implying that in later childhood the BPA concentration may have significant effect on child behavior.

Finally, Evans et al. examined the mothers and their children from the Study for Future Families II (SFFII) (Evans et al. 2014). BPA concentration was measured in spot urine samples collected from women at mean 27 weeks of pregnancy in relation to child behavior assessed at age 6–10 years. This study showed that increased prenatal BPA was associated with increased internalizing and externalizing behaviors, depressed behavior, somatic problems, and oppositional/defiant traits in boys.

Of note is also that two studies reported no association between prenatal BPA exposure and social impairment in children aged 5–9 years (Braun et al. 2014; Miodovnik et al. 2011).

In summary, there are six studies in humans from four independent US birth cohorts showing that high prenatal BPA exposure may contribute to sex-specific behavioral problems in children. It is clear, though, that the studies are not entirely consistent as affected sex and behavioral domains vary between different studies. There are several possible explanations that can account for the discrepancies between the studies, including the differences in timing when the urine samples were collected, instruments that are used for behavioral assessment, child age, as well as differences in ethnicity/genetics and socioeconomic status (low-income minority population vs. white average-income population), among others. As always in epidemiological studies, there are other possible confounders, such as other exposures, that were not measured and, hence, could not have been accounted for but could have contributed to differences in findings. One of the limitations of BPA studies in humans is also the reliability of BPA measurements in urine. BPA has a short half-life, and exposures are likely episodic in nature so BPA concentrations can vary significantly throughout the day. Therefore, the single spot urine samples may not accurately describe a subject's chronic exposure, and it was suggested that the future studies should consider collecting multiple or integrated urine concentration measurements to improve exposure classification (Braun et al. 2011).

Regardless of some limitations, increasing evidence from animal and human studies on BPA neurotoxicity suggests that high maternal BPA exposure may put a developing child at risk for later-life behavioral problems.

3.2.3 *Sex-Specific Effects: A Common Theme*

In both animal and human studies, sex specificity is one of the most common themes regarding BPA-associated effects. It is interesting to note that in some cases BPA has been shown to induce reduction, loss, or reversal of sexually dimorphic phenotypes. Yet, BPA can induce sex differences in other phenotypes that are not sexually dimorphic (Kundakovic et al. 2015). This is not that surprising considering that this compound acts through sex hormone receptors and is well known as an endocrine disruptor. However, the mechanism of this effect is not really clear. In the following section, I will review BPA mechanisms of action and will provide a rationale and experimental evidence suggesting that epigenetic mechanisms may underlie lasting effects of BPA on brain and behavior and explain, at least in part, its sex-specific effects.

3.3 BPA: Epigenetic Mechanisms of Action

The mechanisms that underlie lasting effects of gestational BPA exposure on neurobehavioral outcomes, although not exactly clear, are now emerging. BPA is a selective estrogen receptor (ER) modulator that binds both classic estrogen receptors, ERα and ERβ, although with an affinity that is 10,000–100,000 lower than that of endogenous estradiol (Andersen et al. 1999; Kuiper et al. 1998). BPA also interacts with other estrogen receptors and androgen receptors and may interfere with thyroid hormone signaling (Sohoni and Sumpter 1998; Takayanagi et al. 2006; Xu et al. 2005). BPA effects are typically attributed to its estrogenic or anti-estrogenic action; however, it is not clear how the low potency at ERs could account for the effects of low-dose BPA exposures. Recent studies have shown that epigenetic mechanisms contribute to lasting BPA effects, and these effects may or may not be mediated by ERs (Kundakovic and Champagne 2011).

3.3.1 Epigenetic Mechanisms and Prenatal Programming of Behavioral Outcomes

Prenatal exposure to many environmental toxins has been associated with later behavioral and psychiatric disorders (Kundakovic 2013). Epigenetic mechanisms represent the plausible molecular substrate through which environmental exposures can change the patterns of gene expression associated with normal neurodevelopment, leading to lasting consequences for later behavior and cognitive function. Epigenetic modifications, DNA methylation and histone modifications, control chromatin state in the vicinity of gene regulatory regions and thereby directly regulate gene expression. DNA methylation occurs at position 5 of cytosine residues and predominantly in the context of CpG dinucleotides; it is catalyzed by a family of enzymes called DNA methyltransferases (DNMTs) and is typically involved in gene silencing (Klose and Bird 2006). Histone modifications, such as acetylation and methylation, can shift local chromatin configuration toward a more open or more closed state, leading to either increased or suppressed gene expression, respectively. It is also known that DNA methylation and histone marks often work in concert to regulate gene expression (Fuks 2005).

During development, epigenetic mechanisms are essential for the establishment of cell type-specific gene expression patterns and cellular differentiation, including differentiation of brain cells (Golebiewska et al. 2009; Liu and Casaccia 2010; Miller and Gauthier 2007). Thus, organisms are likely vulnerable to epigenetic disruption across the entire developmental period. Various environmental exposures, including diet, toxins, and stress, can affect DNA methylation and gene expression programming during in utero development (Jirtle and Skinner 2007). Once established, DNA methylation patterns can be passed from one cell generation to another and persist into adulthood, thus providing the mechanism through which the early life environment can exert lasting effects on gene expression and phenotype (Weaver et al. 2004). Animal studies have shown that maternal exposure to drugs (Kaminen-Ahola et al. 2010; Novikova et al. 2008), stress (Mueller and Bale 2008), and endocrine disruptors (Skinner et al. 2008) can alter epigenetic gene programming in the brain and contribute to neurodevelopmental and behavioral deficits in the offspring.

In the following two sections, I will present available evidence from both animal and human studies suggesting that prenatal BPA exposure induces lasting changes in the brain and behavior, at least in part, via epigenetic disruption.

3.3.2 BPA and Disruption of Epigenetic Mechanisms in the Brain: Evidence from Animal Studies

The first evidence that BPA can induce epigenetic changes was provided using the Agouti viable yellow (A^{vy}) mice, which is an established model to study the effects of developmental environmental exposures on the epigenome (Cooney et al. 2002; Dolinoy et al. 2006). This model provides a direct link between environmentally induced changes in DNA methylation and easily observed changes in phenotype. A^{vy} mice (the A^{vy}/a genotype) are genetically identical, but they contain a metastable, DNA methylation sensitive, A^{vy} allele in the *Agouti* gene locus that determines coat color. The expression of the *Agouti* gene varies depending on the DNA methylation levels within a retrotransposon intracisternal A particle (IAP) inserted upstream of the *Agouti* gene. The establishment of epigenetic marks at A^{vy} occurs during early embryogenesis and is a probabilistic event, resulting in a wide coat color distribution of A^{vy} mice ranging from pure yellow (hypomethylated A^{vy} IAP) to pseudoagouti brown (hypermethylated A^{vy} IAP). Maternal exposure to BPA during pregnancy was shown to shift the coat color distribution of offspring toward yellow by decreasing methylation at specific CpG sites in A^{vy} allele (Dolinoy et al. 2007). Methylation levels observed in tail DNA samples were highly correlated with methylation levels in the brain, kidney, and liver from the same animals, implying that BPA-induced changes in A^{vy} methylation were established before germ layer differentiation in the embryonic stem cells and persisted into adulthood. Importantly, these BPA effects can be counteracted by maternal exposure to the diet rich in methyl group donors, including folic acid, further implying that DNA methylation-dependent mechanisms underlie BPA-induced effects.

In addition to this study, several other studies have linked gestational (Anderson et al. 2012; Bromer et al. 2010; Dolinoy et al. 2007) or neonatal (Doshi et al. 2011; Ho et al. 2006; Tang et al. 2012) BPA exposure to long-lasting changes in DNA methylation and altered gene expression in non-neuronal tissues. There is now emerging evidence that BPA also has important effects on the regulation of DNA methylation in the brain. An early genome-wide study by Yaoi et al. (2008) showed that low-dose maternal BPA exposure induced either hypo- or hyper-methylation of the multiple gene loci in the fetal mouse forebrain. Though informative, this study involved methodology that has limited sensitivity, did not reveal specific BPA target genes, and did not examine the persistence of DNA methylation changes into later life. Another study showed that grand-offspring of the dams exposed to BPA during pregnancy had impaired hippocampal neurogenesis and cognition, associated with altered DNA methylation of a gene implicated in neurogenesis (Jang et al. 2012). However, the epigenetic effects were very subtle and associated only with a very high dose of BPA; the effects in the first generation offspring were not reported.

Recently, we reported two rodent studies providing strong experimental evidence that low-dose prenatal BPA exposure induces lasting epigenetic disruption in the brain that may contribute to changes in behavior and learning (Kundakovic et al. 2013a, 2015). Our starting hypothesis is presented in Fig. 3.2. We anticipated that maternal BPA exposure would interfere with gene expression programming in the

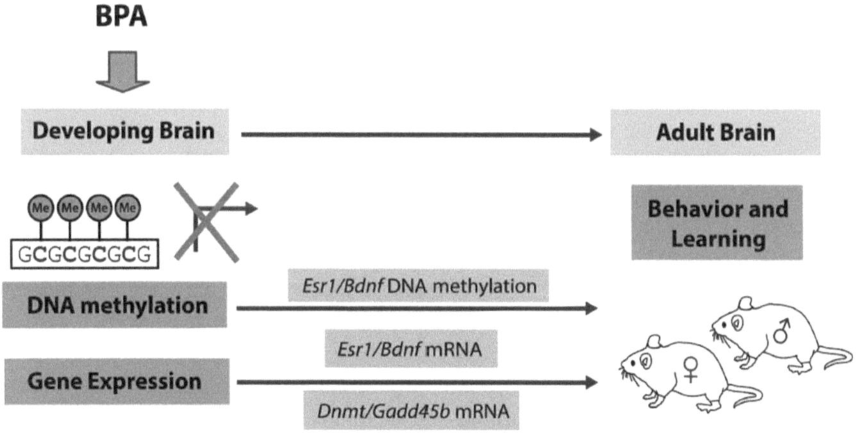

Fig. 3.2 Epigenetic hypothesis for long-term effects of in utero BPA exposure on neurobehavioral outcomes. Maternal BPA exposure disrupts epigenetic gene expression programming in the offspring's developing brain, which at least in part is controlled by DNA methylation (typically associated with transcriptional repression). This induces lasting DNA methylation and gene expression changes (e.g., in the *Esr1* and *Bdnf* genes) in the brain of offspring that may contribute to enduring BPA effects on behavior and learning. BPA-induced epigenetic disruption is likely mediated via the altered regulation of genes encoding proteins involved in DNA methylation (e.g., *Dnmt*) and DNA demethylation (e.g., *Gadd45*). The observed BPA effects are largely sex-specific and may, at least in part, account for BPA-induced disruption of sexually dimorphic phenotypes or, in general, may contribute to sex-specific changes in behavior and learning

offspring's developing brain, which at least in part is controlled by DNA methylation. If BPA disrupts epigenetic regulation during development, this could induce lasting DNA methylation and gene expression changes in the brain that may contribute to enduring BPA effects on brain structure and behavior. We also assumed that those BPA-induced effects may be sex-specific to account for sex-specific changes in behavior.

In the first study, we examined the effect of maternal BPA exposure on brain gene expression and the epigenome in juvenile offspring in the context of social and anxiety-like behavior (Kundakovic et al. 2013a). Three brain regions known to be important for those behaviors were analyzed: the prefrontal cortex, hypothalamus, and hippocampus. Importantly, we also included three doses of BPA (all three of which have been shown to produce environmentally relevant BPA concentrations in mice) and examined BPA effects in both sexes. There are several important findings of this study. First, we showed that BPA induces concordant changes in the expression of genes encoding estrogen receptors (ERs) and DNA methyltransferases (DNMTs) in the cortex and the hypothalamus. ERs are essential for sexual differentiation of the brain and behavior, and they have been shown to play important roles in social and anxiety-like behaviors (McCarthy and Arnold 2011; Patisaul et al. 2006; Tetel and Pfaff 2010). On the other hand, changes in the expression of DNA methylation regulators show that BPA induces lasting disruption in epigenetic regulatory mechanisms and that this could trigger changes in the expression of multiple genes, including ERs. And, indeed, we found that changes in ER alpha mRNA levels were associated with changes in DNA methylation in the corresponding (*Esr1*) gene promoter. This clearly shows that prenatal BPA treatment can induce lasting changes in the brain epigenome, and that these changes, at least in part, may underlie gene expression and behavioral changes. This was further confirmed with behavioral studies; we showed that BPA-induced changes in social and anxiety-like behavior mimic the molecular changes that we observed (Kundakovic et al. 2013a). Importantly, the majority of BPA-induced changes were sex-specific and dose-dependent (following non-monotonous or linear dose–response curves). In many cases, BPA induced loss or reversal of sexually dimorphic phenotypes, particularly at lower doses, and this implied that epigenetic mechanisms may underlie the disruption of sexual dimorphism seen in many BPA studies, at least with regard to neurobehavioral outcomes.

In the second study, we examined the long-term effects of prenatal BPA exposure on the phenotypes relevant to learning and memory (Kundakovic et al. 2015). One of the genes of a particular interest was brain-derived neurotrophic factor (*Bdnf*). *Bdnf* is very important for neurodevelopment and synaptic plasticity, and several early-life exposures have been shown to affect *Bdnf* gene expression and epigenetic regulation (Boersma et al. 2014; Lubin et al. 2008; Onishchenko et al. 2008; Roth et al. 2009). Importantly, *BDNF* is shown to be downregulated in many psychiatric disorders that are associated with early life adversity, such as depression, schizophrenia, and autism, among others (Boulle et al. 2012). Therefore, the interference with the regulation and expression of this gene provides one of the plausible mechanisms through which early life adverse environments can disrupt neurodevelopment leading to lasting consequences for brain plasticity, learning, and behavior.

And, indeed, we were able to show that maternal BPA exposure results in a lasting downregulation of *Bdnf* gene expression in the offspring's hippocampus evident in both juvenile and adult animals, but only in males (Kundakovic et al. 2015). Changes in *Bdnf* mRNA expression were associated with DNA methylation changes in the transcriptionally relevant region of the *Bdnf* gene, which harbors a methylation-sensitive binding site for the transcription factor CREB (Martinowich et al. 2003). We further showed that those changes were associated with changes in the expression of the DNA methylation machinery at both time points as well as with memory deficits in young adult male mice (Kundakovic et al. 2015). These findings strongly imply that prenatal BPA can induce lasting disruption in learning and memory through the mechanisms that, at least in part, include epigenetic disruption of the genes involved in synaptic plasticity, such as *Bdnf*.

In summary, our studies of maternal BPA exposure in mice provide strong evidence that maternal exposure leads to lasting effects on behavior and learning in the offspring that are, at least in part, mediated by epigenetic mechanisms (Fig. 3.2). Importantly, those changes are sex-specific and dose-dependent and can either lead to disruption of sexually dimorphic behaviors or induce sex difference in behaviors that do not significantly differ between sexes. The low-dose effects of BPA strongly imply that epigenetic effects of this compound may occur at environmentally relevant levels of exposure.

Future Directions Our studies focused on candidate genes that are known to be involved in social and anxiety-like behavior (estrogen receptors) as well as in learning and memory (*Bdnf* and NMDA receptor 2B). However, as BPA affects the expression of genes that encode proteins involved in epigenetic regulation, it is very likely that prenatal BPA exposure affects many genes that are epigenetically regulated in the brain and this may well contribute to behavioral phenotypes that we and others have observed. Therefore, it would be extremely important to perform genome-wide epigenetic studies in the brain tissue derived from animals prenatally exposed to BPA to find additional BPA target genes. In addition, we do not understand the upstream mechanisms through which BPA affects the epigenome. There is a possibility that BPA may, in part, act through steroid hormone receptors but other receptors may also be involved, and this still needs to be established. Better understanding of the mechanisms that induce epigenetic disruption may lead to novel preventive approaches and early interventions regarding the adverse effects induced by BPA and other endocrine disrupting chemicals.

3.3.3 BPA Epigenetic Effects in Humans: Current Evidence and Implications for Possible Biomarkers

It has been proposed that epigenetic mechanisms may contribute to lasting effects of environmental exposures on human health outcomes (Jirtle and Skinner 2007). Therefore, there has been a great interest to establish the link between various toxicological exposures, epigenetic variation, and health outcomes, although this has been quite challenging. First, it is very difficult to establish a direct link between a single environmental exposure and epigenetic variation in humans considering that human populations are genetically variable and exposed to mixtures of environmental agents. Second, the epigenome is by definition tissue- and cell type-specific and this is particularly important when studying the effects of environmental exposures in relation to neurobehavioral outcomes since the brain as a target tissue is not accessible in living individuals. However, in utero exposures can leave lasting marks on the epigenome of various tissues (Jirtle and Skinner 2007), and, therefore, there is a possibility that epigenetic signatures of peripheral tissues may, at least in part, be predictive of these signatures in the brain.

There are several lines of evidence implying that epigenetic profiling of peripheral tissue may be useful for studies of behavioral and psychiatric outcomes. First, a study by Davies et al. showed that, in spite of significant between-tissue variation in DNA methylation, some interindividual differences in DNA methylation are correlated across the brain and blood of healthy subjects (Davies et al. 2012). Second, DNA methylation changes have been found both in psychiatric postmortem brain samples (Keller et al. 2010; Mill et al. 2008) and in the peripheral blood of the living psychiatric patients (D'Addario et al. 2012; Fuchikami et al. 2011; Ikegame et al. 2013). However, a challenge to this field of study has been to provide evidence that epigenetic markers in peripheral tissues can indeed predict functionally relevant epigenetic changes in the brain and consequent behavioral and psychiatric outcomes.

To address the question whether BPA affects the epigenome in humans and whether epigenetic biomarkers in the blood can predict functionally relevant epigenetic changes in the brain and behavioral outcomes, we combined an animal and human study (Kundakovic et al. 2015). We again used *BDNF* gene as an excellent candidate for an epigenetic biomarker for behavioral vulnerability. DNA methylation of this gene has been shown to be altered both in the blood (D'Addario et al. 2012; Fuchikami et al. 2011; Ikegame et al. 2013) and brain (Keller et al. 2010) in several psychiatric disorders, including depression, schizophrenia, and bipolar disorder, among others. In addition, this gene is epigenetically regulated during development, and this regulation is sensitive to environmental agents that induce behavioral and learning deficits (Boulle et al. 2012). In our animal study, we showed that prenatal BPA exposure induces sex-specific *Bdnf* DNA methylation changes in the blood, which can predict DNA methylation and gene expression changes in the brain as well as cognitive deficits in male mice (Kundakovic et al. 2015). To explore the translational value of our results, we examined *BDNF* DNA methylation in cord blood samples from the human cohort of Columbia Center for Children's Environmental Health (CCCEH) (Perera et al. 2012). Importantly, as Perera et al. found more behavioral problems in boys than in girls, the behavioral profile of this cohort was consistent with the BPA-induced male-specific learning deficits in our animal study (Kundakovic et al. 2015). To test our hypothesis, we selected cord blood samples of participants that correspond to approximately the lowest quintile (BPA<1 μg/L) and highest quintile (BPA>4 μg/L) of prenatal BPA exposure. The human *BDNF* IV region that we examined has 96 % homology with the

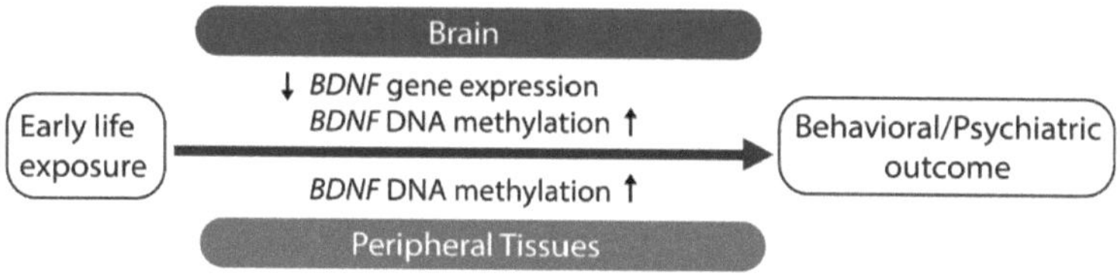

Fig. 3.3 Candidate epigenetic biomarkers for the prediction of behavioral and psychiatric vulnerability. Early life exposures, such as in utero BPA exposure, can leave lasting DNA methylation signatures in peripheral tissues (e.g., *BDNF* DNA methylation changes in the blood) that may reflect functionally relevant changes in brain DNA methylation and gene expression and could be used to predict behavioral and psychiatric outcomes. This figure is adapted from Kundakovic et al. (2015), DNA methylation of BDNF as a biomarker of early-life adversity, PNAS, 112(22), 6807–6813, Fig. 5, p. 6811; used with permission from *Proceedings of the National Academy of Sciences, USA*

sequence of the mouse *Bdnf* IV region with the conserved CREB binding region. Importantly, prenatal exposure to high BPA levels was associated with altered DNA methylation of 2 CpG sites in the human cord blood, in a sex-specific way resembling the animal data (Kundakovic et al. 2015). One of those sites lies exactly within the CREB binding site, and our data suggest that methylation of this site may represent a biomarker in humans predictive of *BDNF* gene expression in the brain and consequent alterations in brain function and behavior.

Based on our findings, we have suggested a model in which the *BDNF* DNA methylation status could be used to predict behavioral vulnerability induced by early environmental exposures (Fig. 3.3). Importantly, we also showed that other early life adversities, such as maternal depression during pregnancy, could also affect the offspring's *BDNF* DNA methylation levels in other peripheral tissues, such as buccal cells (Braithwaite et al. 2015) Together, these studies imply that peripheral epigenetic signatures could be used to predict behavioral and psychiatric vulnerability in response to early life adversity (Kundakovic et al. 2015).

Future Directions As noted above for the animal study, it is very unlikely that *BDNF* gene is the only gene epigenetically altered in response to BPA treatment. Therefore, we are awaiting genome-wide DNA methylation study in human cord blood (or other peripheral tissues) that may reveal novel gene targets and potential biomarkers of BPA exposure that may be relevant for predicting behavioral vulnerability. In addition, to improve our ability to find the epigenetic markers that are truly associated with BPA exposure, human studies need to improve in the same way that was discussed for studies concerning behavioral outcomes, including but not limited to improved BPA exposure classification, controlling for the effects of other exposures and genetic effects, etc. However, combining a human study with an animal study, which can much better control for genetic factors and other exposures, and finding comparable results is always reassuring that the effects found in humans can, indeed, be biologically relevant.

3.4 Multigenerational Effects of BPA and Other Endocrine Disruptors

3.4.1 Experimental Evidence

There is now increasing concern that gestational exposure to endocrine disruptors may not only affect the first generation of developing offspring (F1) but could possibly extend to several subsequent generations. For instance, perinatal exposure to an estrogenic compound diethylstilbestrol (DES) in mice results in genital tract abnormalities and cancers in the first (F1) and second (F2) generation offspring (Newbold et al. 1998, 2000; Veurink et al. 2005). It is important to keep in mind, though, that when mothers are exposed to a toxicant during pregnancy, the mother (F0 generation), the developing embryo (F1 generation), and the developing germ line of the F2 generation experience direct exposure to the compound (Jirtle and Skinner 2007). However, the adverse effects of endocrine disruptors have also been shown to extend beyond the F2 generation, inducing changes in offspring that have not had direct exposure to the chemical. Gestational exposure to the antiandrogenic fungicide, vinclozolin, during the period of gonadal sex determination, resulted in decreased spermatogenic capacity and male subfertility that were observed in F1 offspring and in three subsequent generations (F2–F4) (Anway et al. 2005). In addition, animals from all examined generations (F1–F4) developed pathological states and tissue abnormalities, such as tumors, prostate and kidney disease, and immune cell defects (Anway et al. 2006).

There is now emerging evidence that BPA can also induce transgenerational effects. A recent study by Ziv-Gal et al. has shown that BPA can alter female reproductive function for up to three generations (Ziv-Gal et al. 2015). They administered three different low doses of BPA to pregnant mice from gestational day 11 until birth and then monitored female offspring in three subsequent generations for different developmental and reproductive outcomes. In addition to the F1 generation, the F2 generation mice showed a significantly reduced gestational index, and mice of the F3 generation showed a lower fertility index, compared to controls. Interestingly, some of the adverse effects appeared in the F3 generation only. The F3 generation females showed delays in sexual development that were not seen in the F1 or F2 females that had the direct exposure to BPA. It is important to note that some of the most profound reproductive effects in the F3 generation (e.g., compromised fertility index and delayed vaginal opening) were seen in the mice that belonged to the lowest-dose BPA (50 ng/kg/day) lineage.

Important for the current discussion, another study has shown that transgenerational effects of BPA can extend beyond reproductive outcomes and can include effects on behavior. Wolstenholme et al. fed pregnant mice control or BPA-containing diets that produced plasma BPA concentrations similar to concentrations in humans and subsequently examined the juvenile F1 and F3 generations on social recognition and activity tasks (Wolstenholme et al. 2013). In both generations, BPA exposed mice showed increased levels of social investigation compared to controls.

However, there were effects that were observed only in the F3 generation but not in the F1 generation. For example, only the F3 generation BPA mice showed deficits in social recognition task, and they were more active than controls in the open field test. These results suggest that environmental levels of BPA exposure during gestation may have lasting, transgenerational effects on behavioral phenotypes such as social recognition and activity that might not even be obvious in the first generation offspring.

3.4.2 Possible Mechanisms for Multigenerational Effects

The mechanisms for multigenerational effects of endocrine disruptors can vary depending on the affected generation that is being considered. As previously discussed, the F2 generation still experiences direct exposure to the toxicant, and it seems plausible that the endocrine disruptors may exert their effects by directly affecting the F2 epigenome. For instance, in studies of DES reproductive and carcinogenic effects, these effects were linked to aberrant DNA methylation in developmental and cancer-related genes, implying that epigenetic alterations might underlie multigenerational adverse effects of DES (Newbold et al. 1998, 2000; Veurink et al. 2005). This is a very plausible mechanism as the F2 epigenome may be particularly sensitive to epigenetic dysregulation during the dynamic reprogramming of parental gene imprints occurring at the time of gametogenesis, which coincides with the exposure. Moreover, BPA has indeed been shown to change DNA methylation of gene imprints in the fetal germ cells (Zhang et al. 2012).

However, the term transgenerational inheritance usually refers to the effects observed in the F3 and subsequent generations that did not have the direct exposure to the chemical. Transgenerational inheritance of phenotypes has long been thought to solely occur via genetic transmission through the germ line. It was believed that all epigenetic marks acquired by the previous generation are erased in primordial germ cells and early embryo during epigenetic reprogramming, and that a new organism develops exclusively based on inherited genetic information (Reik 2007). However, endocrine disruptors have not been shown to induce genetic mutations in the germ cells that could explain observed transgenerational effects, and emerging evidence suggests that epigenetic mechanisms may be involved in those effects. Accordingly, there is now evidence that DNA methylation at certain loci can escape reprogramming during development, providing the basis for the hypothesis that transgenerational inheritance may occur through epigenetic mechanisms in rodents and possibly even in humans (Lange and Schneider 2010; Reik 2007).

The transmission of transgenerational effects of vinclozolin on male fertility was proposed to occur epigenetically through the male germ line, as vinclozolin exposure was associated with aberrant DNA methylation patterns in the sperm of F1–F3 generations (Anway et al. 2005; Guerrero-Bosagna et al. 2010). A more recent study showed that the mixture of plastic-derived endocrine disruptors, including BPA, induces DNA methylation changes in the sperm epigenome implying that

BPA may also exert transgenerational effects through DNA methylation-dependent mechanisms (Manikkam et al. 2013). However, another interesting study partly counteracts this hypothesis by showing that endocrine disruptor-induced DNA methylation changes in exposed fetal germ cells are regularly corrected by reprogramming events in the next generation (Iqbal et al. 2015). It is clear that this field warrants further studies and, beyond DNA methylation, other epigenetic mechanisms may be shown to be involved in transgenerational epigenetic inheritance, including histone modifications or noncoding RNAs (Gapp et al. 2014; Vassoler et al. 2013).

Lastly, multigenerational effects of endocrine disruptors may also be explained by behavioral mode of transmission, which again could involve epigenetic mechanisms. This phenomenon is also known as experience-dependent epigenetic inheritance (Danchin et al. 2011). For instance, animal studies have shown that the quality of maternal care can be transmitted from mothers to daughters and granddaughters; the mechanisms involve epigenetic changes in steroid receptor genes that produce long-term changes in gene expression and behavior (Champagne 2008). In this case, epigenetic marks are induced by experience (the level of maternal care) and have to be recreated de novo in each generation. Importantly, the quality of maternal care can also influence other phenotypes in the offspring, such as stress responses, anxiety-like behavior, and learning; these effects have also been associated with epigenetic changes in the offspring (Szyf et al. 2005). In this regard, we and others have shown that BPA induces changes in maternal behavior in mothers (F0 generation) that were directly exposed to BPA during pregnancy (Kundakovic et al. 2013a; Palanza et al. 2002) as well as in the F1 generation females exposed to BPA prenatally (Palanza et al. 2002). This suggests that epigenetically mediated behavioral transmission of maternal care may provide one of the mechanisms for the multigenerational transmission of BPA-induced neurobehavioral effects. While our understanding of this mechanism is limited to studies in rodents, it seems plausible that similar mechanisms may underlie experience-dependent transmission in humans.

3.4.3 Is There Evidence for Multigenerational Effects of Endocrine Disruptors in Humans?

There is now some evidence in humans that endocrine disruptors can induce multigenerational effects. For instance, DES is a synthetic estrogen that was prescribed from the late 1940s to early 1970s for the prevention of miscarriages in pregnant women. The use of DES had tragic consequences as prenatal DES treatment has been linked to an increased risk of reproductive anomalies and tumors, not only in individuals exposed to DES in utero but also in the subsequent generation of offspring (Veurink et al. 2005). The tragic DES story strongly implies that animal studies can be very informative and predictive of human health outcomes in response to toxicological exposures. And, although there is currently no human data showing association between toxicological exposures and adverse health effects that extend beyond the G2 generation, this seems intuitive considering the evidence from animal studies.

In summary, an increasing number of studies imply that exposures to BPA and other endocrine disruptors may have cumulative adverse effects on future generations, and that these effects could be mediated through epigenetic mechanisms.

Future Directions: In studies on the adverse effects of BPA, it would be appropriate and informative to incorporate a multigenerational study design that would enable further exploration of the extent and possible mechanisms through which exposure to this compound could affect multiple generations.

3.5 Conclusions

There is substantial evidence that low-dose prenatal BPA exposure can disrupt neurodevelopment, sexual dimorphism, behavior, and learning in animals, and the consequences may extend to future generations. While the studies in humans are much more limited and more difficult to interpret, increasing evidence suggests that high maternal BPA exposure may contribute to lasting behavioral problems in children. BPA has been shown to disrupt the brain epigenome in rodents and it is likely that, at least in part, epigenetic mechanisms underlie lasting BPA effects on neurobehavioral outcomes. Early studies in humans show that prenatal BPA exposure may leave DNA methylation signatures in peripheral blood that could possibly be used to predict behavioral vulnerability. More detailed studies in both animals and humans are needed to understand BPA effects on the epigenome and its consequences for the brain function and behavior of multiple generations. Epigenetic findings could be used to shape the public policies on BPA use and to open new opportunities for early detection and interventions for environmentally contributed behavioral disorders.

Abbreviations

APVP	Anteroventral periventricular preoptic area
A^{vy}	Agouti viable yellow
Bdnf/*BDNF*	Brain-derived neurotrophic factor (mouse/human gene symbol)
BPA	Bisphenol A
CCCEH	Columbia Center for Children's Environmental Health
CHAMACOS	Center for the Health Assessment of Mothers and Children of Salinas
CREB	Cyclic AMP response element-binding protein
DES	Diethylstilbestrol
DNMTs	DNA methyltransferases

EFSA	European Food Safety Authority
ER	Estrogen Receptor
FDA	U.S. Food and Drug Administration
IAP	Intracisternal A particle retrotransposon
LOAEL	Lowest-observed-adverse-effects level
NMDA	*N*-methyl-D-aspartic acid
SFFII	Study for Future Families II
TDI	Tolerable daily intake

References

Adriani, W., Seta, D. D., Dessi-Fulgheri, F., Farabollini, F., & Laviola, G. (2003). Altered profiles of spontaneous novelty seeking, impulsive behavior, and response to D-amphetamine in rats perinatally exposed to bisphenol A. *Environmental Health Perspectives, 111*, 395–401.

Andersen, H. R., Andersson, A. M., Arnold, S. F., Autrup, H., Barfoed, M., Beresford, N. A., Bjerregaard, P., Christiansen, L. B., Gissel, B., Hummel, R., et al. (1999). Comparison of short-term estrogenicity tests for identification of hormone-disrupting chemicals. *Environmental Health Perspectives, 107*(Suppl 1), 89–108.

Anderson, O. S., Nahar, M. S., Faulk, C., Jones, T. R., Liao, C., Kannan, K., Weinhouse, C., Rozek, L. S., & Dolinoy, D. C. (2012). Epigenetic responses following maternal dietary exposure to physiologically relevant levels of bisphenol A. *Environmental and Molecular Mutagenesis, 53*, 334–42.

Anway, M. D., Cupp, A. S., Uzumcu, M., & Skinner, M. K. (2005). Epigenetic transgenerational actions of endocrine disruptors and male fertility. *Science, 308*, 1466–1469.

Anway, M. D., Leathers, C., & Skinner, M. K. (2006). Endocrine disruptor vinclozolin induced epigenetic transgenerational adult-onset disease. *Endocrinology, 147*, 5515–5523.

Biedermann, S., Tschudin, P., & Grob, K. (2010). Transfer of bisphenol A from thermal printer paper to the skin. *Analytical and Bioanalytical Chemistry, 398*, 571–576.

Boersma, G. J., Lee, R. S., Cordner, Z. A., Ewald, E. R., Purcell, R. H., Moghadam, A. A., & Tamashiro, K. L. (2014). Prenatal stress decreases Bdnf expression and increases methylation of Bdnf exon IV in rats. *Epigenetics, 9*, 437–447.

Boulle, F., van den Hove, D. L., Jakob, S. B., Rutten, B. P., Hamon, M., van Os, J., Lesch, K. P., Lanfumey, L., Steinbusch, H. W., & Kenis, G. (2012). Epigenetic regulation of the BDNF gene: Implications for psychiatric disorders. *Molecular Psychiatry, 17*, 584–596.

Braithwaite, E. C., Kundakovic, M., Ramchandani, P. G., Murphy, S. E., & Champagne, F. A. (2015). Maternal prenatal depressive symptoms predict infant NR3C1 1F and BDNF IV DNA methylation. *Epigenetics, 10*, 408–417.

Braun, J. M., Kalkbrenner, A. E., Calafat, A. M., Yolton, K., Ye, X., Dietrich, K. N., & Lanphear, B. P. (2011). Impact of early-life bisphenol A exposure on behavior and executive function in children. *Pediatrics, 128*, 873–882.

Braun, J. M., Kalkbrenner, A. E., Just, A. C., Yolton, K., Calafat, A. M., Sjodin, A., Hauser, R., Webster, G. M., Chen, A., & Lanphear, B. P. (2014). Gestational exposure to endocrine-disrupting chemicals and reciprocal social, repetitive, and stereotypic behaviors in 4- and 5-year-old children: The HOME study. *Environmental Health Perspectives, 122*, 513–520.

Braun, J. M., Yolton, K., Dietrich, K. N., Hornung, R., Ye, X., Calafat, A. M., & Lanphear, B. P. (2009). Prenatal bisphenol A exposure and early childhood behavior. *Environmental Health Perspectives, 117*, 1945–1952.

Bromer, J. G., Zhou, Y., Taylor, M. B., Doherty, L., & Taylor, H. S. (2010). Bisphenol-A exposure in utero leads to epigenetic alterations in the developmental programming of uterine estrogen response. *FASEB Journal, 24*, 2273–2280.

Calafat, A. M., Kuklenyik, Z., Reidy, J. A., Caudill, S. P., Ekong, J., & Needham, L. L. (2005). Urinary concentrations of bisphenol A and 4-nonylphenol in a human reference population. *Environmental Health Perspectives, 113*, 391–395.

Calafat, A. M., Ye, X., Wong, L. Y., Reidy, J. A., & Needham, L. L. (2008). Exposure of the U.S. population to bisphenol A and 4-tertiary-octylphenol: 2003–2004. *Environmental Health Perspectives, 116*, 39–44.

Champagne, F. A. (2008). Epigenetic mechanisms and the transgenerational effects of maternal care. *Frontiers in Neuroendocrinology, 29*, 386–397.

Chapin, R. E., Adams, J., Boekelheide, K., Gray, L. E., Jr., Hayward, S. W., Lees, P. S., McIntyre, B. S., Portier, K. M., Schnorr, T. M., Selevan, S. G., et al. (2008). NTP-CERHR expert panel report on the reproductive and developmental toxicity of bisphenol A. Birth defects research. *Developmental and Reproductive Toxicology, 83*(Pt. B), 157–395.

Clayton, E. M., Todd, M., Dowd, J. B., & Aiello, A. E. (2010). The impact of bisphenol A and triclosan on immune parameters in the US population, NHANES 2003–2006. *Environmental health perspectives, 119*, 390–6.

Cooney, C. A., Dave, A. A., & Wolff, G. L. (2002). Maternal methyl supplements in mice affect epigenetic variation and DNA methylation of offspring. *Journal of Nutrition, 132*, 2393S–2400S.

D'Addario, C., Dell'Osso, B., Palazzo, M. C., Benatti, B., Lietti, L., Cattaneo, E., Galimberti, D., Fenoglio, C., Cortini, F., & Scarpini, E. (2012). Selective DNA methylation of BDNF promoter in bipolar disorder: Differences among patients with BDI and BDII. *Neuropsychopharmacology, 37*, 1647–1655.

Danchin, E., Charmantier, A., Champagne, F. A., Mesoudi, A., Pujol, B., & Blanchet, S. (2011). Beyond DNA: Integrating inclusive inheritance into an extended theory of evolution. *Nature Reviews Genetics, 12*, 475–486.

Davies, M. N., Volta, M., Pidsley, R., Lunnon, K., Dixit, A., Lovestone, S., Coarfa, C., Harris, R. A., Milosavljevic, A., Troakes, C., et al. (2012). Functional annotation of the human brain methylome identifies tissue-specific epigenetic variation across brain and blood. *Genome Biology, 13*, R43.

Dessi-Fulgheri, F., Porrini, S., & Farabollini, F. (2002). Effects of perinatal exposure to bisphenol A on play behavior of female and male juvenile rats. *Environmental Health Perspectives, 110*(Suppl 3), 403–407.

Dolinoy, D. C., Huang, D., & Jirtle, R. L. (2007). Maternal nutrient supplementation counteracts bisphenol A-induced DNA hypomethylation in early development. *Proceedings of the National Academy of Sciences of the United States of America, 104*, 13056–13061.

Dolinoy, D. C., Weidman, J. R., Waterland, R. A., & Jirtle, R. L. (2006). Maternal genistein alters coat color and protects Avy mouse offspring from obesity by modifying the fetal epigenome. *Environmental Health Perspectives, 114*, 567–572.

Doshi, T., Mehta, S. S., Dighe, V., Balasinor, N., & Vanage, G. (2011). Hypermethylation of estrogen receptor promoter region in adult testis of rats exposed neonatally to bisphenol A. *Toxicology, 289*, 74–82.

Evans, S. F., Kobrosly, R. W., Barrett, E. S., Thurston, S. W., Calafat, A. M., Weiss, B., Stahlhut, R., Yolton, K., & Swan, S. H. (2014). Prenatal bisphenol A exposure and maternally reported behavior in boys and girls. *Neurotoxicology, 45*, 91–99.

Farabollini, F., Porrini, S., Della Seta, D., Bianchi, F., & Dessi-Fulgheri, F. (2002). Effects of perinatal exposure to bisphenol A on sociosexual behavior of female and male rats. *Environmental Health Perspectives, 110*(Suppl 3), 409–414.

Fuchikami, M., Morinobu, S., Segawa, M., Okamoto, Y., Yamawaki, S., Ozaki, N., Inoue, T., Kusumi, I., Koyama, T., & Tsuchiyama, K. (2011). DNA methylation profiles of the brain-derived neurotrophic factor (BDNF) gene as a potent diagnostic biomarker in major depression. *PLoS One, 6*, e23881.

Fuks, F. (2005). DNA methylation and histone modifications: Teaming up to silence genes. *Current Opinion in Genetics and Development, 15*, 490–495.

Funabashi, T., Kawaguchi, M., Furuta, M., Fukushima, A., & Kimura, F. (2004). Exposure to bisphenol A during gestation and lactation causes loss of sex difference in corticotropin-releasing hormone-immunoreactive neurons in the bed nucleus of the stria terminalis of rats. *Psychoneuroendocrinology, 29*, 475–485.

Gapp, K., Jawaid, A., Sarkies, P., Bohacek, J., Pelczar, P., Prados, J., Farinelli, L., Miska, E., & Mansuy, I. M. (2014). Implication of sperm RNAs in transgenerational inheritance of the effects of early trauma in mice. *Nature Neuroscience, 17*, 667–669.

Gioiosa, L., Fissore, E., Ghirardelli, G., Parmigiani, S., & Palanza, P. (2007). Developmental exposure to low-dose estrogenic endocrine disruptors alters sex differences in exploration and emotional responses in mice. *Hormones and Behavior, 52*, 307–316.

Golebiewska, A., Atkinson, S. P., Lako, M., & Armstrong, L. (2009). Epigenetic landscaping during hESC differentiation to neural cells. *Stem Cells, 27*, 1298–1308.

Goncalves, C. R., Cunha, R. W., Barros, D. M., & Martinez, P. E. (2010). Effects of prenatal and postnatal exposure to a low dose of bisphenol A on behavior and memory in rats. *Environmental Toxicology and Pharmacology, 30*, 195–201.

Guerrero-Bosagna, C., Settles, M., Lucker, B., & Skinner, M. K. (2010). Epigenetic transgenerational actions of vinclozolin on promoter regions of the sperm epigenome. *PLoS One, 5*, e13100.

Harley, K. G., Gunier, R. B., Kogut, K., Johnson, C., Bradman, A., Calafat, A. M., & Eskenazi, B. (2013). Prenatal and early childhood bisphenol A concentrations and behavior in school-aged children. *Environmental Research, 126*, 43–50.

Ho, S. M., Tang, W. Y., Belmonte de Frausto, J., & Prins, G. S. (2006). Developmental exposure to estradiol and bisphenol A increases susceptibility to prostate carcinogenesis and epigenetically regulates phosphodiesterase type 4 variant 4. *Cancer Research, 66*, 5624–5632.

Huygh, J., Clotman, K., Malarvannan, G., Covaci, A., Schepens, T., Verbrugghe, W., Dirinck, E., Van Gaal, L., & Jorens, P. G. (2015). Considerable exposure to the endocrine disrupting chemicals phthalates and bisphenol-A in intensive care unit (ICU) patients. *Environment International, 81*, 64–72.

Ikegame, T., Bundo, M., Sunaga, F., Asai, T., Nishimura, F., Yoshikawa, A., Kawamura, Y., Hibino, H., Tochigi, M., Kakiuchi, C., et al. (2013). DNA methylation analysis of BDNF gene promoters in peripheral blood cells of schizophrenia patients. *Neuroscience Research, 77*, 208–214.

Ikezuki, Y., Tsutsumi, O., Takai, Y., Kamei, Y., & Taketani, Y. (2002). Determination of bisphenol A concentrations in human biological fluids reveals significant early prenatal exposure. *Human Reproduction, 17*, 2839–2841.

Iqbal, K., Tran, D. A., Li, A. X., Warden, C., Bai, A. Y., Singh, P., Wu, X., Pfeifer, G. P., & Szabo, P. E. (2015). Deleterious effects of endocrine disruptors are corrected in the mammalian germline by epigenome reprogramming. *Genome Biology, 16*, 59.

Jang, Y. J., Park, H. R., Kim, T. H., Yang, W. J., Lee, J. J., Choi, S. Y., Oh, S. B., Lee, E., Park, J. H., Kim, H. P., et al. (2012). High dose bisphenol A impairs hippocampal neurogenesis in female mice across generations. *Toxicology, 296*, 73–82.

Jasarevic, E., Williams, S. A., Vandas, G. M., Ellersieck, M. R., Liao, C., Kannan, K., Roberts, R. M., Geary, D. C., & Rosenfeld, C. S. (2013). Sex and dose-dependent effects of developmental exposure to bisphenol A on anxiety and spatial learning in deer mice (Peromyscus maniculatus bairdii) offspring. *Hormones and Behavior, 63*, 180–189.

Jirtle, R. L., & Skinner, M. K. (2007). Environmental epigenomics and disease susceptibility. *Nature Reviews Genetics, 8*, 253–262.

Kaminen-Ahola, N., Ahola, A., Maga, M., Mallitt, K. A., Fahey, P., Cox, T. C., Whitelaw, E., & Chong, S. (2010). Maternal ethanol consumption alters the epigenotype and the phenotype of offspring in a mouse model. *PLoS Genetics, 6*, e1000811.

Kawai, K., Nozaki, T., Nishikata, H., Aou, S., Takii, M., & Kubo, C. (2003). Aggressive behavior and serum testosterone concentration during the maturation process of male mice: The effects of fetal exposure to bisphenol A. *Environmental Health Perspectives, 111*, 175–178.

Keller, S., Sarchiapone, M., Zarrilli, F., Videtic, A., Ferraro, A., Carli, V., Sacchetti, S., Lembo, F., Angiolillo, A., Jovanovic, N., et al. (2010). Increased BDNF promoter methylation in the Wernicke area of suicide subjects. *Archives of General Psychiatry, 67*, 258–267.

Kimura, E., Matsuyoshi, C., Miyazaki, W., Benner, S., Hosokawa, M., Yokoyama, K., et al. (2015). Prenatal exposure to bisphenol A impacts neuronal morphology in the hippocampal CA_1 region in developing and aged mice. Archives of Toxicology. doi:10.1007/s00204-015-1485-x, [Epub ahead of print].

Klose, R. J., & Bird, A. P. (2006). Genomic DNA methylation: The mark and its mediators. *Trends in Biochemical Sciences, 31*, 89–97.

Kubo, K., Arai, O., Omura, M., Watanabe, R., Ogata, R., & Aou, S. (2003). Low dose effects of bisphenol A on sexual differentiation of the brain and behavior in rats. *Neuroscience Research, 45*, 345–356.

Kuiper, G. G., Lemmen, J. G., Carlsson, B., Corton, J. C., Safe, S. H., van der Saag, P. T., van der Burg, B., & Gustafsson, J. A. (1998). Interaction of estrogenic chemicals and phytoestrogens with estrogen receptor beta. *Endocrinology, 139*, 4252–4263.

Kundakovic, M. (2013). Prenatal programming of psychopathology: The role of epigenetic mechanisms. *Journal of Medical Biochemistry, 32*, 313–324.

Kundakovic, M., & Champagne, F. A. (2011). Epigenetic perspective on the developmental effects of bisphenol A. *Brain, Behavior, and Immunity, 25*, 1084–1093.

Kundakovic, M., Gudsnuk, K., Franks, B., Madrid, J., Miller, R. L., Perera, F. P., & Champagne, F. A. (2013a). Sex-specific epigenetic disruption and behavioral changes following low-dose in utero bisphenol A exposure. *Proceedings of the National Academy of Sciences of the United States of America, 110*, 9956–9961.

Kundakovic, M., Gudsnuk, K., Herbstman, J. B., Tang, D., Perera, F. P., & Champagne, F. A. (2015). DNA methylation of BDNF as a biomarker of early-life adversity. *Proceedings of the National Academy of Sciences of the United States of America, 112*, 6807–6813.

Kundakovic, M., Lim, S., Gudsnuk, K., & Champagne, F. A. (2013b). Sex-specific and strain-dependent effects of early life adversity on behavioral and epigenetic outcomes. *Frontiers in Psychiatry, 4*, 78.

Lange, U. C., & Schneider, R. (2010). What an epigenome remembers. *Bioessays, 32*, 659–668.

Liu, J., & Casaccia, P. (2010). Epigenetic regulation of oligodendrocyte identity. *Trends in Neurosciences, 33*, 193–201.

Lubin, F. D., Roth, T. L., & Sweatt, J. D. (2008). Epigenetic regulation of BDNF gene transcription in the consolidation of fear memory. *The Journal of Neuroscience, 28*, 10576–10586.

Manikkam, M., Tracey, R., Guerrero-Bosagna, C., & Skinner, M. K. (2013). Plastics derived endocrine disruptors (BPA, DEHP and DBP) induce epigenetic transgenerational inheritance of obesity, reproductive disease and sperm epimutations. *PLoS One, 8*, e55387.

Martinowich, K., Hattori, D., Wu, H., Fouse, S., He, F., Hu, Y., Fan, G., & Sun, Y. E. (2003). DNA methylation-related chromatin remodeling in activity-dependent BDNF gene regulation. *Science, 302*, 890–893.

McCarthy, M. M., & Arnold, A. P. (2011). Reframing sexual differentiation of the brain. *Nature Neuroscience, 14*, 677–683.

Mill, J., Tang, T., Kaminsky, Z., Khare, T., Yazdanpanah, S., Bouchard, L., Jia, P., Assadzadeh, A., Flanagan, J., Schumacher, A., et al. (2008). Epigenomic profiling reveals DNA-methylation changes associated with major psychosis. *American Journal of Human Genetics, 82*, 696–711.

Miller, F. D., & Gauthier, A. S. (2007). Timing is everything: Making neurons versus glia in the developing cortex. *Neuron, 54*, 357–369.

Miodovnik, A., Engel, S. M., Zhu, C., Ye, X., Soorya, L. V., Silva, M. J., Calafat, A. M., & Wolff, M. S. (2011). Endocrine disruptors and childhood social impairment. *Neurotoxicology, 32*, 261–267.

Miyagawa, K., Narita, M., Narita, M., Akama, H., & Suzuki, T. (2007). Memory impairment associated with a dysfunction of the hippocampal cholinergic system induced by prenatal and neonatal exposures to bisphenol-A. *Neuroscience Letters, 418*, 236–241.

Mueller, B. R., & Bale, T. L. (2008). Sex-specific programming of offspring emotionality after stress early in pregnancy. *The Journal of Neuroscience, 28*, 9055–9065.

Nakamura, K., Itoh, K., Sugimoto, T., & Fushiki, S. (2007). Prenatal exposure to bisphenol A affects adult murine neocortical structure. *Neuroscience Letters, 420*, 100–105.

Nakamura, K., Itoh, K., Yaoi, T., Fujiwara, Y., Sugimoto, T., & Fushiki, S. (2006). Murine neocortical histogenesis is perturbed by prenatal exposure to low doses of Bisphenol A. *Journal of Neuroscience Research, 84*, 1197–1205.

Negishi, T., Kawasaki, K., Suzaki, S., Maeda, H., Ishii, Y., Kyuwa, S., Kuroda, Y., & Yoshikawa, Y. (2004). Behavioral alterations in response to fear-provoking stimuli and tranylcypromine induced by perinatal exposure to bisphenol A and nonylphenol in male rats. *Environmental Health Perspectives, 112*, 1159–1164.

Newbold, R. R., Hanson, R. B., Jefferson, W. N., Bullock, B. C., Haseman, J., & McLachlan, J. A. (1998). Increased tumors but uncompromised fertility in the female descendants of mice exposed developmentally to diethylstilbestrol. *Carcinogenesis, 19*, 1655–1663.

Newbold, R. R., Hanson, R. B., Jefferson, W. N., Bullock, B. C., Haseman, J., & McLachlan, J. A. (2000). Proliferative lesions and reproductive tract tumors in male descendants of mice exposed developmentally to diethylstilbestrol. *Carcinogenesis, 21*, 1355–1363.

Novikova, S. I., He, F., Bai, J., Cutrufello, N. J., Lidow, M. S., & Undieh, A. S. (2008). Maternal cocaine administration in mice alters DNA methylation and gene expression in hippocampal neurons of neonatal and prepubertal offspring. *PLoS One, 3*, e1919.

Onishchenko, N., Karpova, N., Sabri, F., Castren, E., & Ceccatelli, S. (2008). Long-lasting depression-like behavior and epigenetic changes of BDNF gene expression induced by perinatal exposure to methylmercury. *Journal of Neurochemistry, 106*, 1378–1387.

Palanza, P. L., Howdeshell, K. L., Parmigiani, S., & vom Saal, F. S. (2002). Exposure to a low dose of bisphenol A during fetal life or in adulthood alters maternal behavior in mice. *Environmental Health Perspectives, 110*(Suppl 3), 415–422.

Patisaul, H. B., Fortino, A. E., & Polston, E. K. (2006). Neonatal genistein or bisphenol-A exposure alters sexual differentiation of the AVPV. *Neurotoxicology and Teratology, 28*, 111–118.

Perera, F., Vishnevetsky, J., Herbstman, J. B., Calafat, A. M., Xiong, W., Rauh, V., & Wang, S. (2012). Prenatal bisphenol A exposure and child behavior in an inner city cohort. *Environmental Health Perspectives, 120*, 1190–1194.

Porrini, S., Belloni, V., Della Seta, D., Farabollini, F., Giannelli, G., & Dessi-Fulgheri, F. (2005). Early exposure to a low dose of bisphenol A affects socio-sexual behavior of juvenile female rats. *Brain Research Bulletin, 65*, 261–266.

Reik, W. (2007). Stability and flexibility of epigenetic gene regulation in mammalian development. *Nature, 447*, 425–432.

Richter, C. A., Birnbaum, L. S., Farabollini, F., Newbold, R. R., Rubin, B. S., Talsness, C. E., Vandenbergh, J. G., Walser-Kuntz, D. R., & vom Saal, F. S. (2007). In vivo effects of bisphenol A in laboratory rodent studies. *Reproductive Toxicology, 24*, 199–224.

Roen, E.L., Wang, Y., Calafat, A.M., Wang, S., Margolis, A., Herbstman, J., et al. (2015). Bisphenol A exposure and behavioral problems among inner city children at 7–9 years of age. *Environmental Research, 142*, 739–45

Roth, T. L., Lubin, F. D., Funk, A. J., & Sweatt, J. D. (2009). Lasting epigenetic influence of early-life adversity on the BDNF gene. *Biological Psychiatry, 65*, 760–769.

Rubin, B. S., Lenkowski, J. R., Schaeberle, C. M., Vandenberg, L. N., Ronsheim, P. M., & Soto, A. M. (2006). Evidence of altered brain sexual differentiation in mice exposed perinatally to low, environmentally relevant levels of bisphenol A. *Endocrinology, 147*, 3681–3691.

Schonfelder, G., Wittfoht, W., Hopp, H., Talsness, C. E., Paul, M., & Chahoud, I. (2002). Parent bisphenol A accumulation in the human maternal-fetal-placental unit. *Environmental Health Perspectives, 110*, A703–707.

Skinner, M. K., Anway, M. D., Savenkova, M. I., Gore, A. C., & Crews, D. (2008). Transgenerational epigenetic programming of the brain transcriptome and anxiety behavior. *PLoS One, 3*, e3745.

Sohoni, P., & Sumpter, J. P. (1998). Several environmental oestrogens are also anti-androgens. *Journal of Endocrinology, 158*, 327–339.

Szyf, M., Weaver, I. C., Champagne, F. A., Diorio, J., & Meaney, M. J. (2005). Maternal programming of steroid receptor expression and phenotype through DNA methylation in the rat. *Frontiers in Neuroendocrinology, 26*, 139–162.

Takayanagi, S., Tokunaga, T., Liu, X., Okada, H., Matsushima, A., & Shimohigashi, Y. (2006). Endocrine disruptor bisphenol A strongly binds to human estrogen-related receptor gamma (ERRgamma) with high constitutive activity. *Toxicology Letters, 167*, 95–105.

Tando, S., Itoh, K., Yaoi, T., Ikeda, J., Fujiwara, Y., & Fushiki, S. (2007). Effects of pre-and neonatal exposure to bisphenol A on murine brain development. *Brain and Development, 29*, 352–356.

Tang, W. Y., Morey, L. M., Cheung, Y. Y., Birch, L., Prins, G. S., & Ho, S. M. (2012). Neonatal exposure to estradiol/bisphenol A alters promoter methylation and expression of Nsbp1 and Hpcal1 genes and transcriptional programs of Dnmt3a/b and Mbd2/4 in the rat prostate gland throughout life. *Endocrinology, 153*, 42–55.

Taylor, J. A., Welshons, W. V., & Vom Saal, F. S. (2008). No effect of route of exposure (oral; subcutaneous injection) on plasma bisphenol A throughout 24h after administration in neonatal female mice. *Reproductive Toxicology, 25*, 169–176.

Tetel, M. J., & Pfaff, D. W. (2010). Contributions of estrogen receptor-alpha and estrogen receptor-ss to the regulation of behavior. *Biochimica et Biophysica Acta, 1800*, 1084–1089.

Tian, Y. H., Baek, J. H., Lee, S. Y., & Jang, C. G. (2010). Prenatal and postnatal exposure to bisphenol a induces anxiolytic behaviors and cognitive deficits in mice. *Synapse, 64*, 432–439.

Vandenberg, L. N., Chahoud, I., Heindel, J. J., Padmanabhan, V., Paumgartten, F. J., & Schoenfelder, G. (2010). Urinary, circulating, and tissue biomonitoring studies indicate widespread exposure to bisphenol A. *Environmental Health Perspectives, 118*, 1055–1070.

Vandenberg, L. N., Colborn, T., Hayes, T. B., Heindel, J. J., Jacobs, D. R., Jr., Lee, D. H., Shioda, T., Soto, A. M., vom Saal, F. S., Welshons, W. V., et al. (2012). Hormones and endocrine-disrupting chemicals: Low-dose effects and nonmonotonic dose responses. *Endocrine Reviews, 33*, 378–455.

Vandenberg, L. N., Hauser, R., Marcus, M., Olea, N., & Welshons, W. V. (2007). Human exposure to bisphenol A (BPA). *Reproductive Toxicology, 24*, 139–177.

Vandenberg, L. N., Maffini, M. V., Sonnenschein, C., Rubin, B. S., & Soto, A. M. (2009). Bisphenol-A and the great divide: A review of controversies in the field of endocrine disruption. *Endocrine Reviews, 30*, 75–95.

Vassoler, F. M., White, S. L., Schmidt, H. D., Sadri-Vakili, G., & Pierce, R. C. (2013). Epigenetic inheritance of a cocaine-resistance phenotype. *Nature Neuroscience, 16*, 42–47.

Veurink, M., Koster, M., & Berg, L. T. (2005). The history of DES, lessons to be learned. *Pharmacy World & Science, 27*, 139–143.

vom Saal, F. S., & Welshons, W. V. (2006). Large effects from small exposures. II. The importance of positive controls in low-dose research on bisphenol A. *Environmental Research, 100*, 50–76.

Weaver, I. C., Cervoni, N., Champagne, F. A., D'Alessio, A. C., Sharma, S., Seckl, J. R., Dymov, S., Szyf, M., & Meaney, M. J. (2004). Epigenetic programming by maternal behavior. *Nature Neuroscience, 7*, 847–854.

Welshons, W. V., Nagel, S. C., & vom Saal, F. S. (2006). Large effects from small exposures. III. Endocrine mechanisms mediating effects of bisphenol A at levels of human exposure. *Endocrinology, 147*, S56–69.

Wolstenholme, J. T., Edwards, M., Shetty, S. R., Gatewood, J. D., Taylor, J. A., Rissman, E. F., & Connelly, J. J. (2012). Gestational exposure to bisphenol a produces transgenerational changes in behaviors and gene expression. *Endocrinology, 153*, 3828–3838.

Wolstenholme, J. T., Goldsby, J. A., & Rissman, E. F. (2013). Transgenerational effects of prenatal bisphenol A on social recognition. *Hormones and Behavior, 64*, 833–839.

Wolstenholme, J. T., Taylor, J. A., Shetty, S. R., Edwards, M., Connelly, J. J., & Rissman, E. F. (2011). Gestational exposure to low dose bisphenol A alters social behavior in juvenile mice. *PLoS One, 6*, e25448.

Xu, L. C., Sun, H., Chen, J. F., Bian, Q., Qian, J., Song, L., & Wang, X. R. (2005). Evaluation of androgen receptor transcriptional activities of bisphenol A, octylphenol and nonylphenol in vitro. *Toxicology, 216*, 197–203.

Xu, X. H., Zhang, J., Wang, Y. M., Ye, Y. P., & Luo, Q. Q. (2010). Perinatal exposure to bisphenol-A impairs learning-memory by concomitant down-regulation of N-methyl-D-aspartate receptors of hippocampus in male offspring mice. *Hormones and Behavior, 58*, 326–333.

Yaoi, T., Itoh, K., Nakamura, K., Ogi, H., Fujiwara, Y., & Fushiki, S. (2008). Genome-wide analysis of epigenomic alterations in fetal mouse forebrain after exposure to low doses of bisphenol A. *Biochemical and Biophysical Research Communications, 376*, 563–567.

Zhang, X. F., Zhang, L. J., Feng, Y. N., Chen, B., Feng, Y. M., Liang, G. J., Li, L., & Shen, W. (2012). Bisphenol A exposure modifies DNA methylation of imprint genes in mouse fetal germ cells. *Molecular Biology Reports, 39*, 8621–8628.

Ziv-Gal, A., Wang, W., Zhou, C., & Flaws, J. A. (2015). The effects of in utero bisphenol A exposure on reproductive capacity in several generations of mice. *Toxicology and Applied Pharmacology, 284*, 354–362.

Chapter 4
The Epigenetic Biomarker γH2AX: From Bench to Clinical Trials

Emmy P. Rogakou, Vassilios Papadakis, and George P. Chrousos

4.1 The γH2AX Paradigm

Epigenetics has emerged as a critical player in human physiology, evolution, and disease in the postgenomic era. The term "epigenetics" has a Greek origin (Greek_ επί: over, on top, Greek_ γενετική: genetics) and depicts the cellular, phenotypic, and physiological traits that are not caused by changes in the DNA sequence. Even though the genome defines the complete set of genetic information contained in the DNA of each individual, the epigenetic mechanisms give plasticity to cells, in order to differentiate during development, and to adapt differently in response to environmental stimuli. Although the same set of genes exists in every cell, different cells in the human body activate different subsets of genes. Misregulation of epigenetic mechanisms can lead to the initiation of a disease. ENCODE, the epigenetic project that aimed to study the epigenomic signatures of human cells grown in culture, has linked epigenomic data to the corresponding genetic information (Dunham et al. 2012; Romanoski et al. 2015; 82) and has produced reference epigenomes for cell types and tissues (Consortium et al. 2015; Dixon et al. 2015; Koren et al. 2015).

Epigenetic factors exert their function by means of three mechanisms: (1) DNA methylation, (2) noncoding RNAs, and (3) histone posttranslational modifications. In this chapter, we are going to focus on a histone posttranslational modification, the γ-phosphorylation of the histone H2AX.

Histone proteins form an octamer, a bead-like structure where DNA is wrapped around, to assemble the nucleosome. Nucleosomes are the fundamental building

E.P. Rogakou, PhD (✉) • G.P. Chrousos, MD
First Department of Pediatrics "Aghia Sophia" Children's Hospital, Medical School, University of Athens, Athens, Greece
e-mail: emrogakou@med.uoa.gr

V. Papadakis, MD, PhD
Department of Pediatric Hematology—Oncology, Marianna V. Vardinoyannis—ELPIDA Oncology Unit, "Aghia Sophia" Children's Hospital, Athens, Greece

D. Hollar (ed.), *Epigenetics, the Environment, and Children's Health Across Lifespans*, DOI 10.1007/978-3-319-25325-1_4

blocks of eukaryotic chromatin. Histone "tails" are the ends of the histone molecules that protrude out from the nucleosomes, providing an exposed surface for potential interactions with other proteins (Luger et al. 1997). The number of posttranslational modifications that have been identified on histone tails is increasing as new technologies are employed in this field and include acetylation, phosphorylation, methylation, ubiquitination, ADP-ribosylation, and many others. It has been proposed that distinct histone modifications on one or more tails act sequentially or in combination to form a "histone code." According to the "histone code" hypothesis, histone modifications are recognized by specialized proteins that initiate a cascade of downstream events. The outcome of these events has an impact on different cell functions: transcription, mitotic and meiotic chromatin condensation, euchromatinization, heterochromatinization, and others. The "histone code" is likely to impact all chromatin-template processes with far-reaching consequences on cell fate decisions toward normal or pathological development (Jenuwein and Allis 2001; Strahl and Allis 2000; Van Attikum and Gasser 2005; Williamson et al. 2012).

The γ-phosphorylation of the histone H2AX has been established as the first paradigm of the "histone code" hypothesis that is implicated in the DNA damage response. To date, a large volume of studies have highlighted γH2AX to be the most sensitive and specific epigenetic biomarker for double-strand break detection in clinical trials and therapeutics.

4.2 The Biology of the γH2AX: Megabase-Long Chromatin Domains

4.2.1 γ-Phosphorylation of Histone H2AX

H2AX is a mammalian variant that belongs to the H2A histone family. In contrast with the other members of the family, H2AX has a unique phosphorylation site on its carboxy-terminal tail, a serine at position 139. This site becomes rapidly phosphorylated when double-strand breaks (DSB) are generated into DNA. It has been well documented that this phosphorylation is specific to DSB and is induced even when a single DBS is present in the nucleus. This specific phosphorylation is denoted as "γ-phosphorylation," and the term"γH2AX" indicates the phosphorylation at serine 139 of the histone H2AX (Rogakou et al. 1998).

The H2AX serine 139 is embedded in a specific SQ motif. This motif is recognized by kinases that are members of the phosphatidylinositol-3 family (PI3), namely, ATM (ataxia telangiectasia mutated), ATR (ATM and Rad3 related), and DNA-PK (DNA-dependent protein kinase) (McKinnon 2004; Paull et al. 2000; Rogakou et al. 1999). In human cells, ATM is the major kinase to control γ-phosphorylation (Daniel et al. 2012). Nevertheless, due to lack of ATM activity the other kinases take over, indicating overlapping roles between ATM and ATR

(Cimprich and Cortez 2008; Falck et al. 2005). However, these kinases perform in different cellular pathways. It has been demonstrated that ATR is the main kinase to γ-phosphorylate H2AX during replication arrest and under hypoxic conditions. The role of DNA-PK in γ-phosphorylation of H2AX is not well understood yet (Chronis & Rogakou, 2007; Downs and Jackson 2004).

4.2.2 *γ-Phosphorylation of Histone H2AX Extents Megabase-Long Domains in Chromatin: The "γ-Phosphorylated Chromatin Platform" Model*

γ-phosphorylation of H2AX is evident within minutes after the generation of DSBs. Most interestingly, γ -phosphorylation is not restricted to the vicinity of sites of the DSB, but extends along both sites of the damage and reaches megabase-long domains in chromatin (Rogakou et al. 1999) (Fig. 4.1). The significance of this fact is of great importance; it depicts a biological amplification mechanism where one

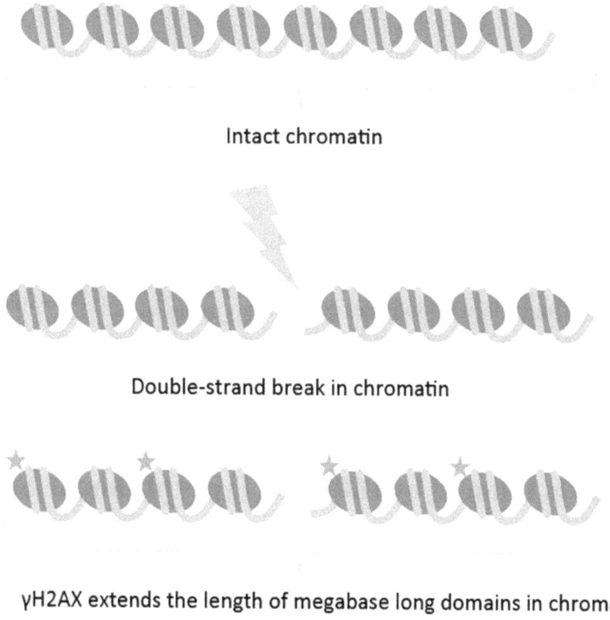

Fig. 4.1 Schematic depiction of the γH2AX induction. (**a**) In intact chromatin, there is no γ-phosphorylation on histone H2AX. Upon generation of a double-stranded break in chromatin (**b**), histone H2AX becomes γ-phosphorylated (depicted by a *star*), and this phosphorylation extends for the lengths of the megabase long domains in chromatin (**c**). Note that histone H2AX is incorporated into nucleosomes, and γ-phosphorylation occurs in situ. For the purpose of this illustration, H2AX is shown to be incorporated into every two nucleosomes. In fact, the percentage of H2AX in respect to the total H2A is different between cell types, spanning from 2.5 to 30 %

double-strand break induces the γ-phosphorylation of thousands of H2AX molecules along megabase-long domains of chromatin that are adjusted to the sites of double-strand breaks.

The physiological benefit of such an epigenetic amplification mechanism is vital to cells. It has been demonstrated that there is a time-dependent sequential recruitment of repair and signal transduction factors on γ-phosphorylated chromatin (Paull et al. 2000). Factors such as the MRN complex (Mre11-Rad50-Nbs1), 53BP1, BRCA1, MDC1, TopBP1, hRad9, and others are recruited to the γ-phosphorylated chromatin (Paull et al. 2000; Daniel et al. 2012; Fatouros & Rogakou, 2007; Lukas et al. 2004; Kobayashi et al. 2004; Fernandez-Capetillo et al. 2004; Stucki and Jackson 2004; Bekker-Jensen et al. 2005). These factors participate in double-strand repair pathways. It has been proposed that the megabase-long γ-phosphorylated chromatin serves as a "chromatin platform," where these factors are recruited to interact with each other (Fernandez-Capetillo et al. 2004).

Subsequently, signal transduction factors depart from the "γ-phosphorylated chromatin platform" to the cytoplasm to mediate cell-cycle arrest and/or apoptosis. As a result, the signal from one double-strand break (one dimension) is amplified on the γ-phosphorylated chromatin (two dimensions) via positive feedback loops, to become adequate to fill the cytoplasm (three dimensions). Remarkably, γH2AX is induced in all phases of the cell cycle and is present in all DSB repair pathways known so far. Among them, the role of γH2AX in homologous recombination (HR) and nonhomologous end-joining (NHEJ) is extensively studied.

The above-described "γ-phosphorylated chromatin platform" model has not only biological significance but also technological implications. As one double-strand break is surrounded by thousands of γ-phosphorylated H2AX nucleosomes, specific antibodies make possible the observation by immunocytochemistry of only one DSB per nucleus, rendering this technology the most sensitive assay for the detection of DSBs. By confocal microscopy, γH2AX foci appear as large, roughly spherical conformations and adopt a band-like conformation in deer mitotic cells (Rogakou et al. 1999), resembling perhaps the known bands in human mitotic chromosomes as seen in routine karyotype tests.

4.3 γH2AX Is Induced by Different Types of Double-Strand Breaks

4.3.1 Typical Kinetics of γH2AX Induction by Ionizing Radiation

Cells subjected to ionizing radiation (IR) exhibit γH2AX formation in minutes. Kinetics followed by immunocytochemistry coupled with confocal or epifluorescent microscopy exhibit the following characteristic: (1) γ-H2AX foci appear rapidly within 1 min, (2) foci increase in number and size, where half-maximal amounts are attained in less than 10 min, (3) foci number and size plateau are reached in 15–60 min, (4) decrease in foci number and size follows at 180 min, and (5) foci finally disappear within 48 h in most of the cases (Rogakou et al. 1998, 1999; Sedelnikova et al. 2002).

This kinetics correlates with the generation and repair of DSBs in the DNA of the cells. Moreover, γH2AX foci correlate one-by-one with the sites of DSB as it has been demonstrated in many quantitative studies and in cells that undergo V(D)J recombination (Chen et al. 2000; Yin et al. 2009). The formation of γH2AX foci is universal, as all normal and cancer cells in eukaryotes respond to the formation of γH2AX at lethal and sublethal amounts of ionizing radiation.

Ionizing radiation is known to produce multiple damaged sites that contain base and sugar alterations as well as single-strand breaks, in addition to double-strand breaks. Nevertheless, a series of experiments have attributed the γH2AX formation specifically to the double-strand breaks and have excluded other types of DNA damage. Neither treatment with H_2O_2, which produces hydroxyl radicals and DNA single-strand breaks at a low spatial distribution, nor low-dose irradiation with ultraviolet light induces directly γH2AX formation (Dixon et al. 2015).

In contrast, double-strand breaks generated by every different mechanism tested so far have shown to induce γH2AX formation (Fig. 4.2). The current dogma is that the generation of double-strand breaks in living eukaryotic cells always induces γH2AX formation. Nevertheless, other types of lesions may also induce γH2AX formation indirectly, as these lesions are converted to double-strand breaks during later steps in DNA repair (Table 4.1, Fig. 4.2) (see later in this chapter).

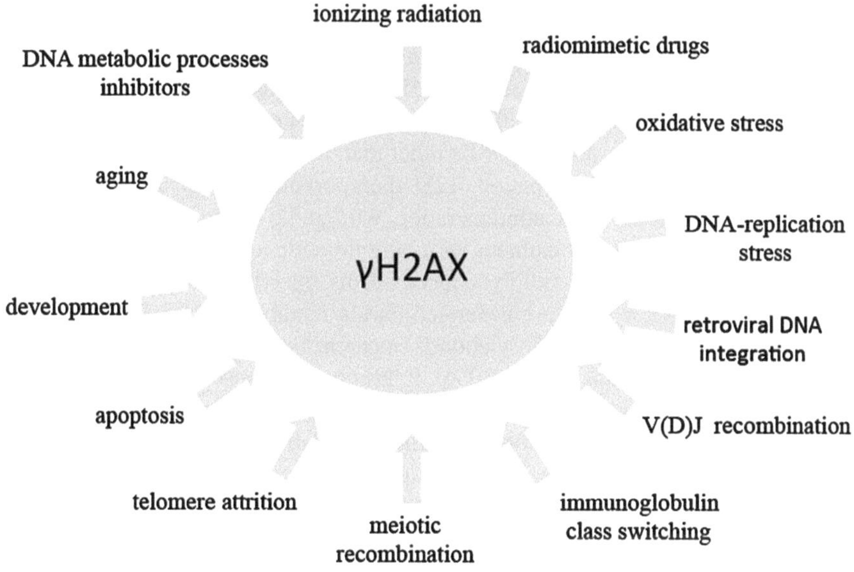

Fig. 4.2 DNA double-strand breaks generated by different environmental, cellular, or developmental events induce γH2AX

Table 4.1 Summary of well-known environmental challenges or cellular processes that induce γH2AX and their mechanism of action to generate DNA double-strand breaks

γH2AX induction	Environmental challenge or cellular process	Mechanism of action	
Direct	Radiation	ROS, α particles, etc.	Radio-chemistry chemistry
	Radionucleotides	ROS	
	Oxidative stress	ROS	
	Radiomimetic drugs	Non-radiolytic mechanisms	
Direct	Topoisomerase II inhibition	Topo II-cleavage complexes	Enzyme-mediated mechanisms
	V(D)J recombination	RAG endonuclease	
	Class switch recombination	AID triggered mechanism	
	Meiotic recombination	Spo11 topoisomerase	
	Apoptosis	CAD or DFF endonuclease	
	Stem cell developmental process	Unknown	
	Telomere attrition	DSB recognition apparatus	
	Retroviral integration	Retroviral integrase	
Indirect	Topoisomerase I inhibition	Topo I-cleavage complexes/ replication fork collision	
	DNA-replication stress	Several endonucleases	
	Intermediates in processing of DNA lesions	Endonucleases, topoisomerases, DNA processing enzymes	
	Aberrant RNA transcription	Non-specified yet	

4.3.2 Direct Generation of Double-Strand Breaks

The original observation of γH2AX formation was induced in experiments where ionizing irradiation was used to generate DSBs in cells, and since then, it has been studied extensively (Rogakou et al. 1998, 1999; Celeste et al. 2002, 2003; Redon et al. 2012). Direct generation of DSBs other than ionizing radiation has also been shown to induce γH2AX formation: 125I incorporation into DNA (Sedelnikova et al. 2002), alpha-particles administration with radio-immunotherapy to solid tumors (Song et al. 2013), treatment with radiomimetic agents such as bleomycin that introduce DSBs by non-radiolytic mechanisms, etc. γH2AX is also induced by reactive oxygen species that are generated during metabolism. It has been shown that constitutive H2AX phosphorylation is apparent in cells that undergo endogenous oxidative stress (Tanaka et al. 2006). It has been demonstrated that cumulative oxidative DNA damage is a key factor in aging and cellular senescence (Celeste et al. 2002; Sedelnikova et al. 2004). In addition, oxidative DNA damage is apparent in inflammation and is considered to predispose cells to neoplastic transformations (Murata et al. 2012).

Apart from irradiation and agents able to break DNA, DSBs are generated during a variety of specialized cellular processes, mediated by endonucleases or topoisomerases.

During V(D)J light chain recombination, RAG endonuclease generates double-strand-breaks between the immunoglobulin and the T cell receptor locus (Chen et al. 2000; Yin et al. 2009; Celeste et al. 2003). In developing thymocytes, γH2AX forms nuclear foci that co-localize with the T cell receptor γ-locus, as it has been shown by immunofluorescence in situ hybridization. These results also demonstrate that γH2AX-immunocytochemistry is such a powerful tool that is able to detect the presence of only one double-strand break in the human nucleus (Chen et al. 2000).

Antibody class switching occurs in mature B cells in response to antigen stimulation and co-stimulatory signals. It is mediated by a unique type of intrachromosomal deletional recombination within special G-rich tandem repeated DNA sequences. The process of class switch recombination (CSR) is initiated by the activation-induced cytidine deaminase (AID), which converts cytosines to uracils by deamination (removal of the amino group of cytosine). Subsequent repair of the deoxy-uracil (dU) residues leads finally to DSBs in order to initiate the process of DNA recombination. γH2AX foci co-localize with the recombinational region in wild type but not in $AID^{-/-}$ mice, indicating that γH2AX forms at sites of CSR and is dependent on the AID activity (Celeste et al. 2003; Panier and Boulton 2014).

In meiotic recombination, γH2AX formation follows the Spo11 activation. Spo11 is a topoisomerase-related protein that introduces DSBs into meiotic chromosomes to initiate meiotic recombination (Celeste et al. 2003; Fernandez-Capetillo et al. 2003; Mahadevaiah et al. 2001). During apoptosis, caspases orchestrate a cascade of cell reactions that result in the typical apoptotic morphology, including DNA fragmentation. During this process, apoptotic endonucleases are activated to produce the typical "apoptotic DNA ladder." It has been demonstrated that γH2AX is induced in apoptosis, following the activation of the caspase-activated DNase, called CAD or DFF (Rogakou et al. 2000; Lu et al. 2006; Solier and Pommier 2009). The biological significance of the γ-phosphorylation during the execution phase of apoptosis is not fully understood yet.

Early events during retroviral replication after the entry of the viral capsid include synthesis of a DNA copy of the viral RNA genome to form a pre-integration complex. This complex enters the nucleus and integrates with the host DNA. Retroviral integration is mechanistically similar to the V(D)J recombination; it is catalyzed by RAG proteins and promotes transient formation of γH2AX at retroviral integration sites (Daniel et al. 2004).

Telomeres are structures located at the termini of linear chromosomes of most eukaryotic organisms. Their major role is to facilitate DNA replication and the protection of chromosome ends from DNA repair enzymes by means of a protein complex known as shelterin. In human fibroblasts that exhibit telomerase-depended senescence, telomeric γH2AX foci are increased in number and extent more than 270 kilobases inward the end of the chromosome (Takai et al. 2003; d'Adda di Fagagna et al. 2003). γH2AX foci, other than telomerase related, also increase in number during aging. They co-localize with 53BP1, MDC1, NBS1, and the phosphorylated form of SMC1, associating DNA repair mechanisms with cellular senescence (Sedelnikova et al. 2004).

The involvement of γH2AX in cell differentiation has recently been implicated. According to a proposed model on differentiation, oxidative stress and DNA damage in undifferentiated cells induce cell-cycle arrest. As a consequence, the duration of the cell cycle is prolonged. During this arrest, transcription factors may accumulate, leading cells to exit the cell-cycle and proceed to differentiation (Santos et al. 2014).

4.3.3 Indirect Generation of Double-Strand Breaks

In addition to DNA double-strand breaks generated directly, there are also DNA double-strand breaks that are generated indirectly. This category includes DNA lesions other than DSBs, which are converted into double-strand breaks during subsequent biological steps. DNA lesions (such as single-strand breaks, base excision repair lesions, aberrant topological DNA formations that obstruct DNA-replication or RNA transcription, etc.) when interfere with DNA or RNA synthesis, cell cycle progression, or other cellular functions may transform to double-stand breaks.

Many pharmaceuticals exhibit the ability to generate indirect double-strand breaks. Enzyme inhibitors, such as PARP, topoisomerases, and other DNA processing enzyme inhibitors, freeze DNA lesions and produce DNA intermediates that may convert to double-strand breaks downstream the cell cycle. A typical example is the topoisomerase I inhibitor that originally introduces single-strand breaks. These lesions are converted to double-strand breaks during the S phase of the cell cycle as DNA-replication forks collide with the damage site (Furuta et al. 2003).

DNA intercalators and nucleotide analogs are used as powerful drugs. DNA intercalators create obstacles to DNA and RNA polymerases, and their repair may result to the generation of double-strand breaks. Nucleotide analogs lead to premature DNA-strand termination that may transform to double-strand breaks.

We are going to refer to these mechanisms later in this chapter, as we are going to examine the mechanisms of action of therapeutics that have been administered during clinical trials where γH2AX has served as a biomarker.

4.4 γH2AX as an Epigenetic Biomarker to Detect Double-Strand Breaks in Clinical Trials

4.4.1 Immunocytochemical Detection of γH2AX Is a Powerful Tool to Locate DNA Double-Strand Breaks in the Cell Nucleus

Since its discovery in 1998 (Rogakou et al. 1998), γH2AX has proven to be an ideal marker to detect double-strand breaks. Since then, γH2AX has been studied thoroughly and has been used as a tool in multiple biomedical fields. The number of scientific publications shows an exponential phase of growth (Fig. 4.3). Today, the volume produced by these studies support with confidence that γH2AX is an

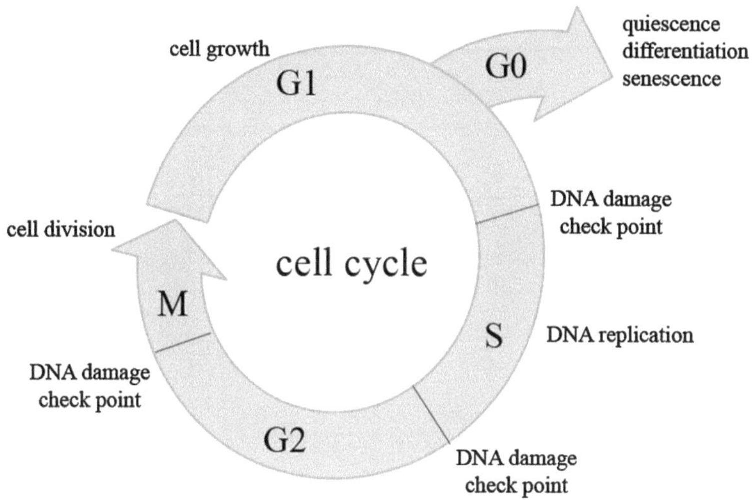

Fig. 4.3 Cell cycle and DNA damage check points. The progression from one phase of the cell cycle to the next is regulated by different control systems, called check points. The major check point systems that sense DNA damage and allow time for repair are located at G1 before entry into S phase, at S phase before the entry into G2 phase, and at the G2 before the entry into mitosis. The G0 phase is a stage where cells do not progress toward the next phase of the cell cycle (typically the S phase), although they exhibit full functionality. This stage is also referred as quiescence, and it is characteristic for both differentiated and senescent cells

excellent marker to detect double-strand breaks, generated from a variety of different environmental challenges or cellular processes. Immunocytochemical detection became the primary method of detection, as it is several orders of magnitude more sensitive than other methods and has the potential for quantification (Rogakou et al. 1999; Sedelnikova et al. 2002).

Assays based on specific antibodies against the characteristic γH2AX epitope (e.g., confocal and epifluorescent microscopy, flow cytometry, and immunoprecipitation) have been incomparably successful for the detection of double-strand breaks (Redon et al. 2010a, b, 2012; Ivashkevich et al. 2012). These assays share four important technical features: (1) specificity, (2) sensitivity, (3) quantification, and (4) repeatability and reproducibility.

1. Specificity: The γH2AX has been proven to detect specifically double-strand breaks in contrast to other DNA damages (Rogakou et al. 1998).
2. Sensitivity: Immunoassays utilizing specific antibodies for γH2AX show the highest score in sensitivity (Chen et al. 2000). The biological model of γ-phosphorylation provides the explanation for this remarkable sensitivity; visualization of only one double-strand break in the whole nucleus is feasible, as γ-phosphorylation spans megabase-long domains in chromatin juxtaposed to the break.
3. Quantification: The presence of γH2AX detected by antibody-based techniques can be quantified by various methods: confocal and epifluorescence microscopy (measured manually or automatically), flow cytometry, western blot quantification, etc. (Redon et al. 2010a, b).

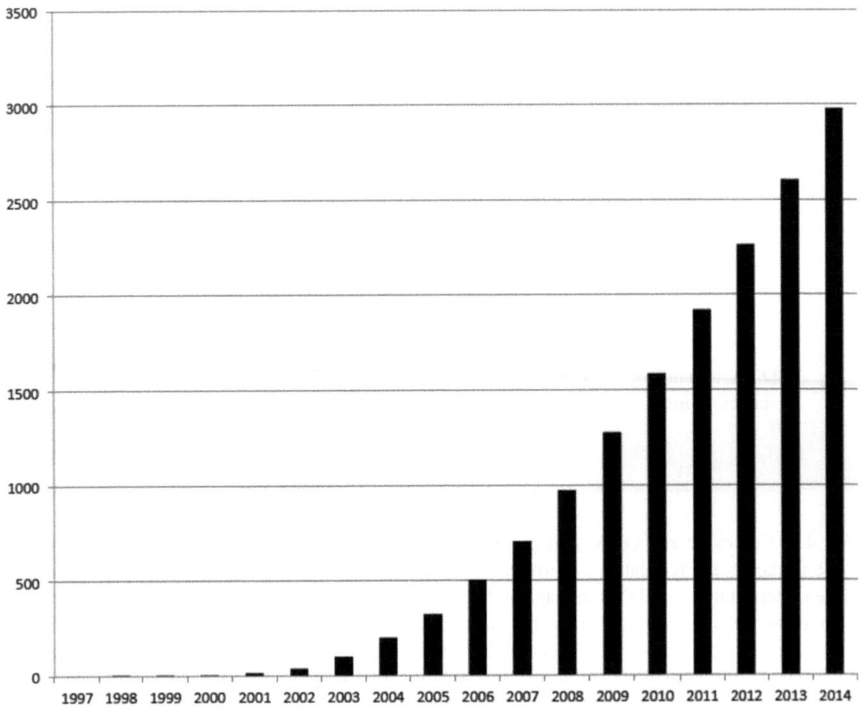

Fig. 4.4 Cumulative diagram of scientific papers that refer to γH2AX. PubMed-NCBI database was searched for γH2AX per calendar year. The cumulative number of these research publications and review articles indicates the available bibliographic information for the γH2AX per calendar year

4. Repeatability and reproducibility: To date, the repeatability and reproducibility of the method have been demonstrated by numerous diverse research laboratories all over the world, as demonstrated by the number of scientific publications (Fig. 4.4).

The above grounds highlight γH2AX to be a reliable assay to use in clinical trials. To date, the research on γH2AX has matured to the point to inflict its utilization in clinical trials as an informative biomarker. Clinical trials on cancer have used γH2AX to monitor and evaluate cancer therapeutics. A representative list of 43 clinical trials on cancer treatment is displayed in Table 4.2.

4.4.2 Cancer Therapeutics and Their Mechanism of Action in Relation to γH2AX Assays

Traditional cancer therapy has progressed through the development of drugs that either introduce DNA lesions or obstruct regular molecular mechanisms that result to apoptosis. Anticancer drugs may induce directly DNA damage or interfere with cellular functions as DNA repair, DNA replication, transcription, cell division, apoptosis, and other. During the inhibition of these processes, DNA intermediates that expose double-strand breaks may be created and detected by γH2AX assays. If apoptosis is induced, γH2AX assays can also be in use, although further development of apoptosis specific assays is needed.

Table 4.3 shows the list of drugs/agents that have been employed in the 43 clinical trials on cancer treatment (displayed in Table 4.2). These drugs/agents may have been administrated as monotherapy or as part of combinational chemotherapy regimens.

The mechanisms of action of these drugs/agents fall into seven categories: The first three categories are (1) DNA damage, (2) DNA-repair inhibition, and (3) DNA-replication stress. There are three additional categories that do not generate double-strand breaks by their mechanism of action: (4) virus-replication inhibition, (5)

Table 4.2 List of selected clinical trials where γH2AX has been utilized as biomarker

Clinical trial registry number	Condition	Age groups	Clinical phase
NCT02207010	Glioblastoma	Adult|Senior	Phase 0
NCT01275183	Head and neck cancer	Adult|Senior	Phase 0
NCT01390571	Brain and central nervous system tumors	Adult|Senior	Phase 1
NCT01076530	Brain tumor; spinal cord tumors	Child	Phase 1
NCT01445418	Breast; ovarian cancer	Adult|Senior	Phase 1
NCT00892736	Breast; ovarian; pancreatic; peritoneal; prostate carcinoma	Adult|Senior	Phase 1
NCT01281150	Solid neoplasm; breast; male breast carcinoma	Adult|Senior	Phase 1
NCT01132573	Acute leukemia	Adult|Senior	Phase 1
NCT00810966	Leukemia; lymphoma; unspecified solid tumor	Adult|Senior	Phase 1
NCT00576654	Lymphoma; breast carcinoma; malignant neoplasm	Adult|Senior	Phase 1
NCT01658319	Lymphoma; leukemia	Adult|Senior	Phase 1
NCT01139970	Myeloid leukemia; acute lymphoblastic leukemia	Adult|Senior	Phase 1
NCT01748825	Neoplasm; lymphoma	Adult|Senior	Phase 1
NCT00740805	Neoplasm; lymphoma; leukemia	Adult|Senior	Phase 1
NCT01455532	Neoplasm	Adult|Senior	Phase 1
NCT00678132	Neoplasm	Adult|Senior	Phase 1
NCT00450502	Neoplasm	Adult|Senior	Phase 1
NCT01202370	Solid neoplasm	Adult|Senior	Phase 1
NCT01017640	Solid neoplasm	Adult|Senior	Phase 1
NCT01213381	Advance solid tumors	Adult|Senior	Phase 1
NCT00422682	Tumors	Adult|Senior	Phase 1
NCT00298675	Tumors	Adult|Senior	Phase 1
NCT00800865	Tumors	Adult|Senior	Phase 1

(continued)

Table 4.2 (continued)

Clinical trial registry number	Condition	Age groups	Clinical phase
NCT00687765	Glioblastoma	Adult|Senior	Phase 1/2
NCT01989546	Advanced ovarian; peritoneal; breast cancer; solid tumors	Adult|Senior	Phase 1/2
NCT01305499	Acute myeloid leukemia	Adult|Senior	Phase 2
NCT01173497	Breast cancer; brain metastases	Adult|Senior	Phase 2
NCT01403610	High grade clioma	Adult|Senior	Phase 2
NCT02038816	Myelodysplastic syndromes	Adult|Senior	Phase 2
NCT01033123	Ovarian cancer	Adult|Senior	Phase 2
NCT01033292	Ovarian cancer	Adult|Senior	Phase 2
NCT00813956	Triple negative breast cancer	Adult|Senior	Phase 2
NCT01204125	Triple negative breast cancer	Adult|Senior	Phase 2
NCT00540358	Triple negative metastatic breast cancer	Adult|Senior	Phase 2
NCT00687687	Uterine carcinosarcoma	Adult|Senior	Phase 2
NCT00938652	Breast cancer	Adult|Senior	Phase 3
NCT01082549	Squamous cell lung cancer	Adult|Senior	Phase 3
NCT02235051	Breast cancer	Adult|Senior	NP
NCT01130259	Breast cancer	Adult|Senior	NP
NCT00369109	Breast; colorectal; pancreatic cancer	Adult|Senior	NP
NCT00493376	Cervix cancer	Adult|Senior	NP
NCT01899391	Prostate cancer radiotherapy	Adult|Senior	NP
NCT00523471	Prostate cancer radiotherapy	Adult|Senior	NP
NCT01518673	Pediatric patients undergoing CT examination	Child	NP

Source: http://clinicaltrials.gov/

checkpoint abrogation, and (6) transcriptional reprogramming. Nevertheless, these drugs prolong the DNA damage response as monitored by γH2AX, if co-administrated with DNA damage agents. Finally, the last category includes (7) aneugens that induce aneuploidy and polyploidy, but do not induce γH2AX by themselves.

4.4.2.1 DNA Damage

DNA damage can be generated directly to cancer cells either by radiotherapy or by drugs. Irradiation and the formation of γH2AX have been extensively studied in animal models (Redon et al. 2010a, b; Moroni et al. 2013). Tirapazamine is an experimental anticancer drug that is activated to a toxic radical within tumor cells in hypoxia state. In this situation, the formation of double-strand breaks has been monitored by γH2AX (Meng et al. 2012; Evans et al. 2008).

Table 4.3 List of therapeutics used in clinical trials that their biological activity has been monitored by the γH2AX biomarker

Therapeutic	Biological target	Mechanism of action	γH2AX induction	
Irradiation	Generation of ROS	DNA damage (DSB)	+	Redon et al. (2010a), Moroni et al. (2013)
Tirapazamine	Generation of ROS in hypoxia	DNA damage (DSB)	+	Meng et al. (2012), Evans et al. (2008)
Methoxyamine	DNA apurinic/ apyrimidinic binder	DNA repair inhibition (BER)	+	Yan et al. (2006, 2007)
Olaparib	PARP inhibitor	DNA repair inhibition (SSB)	+	Redon et al. (2010a, b), Van Vuurden et al. (2011), Fong et al. (2009)
Veliparib	PARP inhibitor	DNA repair inhibition (SSB)	+	Chuang et al. (2012)
BMN 673	PARP inhibitor	DNA repair inhibition (SSB)	+	Shen et al. (2013)
Iniparib	PARP1 inhibitor[a]	DNA repair inhibition (SSB)	+	Sinha (2014), Patel et al. (2012)
5-Azacytidine	DNA-methyltransferase inhibitor	DNA-replication stress	+	Orta et al. (2013), Palii et al. (2008)
Evofosfamide	DNA alkylating agent in hypoxia	DNA-replication stress	+	Meng et al. (2012), Sun et al. (2012)
Mitomycin C	DNA alkylating agent	DNA-replication stress	+	Roh et al. (2008), Niedernhofer et al. (2004)
Cyclophosphamide	DNA alkylating agent	DNA-replication stress	+	Niedernhofer et al. (2004), Cheng et al. (2008)
Carboplatin, cisplatin	DNA alkylating agent	DNA-replication stress	+	Olive and Banáth (2009), Cruet-Hennequart et al. (2014)
Temozolomide	DNA alkylating agent	DNA-replication stress	+	Mirzoeva et al. (2006), Trivedi et al. (2005)
Doxorubicin	DNA intercalation	DNA-replication stress	+	Kurz et al. (2004), Pang et al. (2013)
Gemcitabine	DNA polymerase inhibitor	DNA-replication stress	+	Parsels Leslie et al. (2009), Ewald et al. (2007)
Fludarabine	DNA polymerase inhibitor	DNA-replication stress	+	Ewald et al. (2008), Seedhouse et al. (2009)
Clofarabine	DNA polymerase inhibitor	DNA-replication stress	+	Seedhouse et al. (2009)
Batracylin	DNA topo I and II inhibitor	DNA-replication stress	+	Rao et al. (2007)
Irinotecan	DNA topo I inhibitor	DNA-replication stress	+	Ashour et al. (2015), Petitprez et al. (2013)
Camptothecin	DNA topo I inhibitor	DNA-replication stress	+	Furuta et al. (2003), Ashour et al. (2015)

(continued)

Table 4.3 (continued)

Therapeutic	Biological target	Mechanism of action	γH2AX induction	
Raltegravir	Integrase inhibitor	Virus replication inhibition	–	Cooper et al. (2013)
Entinostat	Histone deacetylase inhibitor	Transcriptional reprogramming	–	Fandy et al. (2009)
MK-1775	Wee1 kinase inhibitor	Checkpoint abrogation	–	Bridges et al. (2011)
Paclitaxel	Microtubule stabilization	Mitosis inhibition	–	Matsuzaki et al. (2010)

[a]This mechanism of action is no longer supported by scientific evidence

4.4.2.2 DNA Repair Inhibition

DNA repair inhibition is a strategy for the development of many cancer drugs. Depending on the specific pathway the drugs target, there are many subcategories. Methoxyamine is a drug that binds covalently to apurinic/apyrimidinic DNA damage sites and inhibits base excision repair. In turn, this may result in an increase in DNA-strand breaks and apoptosis (Yan et al. 2006, 2007).

PARP inhibitors are a novel group of chemotherapeutics that inhibit poly-ADP ribose polymerases (PARP), an enzyme family whose main role is to detect and signal DNA single-strand breaks to the enzymatic machinery involved in the repair. Olaparib (Van Vuurden et al. 2011; Fong et al. 2009), veliparib (Chuang et al. 2012), and BMN 673 (Van Vuurden et al. 2011; Shen et al. 2013) enhance the accumulation of DNA-strand breaks, generating eventually double-strand breaks. Iniparib was originally believed to act as an irreversible inhibitor of PARP1, but its effects against PARP were later disproven (Sinha 2014; Patel et al. 2012).

Topoisomerase I inhibitors freeze the enzyme that is bound covalently to DNA, generating the so-called "topo I-cleavage complexes," establishing single-strand breaks. In S phase, the single-strand breaks are converted to double-strand breaks, as the replication fork that runs DNA collide with the topo I-cleavage complex. Consequently, γH2AX foci co-localize with the topo I-cleavage complexes exclusively in the S phase of the cell cycle. Irinotecan (Ashour et al. 2015; Petitprez et al. 2013) and camptothecin (Furuta et al. 2003; Ashour et al. 2015) are drugs with DNA topoisomerase I inhibitor activity.

Topoisomerase II inhibitors form covalent bonds between the topoisomerase II molecules and the DNA, the so-called topo II-cleavage complexes. Cleavage complexes generated by topoisomerase II inhibitors affect both DNA strands and introduce directly double-strand breaks in DNA. γH2AX foci also co-localize with topo

II-cleavage complexes, but contrary to the "topo I-cleavage complexes," they do appear in all phases of the cell cycle. Batracylin is a dual inhibitor of DNA topoisomerases I and II and is currently used in clinical trials. Its efficacy is evaluated by γH2AX measurements (Rao et al. 2007).

4.4.2.3 DNA-Replication Stress

DNA-replication stress is encountered when DNA-replication forks become stalled by bulky formations and DNA lesions, generated by intracellular flaws or environmental insults.

Several DNA agents form DNA crosslinks, interfering with DNA replication. Mitomycin C, cyclophosphamide, and carboplatin/cisplatin (platinum complex based drugs) act as DNA alkylating agents. As stalled forks eventually collapse, DNA double-strand breaks are generated and can be measured by γH2AX quantification assays (Roh et al. 2008; Niedernhofer et al. 2004; Cheng et al. 2008; Olive and Banáth 2009; Cruet-Hennequart et al. 2014). Temozolomide is an alkylating agent that is converted chemically at physiologic pH to the short-lived active compound MTIC, resulting in obstruction of DNA replication. These lesions have been reported to be repaired by the base excision repair system and eventually induce γH2AX foci (Mirzoeva et al. 2006; Trivedi et al. 2005). Evofosfamide (TH-302) is administered as a prodrug that is activated only in hypoxic regions of tumor, and it becomes a potent DNA alkylating agent. Its activity has also been measured by γH2AX (Meng et al. 2012; Sun et al. 2012).

Agents that intercalate DNA result in the inhibition of macromolecular biosynthesis. Doxorubicin intercalates DNA and inhibits the progression of the enzyme topoisomerase II. Doxorubicin stabilizes the topoisomerase II complex after it has broken the DNA chain. It has also been postulated that Doxorubicin increases free radical production as well. γH2AX quantification assays has been used to monitor these effects (Ewald et al. 2007).

Drugs that can be metabolized in the cell environment into deoxynucleotide analogues have been used as anticancer drugs. Gemcitabine is converted intracellularly to active metabolites that inhibit the production of deoxynucleotides. It also replaces cytidine during DNA replication. Both mechanisms of action result in premature DNA-strand termination and apoptosis. These effects have been monitored by γH2AX measurements (Ewald et al. 2008; Parsels Leslie et al. 2009; Seedhouse et al. 2009). Fludarabine metabolites inhibit DNA polymerases as well as ribonucleotide reductase, thereby interrupting DNA synthesis and inhibiting tumor cell growth. The generation of double-strand breaks has been monitored by γH2AX. Clofarabine belongs to a second-generation nucleoside analog that inhibits the enzymatic activities of ribonucleotide reductase and DNA polymerases. This nucleoside analog also disrupts mitochondrial function and membrane integrity, resulting in the release of pre-apoptotic factors, which activate apoptosis.

5-Azacitidine is incorporated into DNA, where it inhibits DNA methyltransferase reversibly, thereby blocking DNA methylation. Hypomethylation may activate tumor suppressor genes silenced by hypermethylation, also resulting in an antitumor effect. It has also been reported that 5-azacitidine induces γH2AX foci due to replication forks collisions (Orta et al. 2013; Palii et al. 2008).

4.4.2.4 DNA-Damage Checkpoint Abrogation

Another strategy to cancer therapeutics is the development of inhibitors to abrogate the DNA damage checkpoints. MK-1775 is a selective Weel kinase inhibitor that results in G2 DNA-damage checkpoint abrogation. MK-1775 has no effect on γH2AX itself. Nevertheless, if DNA damage has been generated to cells by different means, administration of MK-1775 can lead to mitotic catastrophe and apoptosis. Monitoring the unrepaired double-strand breaks with γH2AX has also been reported for MK-1775 (Bridges et al. 2011).

4.4.2.5 Virus Replication Inhibition

Integrase inhibitors are molecules that are designed to block the action of integrase, a viral enzyme that inserts the viral genome into the DNA of the host cell. Raltegravir has been reported to diminish proviral DNA integration, as monitored by γH2AX assays (Cooper et al. 2013).

4.4.2.6 Transcriptional Reprogramming

Histone deacetylase (HDACs) inhibitors are a class of compounds that interfere with the function of histone deacetylases. As DNA expression is regulated positively by acetylation, HDACs inhibitors sustain transcription. Entinostat is a HDAC inhibitor, with specificity to HDAC1 and HDAC3 inhibition, which potentiates the DNA damage response generated by other agents. γH2AX is induced in peripheral blood mononuclear cells after treatment with entinostat combined with 5-azacitidine (Fandy et al. 2009).

4.4.2.7 Mitosis Inhibition

Aneugens impair the mitotic spindle assembly and are widely used as anticancer drugs, inducing aneuploidy and polyploidy. In contrast to other cancer therapeutics, γH2AX is not induced in paclitaxel or other aneugen-treated cells (Matsuzaki et al. 2010). Most probably, this is due to the fact that the DNA structure remains intact, although the mitotic process is disrupted.

4.5 Future Perspectives

In the present chapter, we have analyzed the biology of the γ-phosphorylation of histone H2AX and its recent applications in clinical trials. γH2AX analysis is already dynamically applied to cancer treatment evaluation and is expected to expand further in this field Shah et al. (2013). As H2AX assays can monitor the fidelity of two major pathways in DSB repair, the homologous repair (HR), and the nonhomologous end joining (NHEJ) (Scully and Xie 2013); a wide range of cancer therapeutics, other than that included in the list of Table 4.3, could also be monitored by γH2AX assays in clinical trials.

Another major progress in the use of γH2AX as a clinical marker will come from the application of these methods to children. Table 4.2 also illustrates that the vast majority of the studies, where γH2AX is used as a biomarker, involve adults and seniors, where studies on children are underrepresented. This reflects the fact that new studies that evaluate novel treatments and methods are initially approved for adult populations and then expand to children, due to regulatory mandates. The fact that a child is a minor and the parents or legal guardians decide for the children's participation in experimental studies has limited the number of studies involving pediatric cohorts. Nevertheless, specifically designed clinical studies for this age group are a necessity.

Although pediatric patients usually exhibit better tolerance in cancer treatment and their survival rate has steadily improved, long-term morbidity and mortality that is associated with currently successful treatments must be addressed (Robison and Hudson 2014; Norris and Adamson 2012). Toward this direction, γH2AX will become an indispensable biomarker for the evaluation of chemotherapy and radiotherapy for identifying pediatric patients with double-strand break repair deficiencies, who may overreact to DNA-damaging cancer therapy (Rübe et al. 2015). This issue has also been addressed in the epidemiology study to quantify risks for pediatric computerized tomography and to optimize doses (EPI-CT). The EPI-CT study has utilized γH2AX as a biomarker for DNA damage to indicate radiation risk (Thierry-Chef et al. 2013). Results are expected to be announced later this year.

Although major research has focused on malignant disease, we are going to witness an increasing body of evidence on γH2AX coming from nonmalignant disease entities. Basic research on γH2AX has demonstrated many different sources of double-strand breaks that induce γ-phosphorylation other than those involved in cancer. Till now, only a few clinical trials have explored the utility of this epigenetic marker to evaluate the burden of nonmalignant diseases. The clinical trial "IRIS" investigates the accelerated immunosenescence in chronic kidney disease patients (NCT02116270). For this reason, T cell senescent populations are evaluated by γH2AX immunostaining. In another clinical study, the effect of stress on women undergoing in vitro fertilization (IVF) is examined (NCT00685282).

The general hypothesis of this research is that stress decreases fertility; counteracting stress with cognitive behavioral therapy will increase fertility in women undergoing IVF. To monitor stress, DNA integrity will be evaluated by γH2AX.

The “DORIAN” study is investigating the role of maternal obesity in healthy and unhealthy aging. Subjects will undergo characterization by biohumoral, imaging, and inflammatory biomarkers. γH2AX in relation to telomere damage will also be evaluated in aging (Iozzo et al. 2014).

The utilization of γH2AX as a biomarker in nonmalignant disease entities will further accelerate toward different physiologies. There is a particularly exciting prospect that the γH2AX analysis may be adapted to serve the evaluation of the efficacy of developmental therapeutics. The outcome of basic research has demonstrated many different sources of double-strand breaks that have not been exploited yet (Fig. 4.2). For this reason, translational research on the γH2AX epigenetic biomarker is expected to develop further, in order to serve the monitoring for diverse physiologies and pathologies and the evaluation of their therapies.

Abbreviations

AID	Activation-induced cytidine deaminase
ATM	Ataxia telangiectasia mutated
CAD	Caspase-activated DNase
CSR	Class switch recombination
DSB	Double-stranded break
dU	Deoxy-uracil
HDAC	Histone deacetylase
NHEJ	Nonhomologous End-Joining
PARP	ADP ribose polymerases
Topo I	Topoisomerase I
Topo II	Topoisomerase II

References

Ashour, M. E., Atteya, R., & El-Khamisy, S. F. (2015). Topoisomerase-mediated chromosomal break repair: An emerging player in many games. *Nature Reviews Cancer, 15*, 137–151.

Bekker-Jensen, S., Lukas, C., Melander, F., Bartek, J., & Lukas, J. (2005). Dynamic assembly and sustained retention of 53BP1 at the sites of DNA damage are controlled by Mdc1/NFBD1. *Journal of Cell Biology, 170*(2), 201–11.

Bridges, K. A., Hirai, H., Buser, C. A., Brooks, C., Liu, H., Buchholz, T. A., et al. (2011). MK-1775, a novel wee1 kinase inhibitor, radiosensitizes p53-defective human tumor cells. *Clinical Cancer Research, 17*(2), 5638–48.

Celeste, A., Difilippantonio, S., Difilippantonio, M. J., Fernandez-Capetillo, O., Pilch, D. R., Sedelnikova, O. A., et al. (2003). H2AX haploinsufficiency modifies genomic stability and tumor susceptibility. *Cell, 114*, 371–83.

Celeste, A., Petersen, S., Romanienko, P. J., Fernandez-Capetillo, O., Chen, H. T., Sedelnikova, O. A., et al. (2002). Genomic instability in mice lacking histone H2AX. *Science, 296*(5569), 922–7.

Chen, H. T., Bhandoola, A., Difilippantonio, M. J., Zhu, J., Brown, M. J., Tai, X., et al. (2000). Response to RAG-mediated VDJ cleavage by NBS1 and gamma-H2AX. *Science, 290*(5498), 1962–5.

Cheng, W.-H., Muftic, D., Muftuoglu, M., Dawut, L., Morris, C., Helleday, T., et al. (2008). WRN is required for ATM activation and the S-phase checkpoint in response to interstrand cross-link-induced DNA double-strand breaks. *Molecular Biology of the Cell, 19*(September), 3923–33.

Chronis, F., & Rogakou, E. P. (2007). Interplay between H2AX and 53BP1 pathways in DNA double-strand break repair response. In D. Gewirtz, S. E. Holt, & S. Grant (Eds.), *Apoptosis, senescence, and cancer* (pp. 243–263). Totowa, NJ: Humana. http://www.springer.com/gp/book/9781588295279.

Chuang, H. C., Kapuriya, N., Kulp, S. K., Chen, C. S., & Shapiro, C. L. (2012). Differential anti-proliferative activities of poly(ADP-ribose) polymerase (PARP) inhibitors in triple-negative breast cancer cells. *Breast Cancer Research and Treatment, 134*(2), 649–59.

Cimprich, K. A., & Cortez, D. (2008). ATR: an essential regulator of genome integrity. *Nature Reviews Molecular Cell Biology, 9*(8), 616–27.

Cooper, A., García, M., Petrovas, C., Yamamoto, T., Koup, R. A., & Nabel, G. J. (2013). HIV-1 causes CD4 cell death through DNA-dependent protein kinase during viral integration. *Nature, 498*, 376–9.

Cruet-Hennequart, S., Villalan, S., Kaczmarczyk, A., O'Meara, E., Sokol, A. M., & Carty, M. P. (2014). Characterization of the effects of cisplatin and carboplatin on cell cycle progression and DNA damage response activation in DNA polymerase eta-deficient human cells. *Cell Cycle, 8*(18), 3043–54.

d'Adda di Fagagna, F., Reaper, P. M., Clay-Farrace, L., Fiegler, H., Carr, P., Von Zglinicki, T., et al. (2003). A DNA damage checkpoint response in telomere-initiated senescence. *Nature, 426*, 194–8.

Daniel, J. A., Pellegrini, M., Lee, B. S., Guo, Z., Filsuf, D., Belkina, N. V., et al. (2012). Loss of ATM kinase activity leads to embryonic lethality in mice. *The Journal of Cell Biology, 198*(3), 295–304.

Daniel, R., Ramcharan, J., Rogakou, E., Taganov, K. D., Greger, J. G., Bonner, W., et al. (2004). Histone H2AX is phosphorylated at sites of retroviral DNA integration but is dispensable for postintegration repair. *The Journal of Biological Chemistry, 279*(44), 45810–4.

Dixon, J. R., Jung, I., Selvaraj, S., Shen, Y., Antosiewicz-Bourget, J. E., Lee, A. Y., et al. (2015). Chromatin architecture reorganization during stem cell differentiation. *Nature, 518*(7539), 331–6.

Downs, J. A., & Jackson, S. P. (2004). A means to a DNA end: The many roles of Ku. *Nature Reviews Molecular Cell Biology, 5*, 367–78.

Dunham, I., Kundaje, A., Aldred, S. F., Collins, P. J., Davis, C. A., Doyle, F., et al. (2012). An integrated encyclopedia of DNA elements in the human genome. *Nature, 489*, 57–74.

Evans, J. W., Chernikova, S. B., Kachnic, L. A., Banath, J. P., Sordet, O., Delahoussaye, Y. M., et al. (2008). Homologous recombination is the principal pathway for the repair of DNA damage induced by tirapazamine in mammalian cells. *Cancer Research, 68*(15), 257–65.

Ewald, B., Sampath, D., & Plunkett, W. (2007). H2AX phosphorylation marks gemcitabine-induced stalled replication forks and their collapse upon S-phase checkpoint abrogation. *Molecular Cancer Therapeutics, 6*(4), 1239–48.

Ewald, B., Sampath, D., & Plunkett, W. (2008). Nucleoside analogs: Molecular mechanisms signaling cell death. *Oncogene, 27*(50), 6522–37.

Falck, J., Coates, J., & Jackson, S. P. (2005). Conserved modes of recruitment of ATM, ATR and DNA-PKcs to sites of DNA damage. *Nature, 434*(March), 605–11.

Fandy, T. E., Herman, J. G., Kerns, P., Jiemjit, A., Sugar, E. A., Choi, S. H., et al. (2009). Early epigenetic changes and DNA damage do not predict clinical response in an overlapping schedule of 5-azacytidine and entinostat in patients with myeloid malignancies. *Blood, 114*, 2764–73.

Fernandez-Capetillo, O., Lee, A., Nussenzweig, M., & Nussenzweig, A. (2004). H2AX: The histone guardian of the genome. *DNA Repair (Amst), 3*, 959–67.

Fernandez-Capetillo, O., Liebe, B., Scherthan, H., & Nussenzweig, A. (2003). H2AX regulates meiotic telomere clustering. *Journal of Cell Biology, 163*, 15–20.

Fong, P. C., Boss, D. S., Yap, T. A., Tutt, A., Wu, P., Mergui-Roelvink, M., et al. (2009). Inhibition of poly(ADP-ribose) polymerase in tumors from BRCA mutation carriers. *The New England Journal of Medicine, 361*, 123–34.

Furuta, T., Takemura, H., Liao, Z.-Y., Aune, G. J., Redon, C., Sedelnikova, O. A., et al. (2003). Phosphorylation of histone H2AX and activation of Mre11, Rad50, and Nbs1 in response to replication-dependent DNA double-strand breaks induced by mammalian DNA topoisomerase I cleavage complexes. *The Journal of Biological Chemistry, 278*(22), 20303–12.

Iozzo, P., Holmes, M., Schmidt, M. V., Cirulli, F., Guzzardi, M. A., Berry, A., et al. (2014). Developmental ORIgins of healthy and unhealthy AgeiNg: The role of maternal obesity: Introduction to DORIAN. *Obesity Facts, 7*, 130–51.

Ivashkevich, A., Redon, C. E., Nakamura, A. J., Martin, R. F., & Martin, O. A. (2012). Use of the γ-H2AX assay to monitor DNA damage and repair in translational cancer research. *Cancer Letters, 327*, 123–33.

Jenuwein, T., & Allis, C. D. (2001). Translating the histone code. *Science, 293*(5532), 1074–80.

Kobayashi, J., Antoccia, A., Tauchi, H., Matsuura, S., & Komatsu, K. (2004). NBS1 and its functional role in the DNA damage response. *DNA Repair (Amst), 3*, 855–61.

Koren, A., Thurman, R., Sandstrom, R., Lawrence, M. S., Reynolds, A., Polak, P., et al. (2015). Cell-of-origin chromatin organization shapes the mutational landscape of cancer. *Nature, 518*(0316), 360–4.

Kurz, E. U., Douglas, P., & Lees-Miller, S. P. (2004). Doxorubicin activates ATM-dependent phosphorylation of multiple downstream targets in part through the generation of reactive oxygen species. *The Journal of Biological Chemistry, 279*(51), 53272–81.

Lu, C., Zhu, F., Cho, Y. Y., Tang, F., Zykova, T., & Ma, W. Y. (2006). Cell apoptosis: Requirement of H2AX in DNA ladder formation, but not for the activation of caspase-3. *Molecular Cell, 23*(1), 121–32.

Luger, K., Mäder, A. W., Richmond, R. K., Sargent, D. F., & Richmond, T. J. (1997). Crystal structure of the nucleosome core particle at 2.8 A resolution. *Nature, 389*(6648), 251–60.

Lukas, C., Melander, F., Stucki, M., Falck, J., Bekker-Jensen, S., Goldberg, M., et al. (2004). Mdc1 couples DNA double-strand break recognition by Nbs1 with its H2AX-dependent chromatin retention. *EMBO Journal, 23*(13), 2674–83.

Mahadevaiah, S. K., Turner, J. M., Baudat, F., Rogakou, E. P., de Boer, P., Blanco-Rodríguez, J., et al. (2001). Recombinational DNA double-strand breaks in mice precede synapsis. *Nature Genetics, 27*(3), 271–6.

Matsuzaki, K., Harada, A., Takeiri, A., Tanaka, K., & Mishima, M. (2010). Whole cell-ELISA to measure the γH2AX response of six aneugens and eight DNA-damaging chemicals. *Mutation Research: Genetic Toxicology and Environmental Mutagenesis, 700*(1–2), 71–9.

McKinnon, P. J. (2004). ATM and ataxia telangiectasia. *EMBO Reports, 5*(8), 772–6.

Meng, F., Evans, J. W., Bhupathi, D., Banica, M., Lan, L., Lorente, G., et al. (2012). Molecular and cellular pharmacology of the hypoxia-activated prodrug TH-302. *Molecular Cancer Therapeutics, 11*, 740–51.

Mirzoeva, O. K., Kawaguchi, T., & Pieper, R. O. (2006). The Mre11/Rad50/Nbs1 complex interacts with the mismatch repair system and contributes to temozolomide-induced G2 arrest and cytotoxicity. *Molecular Cancer Therapeutics, 5*(11), 2757–66.

Moroni, M., Maeda, D., Whitnall, M. H., Bonner, W. M., & Redon, C. E. (2013). Evaluation of the gamma-H2AX assay for radiation biodosimetry in a swine model. *International Journal of Molecular Sciences, 14*, 14119–35.

Murata, M., Thanan, R., Ma, N., & Kawanishi, S. (2012). Role of nitrative and oxidative DNA damage in inflammation-related carcinogenesis (Figure 1). *Journal of Biomedicine & Biotechnology, 2012*, 623019.

Niedernhofer, L. J., Odijk, H., Budzowska, M., van Drunen, E., Maas, A., Theil, A. F., et al. (2004). The structure-specific endonuclease Ercc1-Xpf is required to resolve DNA interstrand cross-link-induced double-strand breaks. *Molecular and Cellular Biology, 24*(13), 5776–87.

Norris, R. E., & Adamson, P. C. (2012). Challenges and opportunities in childhood cancer drug development. *Nature Reviews Cancer, 12*(11), 776–82.

Olive, P. L., & Banáth, J. P. (2009). Kinetics of H2AX phosphorylation after exposure to cisplatin. *Cytometry. Part B, Clinical Cytometry, 76*(2), 79–90.

Orta, M. L., Calderón-Montaño, J. M., Domínguez, I., Pastor, N., Burgos-Morón, E., López-Lázaro, M., et al. (2013). 5-Aza-2′-deoxycytidine causes replication lesions that require Fanconi anemia-dependent homologous recombination for repair. *Nucleic Acids Research, 41*(11), 5827–36.

Palii, S. S., Van Emburgh, B. O., Sankpal, U. T., Brown, K. D., & Robertson, K. D. (2008). DNA methylation inhibitor 5-Aza-2′-deoxycytidine induces reversible genome-wide DNA damage that is distinctly influenced by DNA methyltransferases 1 and 3B. *Molecular and Cellular Biology, 28*(2), 752–71.

Pang, B., Qiao, X., Janssen, L., Velds, A., Groothuis, T., Kerkhoven, R., et al. (2013). Drug-induced histone eviction from open chromatin contributes to the chemotherapeutic effects of doxorubicin. *Nature Communications, 4*, 1908.

Panier, S., & Boulton, S. J. (2014). Double-strand break repair: 53BP1 comes into focus. *Nature Reviews Molecular Cell Biology, 15*(December), 7–18.

Parsels Leslie, A., Morgan, M. A., Tanska, D. M., Parsels, J. D., Palmer, B. D., Booth, R. J., et al. (2009). Gemcitabine sensitization by checkpoint kinase 1 inhibition correlates with inhibition of a Rad51 DNA damage response in pancreatic cancer cells. *Molecular Cancer Therapeutics, 8*(1), 45–54.

Patel, A. G., De Lorenzo, S. B., Flatten, K. S., Poirier, G. G., & Kaufmann, S. H. (2012). Failure of iniparib to inhibit poly(ADP-ribose) polymerase in vitro. *Clinical Cancer Research, 18*(14), 1655–62.

Paull, T. T., Rogakou, E. P., Yamazaki, V., Kirchgessner, C. U., Gellert, M., & Bonner, W. M. (2000). A critical role for histone H2AX in recruitment of repair factors to nuclear foci after DNA damage. *Current Biology, 10*(15), 886–95.

Petitprez, A., Poindessous, V., Ouaret, D., Regairaz, M., Bastian, G., Guérin, E., et al. (2013). Acquired irinotecan resistance is accompanied by stable modifications of cell cycle dynamics independent of MSI status. *International Journal of Oncology, 42*, 1644–53.

Rao, V. A., Agama, K., Holbeck, S., & Pommier, Y. (2007). Batracylin (NSC 320846), a dual inhibitor of DNA topoisomerases I and II induces histone γ-H2AX as a biomarker of DNA damage. *Cancer Research, 67*, 9971–9.

Redon, C. E., Nakamura, A. J., Gouliaeva, K., Rahman, A., Blakely, W. F., & Bonner, W. M. (2010a). The use of gamma-H2AX as a biodosimeter for total-body radiation exposure in non-human primates. *PLoS One, 5*(11), e15544.

Redon, C. E., Nakamura, A. J., Zhang, Y.-W., Ji, J. J., Bonner, M. W., Kinders, R. J., et al. (2010b). Histone gammaH2AX and poly(ADP-ribose) as clinical pharmacodynamic biomarkers. *Clinical Cancer Research, 16*(18), 4532–42.

Redon, C. E., Weyemi, U., Parekh, P. R., Huang, D., Burrell, A. S., & Bonner, W. M. (2012). γ-H2AX and other histone post-translational modifications in the clinic. *Biochimica et Biophysica Acta (BBA) - Gene Regulatory Mechanisms, 1819*(7), 743–56.

Roadmap Epigenomics Consortium, Kundaje, A., Meuleman, W., Ernst, J., Bilenky, M., Yen, A., et al. (2015). Integrative analysis of 111 reference human epigenomes. *Nature, 518*, 317–30.

Robison, L. L., & Hudson, M. M. (2014). Survivors of childhood and adolescent cancer: Life-long risks and responsibilities. *Nature Reviews Cancer, 14*(1), 61–70.

Rogakou, E. P., Boon, C., Redon, C. E., & Bonner, M. W. (1999). Megabase chromatin domains involved in DNA double-strand breaks in vivo. *The Journal of Cell Biology, 146*(5), 905–16.

Rogakou, E. P., Nieves-Neira, W., Boon, C., Pommier, Y., & Bonner, M. W. (2000). Initiation of DNA fragmentation during apoptosis induces phosphorylation of H2AX histone at serine 139. *The Journal of Biological Chemistry, 275*(13), 9390–5.

Rogakou, E. P., Pilch, D. R., Orra, H., Ivanova, V. S., & Bonner, M. W. (1998). DNA double-stranded breaks induce histone H2AX phosphorylation on serine 139. *The Journal of Biological Chemistry, 273*(10), 5858–68.

Roh, D. S., Cook, A. L., Rhee, S. S., Joshi, A., Kowalski, R., Dhaliwal, D. K., et al. (2008). DNA cross-linking, double-strand breaks, and apoptosis in corneal endothelial cells after a single exposure to mitomycin C. *Investigative Ophthalmology & Visual Science, 49*(11), 4837–43.

Romanoski, C. E., Glass, C. K., Stunnenberg, H. G., Wilson, L., & Almouzni, G. (2015). Epigenomics: Roadmap for regulation. *Nature, 518*(7539), 314–316.

Rübe, C. E., Fricke, A., Schneider, R., Simon, K., Kühne, M., Fleckenstein, J., et al. (2015). DNA repair alterations in children with pediatric malignancies: Novel opportunities to identify patients at risk for high-grade toxicities. *International Journal of Radiation Oncology Biology Physics, 78*(2), 359–69.

Santos, M. A., Faryabi, R. B., Ergen, A. V., Day, A. M., Malhowski, A., Canela, A., et al. (2014). DNA-damage-induced differentiation of leukaemic cells as an anti-cancer barrier. *Nature, 514*, 107–11.

Scully, R., & Xie, A. (2013). Double strand break repair functions of histone H2AX. *Mutation Research, 750*, 5–14.

Sedelnikova, O. A., Horikawa, I., Zimonjic, D. B., Popescu, N. C., Bonner, W. M., & Barrett, J. C. (2004). Senescing human cells and ageing mice accumulate DNA lesions with unrepairable double-strand breaks. *Nature Cell Biology, 6*(1), 168–70.

Sedelnikova, O., Rogakou, E. P., Panyutin, I. G., & Bonner, W. M. (2002). Quantitative detection of 125IdU-induced DNA double-strand breaks with γ-H2AX antibody. *Radiation Research, 158*, 486–92.

Seedhouse, C., Grundy, M., Shang, S., Ronan, J., Pimblett, H., Russell, N., et al. (2009). Impaired S-phase arrest in acute myeloid leukemia cells with a FLT3 internal tandem duplication treated with clofarabine. *Clinical Cancer Research, 15*(23), 7291–8.

Shah, K., Cornelissen, B., Kiltie, A. E., & Vallis, K. A. (2013). Can H2AX be used to personalise cancer treatment? *Current Molecular Medicine, 13*, 1–12.

Shen, Y., Rehman, F. L., Feng, Y., Boshuizen, J., Bajrami, I., Elliott, R., et al. (2013). BMN 673, a novel and highly potent PARP1/2 inhibitor for the treatment of human cancers with DNA repair deficiency. *Clinical Cancer Research, 19*, 5003–15.

Sinha, G. (2014). Downfall of iniparib: A PARP inhibitor that doesn't inhibit PARP after all. *Journal of the National Cancer Institute, 106*, 1–2.

Solier, S., & Pommier, Y. (2009). The apoptotic ring: A novel entity with phosphorylated histones H2AX and H2B and activated DNA damage response kinases. *Cell Cycle, 8*, 1853–9.

Song, H., Hedayati, M., Hobbs, R. F., Shao, C., Bruchertseifer, F., Morgenstern, A., et al. (2013). Targeting aberrant DNA double-strand break repair in triple-negative breast cancer with alpha-particle emitter radiolabeled anti-EGFR antibody. *Molecular Cancer Therapeutics, 12*(10), 2043–54.

Strahl, B. D., & Allis, C. D. (2000). The language of covalent histone modifications. *Nature, 403*(January), 41–5.

Stucki, M., & Jackson, S. P. (2004). MDC1/NFBD1: A key regulator of the DNA damage response in higher eukaryotes. *DNA Repair (Amst), 3*, 953–7.

Sun, J. D., Liu, Q., Wang, J., Ahluwalia, D., Ferraro, D., Wang, Y., et al. (2012). Selective tumor hypoxia targeting by hypoxia-activated prodrug TH-302 inhibits tumor growth in preclinical models of cancer. *Clinical Cancer Research, 18*(3), 758–70.

Takai, H., Smogorzewska, A., & De Lange, T. (2003). DNA damage foci at dysfunctional telomeres. *Current Biology, 13*, 1549–56.

Tanaka, T., Halicka, H. D., Huang, X., Traganos, F., & Darzynkiewicz, Z. (2006). Constitutive histone H2AX phosphorylation and ATM activation, the reporters of DNA damage by endogenous oxidants. *Cell Cycle, 5*(17), 1940–5.

Thierry-Chef, I., Dabin, J., Friberg, E. G., Hermen, J., Istad, T. S., Jahnen, N., et al. (2013). Assessing organ doses from paediatric CT scans-A novel approach for an epidemiology study (the EPI-CT study). *International Journal of Environmental Research and Public Health, 10*, 717–28.

Trivedi, R. N., Almeida, K. H., Fornsaglio, J. L., Schamus, S., & Sobol, R. W. (2005). The role of base excision repair in the sensitivity and resistance to temozolomide-mediated cell death. *Cancer Research, 14*, 6394–401.

Van Attikum, H., & Gasser, S. M. (2005). The histone code at DNA breaks: A guide to repair? *Nature Reviews Molecular Cell Biology, 6*, 757–65.

Van Vuurden, D. G., Hulleman, E., Meijer, O. L. M., Wedekind, L. E., Kool, M., Witt, H., et al. (2011). PARP inhibition sensitizes childhood high grade glioma, medulloblastoma and ependymoma to radiation. *Oncotarget, 2*(12), 984–96.

Williamson, E. A., Wray, J. W., Bansal, P., & Hromas, R. (2012). Overview for the histone codes for DNA repair. *Progress in Molecular Biology and Translational Science, 110*, 207–27.

Yan, L., Bulgar, A., Miao, Y., Mahajan, V., Donze, J. R., Gerson, S. L., et al. (2007). Combined treatment with temozolomide and methoxyamine: Blocking apurininc/pyrimidinic site repair coupled with targeting topoisomerase II?? *Clinical Cancer Research, 13*(14), 1532–9.

Yan, T., Seo, Y., Schupp, J. E., Zeng, X., Desai, A. B., & Kinsella, T. J. (2006). Methoxyamine potentiates iododeoxyuridine-induced radiosensitization by altering cell cycle kinetics and enhancing senescence. *Molecular Cancer Therapeutics, 5*(April), 893–902.

Yin, B., Savic, V., Juntilla, M. M., Bredemeyer, A. L., Yang-Iott, K. S., & Helmink, B. (2009). a, et al. Histone H2AX stabilizes broken DNA strands to suppress chromosome breaks and translocations during V(D)J recombination. *Journal of Experimental Medicine, 206*(12), 2625–39.

Chapter 5
Role of Environmental Epigenetics in Perinatal and Neonatal Development

Sripriya Sundararajan and Cynthia F. Bearer

5.1 General Overview

Conrad Waddington (1905–1975) coined the term "epigenetics" and defined it as the "the branch of biology which studies the causal interaction between genes and their products, which bring the phenotype into being" (Goldberg et al. 2007; Rando and Verstrepen 2007; Devaskar and Raychaudhuri 2007; Waddington, 1942). Epigenetics may also be defined as the study of heritable changes in gene activities and phenotype that are not caused by alterations in the DNA nucleotide sequence (Kouzarides 2007; Berger et al. 2009; Bird 2007). Epigenetic alterations induce changes in eukaryotic cellular phenotype by altering gene expression without modifying the DNA sequence. Epigenetics include long-lived and reversible modifications to nucleotides or chromosomes that do not change the sequence but can alter gene expression and phenotype (LaSalle et al. 2013). The mechanisms of gene regulation or epigenetic modifications include DNA methylation, histone modification, and many others that include noncoding ribonucleic acids (RNAs) such as microRNAs (miRNAs) (Maccani and Marsit 2009; Miller and Grant 2013; Devaskar and Raychaudhuri 2007).

The early focus of the field of fetal programming was on associations between prenatal exposures and metabolic diseases and related disorders such as coronary heart disease, obesity, and diabetes (Maccani et al. 2013). There is significant evidence that the risk of chronic adult diseases and disorders results from exposure to

S. Sundararajan, M.D. (✉) • C.F. Bearer, M.D., Ph.D.
Division of Neonatology, University of Maryland School of Medicine, 110 S. Paca Street, 8th floor, Baltimore, MD 21201, USA
e-mail: ssundararajan@peds.umaryland.edu; cbearer@peds.umaryland.edu

D. Hollar (ed.), *Epigenetics, the Environment, and Children's Health Across Lifespans*, DOI 10.1007/978-3-319-25325-1_5

environmental factors early in development (Bateson et al. 2004; McMillen and Robinson 2005). Epigenetic modifications are established early in development, are susceptible to changes during gestation, and remain vulnerable targets to environmental disruption and subsequently may lead to disease manifestations (Rakyan et al. 2002; Waterland and Jirtle 2003). Epigenetic changes can occur at the parental, prenatal, and postnatal stages of life (Pickard et al. 2001; Reik et al. 2001; McMillen and Robinson 2005; Robins et al. 2011; Pembrey et al. 2014). For example, early developmental (in utero) exposure to a high dose of bisphenol A, a high-production-volume chemical used in the manufacture of polycarbonate plastic, shifted coat color distribution of agouti (A^{vy}) mouse offspring toward yellow by decreasing cytosine-phosphorylated-guanine (CpG) dinucleotide methylation at the A^{vy} and CDK5 activator binding protein ($Cabp^{IAP}$) metastable epialleles (Dolinoy et al. 2007). Moreover, restoration of normal methylation patterns and coat color distribution occurred with maternal supplementation of genistein or methyl donors such as folate, choline, betaine, and vitamin B_{12} (Dolinoy et al. 2006, 2007). Furthermore, animal studies have demonstrated epigenetic changes in rat pups resulting from differences in maternal care in the postnatal extrauterine environment as elaborated at a later section in this chapter (Meaney and Szyf 2005).

Epigenetic variations established in response to in utero environmental exposures could be considered as "perinatal programming." These seminal studies highlight the importance of environmental exposures that could potentially affect perinatal programming and subsequent development of adult-onset diseases. Figure 5.1 summarizes the concept of perinatal programming and the contributions

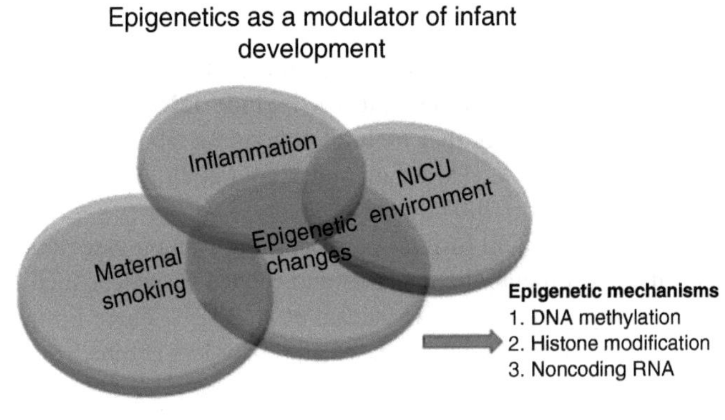

Fig. 5.1 Concept of perinatal programming: Schematic representation of the prenatal, natal, and postnatal environmental factors that could potentially impact the epigenetic machinery of the developing fetus and young infant with subsequent consequences for neurodevelopment and later onset of childhood and adult diseases. There are potential applications that can favorably alter maternal and child health care, and some include cessation of smoking, abstinence from alcohol, the use of maternal dietary supplements during pregnancy, reduction of inflammation in the perinatal period by minimizing infections, and provision of single family room NICU and thus giving an opportunity for more family time with the premature infant in the immediate postnatal life

from the various stimuli during intrauterine and extrauterine life. Overall, this chapter principally highlights the impact of environmental exposures on the developing fetus and neonate through regulation of gene transcription by epigenetic mechanisms.

Section 5.3 focuses on eminent environmental factors including maternal alcohol and tobacco exposure that could potentially alter the epigenetics of the developing fetus. Section 5.4 elucidates the role of epigenetics in inflammation, allergic immune disease, and organ fibrosis, including wound healing. Section 5.5 elaborates the influence of the NICU environment via possible epigenetic mechanisms on the developing neonate in the postnatal state.

5.2 Mechanisms of Epigenetic Regulation

For more comprehensive review, the different mechanisms of epigenetic regulation are described in depth in Chaps. 2–4. The three major epigenetic mechanisms are briefly enumerated here.

1. *DNA methylation*: DNA methylation refers to addition of methyl groups at the fifth ring position (a carbon atom) of cytosine, most frequently at CpG sites with the help of enzymes such as DNA methyltransferase (DNMT). Around 3 % of cytosines are methylated in the human genome almost exclusively in the context of the dinucleotide, CpG (Miller and Grant 2013). In general, such methylation of gene promoters suppresses gene expression. Promoter methylation is associated with gene silencing via transcriptional repression without alteration of the underlying DNA sequence. Highly methylated areas therefore tend to be less transcriptionally active (Jaenisch and Bird 2003; Bird 2002). DNA methylation also regulates chromatin condensation via (1) the occlusion of DNA-binding proteins that act as or recruit transcriptional activators or (2) the recruitment of Methyl-CpG-binding domain protein 2 (MBD) which recruits transcriptional corepressor complexes (Miller and Grant 2013). MBD proteins are involved in the repression of transcription from methylated gene promoters.
2. *Histone modification*: Histone (proteins around which DNA is wrapped) modification occurs largely, but not exclusively, in the N-terminal tail region and includes acetylation and methylation by chromatin-modifying enzymes (Kouzarides 2007). The effects of individual histone modifications on transcription are not straightforward and provide a complex means of regulating transcription. Generalities such as methylation of histone H3 lysine 9 (K9) contributing to gene inactivation, and methylation of histone H3 lysine 4 (K4) increasing gene expressions, are complicated by the compound effect of multiple modifications.
3. *miRNAs*: Non-protein-coding RNAs such as miRNAs are small (~22) nucleotides that act by interfering with translation of protein from messenger RNA (mRNA) during the process of posttranscriptional gene regulation and therefore

reduce gene function. Excess of 60 % of protein-coding genes are thought to be regulated by miRNA, highlighting their functional importance in physiologic and pathologic disease states (Puumala and Hoyme 2015). miRNAs have been shown to regulate a number of key cellular processes including cell proliferation, differentiation, carcinogenesis, and cellular death (Maccani and Marsit 2009; Miska 2005). Thus different epigenetic mechanisms could potentially alter gene expression.

5.3 Epigenetics and the Prenatal Environment

The environment plays a very important role in the proper functioning of the cell. Environmental epigenetics studies how cells respond to the environment by changing DNA methylation and chromatin structure or expressing small RNAs. This response modulates the gene expression and the ultimate phenotype (Lane 2014). The influence of the prenatal environment on neurobehavioral outcomes such as schizophrenia, depression, inhibitory control, and attention deficit/hyperactivity disorder has been investigated (Van den Bergh 2011). Addictive substances, diet, prescriptive drugs, environmental pollutants, emotional stress, maternal smoking, and alcohol all alter epigenetics as one mechanism of their impact (Lo and Zhou 2014). Specific sets of epigenetic modifications can result in specific molecular changes, which can then be involved in epigenetic processes such as gene silencing and X-chromosome inactivation or imprinting. The gene sequence remains unchanged throughout life; however, environmental factors such as stress (McGowan et al. 2009), diet (Weaver et al. 2005), or maternal care (Weaver et al. 2004) act through certain chemical reactions, referred to as epigenetic signal cascade, to influence the chromatin state.

Briefly, the epigenetic cascade starts with an environmental signal that recruits an epigenetic initiator such as cAMP response element-binding protein (CREB) or RE1-silencing transcription factor (REST) to directly act on the chromatin to invoke modification through different pathways (such as DNA methylation and histone modification) and establish pattern of epigenetic markers (Berger et al. 2009; Stankiewicz et al. 2013). These reactions can unravel the chromatin and cause stretches of DNA containing a gene to be exposed for longer or shorter periods of time, essentially turning the gene on or off and allowing for changes in protein production (Stankiewicz et al. 2013). This change in protein production, in turn, can affect physiological and behavioral traits and can be passed from one cell to the next as the cells multiply within an organism and can even be passed from parents to children (Rothstein et al. 2009). Epigenetic mechanisms such as DNA methylation and histone modifications are strongly suspected both in long-term and in rapid, dynamic gene expression regulation during stress (Tsankova et al. 2007). The role of epigenetics is thought to underlie synaptic plasticity, memory, and cognitive processes as well as in shaping stress-vulnerable phenotypes and behavioral adaptations to chronic stress (Siegmund and Wotjak 2007).

Alcohol Maternal alcohol use during pregnancy can result in fetal alcohol spectrum disorder (FASD). FASD is a continuum of various permanent birth defects that are caused by maternal consumption of alcohol during pregnancy. FASD includes fetal alcohol syndrome (FAS), partial FAS, alcohol-related neurodevelopmental disorders, alcohol-related birth defects, and fetal alcohol effects. FAS, the most severe form of FASD, is a diagnosis made on a constellation of findings including growth retardation, facial dysmorphology, and central nervous system dysfunction where heavy maternal alcohol exposure is confirmed. The mammalian one-carbon metabolism provides the methyl groups for all biologic methylation reactions that in turn are dependent on methyl donors such as methionine and choline and cofactors, folic acid, vitamin B12, and pyridoxal phosphate (Devaskar and Raychaudhuri 2007). Alcohol use in pregnancy could induce epigenetic alterations through interference in the methionine and folic acid metabolic pathway (Puumala and Hoyme 2015). It is well known that alcohol interferes with folate metabolism, but alcohol has been shown to alter methylation patterns in animal models exposed to alcohol in utero. Multiple epigenetic mechanisms have been postulated. One is excessive alcohol exposure induces DNA methylation changes in sperm, embryos, or developing fetal brain (Garro et al. 1991; Otero et al. 2012; Ouko et al. 2009). In a murine model, alcohol exposure during the early neurulation period induced aberrant changes in DNA methylation patterns with associated alterations in gene expression. These alterations were thought to contribute to the observed abnormal fetal development (Liu et al. 2009). Altered and aberrant methylation patterns including hypo- and hypermethylation were observed after alcohol exposure coincident with the retardation of migration, neuronal formation, and growth processes of neural stem cells (Zhou et al. 2011a). Prenatal alcohol exposure interferes with the intrinsic DNA methylation program and gene expression of the offspring (Ouko et al. 2009; Zhou et al. 2011b, c).

Promising animal studies have shown that choline supplementation could potentially ameliorate some of the harmful effects of alcohol in FASD. Choline, an essential nutrient for synthesis of acetylcholine, membrane phospholipids, methionine, and DNA, RNA, and histone methylation, is critical for fetal neural development. Choline supplementation through alteration in DNA methylation during the critical period of organogenesis in utero strongly influences neural precursor cell proliferation and apoptosis (Zeisel 2007, 2011). Prenatal choline supplementation to alcohol-exposed pregnant dams has improved behavioral measures in their offspring, although the mechanism underlying this effect is unknown (Thomas et al. 2009). Prenatal choline supplementation to dams exposed or unexposed to alcohol resulted in both hypomethylation and hypermethylation, making the mechanism of increasing availability of choline unlikely (Otero et al. 2012). Human studies to evaluate the functional efficacy of choline supplementation in children with FASD are currently being done (Wozniak et al. 2013).

Smoking As per the Centers for Disease Control and Prevention (CDC) surveillance study, 12.3 % of women smoke during pregnancy (Tong et al. 2013). Maternal smoking has significant adverse effects to the developing fetus and hence to the long-term health of children and the adults they become. Continued postnatal exposure compounds these detrimental effects (Hackshaw et al. 2011). Previous work from multiple studies has suggested that maternal cigarette smoking during pregnancy is associated with increased risk for spontaneous abortion (Castles et al. 1999; Shah and Bracken 2000), respiratory disease (Cook et al. 1999; Cook and Strachan 1999), immune system problems such as asthma and allergies (Prescott and Clifton 2009a), and cancer in later life (Doherty et al. 2009). Smoking exposure during the crucial period of fetal programming could influence epigenetics. Adverse in utero environment could foster untoward effects in the offspring through epigenetic mechanisms, including DNA methylation and miRNA expression.

Knopik et al. elucidate the epigenetics of maternal cigarette smoking during pregnancy and effects on child development (Knopik et al. 2012). CYP1A1 is an enzyme that catalyzes the metabolic conversion of polycyclic aromatic hydrocarbons into harmful hydrophilic DNA adducts that eventually are metabolized through phase II metabolism into nontoxic compounds. Maternal tobacco use was associated with increased CYP1A1 gene expression in association with hypomethylation of the CYP1A1 promoter CpG sites immediately proximal to the 5′-xenobiotic response element transcription factor-binding element (Suter et al. 2010b). This could signify that maternal smoke exposure through downregulation of methylation of key promoter regions could contribute to specific phenotypes that could adversely affect neonatal development (Suter et al. 2010a). Further, Suter et al. expanded their placental promoter DNA methylation analysis to that of gene expression on a genome-wide scale associated with maternal tobacco exposure. They identified a significant Pearson correlation (≥ 0.7 or ≤ -0.7) between aberrant placental transcriptional regulation and differential CpG methylation from maternal tobacco use, an 18-fold increase among smokers compared to nonsmokers with a dominant effect among oxidative stress pathways (Suter et al. 2011).

DNA methylation patterns are essential both for the growth and maintenance of normal tissue-specific expression profiles in different cell types during development. In addition, some of these patterns are set in the in utero period of fetal and placental development (Hajkova et al. 2002; Jaenisch 1997; Rougier et al. 1998). Methylation patterns altered by in utero tobacco smoke exposure have been linked with adverse placental morphology and birth outcomes (Serman et al. 2007; Sinclair et al. 2007). Wilhelm-Benartzi et al. identified an association between infant growth and in utero tobacco smoke exposure and altered methylation patterns in repetitive elements (stretches of DNA that show a large number of repeated bases) and gene-associated DNA from human term placenta tissue samples. Methylation levels of Alu element AluYb8, a repetitive element in the placenta, were significantly increased by maternal tobacco use during pregnancy (Wilhelm-Benartzi et al. 2012). Thus altered methylation patterns identified in placenta samples from tobacco smoke exposure group suggest the role of epigenetic alterations from in utero environment and a possible link between environmental conditions during fetal development and diseases in later life.

Studies have linked exposure to maternal smoking with alterations of DNA methylation (Suter et al. 2013; Markunas et al. 2014), specifically in the aryl hydrocarbon receptor repressor (AHRR) gene in neonatal blood (Novakovic et al. 2014; Joubert et al. 2012). AHRR is involved in detoxification of chemicals found in tobacco smoke, and hypomethylation of the gene promoter has been associated with both higher AHRR gene expression in cord blood mononuclear cells and maternal smoking (Novakovic et al. 2014). Maternal smoking was associated with intermediate to high methylation changes limited to CpG islands within cord blood mononuclear cells and buccal epithelium. The smoking-associated DNA methylation changes in cord blood persisted for up to 18 months of age, thus highlighting the postnatal stability of these epigenetic changes in response to maternal smoking in pregnancy (Novakovic et al. 2014). Recently, ten additional genes with altered methylation patterns were identified, establishing new associations of maternal smoking to changes in gene expression in newborn blood (Markunas et al. 2014).

The placenta is the essential regulator of the in utero fetal environment and has been described as the third brain linking the mother and the infant (Yen 1994). Recent research suggests similarities between neuronal and placental DNA methylation profiles in regions of genes associated with neuronal development (Schroeder and LaSalle 2013; Schroeder et al. 2013). While epigenetics refers to the study of single genes or sets of genes, epigenomics refers to the global analyses of epigenetic changes across the entire genome. Recent research has looked at the placental epigenome and its dysregulation from the prenatal intrauterine environment of tobacco exposure with infant neurobehavioral outcome (Maccani and Marsit 2009; Maccani et al. 2010, 2013). Expression of placental miRNA 16 was negatively associated with attention score, while miR-146a and miR-182 were positively associated with quality movement scores as assessed by the NICU Network Neurobehavioral Scales. This work suggests that epigenetic dysregulation in the placenta can profoundly influence infant neurobehavior.

Future studies should be aimed at identifying more of such patterns of epigenetic markers within the chromosomes associated with environmental exposures and determine their stability and biological and clinical relevance and whether the identified epigenetic markers could be targets for intervention to alter the phenotype and improve the natural history of the disease.

5.4 Epigenetic Phenomena and Inflammation

The inflammatory response that follows any stressful event such as injury, infection, or tissue stress is very complex and involves a highly organized and sophisticated regulatory network to carry out functions at signal-specific and gene-specific levels (Medzhitov and Horng 2009). Recent studies have highlighted the importance of chromatin modifications in the control of inflammatory gene expression (Bernstein et al. 2007). The inflammatory network involves the transcription factors, transcription regulators, and chromatin modifiers for each specific gene (Medzhitov 2008). Transcription factors of the NF-κB, FOXP3, IRF, and STAT families along with epigenetic phenomena, including DNA methylation and covalent histone modifications, have been shown to be critical in the regulation of inflammatory genes (Medzhitov and Horng 2009; Bayarsaihan 2011).

Fetal immune development involving regulation of Th1/Th2 cells may be under epigenetic influence to alter gene transcription and the resulting phenotype and thereby disease predisposition. One of the most striking examples of epigenetic influence in gene expression and function without alteration in gene sequence is one that involves $CD4^+$ T immune cells in the developing immune system of neonates (Sanders 2006). $CD4^+$ T cells are a family of cells that consists of a naive T-cell precursor and its two effector T-cell subsets, Th1 and Th2 cells. Following antigen exposure, Th1 and Th2 cells secrete IFNγ and IL4, respectively (Mosmann and Sad 1996). Multiple epigenetic modifications of cytokine gene expression have been studied to explain the observed functional differences in these T-cell subsets. Methylation of the gene promoter is the main epigenetic mechanism controlling Th1 expression and the maturation of the Th1 subset of T cells (Prescott and Clifton 2009b). IFNγ gene expression increased after exposure of activated T cells to DNA methylation inhibitors (Young et al. 1994) and when T cells from DNA methyltransferase knockout mice were used (Makar and Wilson 2004). Decreased DNA methylation was found in the IFNγ promoter region of Th1 cells (Melvin et al. 1995), suggesting that hypomethylation correlated with increased gene expression of IFNγ. Hypermethylation (gene silencing) of the IFNγ promoter has been associated with reduced IFNγ expression in immature neonatal $CD4^+$ T precursor cells (White et al. 2002).

IL-4 silencing also occurred as naive T cells developed into Th1 cells in association with DNA methylation/histone deacetylation of the GATA-3 and IL-4 genes (Grogan et al. 2001; Mullen et al. 2001). Similarly, the fetal development of Th2 cells is associated with increased IL-4 gene expression and IFNγ silencing through epigenetic mechanisms (Fields et al. 2002, 2004; Grogan et al. 2001; Hewitt et al. 2004). Therefore, the role of epigenetic mechanism in the commitment of a naive T cell to either Th1 or Th2 cell lineage and the subsequent cytokine expression is striking.

Detailed investigation of the regulatory function of human CD25(+)CD4(+) T cells (T_{reg} cells) at various stages of maturity from thymocytes, cord blood (CB), and adult peripheral blood (APB) has confirmed functional immaturity in neonatal compared to adult T_{reg} cells (Fujimaki et al. 2008; Schaub et al. 2008). FOXP3, a master transcription factor for CD4(+) regulatory T cells (Treg), has been shown to be under the influence of epigenetic mechanism including DNA methylation and histone modification (Lal et al. 2009). Demethylation of a specific CpG site in FOXP3 using the DNA methyltransferase (DNMT) inhibitor 5-aza-2′-deoxycytidine (Aza) induced acetylation of histone 3. The acetylation of histone 3 resulted in strong and stable induction of FOXP3 gene expression. Epigenetic regulation of FOXP3 can be controlled predictably with DNMT inhibitors to generate functional, stable, and specific T_{reg} cells (Lal et al. 2009). Additional studies have also shown that DNA demethylation, an epigenetic phenomenon, controls FOXP3 promoter gene expression and thereby increases FOXP3 and regulates stable T_{reg} cell lineage and differentiation during immune system development (Polansky et al. 2008; Janson et al. 2008).

Experimental studies provide substantial in vitro data indicating that DNA methylation of genes critical to T-helper cell differentiation may induce polarization toward or away from an allergic phenotype (Miller and Ho 2008). Changes in the DNA methylation status during development can affect the gene transcription and subsequent phenotype of the child by increasing the risk of allergic airway diseases such as asthma (Prescott and Clifton 2009b). There is strong evidence to suggest that maternal supplementation with dietary folate in mice is associated with increased methylation of Runt-related transcription factor 3 (Runx3). Runx3, a regulatory gene known to negatively regulate allergic airway disease, was found to be excessively methylated after in utero supplementation with methyl donors. This excessive methylation was associated with decreased transcriptional activity of Runx3 and increased disease severity of allergic airway disease in mice (Hollingsworth et al. 2008). This highlights the importance of epigenetic mechanisms that can modify the heritable risk of allergic diseases such as asthma.

Altered DNA methylation, an epigenetic event, has been linked with acute bacterial infections and toxin exposure (Kondo et al. 2000). The gold standard for the diagnosis of bacterial sepsis in premature neonates is the establishment of the growth of the microorganism in a blood culture. Recent research explored the potential use of the DNA methylation pattern of CpG sites in the promoter region of calcitonin-related polypeptide-α (CALCA) gene as an epigenetic marker for bacterial sepsis in preterm newborns. CALCA is a gene that codes for procalcitonin (PCT). PCT is an acute-phase protein whose level in serum increases with an acute bacterial infection. This retrospective investigation demonstrated altered DNA methylation patterns at eight CpG sites in the promoter region of the CALCA in neonates with bacterial sepsis (Tendl et al. 2013).

DNA methylation plays an important role in the regulation of inflammatory genes. Chronic inflammation has been associated with the induction of aberrant DNA methylation patterns (Hur et al. 2011; Hsieh et al. 1998; Issa et al. 2001; Toyota et al. 2002). Toll-like receptors (TLRs) are major receptors that enable inflammatory cells to recognize invading microbial pathogens. Promoter hypomethylation of the Toll-like receptor 2 (TLR2) gene was found to be responsible for cystic fibrosis (CF)-related upregulation of TLR2. TLR2 expression is epigenetically upregulated and is associated with an increased pro-inflammatory response to bacterial peptidoglycan in CF bronchial epithelial cells (Shuto et al. 2006). In conclusion, inflammation has a high capacity to induce methylation changes in DNA.

Epigenetic mechanisms have been linked with tissue repair and wound healing processes. Besides understanding the regulation of wound healing by growth factors and cytokines, recent focus has been on the epigenetic regulation at the wound site. Following a specific tissue injury, wound healing involves deposition of type 1 collagen and other extracellular matrix proteins. Excessive deposition of collagen, which occurs in some fibrotic disease states, results in organ dysfunction and failure. The promoter region of type 1 collagen protein alpha1 (COL1A1) is being extensively studied to better understand gene regulation of the fibrotic process.

Hypermethylation of CpG sites in the COL1A1 promoter is associated with reduced collagen synthesis at the transcriptional level in myopic scleras (Zhou et al. 2012). TGFβ treatment of cardiac fibroblasts (1) upregulated the expression of collagen type 1 mRNA through demethylation of DNA in the COL1A1 promoter regions and (2) promoted inhibition of DNA methyltransferases (DNMTs) (Pan et al. 2013). Numerous other studies have linked aberrant histone acetylation with the pathogenesis of pulmonary, renal, and cardiac fibrosis (Coward et al. 2009, 2010; Marumo et al. 2010; Kook et al. 2003; Robinson et al. 2012).

Recent investigation has implicated miRNAs in the regulation of inflammation. miR-146a limits TLR signaling by blocking signaling molecule TNF (tumor necrosis factor) receptor-associated factor 6 protein (TRAF6) (Taganov et al. 2006). Studies by Liu et al. found miR-147 to be induced upon stimulation of multiple TLRs and functioned as a negative regulator of TLR-associated signaling events in murine macrophages. Their data demonstrate a negative-feedback loop in which TLR stimulation induces miR-147 to prevent excessive inflammatory responses.

In summary, knowledge about alterations in histone modifications, DNA methylation, and miRNA regulation will provide a better understanding of the molecular basis for various acute and chronic inflammatory diseases. Progress in the studies of epigenetic alteration during an inflammatory response opens opportunities for the development of efficient medications for specific targets. Epigenetic analysis of a gene encoding an acute-phase protein involved in infectious processes has the potential to serve as an epigenetic biomarker in addition to the classical serum markers for the diagnosis of neonatal sepsis. Among the drugs currently proposed for epigenetic therapy are histone deacetylase inhibitors and demethylating agents, which target chromatin in rapidly dividing tumor cells and restore normal cell functions (Karberg 2009). Investigating epigenetic pathways including genomic methylation and understanding the mechanisms of underlying organ fibrosis may lead to effective ways of targeting the epigenetic control pathways to reduce excessive collagen deposition in organs. The use of more recent technologically advanced tools such as whole-genome microarray expression profiling and chromatin immunoprecipitation-based sequencing (ChIP-seq) methods will be extremely useful in the development of epigenetic drugs with greater specificity.

5.5 Epigenetics and the NICU Environment

Epigenetic regulation from both intrauterine and extrauterine environments, known to affect infant outcome, is currently under investigation as an important determinant of outcome in the neonatal intensive care unit (NICU). Epigenetic changes can have profound impacts on phenotypes such as intellectual disability by rendering the transcription of specific genes more or less responsive to later stimulation. This is already evident from multiple human diseases with specific deficiencies (LaSalle et al. 2013). A number of epigenetic changes have been linked with pediatric diseases. These include, but are not limited to, autism, schizophrenia, and intellectual disability (Gluckman et al. 2005; Lesseur et al. 2013, 2014; Van den Bergh 2011). Abnormal epigenetic regulation has been reported as the mechanism underlying the pathogenesis of various human neuronal diseases such as Rett syndrome (Amir et al. 1999), Prader-Willi, Angelman (Sutcliffe et al. 1994; Dittrich et al. 1996), Rubinstein-Taybi (Das et al. 2009), X-linked intellectual disability, and autism (Iwase et al. 2007).

While the period of fetal development is indeed a critical period during which changes in the intrauterine environment can have significant consequences for future health, premature babies are exposed to a radically different environment. A growing body of research suggests that the NICU extrauterine environment plays an equally and if not greater critical role in infant neurodevelopment and overall future health (Lester et al. 2011, 2014).

By providing a single family room NICU and therefore substantially altering the extrauterine environment experienced by these infants, the overall medical and neurobehavioral outcome of the preterm infant could potentially improve. The benefits offered by the single family room NICU include (1) better practice for infection control by avoiding overcrowding and reducing pathogen exposure; (2) more family time, particularly more skin-to-skin holding and more breastfeeding, both of which improve physiological stability, reduce feeding intolerance, and improve growth; and (3) reduction of noise and light (Lester et al. 2011). In addition, more recent studies have indicated that the ability to provide maternal biological sounds and voice improves cardiorespiratory regulation, reduced frequency of apnea and bradycardia episodes, and cognitive outcomes to these extremely preterm infants (Doheny et al. 2012a, b; McMahon et al. 2012). No human studies are currently available linking postnatal epigenetic alterations in the preterm newborn and the extrauterine environmental change in the NICU.

But animal studies have demonstrated epigenetic changes in rat pups resulting from differences in maternal care in the postnatal extrauterine environment. Maternal behavior toward rodents modified the expression of genes associated with behavioral and neuroendocrine responses by altering the methylation status of the glucocorticoid receptor promoter region in the hippocampus of the rat pups (Meaney and Szyf 2005; Weaver et al. 2004). Weaver et al. studied changes in the DNA methylation pattern arising from differences in maternal care between the offspring of high- and low-licking/grooming rat mothers. They specifically noted altered histone acetylation and transcription factor (nerve growth factor-induced clone A) binding to the glucocorticoid receptor promoter which reversed with cross-fostering, and these changes persisted into adulthood. These findings highlight the importance of environmental programming and the subsequent impact on structural modifications of the DNA and thereafter its genetic function from maternal behavior.

Examining epigenctic alterations in the prenatal fetuses, premature infants, and postnatal neonates to make important links between epigenetic markers and childhood/adult diseases will pave the way for better understanding of the pathogenesis of diseases. Ultimately, newer therapies targeting epigenetic markers can aid in preventive strategies and novel therapy for diseases.

References

Amir, R. E., van den Veyver, I. B., Wan, M., Tran, C. Q., Francke, U., & Zoghbi, H. Y. (1999). Rett syndrome is caused by mutations in X-linked MECP2, encoding methyl-CpG-binding protein 2. *Nature Genetics, 23*, 185–188.

Bateson, P., Barker, D., Clutton-Brock, T., Deb, D., D'udine, B., Foley, R. A., Gluckman, P., Godfrey, K., Kirkwood, T., Lahr, M. M., Mcnamara, J., Metcalfe, N. B., Monaghan, P., Spencer, H. G., & Sultan, S. E. (2004). Developmental plasticity and human health. *Nature, 430*, 419–421.

Bayarsaihan, D. (2011). Epigenetic mechanisms in inflammation. *Journal of Dental Research, 90*, 9–17.

Berger, S. L., Kouzarides, T., Shiekhattar, R., & Shilatifard, A. (2009). An operational definition of epigenetics. *Genes and Development, 23*, 781–783.

Bernstein, B. E., Meissner, A., & Lander, E. S. (2007). The mammalian epigenome. *Cell, 128*, 669–681.

Bird, A. (2002). DNA methylation patterns and epigenetic memory. *Genes and Development, 16*, 6–21.

Bird, A. (2007). Perceptions of epigenetics. *Nature, 447*, 396–398.

Castles, A., Adams, E. K., Melvin, C. L., Kelsch, C., & Boulton, M. L. (1999). Effects of smoking during pregnancy. Five meta-analyses. *American Journal of Preventive Medicine, 16*, 208–215.

Cook, D. G., & Strachan, D. P. (1999). Health effects of passive smoking-10: Summary of effects of parental smoking on the respiratory health of children and implications for research. *Thorax, 54*, 357–366.

Cook, D. G., Strachan, D. P., & Carey, I. M. (1999). Health effects of passive smoking. *Thorax, 54*, 469.

Coward, W. R., Watts, K., Feghali-Bostwick, C. A., Jenkins, G., & Pang, L. (2010). Repression of IP-10 by interactions between histone deacetylation and hypermethylation in idiopathic pulmonary fibrosis. *Molecular and Cellular Biology, 30*, 2874–2886.

Coward, W. R., Watts, K., Feghali-Bostwick, C. A., Knox, A., & Pang, L. (2009). Defective histone acetylation is responsible for the diminished expression of cyclooxygenase 2 in idiopathic pulmonary fibrosis. *Molecular and Cellular Biology, 29*, 4325–4339.

Das, C., Lucia, M. S., Hansen, K. C., & Tyler, J. K. (2009). CBP/p300-mediated acetylation of histone H3 on lysine 56. *Nature, 459*, 113–117.

Devaskar, S. U., & Raychaudhuri, S. (2007). Epigenetics—A science of heritable biological adaptation. *Pediatric Research, 61*, 1R–4R.

Dittrich, B., Buiting, K., Korn, B., Rickard, S., Buxton, J., Saitoh, S., Nicholls, R. D., Poustka, A., Winterpacht, A., Zabel, B., & Horsthemke, B. (1996). Imprint switching on human chromosome 15 may involve alternative transcripts of the SNRPN gene. *Nature Genetics, 14*, 163–170.

Doheny, L., Hurwitz, S., Insoft, R., Ringer, S., & Lahav, A. (2012a). Exposure to biological maternal sounds improves cardiorespiratory regulation in extremely preterm infants. *The Journal of Maternal-Fetal & Neonatal Medicine: The Official Journal of the European Association of Perinatal Medicine, the Federation of Asia and Oceania Perinatal Societies, the International Society of Perinatal Obstetricians, 25*, 1591–1594.

Doheny, L., Morey, J. A., Ringer, S. A., & Lahav, A. (2012b). Reduced frequency of apnea and bradycardia episodes caused by exposure to biological maternal sounds. *Pediatrics International: Official Journal of the Japan Pediatric Society, 54*, e1–e3.

Doherty, S. P., Grabowski, J., Hoffman, C., Ng, S. P., & Zelikoff, J. T. (2009). Early life insult from cigarette smoke may be predictive of chronic diseases later in life. *Biomarkers, 14*(Suppl 1), 97–101.

Dolinoy, D. C., Huang, D., & Jirtle, R. L. (2007). Maternal nutrient supplementation counteracts bisphenol A-induced DNA hypomethylation in early development. *Proceedings of the National Academy of Sciences of the United States of America, 104*, 13056–13061.

Dolinoy, D. C., Weidman, J. R., Waterland, R. A., & Jirtle, R. L. (2006). Maternal genistein alters coat color and protects Avy mouse offspring from obesity by modifying the fetal epigenome. *Environmental Health Perspectives, 114*, 567–572.
Fields, P. E., Kim, S. T., & Flavell, R. A. (2002). Cutting edge: Changes in histone acetylation at the IL-4 and IFN-gamma loci accompany Th1/Th2 differentiation. *The Journal of Immunology, 169*, 647–650.
Fields, P. E., Lee, G. R., Kim, S. T., Bartsevich, V. V., & Flavell, R. A. (2004). Th2-specific chromatin remodeling and enhancer activity in the Th2 cytokine locus control region. *Immunity, 21*, 865–876.
Fujimaki, W., Takahashi, N., Ohnuma, K., Nagatsu, M., Kurosawa, H., Yoshida, S., Dang, N. H., Uchiyama, T., & Morimoto, C. (2008). Comparative study of regulatory T cell function of human CD25CD4 T cells from thymocytes, cord blood, and adult peripheral blood. *Clinical & Developmental Immunology, 2008*, 305859.
Garro, A. J., Mcbeth, D. L., Lima, V., & Lieber, C. S. (1991). Ethanol consumption inhibits fetal DNA methylation in mice: Implications for the fetal alcohol syndrome. *Alcoholism: Clinical and Experimental Research, 15*, 395–398.
Gluckman, P. D., Hanson, M. A., & Pinal, C. (2005). The developmental origins of adult disease. *Maternal & Child Nutrition, 1*, 130–141.
Goldberg, A. D., Allis, C. D., & Bernstein, E. (2007). Epigenetics: A landscape takes shape. *Cell, 128*, 635–638.
Grogan, J. L., Mohrs, M., Harmon, B., Lacy, D. A., Sedat, J. W., & Locksley, R. M. (2001). Early transcription and silencing of cytokine genes underlie polarization of T helper cell subsets. *Immunity, 14*, 205–215.
Hackshaw, A., Rodeck, C., & Boniface, S. (2011). Maternal smoking in pregnancy and birth defects: A systematic review based on 173 687 malformed cases and 11.7 million controls. *Human Reproduction Update, 17*, 589–604.
Hajkova, P., Erhardt, S., Lane, N., Haaf, T., El-Maarri, O., Reik, W., Walter, J., & Surani, M. A. (2002). Epigenetic reprogramming in mouse primordial germ cells. *Mechanisms of Development, 117*, 15–23.
Hewitt, S. L., High, F. A., Reiner, S. L., Fisher, A. G., & Merkenschlager, M. (2004). Nuclear repositioning marks the selective exclusion of lineage-inappropriate transcription factor loci during T helper cell differentiation. *European Journal of Immunology, 34*, 3604–3613.
Hollingsworth, J. W., Maruoka, S., Boon, K., Garantziotis, S., Li, Z., Tomfohr, J., Bailey, N., Potts, E. N., Whitehead, G., Brass, D. M., & Schwartz, D. A. (2008). In utero supplementation with methyl donors enhances allergic airway disease in mice. *The Journal of Clinical Investigation, 118*, 3462–3469.
Hsieh, C. J., Klump, B., Holzmann, K., Borchard, F., Gregor, M., & Porschen, R. (1998). Hypermethylation of the p16INK4a promoter in colectomy specimens of patients with long-standing and extensive ulcerative colitis. *Cancer Research, 58*, 3942–3945.
Hur, K., Niwa, T., Toyoda, T., Tsukamoto, T., Tatematsu, M., Yang, H. K., & Ushijima, T. (2011). Insufficient role of cell proliferation in aberrant DNA methylation induction and involvement of specific types of inflammation. *Carcinogenesis, 32*, 35–41.
Issa, J. P., Ahuja, N., Toyota, M., Bronner, M. P., & Brentnall, T. A. (2001). Accelerated age-related CpG island methylation in ulcerative colitis. *Cancer Research, 61*, 3573–3577.
Iwase, S., Lan, F., Bayliss, P., de la Torre-Ubieta, L., Huarte, M., Qi, H. H., Whetstine, J. R., Bonni, A., Roberts, T. M., & Shi, Y. (2007). The X-linked mental retardation gene SMCX/JARID1C defines a family of histone H3 lysine 4 demethylases. *Cell, 128*, 1077–1088.
Jaenisch, R. (1997). DNA methylation and imprinting: Why bother? *Trends in Genetics, 13*, 323–329,
Jaenisch, R., & Bird, A. (2003). Epigenetic regulation of gene expression: How the genome integrates intrinsic and environmental signals. *Nature Genetics, 33*(Suppl), 245–254.
Janson, P. C., Winerdal, M. E., Marits, P., Thorn, M., Ohlsson, R., & Winqvist, O. (2008). FOXP3 promoter demethylation reveals the committed Treg population in humans. *PLoS One, 3*, e1612.

Joubert, B. R., Haberg, S. E., Nilsen, R. M., Wang, X., Vollset, S. E., Murphy, S. K., Huang, Z., Hoyo, C., Midttun, O., Cupul-Uicab, L. A., Ueland, P. M., Wu, M. C., Nystad, W., Bell, D. A., Peddada, S. D., & London, S. J. (2012). 450K epigenome-wide scan identifies differential DNA methylation in newborns related to maternal smoking during pregnancy. *Environmental Health Perspectives, 120*, 1425–1431.

Karberg, S. (2009). Switching on epigenetic therapy. *Cell, 139*, 1029–1031.

Knopik, V. S., Maccani, M. A., Francazio, S., & Mcgeary, J. E. (2012). The epigenetics of maternal cigarette smoking during pregnancy and effects on child development. *Development and Psychopathology, 24*, 1377–1390.

Kondo, Y., Kanai, Y., Sakamoto, M., Mizokami, M., Ueda, R., & Hirohashi, S. (2000). Genetic instability and aberrant DNA methylation in chronic hepatitis and cirrhosis—A comprehensive study of loss of heterozygosity and microsatellite instability at 39 loci and DNA hypermethylation on 8 CpG islands in microdissected specimens from patients with hepatocellular carcinoma. *Hepatology, 32*, 970–979.

Kook, H., Lepore, J. J., Gitler, A. D., Lu, M. M., Wing-Man Yung, W., Mackay, J., Zhou, R., Ferrari, V., Gruber, P., & Epstein, J. A. (2003). Cardiac hypertrophy and histone deacetylase-dependent transcriptional repression mediated by the atypical homeodomain protein Hop. *The Journal of Clinical Investigation, 112*, 863–871.

Kouzarides, T. (2007). Chromatin modifications and their function. *Cell, 128*, 693–705.

Lal, G., Zhang, N., van der Touw, W., Ding, Y., Ju, W., Bottinger, E. P., Reid, S. P., Levy, D. E., & Bromberg, J. S. (2009). Epigenetic regulation of Foxp3 expression in regulatory T cells by DNA methylation. *The Journal of Immunology, 182*, 259–273.

Lane, R. H. (2014). Fetal programming, epigenetics, and adult onset disease. *Clinics in Perinatology, 41*, 815–831.

Lasalle, J. M., Powell, W. T., & Yasui, D. H. (2013). Epigenetic layers and players underlying neurodevelopment. *Trends in Neurosciences, 36*, 460–470.

Lesseur, C., Armstrong, D. A., Paquette, A. G., Koestler, D. C., Padbury, J. F., & Marsit, C. J. (2013). Tissue-specific Leptin promoter DNA methylation is associated with maternal and infant perinatal factors. *Molecular and Cellular Endocrinology, 381*, 160–167.

Lesseur, C., Paquette, A. G., & Marsit, C. J. (2014). Epigenetic regulation of infant neurobehavioral outcomes. *Medical Epigenetics, 2*, 71–79.

Lester, B. M., Hawes, K., Abar, B., Sullivan, M., Miller, R., Bigsby, R., Laptook, A., Salisbury, A., Taub, M., Lagasse, L. L., & Padbury, J. F. (2014). Single-family room care and neurobehavioral and medical outcomes in preterm infants. *Pediatrics, 134*, 754–760.

Lester, B. M., Miller, R. J., Hawes, K., Salisbury, A., Bigsby, R., Sullivan, M. C., & Padbury, J. F. (2011). Infant neurobehavioral development. *Seminars in Perinatology, 35*, 8–19.

Liu, Y., Balaraman, Y., Wang, G., Nephew, K. P., & Zhou, F. C. (2009). Alcohol exposure alters DNA methylation profiles in mouse embryos at early neurulation. *Epigenetics: Official Journal of the DNA Methylation Society, 4*, 500–511.

Lo, C. L., & Zhou, F. C. (2014). Environmental alterations of epigenetics prior to the birth. *International Review of Neurobiology, 115*, 1–49.

Maccani, M. A., Avissar-Whiting, M., Banister, C. E., Mcgonnigal, B., Padbury, J. F., & Marsit, C. J. (2010). Maternal cigarette smoking during pregnancy is associated with downregulation of miR-16, miR-21, and miR-146a in the placenta. *Epigenetics: Official Journal of the DNA Methylation Society, 5*, 583–589.

Maccani, M. A., & Marsit, C. J. (2009). Epigenetics in the placenta. *American Journal of Reproductive Immunology, 62*, 78–89.

Maccani, M. A., Padbury, J. F., Lester, B. M., Knopik, V. S., & Marsit, C. J. (2013). Placental miRNA expression profiles are associated with measures of infant neurobehavioral outcomes. *Pediatric Research, 74*, 272–278.

Makar, K. W., & Wilson, C. B. (2004). DNA methylation is a nonredundant repressor of the Th2 effector program. *The Journal of Immunology, 173*, 4402–4406.

Markunas, C. A., Xu, Z., Harlid, S., Wade, P. A., Lie, R. T., Taylor, J. A., & Wilcox, A. J. (2014). Identification of DNA methylation changes in newborns related to maternal smoking during pregnancy. *Environmental Health Perspectives, 122*, 1147–1153.

Marumo, T., Hishikawa, K., Yoshikawa, M., Hirahashi, J., Kawachi, S., & Fujita, T. (2010). Histone deacetylase modulates the proinflammatory and -fibrotic changes in tubulointerstitial injury. *American Journal of Physiology. Renal Physiology, 298*, F133–F141.

Mcgowan, P. O., Sasaki, A., D'alessio, A. C., Dymov, S., Labonte, B., Szyf, M., Turecki, G., & Meaney, M. J. (2009). Epigenetic regulation of the glucocorticoid receptor in human brain associates with childhood abuse. *Nature Neuroscience, 12*, 342–348.

Mcmahon, E., Wintermark, P., & Lahav, A. (2012). Auditory brain development in premature infants: The importance of early experience. *Annals of the New York Academy of Sciences, 1252*, 17–24.

Mcmillen, I. C., & Robinson, J. S. (2005). Developmental origins of the metabolic syndrome: Prediction, plasticity, and programming. *Physiological Reviews, 85*, 571–633.

Meaney, M. J., & Szyf, M. (2005). Environmental programming of stress responses through DNA methylation: Life at the interface between a dynamic environment and a fixed genome. *Dialogues in Clinical Neuroscience, 7*, 103–123.

Medzhitov, R. (2008). Origin and physiological roles of inflammation. *Nature, 454*, 428–435.

Medzhitov, R., & Horng, T. (2009). Transcriptional control of the inflammatory response. *Nature Reviews Immunology, 9*, 692–703.

Melvin, A. J., Mcgurn, M. E., Bort, S. J., Gibson, C., & Lewis, D. B. (1995). Hypomethylation of the interferon-gamma gene correlates with its expression by primary T-lineage cells. *European Journal of Immunology, 25*, 426–430.

Miller, J. L., & Grant, P. A. (2013). The role of DNA methylation and histone modifications in transcriptional regulation in humans. *Sub-Cellular Biochemistry, 61*, 289–317.

Miller, R. L., & Ho, S. M. (2008). Environmental epigenetics and asthma: Current concepts and call for studies. *American Journal of Respiratory and Critical Care Medicine, 177*, 567–573.

Miska, E. A. (2005). How microRNAs control cell division, differentiation and death. *Current Opinion in Genetics & Development, 15*, 563–568.

Mosmann, T. R., & Sad, S. (1996). The expanding universe of T-cell subsets: Th1, Th2 and more. *Immunology Today, 17*, 138–146.

Mullen, A. C., High, F. A., Hutchins, A. S., Lee, H. W., Villarino, A. V., Livingston, D. M., Kung, A. L., Cereb, N., Yao, T. P., Yang, S. Y., & Reiner, S. L. (2001). Role of T-bet in commitment of TH1 cells before IL-12-dependent selection. *Science, 292*, 1907–1910.

Novakovic, B., Ryan, J., Pereira, N., Boughton, B., Craig, J. M., & Saffery, R. (2014). Postnatal stability, tissue, and time specific effects of AHRR methylation change in response to maternal smoking in pregnancy. *Epigenetics: Official Journal of the DNA Methylation Society, 9*, 377–386.

Otero, N. K., Thomas, J. D., Saski, C. A., Xia, X., & Kelly, S. J. (2012). Choline supplementation and DNA methylation in the hippocampus and prefrontal cortex of rats exposed to alcohol during development. *Alcoholism: Clinical and Experimental Research, 36*, 1701–1709.

Ouko, L. A., Shantikumar, K., Knezovich, J., Haycock, P., Schnugh, D. J., & Ramsay, M. (2009). Effect of alcohol consumption on CpG methylation in the differentially methylated regions of H19 and IG-DMR in male gametes: Implications for fetal alcohol spectrum disorders. *Alcoholism: Clinical and Experimental Research, 33*, 1615–1627.

Pan, X., Chen, Z., Huang, R., Yao, Y., & Ma, G. (2013). Transforming growth factor beta1 induces the expression of collagen type I by DNA methylation in cardiac fibroblasts. *PLoS One, 8*, e60335.

Pembrey, M., Saffery, R., & Bygren, L. O. (2014). Human transgenerational responses to early-life experience: Potential impact on development, health and biomedical research. *Journal of Medical Genetics, 51*, 563–572.

Pickard, B., Dean, W., Engemann, S., Bergmann, K., Fuermann, M., Jung, M., Reis, A., Allen, N., Reik, W., & Walter, J. (2001). Epigenetic targeting in the mouse zygote marks DNA for later methylation: A mechanism for maternal effects in development. *Mechanisms of Development, 103*, 35–47.

Polansky, J. K., Kretschmer, K., Freyer, J., Floess, S., Garbe, A., Baron, U., Olek, S., Hamann, A., von Boehmer, H., & Huehn, J. (2008). DNA methylation controls Foxp3 gene expression. *European Journal of Immunology, 38*, 1654–1663.

Prescott, S. L., & Clifton, V. (2009a). Asthma and pregnancy: Emerging evidence of epigenetic interactions in utero. *Current Opinion in Allergy and Clinical Immunology, 9*, 417–426.
Prescott, S. L., & Clifton, V. (2009b). Asthma and pregnancy: Emerging evidence of epigenetic interactions in utero. *Current Opinion in Allergy and Clinical Immunology, 9*, 417–426.
Puumala, S. E., & Hoyme, H. E. (2015). Epigenetics in pediatrics. *Pediatrics in Review/American Academy of Pediatrics, 36*, 14–21.
Rakyan, V. K., Blewitt, M. E., Druker, R., Preis, J. I., & Whitelaw, E. (2002). Metastable epialleles in mammals. *Trends in Genetics, 18*, 348–351.
Rando, O. J., & Verstrepen, K. J. (2007). Timescales of genetic and epigenetic inheritance. *Cell, 128*, 655–668.
Reik, W., Dean, W., & Walter, J. (2001). Epigenetic reprogramming in mammalian development. *Science, 293*, 1089–1093.
Robins, J. C., Marsit, C. J., Padbury, J. F., & Sharma, S. S. (2011). Endocrine disruptors, environmental oxygen, epigenetics and pregnancy. *Frontiers in Bioscience (Elite Edition), 3*, 690–700.
Robinson, C. M., Watson, C. J., & Baugh, J. A. (2012). Epigenetics within the matrix: A neo-regulator of fibrotic disease. *Epigenetics: Official Journal of the DNA Methylation Society, 7*, 987–993.
Rothstein, M. A., Cai, Y., & Marchant, G. E. (2009). The ghost in our genes: Legal and ethical implications of epigenetics. *Health Matrix, 19*, 1–62.
Rougier, N., Bourc'his, D., Gomes, D. M., Niveleau, A., Plachot, M., Paldi, A., & Viegas-Pequignot, E. (1998). Chromosome methylation patterns during mammalian preimplantation development. *Genes and Development, 12*, 2108–2113.
Sanders, V. M. (2006). Epigenetic regulation of Th1 and Th2 cell development. *Brain, Behavior, and Immunity, 20*, 317–324.
Schaub, B., Liu, J., Schleich, I., Hoppler, S., Sattler, C., & von Mutius, E. (2008). Impairment of T helper and T regulatory cell responses at birth. *Allergy, 63*, 1438–1447.
Schroeder, D. I., Blair, J. D., Lott, P., Yu, H. O., Hong, D., Crary, F., Ashwood, P., Walker, C., Korf, I., Robinson, W. P., & Lasalle, J. M. (2013). The human placenta methylome. *Proceedings of the National Academy of Sciences of the United States of America, 110*, 6037–6042.
Schroeder, D. I., & Lasalle, J. M. (2013). How has the study of the human placenta aided our understanding of partially methylated genes? *Epigenomics, 5*, 645–654.
Serman, L., Vlahovic, M., Sijan, M., Bulic-Jakus, F., Serman, A., Sincic, N., Matijevic, R., Juric-Lekic, G., & Katusic, A. (2007). The impact of 5-azacytidine on placental weight, glycoprotein pattern and proliferating cell nuclear antigen expression in rat placenta. *Placenta, 28*, 803–811.
Shah, N. R., & Bracken, M. B. (2000). A systematic review and meta-analysis of prospective studies on the association between maternal cigarette smoking and preterm delivery. *American Journal of Obstetrics and Gynecology, 182*, 465–472.
Shuto, T., Furuta, T., Oba, M., Xu, H., Li, J. D., Cheung, J., Gruenert, D. C., Uehara, A., Suico, M. A., Okiyoneda, T., & Kai, H. (2006). Promoter hypomethylation of Toll-like receptor-2 gene is associated with increased proinflammatory response toward bacterial peptidoglycan in cystic fibrosis bronchial epithelial cells. *FASEB Journal: Official Publication of the Federation of American Societies for Experimental Biology, 20*, 782–784.
Siegmund, A., & Wotjak, C. T. (2007). A mouse model of posttraumatic stress disorder that distinguishes between conditioned and sensitised fear. *Journal of Psychiatric Research, 41*, 848–860.
Sinclair, K. D., Allegrucci, C., Singh, R., Gardner, D. S., Sebastian, S., Bispham, J., Thurston, A., Huntley, J. F., Rees, W. D., Maloney, C. A., Lea, R. G., Craigon, J., Mcevoy, T. G., & Young, L. E. (2007). DNA methylation, insulin resistance, and blood pressure in offspring determined by maternal periconceptional B vitamin and methionine status. *Proceedings of the National Academy of Sciences of the United States of America, 104*, 19351–19356.
Stankiewicz, A. M., Swiergiel, A. H., & Lisowski, P. (2013). Epigenetics of stress adaptations in the brain. *Brain Research Bulletin, 98*, 76–92.

Sutcliffe, J. S., Nakao, M., Christian, S., Orstavik, K. H., Tommerup, N., Ledbetter, D. H., & Beaudet, A. L. (1994). Deletions of a differentially methylated CpG island at the SNRPN gene define a putative imprinting control region. *Nature Genetics, 8*, 52–58.

Suter, M., Abramovici, A., Showalter, L., Hu, M., Shope, C. D., Varner, M., & Aagaard-Tillery, K. (2010a). In utero tobacco exposure epigenetically modifies placental CYP1A1 expression. *Metabolism, Clinical and Experimental, 59*, 1481–1490.

Suter, M., Abramovici, A., & Aagaard-Tillery, K. (2010b). Genetic and epigenetic influences associated with intrauterine growth restriction due to in utero tobacco exposure. *Pediatric Endocrinology Reviews: PER, 8*, 94–102.

Suter, M. A., Anders, A. M., & Aagaard, K. M. (2013). Maternal smoking as a model for environmental epigenetic changes affecting birthweight and fetal programming. *Molecular Human Reproduction, 19*, 1–6.

Suter, M., Ma, J., Harris, A., Patterson, L., Brown, K. A., Shope, C., Showalter, L., Abramovici, A., & Aagaard-Tillery, K. M. (2011). Maternal tobacco use modestly alters correlated epigenome-wide placental DNA methylation and gene expression. *Epigenetics: Official Journal of the DNA Methylation Society, 6*, 1284–1294.

Taganov, K. D., Boldin, M. P., Chang, K. J., & Baltimore, D. (2006). NF-kappaB-dependent induction of microRNA miR-146, an inhibitor targeted to signaling proteins of innate immune responses. *Proceedings of the National Academy of Sciences of the United States of America, 103*, 12481–12486.

Tendl, K. A., Schulz, S. M., Mechtler, T. P., Bohn, A., Metz, T., Greber-Platzer, S., Kasper, D. C., Herkner, K. R., & Item, C. B. (2013). DNA methylation pattern of CALCA in preterm neonates with bacterial sepsis as a putative epigenetic biomarker. *Epigenetics: Official Journal of the DNA Methylation Society, 8*, 1261–1267.

Thomas, J. D., Abou, E. J., & Dominguez, H. D. (2009). Prenatal choline supplementation mitigates the adverse effects of prenatal alcohol exposure on development in rats. *Neurotoxicology and Teratology, 31*, 303–311.

Tong, V. T., Dietz, P. M., Morrow, B., D'angelo, D. V., Farr, S. L., Rockhill, K. M., England, L. J., & Centers for Disease, Control & Prevention. (2013). Trends in smoking before, during, and after pregnancy—Pregnancy risk assessment monitoring system, United States, 40 sites, 2000–2010. *MMWR Surveillance Summaries, 62*, 1–19.

Toyota, M., Itoh, F., Kikuchi, T., Satoh, A., Obata, T., Suzuki, H., Ishii, S., Endo, T., Tokino, T., & Imai, K. (2002). DNA methylation changes in gastrointestinal disease. *Journal of Gastroenterology, 37*(Suppl 14), 97–101.

Tsankova, N., Renthal, W., Kumar, A., & Nestler, E. J. (2007). Epigenetic regulation in psychiatric disorders. *Nature Reviews. Neuroscience, 8*, 355–367.

van den Bergh, B. R. (2011). Developmental programming of early brain and behaviour development and mental health: A conceptual framework. *Developmental Medicine and Child Neurology, 53*(Suppl 4), 19–23.

Waddington, C. H. (1942) The epigenotype. Endeavour 1: 18–20.

Waterland, R. A., & Jirtle, R. L. (2003). Transposable elements: Targets for early nutritional effects on epigenetic gene regulation. *Molecular and Cellular Biology, 23*, 5293–5300.

Weaver, I. C., Cervoni, N., Champagne, F. A., D'alessio, A. C., Sharma, S., Seckl, J. R., Dymov, S., Szyf, M., & Meaney, M. J. (2004). Epigenetic programming by maternal behavior. *Nature Neuroscience, 7*, 847–854.

Weaver, I. C., Champagne, F. A., Brown, S. E., Dymov, S., Sharma, S., Meaney, M. J., & Szyf, M. (2005). Reversal of maternal programming of stress responses in adult offspring through methyl supplementation: Altering epigenetic marking later in life. *The Journal of Neuroscience: The Official Journal of the Society for Neuroscience, 25*, 11045–11054.

White, G. P., Watt, P. M., Holt, B. J., & Holt, P. G. (2002). Differential patterns of methylation of the IFN-gamma promoter at CpG and non-CpG sites underlie differences in IFN-gamma gene expression between human neonatal and adult CD45RO- T cells. *Journal of Immunology, 168*, 2820–2827.

Wilhelm-Benartzi, C. S., Houseman, E. A., Maccani, M. A., Poage, G. M., Koestler, D. C., Langevin, S. M., Gagne, L. A., Banister, C. E., Padbury, J. F., & Marsit, C. J. (2012). In utero exposures, infant growth, and DNA methylation of repetitive elements and developmentally related genes in human placenta. *Environmental Health Perspectives, 120*, 296–302.

Wozniak, J. R., Fuglestad, A. J., Eckerle, J. K., Kroupina, M. G., Miller, N. C., Boys, C. J., Brearley, A. M., Fink, B. A., Hoecker, H. L., Zeisel, S. H., & Georgieff, M. K. (2013). Choline supplementation in children with fetal alcohol spectrum disorders has high feasibility and tolerability. *Nutrition Research, 33*, 897–904.

Yen, S. S. (1994). The placenta as the third brain. *The Journal of Reproductive Medicine, 39*, 277–280.

Young, H. A., Ghosh, P., Ye, J., Lederer, J., Lichtman, A., Gerard, J. R., Penix, L., Wilson, C. B., Melvin, A. J., Mcgurn, M. E., et al. (1994). Differentiation of the T helper phenotypes by analysis of the methylation state of the IFN-gamma gene. *The Journal of Immunology, 153*, 3603 3610.

Zeisel, S. H. (2007). Gene response elements, genetic polymorphisms and epigenetics influence the human dietary requirement for choline. *IUBMB Life, 59*, 380–387.

Zeisel, S. H. (2011). What choline metabolism can tell us about the underlying mechanisms of fetal alcohol spectrum disorders. *Molecular Neurobiology, 44*, 185–191.

Zhou, F. C., Balaraman, Y., Teng, M., Liu, Y., Singh, R. P., & Nephew, K. P. (2011a). Alcohol alters DNA methylation patterns and inhibits neural stem cell differentiation. *Alcoholism: Clinical and Experimental Research, 35*, 735–746.

Zhou, F. C., Chen, Y., & Love, A. (2011b). Cellular DNA methylation program during neurulation and its alteration by alcohol exposure. *Birth Defects Research. Part A, Clinical and Molecular Teratology, 91*, 703–715.

Zhou, X., Ji, F., An, J., Zhao, F., Shi, F., Huang, F., Li, Y., Jiao, S., Yan, D., Chen, X., Chen, J., & Qu, J. (2012). Experimental murine myopia induces collagen type Ialpha1 (COL1A1) DNA methylation and altered COL1A1 messenger RNA expression in sclera. *Molecular Vision, 18*, 1312–1324.

Zhou, F. C., Zhao, Q., Liu, Y., Goodlett, C. R., Liang, T., Mcclintick, J. N., Edenberg, H. J., & Li, L. (2011c). Alteration of gene expression by alcohol exposure at early neurulation. *BMC Genomics, 12*, 124.

Chapter 6
Nutrition in Early Life, Epigenetics, and Health

Xinyin Jiang

6.1 The Developmental Origins of Health and Disease: History and Epidemiological Evidence

In the 1980s, Barker and colleagues published a series of reports which illustrate the associations between low birth weight and increased risk of type 2 diabetes, cardiovascular disease, and stroke in a British cohort (Barker and Osmond 1988; Barker et al. 1989; Hales and Barker 1992), which gave rise to the hypothesis of Developmental Origins of Health and Disease. According to this hypothesis, which is also referred to as developmental programming, an organism has tremendous developmental plasticity in the intrauterine period, during which limited access to nutrients triggers adaption of body which assumes the same environment will be maintained after birth. However, if the postnatal environment is full of resources (e.g., unlimited calorie access) which the individual is not tuned for, he/she may amass an excessive amount of nutrients which increases the risk of metabolic diseases (Bateson et al. 2004). Later studies corroborate the findings by Barker and colleagues in various cohorts from Scandinavia (Naess et al. 2013; Stuart et al. 2013), the United States (Rich-Edwards et al. 1997), Denmark (Baker et al. 2008), etc.

Nutrition is a critical determinant of birth weight. Large cohort studies which directly assess the influence of maternal nutrition status on short- and long-term development of children are made available by investigating the lasting impacts of several large famines in recent human history. Famines lead to substantial decrease in total caloric intake and global reduction in the consumption of most macronutrients and micronutrients (Roseboom et al. 2006). The Dutch Hunger Winter birth

X. Jiang, Ph.D., R.D. (✉)
Department of Health and Nutrition Sciences, Brooklyn College, The City University of New York, 2900 Bedford Ave., Brooklyn, NY 11210, USA
e-mail: XinyinJiang@brooklyn.cuny.edu

D. Hollar (ed.), *Epigenetics, the Environment, and Children's Health Across Lifespans*, DOI 10.1007/978-3-319-25325-1_6

cohort study is one of the most classic examples of this kind. From November 1944 to April 1945, due to military blockage during World War II, access to food suddenly became very limited and the average caloric intakes plummeted in the Netherlands, leading to a severe famine during that winter (Roseboom et al. 2006). The blockage ended in the spring of 1945, after which nutrition intake resumed to normal swiftly, creating a well-defined, short period of nutrition restriction which enables the study of maternal nutrition restriction in each trimester of pregnancy (Roseboom et al. 2006). Data from this cohort suggests that maternal undernutrition in early gestation results in greater risk of having an atherogenic lipid profile, female obesity, and coronary heart disease, whereas nutrition restriction in mid gestation increases the risk of microalbuminuria, and restriction in late gestation seems to affect birth weight the most severely, resulting in an average birth weight about 200 g lighter than unexposed babies (Roseboom et al. 2006, 2011). The Dutch famine studies suggest that prenatal undernutrition is associated with long-term adverse outcomes of body weight management and metabolic diseases and that the influence is specific to different stages of pregnancy.

Others have also studied famines such as the Chinese famine (Li et al. 2011a; Zheng et al. 2012) from 1957 to 1963 and the Biafran famine (Hult et al. 2010) in Nigeria from 1967 to 1970. These famines lasted for longer periods of time and had more gradual recovery of food supplies compared to the Dutch famine. The longer exposure to famine makes it difficult to discern the effect of prenatal maternal undernutrition and early postnatal nutrition restriction. Nevertheless, both studies corroborate findings from the Dutch Hunger Winter cohort, evidencing that exposure to famine/general nutrition restriction early in life increases the risk of hypertension, diabetes, obesity, and unfavorable lipid profiles.

6.2 Underlying Mechanisms

Nutrition is an indispensable component for the survival and proper functioning of an organism. The remarkable plasticity in early life provides an unparalleled opportunity for dietary components to exert lasting impacts on children's health. The original hypothesis of Developmental Origins of Health and Disease by Barker et al. (2008) narrowly defined the intrauterine period as the key stage of developmental programming, yet later research has gradually revealed that the window of sensitivity can be expanded from preconception to early postnatal life when nutrition interventions demonstrate permanent influence on later health outcomes (Patel and Srinivasan 2011) (Fig. 6.1). Many mechanisms, such as changes in cell numbers in metabolic organs, activation of the neuroendocrine system, and oxidative stress have been proposed in mediating the developmental origins of health and disease, yet the most widely accepted molecular mechanism involves epigenetic modifications. Epigenetics describes the study of heritable traits of gene expression secondary to modifications on the chromatins that are not related to changes in DNA sequence. Common epigenetic modifications include DNA methylation and histone

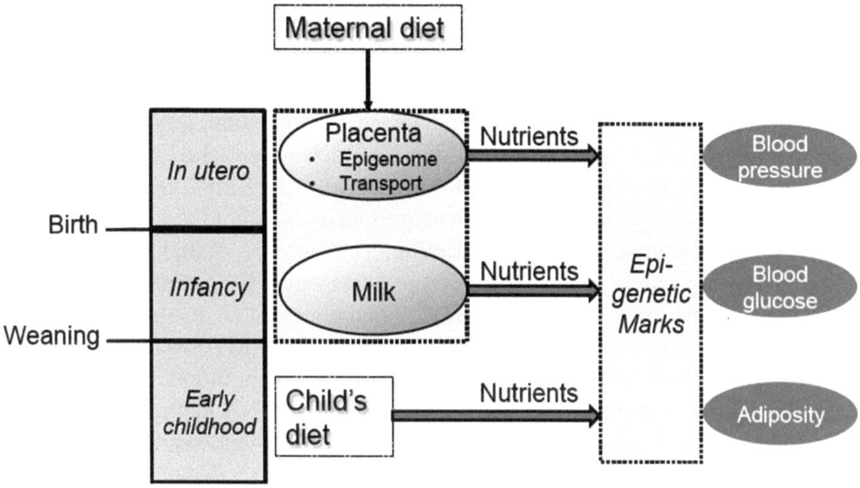

Fig. 6.1 Nutrition exposures in utero and in early childhood modify various epigenetic marks, leading to lasting alterations in metabolic functions

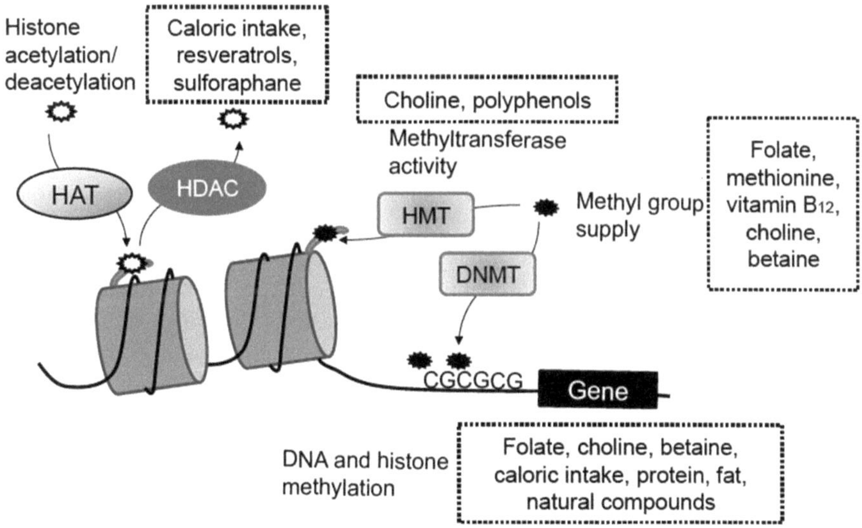

Fig. 6.2 Interaction between nutrients and epigenetic modifications. *DNMT* DNA methyltransferase, *HAT* histone acetyltransferase, *HDAC* histone deacetylase, *HMT* histone methyltransferase

modifications. Additionally, noncoding RNAs may also be loosely defined as a source of epigenetic influence. This section of the chapter will provide an overview of the mechanisms and evidence about nutrition-mediated epigenetic changes during early development with a focus on DNA and histone methylation (Fig. 6.2).

6.2.1 DNA Methylation

DNA cytosine methylation is the most extensively studied epigenetic mark, which usually occurs at the C-5 carbon of the cytosine residue within a cytosine-guanine dinucleotide (CpG). Consecutive CpG dinucleotides are enriched in the promoter regions of genes, forming CpG islands (Illingworth and Bird 2009). Most of the CpG islands are hypomethylated, but increased methylation of these CpG islands is often associated with repressed transcription. The cytosine methylation marks are laid down by a group of highly conserved proteins called DNA methyltransferases (DNMTs). Among these enzymes, DNMT1 maintains DNA methylation pattern during cellular replication across the life span, whereas DNMT-3A, DNMT-3B, and their cofactors DNMT3-like mediate de novo establishment of methylation pattern in early development (Ooi et al. 2009). The pattern of DNA methylation changes dramatically between fertilization and implantation as demonstrated by global active demethylation of the paternal genome and passive demethylation of the maternal genome after fertilization, followed by remethylation of both parental genomes during embryogenesis (Hochberg et al. 2011; Jirtle and Skinner 2007; Mayer et al. 2000). Tissue-specific cytosine methylation is further established during later stages of prenatal and early postnatal development.

The dynamic changes in DNA methylation during early development provide the fetus with high levels of epigenetic plasticity in response to outside nutritional signals which may revise the DNA methylation pattern to one that favors the survival of the fetus (Susser et al. 2012). Many dietary components/nutrients have been demonstrated to affect DNA methylation in early development and pathogenic processes. These nutrients interact with DNA methylation by increasing the availability of substrate methyl groups and activating methyltransferases or by other indirect mechanisms. In the body, the methyl groups added to the cytosine come from methyl donors such as folate, choline, betaine, and methionine (Bertolo and McBreairty 2013). A methyl-devoid diet in ewes alters the methylation status of 72 CpG loci or 4 % of the1400 CpG islands tested (Sinclair et al. 2007) and increases obesity risk of offspring, whereas differing maternal methyl nutrient consumption changes the fur coat color of pups by affecting the CpG methylation of a transposon in mice (Waterland and Jirtle 2003). Section 6.4.4 will further specify the effects of these methyl nutrients on epigenetic modifications. Dietary flavonoids, such as polyphenols found in tea, genistein from soy, and anthocyanin from berries, may inhibit DNMTs, thereby preventing CpG hypermethylation that occurs during the pathogenic process of cancer (Gilbert and Liu 2010; Li and Tollefsbol 2010). Other nutrition manipulations, such as altering total caloric intake or protein content in maternal diet, affect CpG methylation of regulatory genes in nutrient metabolism and somatic growth pathways, such as insulin growth-like factor 2 (*IGF2*). A detailed discussion on the influence of global nutrient availability and dietary macronutrient distribution on epigenetics is included in Sects. 6.4.1–6.4.3.

6.2.2 *Histone Modifications*

Histones are proteins that the DNA wraps around within the cell nucleus. Histones and DNA form nucleosomes, which are the building blocks of chromatin. Histone modifications are more complicated than DNA methylation in that the modifications can occur in multiple lysine, arginine, or other residues of the histone proteins. Moreover, there are a variety of modifications occurring on those residues, such as phosphorylation, acetylation, methylation, ADP-ribosylation, and biotinylation (Choi and Friso 2010). Modifications on the histones affect the structure of chromatin. Histone acetylation is usually associated with an active chromatin structure, whereas histone methylation marks can be either transcriptionally permissive or repressive, depending on the residues that are modified. Histone modifications are critical for early development. After global DNA demethylation, the de novo methyltransferases recognize histone methylation state of different genomic regions which pinpoints them to the locations where DNA methylation marks need to be laid down (Ooi et al. 2007). Embryonic stem cells maintain a bivalent state where both histone activation marks (e.g., trimethylation at histone H3 lysine 4, H3K4me3) and repression marks (e.g., trimethylation at histone H3 lysine 27, H3K27me3) are present in a lot of gene promoters, enabling fast derepression and induction of gene expression as the stem cells differentiate into different lineages (Voigt et al. 2013).

Several nutrients have been related to histone modifications during prenatal development. Methyl nutrients provide substrates for histone methylation and affect activity of histone methyltransferases. For example, feeding rats with a choline-supplemented diet during gestation increases hippocampal histone H3 lysine 9 (H3K9) methylation, whereas a choline-deficient diet increases the expression of the histone methyltransferase G9a (Davison et al. 2009).

A group of proteins called sirtuins function as histone deacetylases (Schemies et al. 2010) and endogenous regulators of histone acetylase P300 activity (Yang et al. 2006). The expression of sirtuins is influenced by total caloric intake and natural compounds such as resveratrol from grape skins, sulforaphane from cruciferous vegetables, and curcumin from turmeric roots (*Curcuma longa*) (Delage and Dashwood 2008). Maternal calorie restriction seems to decrease sirtuin 1 expression of male rat offspring in white adipose tissue, liver, and skeletal muscle (Palou et al. 2013), which may in turn alter histone acetylation. Table 6.1 provides an overview of select studies in which early dietary exposures affect epigenetic markers and functional endpoints of offspring.

6.3 Susceptible Pathways

As is described above, nutrition and epigenetics interact to influence biological pathways and physiological functions. However, biological pathways are not equally sensitive to environmental exposures. In fact, only a small proportion of

Table 6.1 Studies that examined the effects of early dietary exposures on epigenetic modifications and functional outcomes

Study	Dietary exposures	Timing of dietary exposures	Model	Epigenetic-related changes in offspring	Functional or gene expression changes in offspring
Palou et al. (2013)	Caloric restriction	Day 1–day 12 of gestation	Rat	↓ mRNA expression of SIRT1 (histone deacetylase) in white adipose tissue, muscle, and liver	
Zhang et al. (2010)	Caloric restriction	One month before until 6 days after conception	Sheep	↓ *IGF2/H19* methylation in lamb adrenal glands	↓ mRNA expression of *IGF2*; ↑ cortisol in female lambs
Begum et al. (2013)	Caloric restriction	Sixty days before until 30 days of gestation	Sheep	↓ CpG methylation, ↑H3K9 acetylation, and ↓ H3K27me3 of GR in ventral hypothalamus	↑ Hypothalamic GR, ↓ POMC mRNA expression
Heijmans et al. (2008), Tobi et al. (2014)	Famine	Periconceptional period	Human	↓ Blood CpG methylation of IGF2; ↑ blood CpG methylation of *INSR* and *CPT1A*	Positive associations between *INSR* methylation and birth weight and between *CPT1A* methylation and LDL levels
Vucetic et al. (2010)	High-fat diet	From 3 months before gestation to weaning	Mouse	↓ Global and μ-opioid receptor CpG methylation	↑ Body weight at weaning; altered brain dopamine-related gene expression
Gong et al. (2010)	Low protein or low protein with folate	Throughout gestation	Rat	↑ Hepatic CpG methylation of *Igf2* of low-protein male pups; normalized in the low protein-folate group	↑ *Igf2* and *H19* mRNA, ↑ *Dnmt1* expression in liver of low-protein male pups; normalized in the low protein-folate group
Lillycrop et al. (2005)	Low protein or low protein with folate	Throughout gestation	Rat	↓ Hepatic CpG methylation of *PPARA* and GR in low protein pups; normalized in the low protein-folate group	↑ *PPARA* and GR mRNA in liver of low protein pups; normalized in low protein-folate group
Chen et al. (2009)	Protein restriction	Gestation or gestation + lactation	Mouse	↓ Muscle Sirt1 of gestational low-protein group	↓ Body weight, ↓ fasting glucose and insulin of (gestation + lactation) protein restriction

Drake et al. (2012), Herrick et al. (2003)	High meat and fish consumption	Late pregnancy	Human	↑ Glucocorticoid receptor exon 1F CpG methylation	↑ Plasma cortisol, ↑ blood pressure
Sinclair et al. (2007)	Folate, vitamin B12 and methionine restriction	Periconceptional period	Sheep	Methylation of 71 CpG loci was altered	↑ Growth rates in the methyl deficient group; ↑ fat mass, ↑ blood pressure, ↓ glucose tolerance of methyl deficient males
Waterland and Jirtle (2003)	Folate, vitamin B12, choline, and betaine supplementation	From 2 weeks before pregnancy to weaning	Mouse	↑ A^{vy} CpG methylation	Darker fur coat color
Davison et al. (2009)	Choline supplement, control, or deficient diets	Embryonic days E11–E17	Rat	↑ H3K9me2 and H3K27me3 in choline-supplemented rats; ↑ H3K4me2 in choline-deficient rats; ↓ gene expression of histone methyltransferase G9a in choline-deficient rats	
Mehedint et al. (2010a)	Choline deficiency	Embryonic days E11–E17	Mouse	↓ Hippocampal CpG methylation of vascular endothelial growth factor and angiopoietin 2	↑ Angiogenic gene expression in fetal brain; ↓ hippocampal endothelial cell proliferation; ↓ numbers of blood vessels
Jiang et al. (2012)	Choline supplementation	Third trimester of pregnancy	Human	↑ Placental CpG methylation of *CRH* and GR; ↑ placental *DNMT1* and histone methyltransferase expression	↓ Placental *CRH* mRNA expression; ↓ cord blood cortisol
Steegers-Theunissen et al. (2009)	Folic acid supplementation	Periconceptional period	Human	↑ CpG methylation of *IGF2*	
Hoyo et al. (2011)	Folic acid intake	Periconceptional period	Human	↓ CpG methylation of *IGF2* (CTCF-binding region)	

Abbreviations: *CPT1A* carnitine palmitoyltransferase-1A, *CRH* corticotropin-releasing hormone, *CTCF* CCCTC-binding factor, *GR* glucocorticoid receptor, *IGF2* insulin growth-like factor 2, *INSR* insulin receptor, *LDL* low-density lipoprotein, *POMC* Pro-opiomelanocortin genes undergo differential expression after in utero exposure to dietary treatments such as protein restriction (Burdge and Lillycrop 2010). Consistent with the nurturing effect of dietary nutrients, biological pathways that are susceptible to early nutrition exposures are often the ones related to cellular growth and development and, accordingly, the accretion and metabolism of macronutrients. Alterations in these pathways increase the risk of diseases associated with disturbance of energy homeostasis, such as obesity, type 2 diabetes, and cardiovascular disease. This section highlights several biological pathways that are targeted by early nutrition exposures.

6.3.1 The Imprinted Gene IGF2 and Somatic Growth

Imprinted genes are a group of genes expressed in only one parental allele. Their allelic expression is often controlled by imprinting control regions (ICRs) which are enriched with CpGs. The paternal and maternal alleles are differentially methylated in the ICRs, which determines the monoallelic expression pattern of these imprinted genes (Morison et al. 2001). Many imprinted genes are critical regulators of fetal growth and are considered as major targets of developmental programming of somatic size (Lui et al. 2008). The imprinted patterns of these genes at the ICRs are tightly regulated and escape the wave of DNA demethylation after fertilization (Brandeis et al. 1993). However, multiple nutrients have been demonstrated to modify CpG methylation of the ICRs and expression of certain imprinted genes, thereby affecting fetal and neonatal growth. Here we will use the imprinted gene insulin growth-like factor 2 (IGF2) as an example to illustrate this effect.

IGF2 is one of the well-characterized imprinted genes that play a significant role in prenatal somatic growth through a balance of cell proliferation and apoptosis (Huang et al. 2012). The correlation between IGF2 concentrations and body weight has been observed across species, with prenatal IGF2 expression positively linked to larger fetuses (DeChiara et al. 1990) and postnatal IGF2 levels negatively associated with fat mass and obesity (Jones et al. 2001). The monoallelic expression of IGF2 depends on the methylation of its ICRs. One of the ICRs of *IGF2* contains binding sites for the insulator protein CCTCC-binding factor (CTCF), which blocks *IGF2* transcription of the unmethylated maternal allele. The paternal allele of *IGF2* is methylated, which prevents CTCF from binding and enables the expression of *IGF2* (Huang et al. 2012).

IGF2 expression is affected by various nutrition exposures in early life. Gestational caloric restriction increases *IGF2* mRNA expression in the liver and skeletal muscle and alters mRNA (Brameld et al. 2000) and CpG methylation of *IGF2* in the adrenal gland of lambs (Williams-Wyss et al. 2014; Zhang et al. 2010). Similarly, in utero restriction of a single macronutrient, protein, alters *Igf2* methylation in the liver of rat pups. Interestingly, the effect of protein restriction on *Igf2* can be reversed by supplementing the methyl donor folic acid, indicating the involvement of DNA methylation (Gong et al. 2010). The influence of energy consumption and nutrient availability on IGF2 is also evidenced in humans. For example, exposure to the Dutch famine in early gestation is associated with lower *IGF2* methylation after six decades (Heijmans et al. 2008; Tobi et al. 2012), which may mediate in part the higher risk of metabolic diseases of these individuals. The impact of folic acid on IGF2 garners tremendous attention due to the debate over the safety of its mandatory fortification in foods.

Periconceptional folic acid supplementation is associated with 4.5 % higher methylation of a differentially methylated region (DMR) of *IGF2* among 17-month-old toddlers (Steegers-Theunissen et al. 2009). However, another study suggests that such supplementation can also lead to lower level of *IGF2* methylation at a different DMR (Hoyo et al. 2011) which is associated with overweight babies at 1 year of age (Perkins et al. 2012). The association between *IGF2* methylation and adiposity may persist into later life, as Huang et al. have demonstrated that *IGF2* methylation is positively associated with subcutaneous fat mass at the age of 17 and negatively associated with head circumference at birth (Huang et al. 2012). In sum, the programming of IGF2 by early nutrition is complicated, and the phenotypic effect of its methylation changes remains to be further discerned.

6.3.2 The Hypothalamic-Pituitary-Adrenal (HPA) Axis

The HPA axis mediates stress response and macronutrient mobilization. In this axis, the hypothalamus produces corticotropin-releasing hormone (CRH) in response to various stimuli, which increases the release of adrenocorticotropic hormone (ACTH) from the pituitary gland. ACTH increases the production and release of cortisol from the adrenal cortex (Smith and Vale 2006). Cortisol is a major stress hormone, which is received by glucocorticoid receptors in various tissues, such as the liver, kidney, placenta, and brain, and elicits a series of physiological changes such as elevated blood glucose, lipolysis, and protein breakdown. This axis can be programmed by early stress related to antenatal glucocorticoid treatment (Levitt et al. 1996) and suboptimal maternal care during lactation (Korosi and Baram 2010). Studies by Meaney and colleagues have shown in rats that maternal care during lactation (i.e., licking and arched back nursing behavior) reduces mRNA expression of *CRH* and circulating ACTH and increases expression of the glucocorticoid receptor, which leads to reduced corticosterone release during acute stress and improved sensitivity of stress response (Liu et al. 1997). Further study suggests that such changes are due to lower CpG methylation and altered histone acetylation of the glucocorticoid receptor promoter (Meaney and Szyf 2005). Similar difference in glucocorticoid receptor methylation is observed among victims of childhood abuse in humans (Steiger et al. 2013). These findings confirm that epigenetic alterations are involved in the programming of the HPA axis.

The influence of maternal nutrition on the HPA axis components has been tested by multiple animal studies. Both maternal caloric restriction and protein deficiency during gestation increase glucocorticoid receptor expression by decreasing its CpG methylation in either the hypothalamus or liver of offspring (Begum et al. 2013; Lillycrop et al. 2005). Human evidence is still scarce. However, a recent human study reports that a higher intake of choline during the third trimester of pregnancy increases placental *CRH* and *NR3C1* (the gene that encodes the glucocorticoid receptor) methylation (Jiang et al. 2012). The long-term consequences of HPA alteration at birth are still unclear, although a chronically active HPA axis and elevated cortisol levels have been associated with increased obesity, hypertension, diabetes, and mental problems (Levitt et al. 1996; O'Donnell et al. 2009; Smith and Vale 2006). Understanding the interaction between nutrient availability and stress programming may offer a simple way of nutrition manipulation to neutralize the negative influence of antenatal stress (Lucassen et al. 2013).

6.3.3 Peroxisome Proliferator-Activated Receptors

The peroxisome proliferator-activated receptors (PPARs) are key regulators of macronutrient metabolism. There are three isoforms of PPARs, PPARα, PPARβ/δ, and PPARγ, which are all nuclear receptors. PPARs form heterodimers with retinoid X receptor (RXR), which then bind to the PPAR response elements in the genome and activate the transcription of a variety of genes (Grygiel-Górniak 2014). The PPARs regulate several genes involved in lipid catabolism and gluconeogenesis. The activity of PPARs relies on nutrients such as the omega-3 fatty acids, which serve as PPAR ligands. Moreover, the activity of the PPAR signaling pathway seems to be affected by prenatal and early postnatal exposures to macronutrients and energy availability immensely. Receiving excess energy, especially from a maternal high-fat diet, decreases PPARα expression of fetal liver in part due to altered promoter CpG and histone methylation patterns. Such changes are associated with decreased lipid catabolism, increased liver fat accumulation, and later obesity (Magliano et al. 2013). More discussion on the programming of PPARs by early nutrition can be found in the next section.

6.4 Effects of Different Nutrients during Early Life on Health and Development

This section will introduce the influence of different nutrition manipulations in early life on health and development (Table 6.1).

6.4.1 Caloric Restriction

As is mentioned in Sect. 6.1, studies of famine-affected cohorts consistently demonstrate that individuals exposed to famine in utero display abnormal metabolic panels, such as elevated blood glucose and high waist circumference in adulthood (Roseboom et al. 2006). Moreover, exposure to famine as an infant or in early childhood is also associated with higher blood pressure, obesity, and greater risk of metabolic syndrome as an adult (Hult et al. 2010; Li et al. 2011a; Zheng et al. 2012).

Several animal models have been established to delineate the effect of in utero nutrition restriction on fetal growth and development. Sheep have a relatively long pregnancy (~140 days) with only 1–2 fetuses, making it a good proxy of human pregnancy to study the intrauterine influence on fetal health. Surgical procedures can be done to reduce placental blood flow and nutrient supply to the fetus, generating prenatal malnutrition secondary to placental insufficiency. Using this model, studies show that prenatal undernutrition leads to lower birth weight, increased lipogenic and adipogenic activity of adipocytes, lower leptin levels, and altered glucocorticoid signaling, which all result in higher incidence of obesity in later life (Sarr et al. 2012). In rodents, undernutrition in early life is achieved by restricting maternal food intake to 30–50 % of normal *ad libitum* intake during gestation or lactation. Similar to sheep, early undernutrition results in smaller infants, β-cell dysfunction (Fernandez-Twinn and Ozanne 2006), and reduced cell numbers in various tissues (Winick and Noble 1966). Long-lasting impacts include elevated blood pressure and hyperinsulinemia, which also predisposes the animals to diabetes and cardiovascular disease (Carmody et al. 2011; Woodall et al. 1996). Taken together, the adverse effect of global nutrition restriction in early life on adiposity, blood pressure, and glucose hemostasis seems to be universal in both animals and humans.

The involvement of epigenetics in mediating the lasting metabolic changes secondary to early caloric restriction is gradually being unveiled. As is mentioned in Sect. 6.3, pathways related to growth and macronutrient metabolism are especially sensitive to caloric restriction-induced epigenetic modifications. For example, food restriction of pregnant mice or ewes alters promoter CpG methylation of glucocorticoid receptor regulating genes, which is positively associated with obesity of pups (Begum et al. 2013; Ogawa et al. 2014). Observations in the famine studies strongly support that these epigenetic changes are long lasting and partly explain the increased risk of metabolic syndrome of affected individuals. As is mentioned previously, individuals exposed to the Dutch famine prenatally have less CpG methylation of the imprinted *IGF2* compared with their unexposed siblings six decades after the famine was over (Heijmans et al. 2008). Moreover, in the same cohort, periconceptional exposure to famine was associated with lower CpG methylation of the imprinted gene *INSIGF* and higher methylation of *IL10, LEP, ABCA1, GNASAS, and MEG3*, which are genes encoding proteins related to immune function, fat metabolism, body weight control, and somatic growth (Talens et al. 2010). The most recent study from this cohort examined the genome-wide CpG methylation signatures and showed that *INSR*, which mediates early growth and insulin signaling, and *CPT1A*, which mediates fatty acid oxidation, are both hypermethylated. Hypermethylation of these two genes is associated with higher birth weight or LDL cholesterol levels, respectively (Tobi et al. 2014).

6.4.1.1 Catch-Up Growth

Interestingly, global undernutrition results in lower weight and length than unexposed pups or children at birth, which is opposite to their higher body weight and adiposity than unexposed peers observed in later life, suggesting that catch-up growth occurs at some point during their postnatal development. Indeed, compensatory or catch-up growth occurs in most prenatally restricted individuals from 0 to 2 years of age in humans (Brenseke et al. 2013). However, such compensation is often exacerbated, which leads to increased adiposity (Ong et al. 2000). Postnatal overcompensation is more pronounced when the individual is placed in a nutrition excess environment, such as administration of a high-fat diet in animals or a nutrient-enriched diet in humans, resulting in greater weight gain, insulin resistance, hyperglycemia, and other metabolic abnormalities (Brenseke et al. 2013). In this regard, "mismatch" between prenatal and postnatal nutrition access renders an individual additionally susceptible to obesity in postnatal life, which implies that using extra dietary intake to spur catch-up growth of prenatally restricted individuals may not be beneficial, whereas adequate but not excess nutrition may help them slowly grow into the normal trajectory of development.

6.4.2 Overnutrition

Although maternal caloric restriction is associated with increased risk of metabolic disorders for offspring, maternal overnutrition, induced by excess macronutrient or total calorie intake, also exerts adverse influence on the metabolic health of offspring. The concern of maternal overnutrition is increasing due to the epidemic of obesity. Multiple epidemiological studies have associated maternal BMI with their children's, which indicates that maternal obesity increases the risk of childhood obesity (Vohr et al. 1999) and insulin resistance which may persist into adolescence and adulthood (Gillman et al. 2003). Maternal overnutrition also occurs in gestational diabetes mellitus (GDM) complicated pregnancies, where excess glucose and lipid are transported to the fetuses. GDM is associated with fetal overgrowth or macrosomia, and these babies are more likely to develop type 2 diabetes and obesity in the future (Stuart et al. 2013).

In animals, overnutrition can be generated by *ad libitum* high-fat or high-carbohydrate feeding or a cafeteria diet, which contains processed meats, cheese, cookies, cereals, and crackers (Sampey et al. 2011). Studies using rodents, sheep, and nonhuman primates show similar findings of increased risk of macrosomia, hyperglycemia, hyperinsulinemia, and hepatic fat accumulation when exposed to prenatal overnutrition (Li et al. 2011b). Overnutrition during lactation can also program later body weight and food intake of pups (Patel and Srinivasan 2011). When nursed by a diabetic mom, rat pups have increased risk of overeating and obesity, suggesting that the influence of overnutrition extends to early postnatal life.

An underlying mechanism by which early overnutrition alters later metabolic function is related to the surge of the circulating levels of leptin and insulin that regulate energy balance. Moreover, the levels of leptin and insulin are partially regulated by the hypothalamic neurons in the arcuate nucleus and paraventricular nucleus, or the appetite control center. In rodents, both perinatal and early postnatal overnutrition increase the hypothalamic expression of the appetite stimulator neuropeptide Y and the appetite suppressor proopiomelanocortin (POMC) at birth (Morris and Chen 2009). In addition, feeding mice a high-fat diet during lactation impairs the formation of POMC and agouti-related peptide (AgRP) projections to hypothalamic target sites, which may largely affect future appetite control in response to food restriction or excess (Vogt et al. 2014). Epigenetic modifications seem to play a role in the changes of the hypothalamic neurons, as *Pomc* promoter CpG methylation is altered by postweaning high-fat feeding (Marco et al. 2013). Another gene that encodes the μ-opioid receptor (MOR) in the reward-related region of the brain is also hypomethylated in dams fed a high-fat diet during gestation, which may contribute to overeating in later life (Vucetic et al. 2010).

Hepatic responses to a high-fat or cafeteria diet-induced early overnutrition are phenomenal, which favor lipid accumulation and dampen lipid catabolism. In a nonhuman primate model, fetuses from a high-fat dam exhibit signs of fatty liver such as elevated lipogenic activity and accumulation of triglycerides, along with increased expression of gluconeogenic enzymes such as the rate-limiting enzyme phosphoenolpyruvate carboxykinase (*PCK1*) (McCurdy et al. 2009). High-fat feeding in mice during gestation and lactation also activates the expression of lipogenic genes with concomitant decrease in PPAR-α and the lipolytic protein carnitine palmitoyltransferase-1 which PPAR-α regulates (Magliano et al. 2013). Since PPAR-α is a susceptible target of epigenetic regulation, the lasting changes in liver triggered by maternal overnutrition are likely to be mediated by an epigenetic mechanism.

6.4.3 Protein Restriction

Protein restriction in pregnant and lactating animals is a commonly used model to examine the effect of malnutrition in early life on growth and development of pups. Gestational protein restriction results in low birth weight. Restricting protein from gestation to weaning results in better insulin sensitivity in young pups but elevated blood glucose and reduced insulin sensitivity in older pups (Mortensen et al. 2010). Similar to observations from global caloric restriction, recuperation of protein following maternal protein restriction leads to catch-up growth, which exacerbates insulin resistance and decreases expression of Sirt1, a histone deacetylase and determinant of longevity (Chen et al. 2009). Protein restriction in early childhood (weaning—55 days) seems to result in even more pronounced blood glucose elevation and insulin reduction in rats than early time points, whereas protein restriction in adulthood does not affect glucose tolerance, suggesting that susceptibility to protein restriction is limited to the perinatal period and early childhood (Miñana-Solis and Escobar 2008).

Lillycrop and Burdge led a research team which conducted a series of rat studies that examined the physiological and molecular alterations in offspring experiencing maternal protein restriction (Review in Burdge and Lillycrop 2010). Their studies consistently exhibit increased expression of hepatic PPARα and glucocorticoid receptor, which also leads to elevations of the downstream proteins that they regulate, such as the lipid catabolic genes acyl-CoA oxidase and carnitine palmitoyltransferase-1 and the gluconeogenic enzyme PCK1. These gene expression changes are consistent with the higher blood glucose and altered glucose tolerance of pups from the protein-restricted dams (Burdge and Lillycrop 2010). Both glucocorticoid receptor and PPARα promoter CpGs are hypomethylated in the protein-deficient dams, confirming the involvement of an epigenetic mechanism (Lillycrop et al. 2005). In addition, others have also found that maternal protein restriction decreases leptin expression by altering CpG methylation of the leptin promoter (Jousse et al. 2011), which disturbs the control of food intake and adiposity (Altobelli et al. 2013). These findings demonstrate that epigenetic modifications are important mediators of the lasting changes in energy metabolism induced by maternal protein restriction.

In humans, protein restriction during pregnancy and early childhood is seen in the famine studies, the result of which is summarized in the previous sections. Protein supplementation has generated mixed results. When supplementation to pregnant women deficient in protein improves birth weight and reduces the proportion of low-birth-weight babies, additional protein to pregnant women who are protein sufficient inversely affects fetal growth (Liberato et al. 2013). Studies on the long-term influence of protein supplementation are scarce. A study on a presumably protein energy-deficient population in Guatemala shows that receiving protein supplementation in early childhood improved height and weight of the next generation, and such improvement was more pronounced when protein was given to the women when they were at 3–7 years old compared to earlier ages (Stein et al. 2003). Another study shows that children of women who were advised to eat high amounts of red meat and low carbohydrate during pregnancy had higher CpG methylation of the glucocorticoid receptor, which was associated with higher BMI and waist circumference but lower blood pressure at 40 years of age (Drake et al. 2012; Herrick et al. 2003; Shiell et al. 2001). These results seem to corroborate animal studies suggesting that the susceptibility of glucocorticoid receptor to protein intake-mediated epigenetic changes also applies to humans.

6.4.4 Methyl Nutrients

The methyl nutrients are nutrients involved in one-carbon metabolism, which provides methyl groups for various methylation reactions in the body, including DNA and histone methylation. Metabolism of these nutrients in the one-carbon cycle and their provision of methyl groups are illustrated in Fig. 6.3.

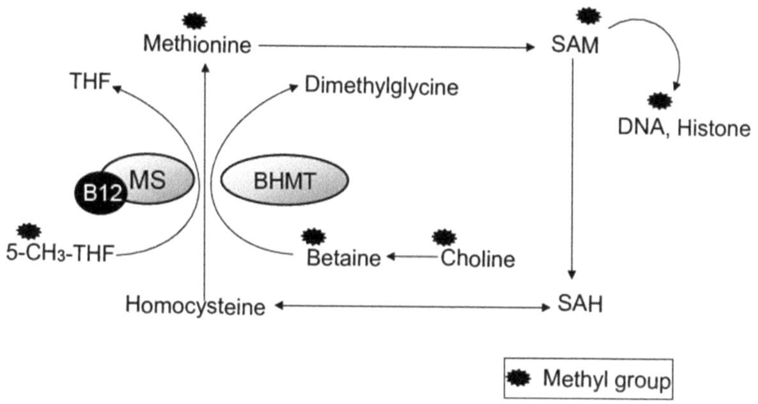

Fig. 6.3 One-carbon metabolism. The universal methyl donor *S*-adenosylmethionine (SAM) is the direct provider of methyl groups for DNA and histone methylation. After the methyl group is transferred to DNA or histones, SAM is converted to *S*-adenosylhomocysteine (SAH). SAH is broken down to homocysteine, which needs to be recycled into methionine to replenish the methyl group supply. There are two complementary pathways that provide methyl groups for homocysteine remethylation, one mediated by folate and the other mediated by choline. Methyltetrahydrofolate (5-CH_3-THF) transfers its methyl group to homocysteine via methionine synthase (MS), with vitamin B_{12} as a cofactor. Choline is oxidized into betaine, which transfers its methyl group to homocysteine by betaine homocysteine methyltransferase (BHMT). While folate-mediated homocysteine remethylation occurs in many tissues, BHMT is mainly expressed in the liver

The most far-reaching evidence showing effects of methyl nutrients on children's health, which leads to massive public policy change, is the observed link between maternal antenatal folate deficiency and the risk of neural tube defects (NTDs). Starting from the 1980s, folate deficiency during early pregnancy has been related to increased risk of NTDs. Later, several large randomized, controlled trials confirmed the effectiveness of maternal folate supplementation in reducing NTDs, which eventually led to the mandated fortification of folic acid to cereal grains in North America in the mid-1990s (Institute of Medicine 1998). Some more recent studies suggest that other methyl nutrients, such as vitamin B_{12} and choline, may have independent effects on reducing NTDs in populations that are folate sufficient (Shaw et al. 2009; Thompson et al. 2009), although null results have also been reported (Mills et al. 2014).

The methyl nutrients are extensively studied for their effects on programming children's health, because they provide substrates for DNA and histone methylation, which may then affect epigenetic modifications of important metabolic pathways. One of the earliest studies that demonstrate the impact of maternal methyl nutrient intake on offspring phenotype through epigenetic programming was conducted in the agouti viable yellow (A^{vy}) mouse model (Waterland and Jirtle 2003). In wild-type mice, the agouti protein is only expressed in the skin. However, in this model, a transposon intracisternal *A* particle (IAP) is stably inserted into the 5′ upstream region of the Agouti gene, leading to constitutive expression of the agouti protein in all tissues, which causes yellow pigmentation of the fur coat. However, the expression of agouti is suppressed by hypermethylation of IAP. As such, A^{vy} is called a metastatic epiallele. Supplementing mice with methyl nutrients during gestation leads to higher methylation of IAP, resulting in more pups with brown fur coat color which resembles wild-type mice without ectopic agouti expression. Later studies of maternal methyl nutrient supplementation also show lower risk of obesity in A^{vy} mice (Waterland et al. 2008) and repression of the kinked tail phenotype occurring in a similar epiallele Axin(Fu) (Waterland et al. 2006). Emerging evidence suggests that the influence of maternal methyl nutrient intake is not only limited to metastatic epialleles but also alleles sensitive to developmental programming such as the imprinted gene *IGF2* (as is specified in Sect. 6.3.1) and metabolism regulators PPARα and glucocorticoid receptor in both human and animal models.

Choline, the less studied nutrient compared to its counterpart folate, garners a growing amount of attention due to emerging studies that demonstrate its potent influence in early life in both animals and humans. In animals, there is consistent evidence that choline supplementation during certain periods of gestation or lactation improves spatial memory and prevents cognitive decline (review in McCann et al. 2006), although in humans, results are less clear (Boeke et al. 2013; Ross et al. 2013). Choline appears to be a potent regulator of epigenetic modifications in early life. Animal studies of maternal supplementation or deficiency of choline demonstrate changes in the CpG methylation of *Igf2* (Kovacheva et al. 2007), as well as alterations in global and site-specific histone marks H3K9 and H3K4 methylation on genes such as Calbindin in the fetal liver or brain (Mehedint et al. 2010b). Moreover, these epigenetic changes may have resulted in functional alterations in fetal brain angiogenesis (Mehedint et al. 2010a) or delayed progression of chemically induced mammary tumors in later life (Kovacheva et al. 2009). In humans, a randomized controlled study which provided third-trimester pregnant women with either 930 mg/day or 480 mg/day choline for 12 weeks suggests that high maternal choline intake may increase the CpG methylation of *CRH* and glucocorticoid receptor of placental tissue and, accordingly, decrease cortisol levels in fetal cord blood. The authors concluded that improving maternal choline intake might help alleviate stress of the child and lower the risks of stress-related diseases (Jiang et al. 2012).

It should be noted that the influence of methyl nutrient supplementation or restriction in early life is not unidirectional. While it is intuitive to think that depriving maternal methyl nutrients decreases methyl group supply and methylation of genes, choline deficiency during gestation unexpectedly increases CpG methylation of *Igf2* in fetal rats (Kovacheva et al. 2007). Conversely, choline supplementation decreases CpG methylation of cortisol-regulating genes in human cord blood (Jiang et al. 2012). These changes occur in an unexpected direction, which suggests that the influence of methyl nutrients on epigenetic programming is more complicated than just providing the methyl groups and may involve other delicate regulatory systems such as compensatory activation of DNMTs and methyl-binding proteins (Kovacheva et al. 2007).

6.5 Sexual Dimorphism of the Influence of Early Nutrition Exposures

Men and women display differential susceptibility to various common diseases, such as obesity, diabetes, cardiovascular disease, and neuropsychological disorders. Interestingly, various influences of early nutrition on phenotypic traits also seem to be dependent on the sex of an individual. Males seem to be more susceptible to altered glucose homeostasis and other metabolic abnormalities compared to females when exposed to the same nutrition insult in early life. For example, maternal high-fat diet increases pancreatic beta-cell oxidative stress and functional deterioration in male but not female mice (Yokomizo et al. 2014). Periconceptional feeding of a methyl-deficient diet disrupts glucose homeostasis in male rats specifically (Maloney et al. 2011). In addition, several maternal dietary restrictions such as protein deficiency, B vitamin deficiency, and caloric restriction impair blood pressure of only the male offspring in rodent models (Aiken and Ozanne 2013).

The sexually dimorphic influence of early nutrition may be explained by several mechanisms, such as different hormonal secretion of the two sexes and random X chromatin inactivation of females (Hochberg et al. 2011). The placenta may play an essential role in mediating the sexually dimorphic responses to prenatal nutrition exposures. The placenta serves as a buffering organ which adapts to adverse maternal environment and retains normal transfer of nutrients to the fetus. Maternal undernutrition is more likely to decrease placental size in males (Aiken and Ozanne 2013), whereas the placenta of the female fetus seems to be more actively adapted to the environment by altering the gene transcription of various pathways in a more coordinated pattern, so as to handle existing environmental stress. Such coordinated adjustment is not observed in males (Mao et al. 2010).

Differential epigenetic responses may be an underlying contributor to sexual dimorphism. For example, when comparing people who were exposed to the Dutch famine periconceptionally with their unexposed siblings, the CpG methylation of several imprinted or growth signaling genes was differentially altered in men and women (Tobi et al. 2009). Histone H3 acetylation in developing mouse brain is different between males and females, when hormone-masculinized females display "male-like" H3 acetylation pattern (Tsai et al. 2009). The sexual dimorphism in prenatal programming indicates that when we are interpreting data about the influence of early nutrition, sex is a potential confounder that requires the outcomes of males and females to be analyzed independently.

6.6 Conclusions and Future Directions

As is illustrated in the previous sections, a large body of studies has demonstrated the potent effect of early nutrition on regulating somatic development and health outcomes via epigenetic modifications. However, gaps of knowledge remain.

More human studies are needed to delineate whether the epigenetic changes induced by early nutrition exposures, such as antenatal folic acid supplementation, actually lead to functional differences in both short and long terms. Since various environmental signals such as maternal stress, toxins, and pollutants are exerted on the same individual during early development in addition to dietary intake, delineating the interactions between nutrition and these environmental factors in shaping the epigenome is warranted. Moreover, these interactions highlight the possibility of manipulating nutritional intake in early life to offset the negative impacts of harmful environmental signals on growth and development.

Abbreviations

ACTH	Adrenocorticotropic hormone
AgRP	Agouti-related peptide
BHMT	Betaine homocysteine methyltransferase
CpG	Cytosine-guanine dinucleotide
CPT1A	Carnitine palmitoyltransferase-1A
CRH	Corticotropin-releasing hormone
CTCF	CCCTC-binding factor
DNMT	DNA methyltransferase
GR	Glucocorticoid receptor
HAT	Histone acetyltransferase
HDAC	Histone deacetylase
HMT	Histone methyltransferase
HPA axis	The hypothalamic-pituitary-adrenal axis
IGF2	Insulin growth-like factor 2
INSR	Insulin receptor
LDL	Low-density lipoprotein
5-CH_3-THF	Methyltetrahydrofolate
NTD	Neural tube defect
PCK1	Phosphoenolpyruvate carboxykinase
POMC	Pro-opiomelanocortin
PPAR	Peroxisome proliferator-activated receptors
SAM	*S*-adenosylmethionine

References

Aiken, C. E., & Ozanne, S. E. (2013). Sex differences in developmental programming models. *Reproduction, 145*, R1–R13.

Altobelli, G., Bogdarina, I. G., Stupka, E., Clark, A. J., & Langley-Evans, S. (2013). Genome-wide methylation and gene expression changes in newborn rats following maternal protein restriction and reversal by folic acid. *PLoS One, 8*, e82989.

Baker, J. L., Olsen, L. W., & Sørensen, T. I. (2008). Weight at birth and all-cause mortality in adulthood. *Epidemiology, 19*, 197–203.
Barker, D. J., & Osmond, C. (1988). Low birth weight and hypertension. *BMJ, 297*, 134–135.
Barker, D. J., Winter, P. D., Osmond, C., Margetts, B., & Simmonds, S. J. (1989). Weight in infancy and death from ischaemic heart disease. *Lancet, 2*, 577–580.
Bateson, P., Barker, D., Clutton-Brock, T., Deb, D., D'Udine, B., Foley, R. A., et al. (2004). Developmental plasticity and human health. *Nature, 430*, 419–421.
Begum, G., Davies, A., Stevens, A., Oliver, M., Jaquiery, A., Challis, J., et al. (2013). Maternal undernutrition programs tissue-specific epigenetic changes in the glucocorticoid receptor in adult offspring. *Endocrinology, 154*, 4560–4569.
Bertolo, R. F., & McBreairty, L. E. (2013). The nutritional burden of methylation reactions. *Current Opinion in Clinical Nutrition and Metabolic Care, 16*, 102–108.
Boeke, C. E., Gillman, M. W., Hughes, M. D., Rifas-Shiman, S. L., Villamor, E., & Oken, E. (2013). Choline intake during pregnancy and child cognition at age 7 years. *American Journal of Epidemiology, 177*, 1338–1347.
Brameld, J. M., Mostyn, A., Dandrea, J., Stephenson, T. J., Dawson, J. M., Buttery, P. J., et al. (2000). Maternal nutrition alters the expression of insulin-like growth factors in fetal sheep liver and skeletal muscle. *The Journal of Endocrinology, 167*, 429–437.
Brandeis, M., Kafri, T., Ariel, M., Chaillet, J. R., McCarrey, J., Razin, A., et al. (1993). The ontogeny of allele-specific methylation associated with imprinted genes in the mouse. *The EMBO Journal, 12*, 3669–3677.
Brenseke, B., Prater, M. R., Bahamonde, J., & Gutierrez, J. C. (2013). Current thoughts on maternal nutrition and fetal programming of the metabolic syndrome. *Journal of Pregnancy, 2013*, 368461.
Burdge, G. C., & Lillycrop, K. A. (2010). Nutrition, epigenetics, and developmental plasticity: Implications for understanding human disease. *Annual Review of Nutrition, 30*, 315–339.
Carmody, J. S., Wan, P., Accili, D., Zeltser, L. M., & Leibel, R. L. (2011). Respective contributions of maternal insulin resistance and diet to metabolic and hypothalamic phenotypes of progeny. *Obesity (Silver Spring), 19*, 492–499.
Chen, J. H., Martin-Gronert, M. S., Tarry-Adkins, J., & Ozanne, S. E. (2009). Maternal protein restriction affects postnatal growth and the expression of key proteins involved in lifespan regulation in mice. *PLoS One, 4*, e4950.
Choi, S. W., & Friso, S. (2010). Epigenetics: A new bridge between nutrition and health. *Advances in Nutrition, 1*, 8–16.
Davison, J. M., Mellott, T. J., Kovacheva, V. P., & Blusztajn, J. K. (2009). Gestational choline supply regulates methylation of histone H3, expression of histone methyltransferases G9a (Kmt1c) and Suv39h1 (Kmt1a), and DNA methylation of their genes in rat fetal liver and brain. *The Journal of Biological Chemistry, 284*, 1982–1989.
DeChiara, T. M., Efstratiadis, A., & Robertson, E. J. (1990). A growth-deficiency phenotype in heterozygous mice carrying an insulin-like growth factor II gene disrupted by targeting. *Nature, 345*, 78–80.
Delage, B., & Dashwood, R. H. (2008). Dietary manipulation of histone structure and function. *Annual Review of Nutrition, 28*, 347–366.
Drake, A. J., McPherson, R. C., Godfrey, K. M., Cooper, C., Lillycrop, K. A., Hanson, M. A., et al. (2012). An unbalanced maternal diet in pregnancy associates with offspring epigenetic changes in genes controlling glucocorticoid action and foetal growth. *Clinical Endocrinology, 77*, 808–815.
Fernandez-Twinn, D. S., & Ozanne, S. E. (2006). Mechanisms by which poor early growth programs type-2 diabetes, obesity and the metabolic syndrome. *Physiology and Behavior, 88*, 234–243.
Gilbert, E. R., & Liu, D. (2010). Flavonoids influence epigenetic-modifying enzyme activity: Structure – function relationships and the therapeutic potential for cancer. *Current Medicinal Chemistry, 17*, 1756–1768.

Gillman, M. W., Rifas-Shiman, S., Berkey, C. S., Field, A. E., & Colditz, G. A. (2003). Maternal gestational diabetes, birth weight, and adolescent obesity. *Pediatrics, 111*, e221–e226.

Gong, L., Pan, Y. X., & Chen, H. (2010). Gestational low protein diet in the rat mediates Igf2 gene expression in male offspring via altered hepatic DNA methylation. *Epigenetics, 5*, 619–626.

Grygiel-Górniak, B. (2014). Peroxisome proliferator-activated receptors and their ligands: Nutritional and clinical implications—a review. *Nutrition Journal, 13*, 17.

Hales, C. N., & Barker, D. J. (1992). Type 2 (non-insulin-dependent) diabetes mellitus: The thrifty phenotype hypothesis. *Diabetologia, 35*, 595–601.

Heijmans, B. T., Tobi, E. W., Stein, A. D., Putter, H., Blauw, G. J., Susser, E. S., et al. (2008). Persistent epigenetic differences associated with prenatal exposure to famine in humans. *Proceedings of the National Academy of Sciences of the United States of America, 105*, 17046–17049.

Herrick, K., Phillips, D. I., Haselden, S., Shiell, A. W., Campbell-Brown, M., & Godfrey, K. M. (2003). Maternal consumption of a high-meat, low-carbohydrate diet in late pregnancy: Relation to adult cortisol concentrations in the offspring. *The Journal of Clinical Endocrinology and Metabolism, 88*, 3554–3560.

Hochberg, Z., Feil, R., Constancia, M., Fraga, M., Junien, C., Carel, J. C., et al. (2011). Child health, developmental plasticity, and epigenetic programming. *Endocrine Reviews, 32*, 159–224.

Hoyo, C., Murtha, A. P., Schildkraut, J. M., Jirtle, R. L., Demark-Wahnefried, W., Forman, M. R., et al. (2011). Methylation variation at IGF2 differentially methylated regions and maternal folic acid use before and during pregnancy. *Epigenetics, 6*, 928–936.

Huang, R. C., Galati, J. C., Burrows, S., Beilin, L. J., Li, X., Pennell, C. E., et al. (2012). DNA methylation of the IGF2/H19 imprinting control region and adiposity distribution in young adults. *Clinical Epigenetics, 4*, 21.

Hult, M., Tornhammar, P., Ueda, P., Chima, C., Bonamy, A. K., Ozumba, B., et al. (2010). Hypertension, diabetes and overweight: Looming legacies of the Biafran famine. *PLoS One, 5*, e13582.

Illingworth, R. S., & Bird, A. P. (2009). CpG islands—'A rough guide'. *FEBS Letters, 583*, 1713–1720.

Institute of Medicine. (1998). *Dietary reference intakes for thiamin, riboflavin, niacin, vitamin B6, folate, vitamin B12, pantothenic acid, biotin and choline*. Washington, DC: National Academy Press.

Jiang, X., Yan, J., West, A. A., Perry, C. A., Malysheva, O. V., Devapatla, S., et al. (2012). Maternal choline intake alters the epigenetic state of fetal cortisol-regulating genes in humans. *The FASEB Journal, 26*, 3563–3574.

Jirtle, R. L., & Skinner, M. K. (2007). Environmental epigenomics and disease susceptibility. *Nature Reviews Genetics, 8*, 253–262.

Jones, B. K., Levorse, J., & Tilghman, S. M. (2001). Deletion of a nuclease-sensitive region between the Igf2 and H19 genes leads to Igf2 misregulation and increased adiposity. *Human Molecular Genetics, 10*, 807–814.

Jousse, C., Parry, L., Lambert-Langlais, S., Maurin, A. C., Averous, J., Bruhat, A., et al. (2011). Perinatal undernutrition affects the methylation and expression of the leptin gene in adults: implication for the understanding of metabolic syndrome. *The FASEB Journal, 25*, 3271–3278.

Korosi, A., & Baram, T. Z. (2010). Plasticity of the stress response early in life: Mechanisms and significance. *Developmental Psychobiology, 52*, 661–670.

Kovacheva, V. P., Davison, J. M., Mellott, T. J., Rogers, A. E., Yang, S., O'Brien, M. J., et al. (2009). Raising gestational choline intake alters gene expression in DMBA-evoked mammary tumors and prolongs survival. *The FASEB Journal, 23*, 1054–1063.

Kovacheva, V. P., Mellott, T. J., Davison, J. M., Wagner, N., Lopez-Coviella, I., Schnitzler, A. C., et al. (2007). Gestational choline deficiency causes global and Igf2 gene DNA hypermethylation by up-regulation of Dnmt1 expression. *The Journal of Biological Chemistry, 282*, 31777–31788.

Levitt, N. S., Lindsay, R. S., Holmes, M. C., & Seckl, J. R. (1996). Dexamethasone in the last week of pregnancy attenuates hippocampal glucocorticoid receptor gene expression and elevates blood pressure in the adult offspring in the rat. *Neuroendocrinology, 64*, 412–418.

Li, Y., Jaddoe, V. W., Qi, L., He, Y., Wang, D., Lai, J., et al. (2011a). Exposure to the Chinese famine in early life and the risk of metabolic syndrome in adulthood. *Diabetes Care, 34*, 1014–1018.

Li, M., Sloboda, D. M., & Vickers, M. H. (2011b). Maternal obesity and developmental programming of metabolic disorders in offspring: Evidence from animal models. *Experimental Diabetes Research, 2011*, 592408.

Li, Y., & Tollefsbol, T. O. (2010). Impact on DNA methylation in cancer prevention and therapy by bioactive dietary components. *Current Medicinal Chemistry, 17*, 2141–2151.

Liberato, S. C., Singh, G., & Mulholland, K. (2013). Effects of protein energy supplementation during pregnancy on fetal growth: A review of the literature focusing on contextual factors. *Food and Nutrition Research 57*. doi: 10.3402/fnr.v57i0.20499. eCollection 2013.

Lillycrop, K. A., Phillips, E. S., Jackson, A. A., Hanson, M. A., & Burdge, G. C. (2005). Dietary protein restriction of pregnant rats induces and folic acid supplementation prevents epigenetic modification of hepatic gene expression in the offspring. *Journal of Nutrition, 135*, 1382–1386.

Liu, D., Diorio, J., Tannenbaum, B., Caldji, C., Francis, D., Freedman, A., et al. (1997). Maternal care, hippocampal glucocorticoid receptors, and hypothalamic-pituitary-adrenal responses to stress. *Science, 277*, 1659–1662.

Lucassen, P. J., Naninck, E. F., van Goudoever, J. B., Fitzsimons, C., Joels, M., & Korosi, A. (2013). Perinatal programming of adult hippocampal structure and function; emerging roles of stress, nutrition and epigenetics. *Trends in Neurosciences, 36*, 621–631.

Lui, J. C., Finkielstain, G. P., Barnes, K. M., & Baron, J. (2008). An imprinted gene network that controls mammalian somatic growth is down-regulated during postnatal growth deceleration in multiple organs. *American Journal of Physiology: Regulatory, Integrative and Comparative Physiology, 295*, R189–R196.

Magliano, D. C., Bargut, T. C., de Carvalho, S. N., Aguila, M. B., Mandarim-de-Lacerda, C. A., & Souza-Mello, V. (2013). Peroxisome proliferator-activated receptors-alpha and gamma are targets to treat offspring from maternal diet-induced obesity in mice. *PLoS One, 8*, e64258.

Maloney, C. A., Hay, S. M., Young, L. E., Sinclair, K. D., & Rees, W. D. (2011). A methyl-deficient diet fed to rat dams during the peri-conception period programs glucose homeostasis in adult male but not female offspring. *The Journal of Nutrition, 141*, 95–100.

Mao, J., Zhang, X., Sieli, P. T., Falduto, M. T., Torres, K. E., & Rosenfeld, C. S. (2010). Contrasting effects of different maternal diets on sexually dimorphic gene expression in the murine placenta. *Proceedings of the National Academy of Sciences of the United States of America, 107*, 5557–5562.

Marco, A., Kisliouk, T., Weller, A., & Meiri, N. (2013). High fat diet induces hypermethylation of the hypothalamic Pomc promoter and obesity in post-weaning rats. *Psychoneuroendocrinology, 38*, 2844–2853.

Mayer, W., Niveleau, A., Walter, J., Fundele, R., & Haaf, T. (2000). Demethylation of the zygotic paternal genome. *Nature, 403*, 501–502.

McCann, J. C., Hudes, M., & Ames, B. N. (2006). An overview of evidence for a causal relationship between dietary availability of choline during development and cognitive function in offspring. *Neuroscience and Biobehavioral Reviews, 30*, 696–712.

McCurdy, C. E., Bishop, J. M., Williams, S. M., Grayson, B. E., Smith, M. S., Friedman, J. E., et al. (2009). Maternal high-fat diet triggers lipotoxicity in the fetal livers of nonhuman primates. *The Journal of Clinical Investigation, 119*, 323–335.

Meaney, M. J., & Szyf, M. (2005). Environmental programming of stress responses through DNA methylation: Life at the interface between a dynamic environment and a fixed genome. *Dialogues in Clinical Neuroscience, 7*, 103–123.

Mehedint, M. G., Craciunescu, C. N., & Zeisel, S. H. (2010a). Maternal dietary choline deficiency alters angiogenesis in fetal mouse hippocampus. *Proceedings of the National Academy of Sciences of the United States of America, 107*, 12834–12839.

Mehedint, M. G., Niculescu, M. D., Craciunescu, C. N., & Zeisel, S. H. (2010b). Choline deficiency alters global histone methylation and epigenetic marking at the Re1 site of the calbindin 1 gene. *The FASEB Journal, 24*, 184–195.

Mills, J. L., Fan, R., Brody, L. C., Liu, A., Ueland, P. M., Wang, Y., et al. (2014). Maternal choline concentrations during pregnancy and choline-related genetic variants as risk factors for neural tube defects. *American Journal of Clinical Nutrition, 100*, 1069–1074.

Miñana-Solis, M. D. C., & Escobar, C. (2008). Post-weaning protein malnutrition in the rat produces short and long term metabolic impairment, in contrast to earlier and later periods. *International Journal of Biological Sciences, 4*, 422–432.

Morison, I. M., Paton, C. J., & Cleverley, S. D. (2001). The imprinted gene and parent-of-origin effect database. *Nucleic Acids Research, 29*, 275–276.

Morris, M. J., & Chen, H. (2009). Established maternal obesity in the rat reprograms hypothalamic appetite regulators and leptin signaling at birth. *International Journal of Obesity, 33*, 115–122.

Mortensen, O. H., Olsen, H. L., Frandsen, L., Nielsen, P. E., Nielsen, F. C., Grunnet, N., et al. (2010). Gestational protein restriction in mice has pronounced effects on gene expression in newborn offspring's liver and skeletal muscle; protective effect of taurine. *Pediatric Research, 67*, 47–53.

Naess, O., Stoltenberg, C., Hoff, D. A., Nystad, W., Magnus, P., Tverdal, A., et al. (2013). Cardiovascular mortality in relation to birth weight of children and grandchildren in 500,000 Norwegian families. *European Heart Journal, 34*, 3427–3436.

O'Donnell, K., O'Connor, T. G., & Glover, V. (2009). Prenatal stress and neurodevelopment of the child: Focus on the HPA axis and role of the placenta. *Developmental Neuroscience, 31*, 285–292.

Ogawa, T., Shibato, J., Rakwal, R., Saito, T., Tamura, G., Kuwagata, M., et al. (2014). Seeking genes responsible for developmental origins of health and disease from the fetal mouse liver following maternal food restriction. *Congenital Anomalies (Kyoto), 54*, 195–219.

Ong, K. K., Ahmed, M. L., Emmett, P. M., Preece, M. A., & Dunger, D. B. (2000). Association between postnatal catch-up growth and obesity in childhood: Prospective cohort study. *BMJ, 320*, 967–971.

Ooi, S. K., O'Donnell, A. H., & Bestor, T. H. (2009). Mammalian cytosine methylation at a glance. *Journal of Cell Science, 122*, 2787–2791.

Ooi, S. K., Qiu, C., Bernstein, E., Li, K., Jia, D., Yang, Z., et al. (2007). DNMT3L connects unmethylated lysine 4 of histone H3 to de novo methylation of DNA. *Nature, 448*, 714–717.

Palou, M., Priego, T., Sánchez, J., Palou, A., & Picó, C. (2013). Metabolic programming of sirtuin 1 (SIRT1) expression by moderate energy restriction during gestation in rats may be related to obesity susceptibility in later life. *The British Journal of Nutrition, 109*, 757–764.

Patel, M. S., & Srinivasan, M. (2011). Metabolic programming in the immediate postnatal life. *Annals of Nutrition and Metabolism, 58*(Suppl 2), 18–28.

Perkins, E., Murphy, S. K., Murtha, A. P., Schildkraut, J., Jirtle, R. L., Demark-Wahnefried, W., et al. (2012). Insulin-like growth factor 2/H19 methylation at birth and risk of overweight and obesity in children. *The Journal of Pediatrics, 161*, 31–39.

Rich-Edwards, J. W., Stampfer, M. J., Manson, J. E., Rosner, B., Hankinson, S. E., Colditz, G. A., et al. (1997). Birth weight and risk of cardiovascular disease in a cohort of women followed up since 1976. *BMJ, 315*, 396–400.

Roseboom, T., de Rooij, S., & Painter, R. (2006). The Dutch famine and its long-term consequences for adult health. *Early Human Development, 82*, 485–491.

Roseboom, T. J., Painter, R. C., van Abeelen, A. F., Veenendaal, M. V., & de Rooij, S. R. (2011). Hungry in the womb: What are the consequences? Lessons from the Dutch famine. *Maturitas, 70*, 141–145.

Ross, R. G., Hunter, S. K., McCarthy, L., Beuler, J., Hutchison, A. K., Wagner, B. D., et al. (2013). Perinatal choline effects on neonatal pathophysiology related to later schizophrenia risk. *The American Journal of Psychiatry, 170*, 290–298.

Sampey, B. P., Vanhoose, A. M., Winfield, H. M., Freemerman, A. J., Muehlbauer, M. J., Fueger, P. T., et al. (2011). Cafeteria diet is a robust model of human metabolic syndrome with liver and adipose inflammation: Comparison to high-fat diet. *Obesity (Silver Spring), 19*, 1109–1117.

Sarr, O., Yang, K., & Regnault, T. R. (2012). In utero programming of later adiposity: The role of fetal growth restriction. *Journal of Pregnancy, 2012*, 134758.

Schemies, J., Uciechowska, U., Sippl, W., & Jung, M. (2010). NAD(+) -dependent histone deacetylases (sirtuins) as novel therapeutic targets. *Medicinal Research Reviews, 30*, 861–889.

Shaw, G. M., Finnell, R. H., Blom, H. J., Carmichael, S. L., Vollset, S. E., Yang, W., et al. (2009). Choline and risk of neural tube defects in a folate-fortified population. *Epidemiology, 20*, 714–719.

Shiell, A. W., Campbell-Brown, M., Haselden, S., Robinson, S., Godfrey, K. M., & Barker, D. J. (2001). High-meat, low-carbohydrate diet in pregnancy: Relation to adult blood pressure in the offspring. *Hypertension, 38*, 1282–1288.

Sinclair, K. D., Allegrucci, C., Singh, R., Gardner, D. S., Sebastian, S., Bispham, J., et al. (2007). DNA methylation, insulin resistance, and blood pressure in offspring determined by maternal periconceptional B vitamin and methionine status. *Proceedings of the National Academy of Sciences of the United States of America, 104*, 19351–19356.

Smith, S. M., & Vale, W. W. (2006). The role of the hypothalamic-pituitary-adrenal axis in neuroendocrine responses to stress. *Dialogues in Clinical Neuroscience, 8*, 383–395.

Steegers-Theunissen, R. P., Obermann-Borst, S. A., Kremer, D., Lindemans, J., Siebel, C., Steegers, E. A., et al. (2009). Periconceptional maternal folic acid use of 400 microg per day is related to increased methylation of the IGF2 gene in the very young child. *PLoS One, 4*, e7845.

Steiger, H., Labonté, B., Groleau, P., Turecki, G., & Israel, M. (2013). Methylation of the glucocorticoid receptor gene promoter in bulimic women: Associations with borderline personality disorder, suicidality, and exposure to childhood abuse. *International Journal of Eating Disorders, 46*, 246–255.

Stein, A. D., Barnhart, H. X., Hickey, M., Ramakrishnan, U., Schroeder, D. G., & Martorell, R. (2003). Prospective study of protein-energy supplementation early in life and of growth in the subsequent generation in Guatemala. *The American Journal of Clinical Nutrition, 78*, 162–167.

Stuart, A., Amer-Wåhlin, I., Persson, J., & Källen, K. (2013). Long-term cardiovascular risk in relation to birth weight and exposure to maternal diabetes mellitus. *International Journal of Cardiology, 168*, 2653–2657.

Susser, E., Kirkbride, J. B., Heijmans, B. T., Kresovich, J. K., Lumey, L. H., & Stein, A. D. (2012). Maternal prenatal nutrition and health in grandchildren and subsequent generations. *Annual Review of Anthropology, 41*, 577–610.

Talens, R. P., Boomsma, D. I., Tobi, E. W., Kremer, D., Jukema, J. W., Willemsen, G., et al. (2010). Variation, patterns, and temporal stability of DNA methylation: Considerations for epigenetic epidemiology. *The FASEB Journal, 24*, 3135–3144.

Thompson, M. D., Cole, D. E., & Ray, J. G. (2009). Vitamin B-12 and neural tube defects: The Canadian experience. *The American Journal of Clinical Nutrition, 89*, 697S–701S.

Tobi, E. W., Goeman, J. J., Monajemi, R., Gu, H., Putter, H., Zhang, Y., et al. (2014). DNA methylation signatures link prenatal famine exposure to growth and metabolism. *Nature Communications, 5*, 5592.

Tobi, E. W., Lumey, L. H., Talens, R. P., Kremer, D., Putter, H., Stein, A. D., et al. (2009). DNA methylation differences after exposure to prenatal famine are common and timing- and sex-specific. *Human Molecular Genetics, 18*, 4046–4053.

Tobi, E. W., Slagboom, P. E., van Dongen, J., Kremer, D., Stein, A. D., Putter, H., et al. (2012). Prenatal famine and genetic variation are independently and additively associated with DNA methylation at regulatory loci within IGF2/H19. *PLoS One, 7*, e37933.

Tsai, H. W., Grant, P. A., & Rissman, E. F. (2009). Sex differences in histone modifications in the neonatal mouse brain. *Epigenetics, 4*, 47–53.

Vogt, M. C., Paeger, L., Hess, S., Steculorum, S. M., Awazawa, M., Hampel, B., et al. (2014). Neonatal insulin action impairs hypothalamic neurocircuit formation in response to maternal high-fat feeding. *Cell, 156*, 495–509.

Vohr, B. R., McGarvey, S. T., & Tucker, R. (1999). Effects of maternal gestational diabetes on offspring adiposity at 4–7 years of age. *Diabetes Care, 22*, 1284–1291.

Voigt, P., Tee, W. W., & Reinberg, D. (2013). A double take on bivalent promoters. *Genes and Development, 27*, 1318–1338.

Vucetic, Z., Kimmel, J., Totoki, K., Hollenbeck, E., & Reyes, T. M. (2010). Maternal high-fat diet alters methylation and gene expression of dopamine and opioid-related genes. *Endocrinology, 151*, 4756–4764.

Waterland, R. A., Dolinoy, D. C., Lin, J. R., Smith, C. A., Shi, X., & Tahiliani, K. G. (2006). Maternal methyl supplements increase offspring DNA methylation at Axin fused. *Genesis, 44*, 401–406.

Waterland, R. A., & Jirtle, R. L. (2003). Transposable elements: Targets for early nutritional effects on epigenetic gene regulation. *Molecular and Cellular Biology, 23*, 5293–5300.

Waterland, R. A., Travisano, M., Tahiliani, K. G., Rached, M. T., & Mirza, S. (2008). Methyl donor supplementation prevents transgenerational amplification of obesity. *International Journal of Obesity, 32*, 1373–1379.

Williams-Wyss, O., Zhang, S., MacLaughlin, S. M., Kleemann, D., Walker, S. K., Suter, C. M., et al. (2014). Embryo number and periconceptional undernutrition in the sheep have differential effects on adrenal epigenotype, growth, and development. *The American Journal of Physiology: Endocrinology and Metabolism, 307*, E141–E150.

Winick, M., & Noble, A. (1966). Cellular response in rats during malnutrition at various ages. *Journal of Nutrition, 89*, 300–306.

Woodall, S. M., Johnston, B. M., Breier, B. H., & Gluckman, P. D. (1996). Chronic maternal undernutrition in the rat leads to delayed postnatal growth and elevated blood pressure of offspring. *Pediatric Research, 40*, 438–443.

Yang, T., Fu, M., Pestell, R., & Sauve, A. A. (2006). SIRT1 and endocrine signaling. *Trends in Endocrinology and Metabolism, 17*, 186–191.

Yokomizo, H., Inoguchi, T., Sonoda, N., Sakaki, Y., Maeda, Y., Inoue, T., et al. (2014). Maternal high-fat diet induces insulin resistance and deterioration of pancreatic β-cell function in adult offspring with sex differences in mice. *The American Journal of Physiology: Endocrinology and Metabolism, 306*, E1163–E1175.

Zhang, S., Rattanatray, L., MacLaughlin, S. M., Cropley, J. E., Suter, C. M., Molloy, L., et al. (2010). Periconceptional undernutrition in normal and overweight ewes leads to increased adrenal growth and epigenetic changes in adrenal IGF2/H19 gene in offspring. *The FASEB Journal, 24*, 2772–2782.

Zheng, X., Wang, Y., Ren, W., Luo, R., Zhang, S., Zhang, J. H., et al. (2012). Risk of metabolic syndrome in adults exposed to the great Chinese famine during the fetal life and early childhood. *European Journal of Clinical Nutrition, 66*, 231–236.

Chapter 7
What Is the Risk? Dental Amalgam, Mercury Exposure, and Human Health Risks Throughout the Life Span

John Kall, Amanda Just, and Michael Aschner

7.1 Introduction

According to the US Food and Drug Administration (USFDA), "Dental amalgam is a mixture of metals, consisting of liquid mercury and a powdered alloy composed of silver, tin, and copper. Approximately 50 % of dental amalgam is elemental mercury by weight. Dental amalgam fillings are also known as 'silver fillings' because of their silver-like appearance" (USFDA 2009b).

All dental amalgam restorations contain this mercury (World Health Organization (WHO) 2005), and the literature is consistent that these fillings emit mercury vapor (Gay et al. 1979; Hahn et al. 1989; Haley 2005; Health Canada 1996; Leistevuo et al. 2001; Mahler et al. 1994; Nylander et al. 1987; Vimy and Lorscheider 1985a, b; Vimy et al. 1986, 1990). In recent years, government officials, scientists, dentists, consumers, and many others have raised serious concerns about the risks this dental mercury poses to humans and to the environment.

In order to identify potential health hazards of dental mercury amalgam throughout the life span, we reviewed the scientific literature pertaining to dental mercury risk. This review was accomplished by assessing scientific documents relevant to this topic collected by the International Academy of Oral Medicine and Toxicology (IAOMT, www.iaomt.org) and by conducting a PubMed (www.ncbi.nlm.nih.gov/

J. Kall, D.M.D. (✉) • A. Just, M.S.
International Academy of Oral Medicine and Toxicology, ChampionsGate, FL 33896, USA
e-mail: jkall2323@gmail.com

M. Aschner, Ph.D.
International Academy of Oral Medicine and Toxicology, ChampionsGate, FL 33896, USA

Department of Molecular Pharmacology, Albert Einstein College of Medicine, Bronx, NY, USA

Vanderbilt Kennedy Center for Research on Human Development, Nashville, TN, USA

D. Hollar (ed.), *Epigenetics, the Environment, and Children's Health Across Lifespans*, DOI 10.1007/978-3-319-25325-1_7

PubMed) literature search on "dental mercury risk." More details on both of these search mechanisms are provided in the Methods portion of this article (Sect. 7.2).

7.1.1 History of Use and Regulations

Over a million dentists around the world routinely use dental mercury amalgam to repair decayed teeth, but controversy has surrounded the use of mercury in dentistry since the neurotoxin was first widely introduced as an ingredient in this type of filling material. The American Society of Dental Surgeons, the predecessor to the American Dental Association, made its members pledge not to use mercury because of its known toxicity (Health Canada 1996). This action was part of what is referred to as the "amalgam war," which began in the 1840s and resulted in a series of arguments among dentists and others over the safety and use of dental mercury (Hyson 2006).

It is now known that exposure to mercury, even in minute amounts, is toxic and poses significant risks to human health. For example, a World Health Organization (WHO) report warned of mercury: "It may cause harmful effects to the nervous, digestive, respiratory, immune systems and to the kidneys, besides causing lung damage. Adverse health effects from mercury exposure can be: tremors, impaired vision and hearing, paralysis, insomnia, emotional instability, developmental deficits during fetal development, and attention deficit and developmental delays during childhood. Recent studies suggest that mercury may have no threshold below which some adverse effects do not occur" (WHO 2005, p. 1).

Scientific evidence confirms that in millions of individuals with dental mercury amalgam fillings, mercury exposure exceeds the reference exposure level (REL) of 0.3 mcg Hg^0/m^3 set by the US Environmental Protection Agency (USEPA) and of 0.03 mcg Hg^0/m^3 set by the California Environmental Protection Agency (Richardson et al. 2011). (REL is a term used to denote the exposure level defined by national and international regulatory agencies at which there is an expectation of *no negative health outcomes* within the population.)

Also, reports from the World Health Organization (WHO) and Canada's federal department of health (Health Canada) conclude that mercury vapor from dental amalgam is the greatest source of human exposure to mercury in nonindustrial settings (Health Canada 1996; WHO 1991, 2005). In 1991, the WHO Environmental Health Criteria 118 concluded that the "[e]stimated average daily intake and retention" from dental amalgam was 3.8–21 (3–17) mcg/day (WHO 1991). In the 2003 Executive Summary of this document, the WHO stated, "Dental amalgam constitutes a potentially significant source of exposure to elemental mercury, with estimates of daily intake from amalgam restorations ranging from 1 to 27 [mcg]/day" (Risher 2003).

The United Nations Environment Programme's Intercessional Negotiating Committee agreed upon the text of a global, legally binding mercury treaty in January 2013, and over 100 nations have since signed the "Minamata Convention on Mercury." The United States was the first country to give its support for ratification of the international treaty, and Annex A, Part II, of the Minamata Convention on Mercury has a number of provisions with regard to phasing down the use of dental mercury amalgam including “setting national objectives aiming at minimizing its use” and “promoting the use of cost-effective and clinically effective mercury-free alternatives for dental restoration” (United Nations Environment Programme [UNEP] 2013, p. 23).

Norway banned dental mercury amalgam in 2008 (Ministry of the Environment 2007), and Sweden banned the use of amalgam for almost all purposes in 2009 (Swedish Chemicals Agency 2008). Japan and Switzerland have restricted or almost banned amalgam (BIO Intelligence (for European Commission report) 2012). France has recommended that alternative mercury-free dental materials be used for pregnant women, and Germany, Finland, Austria, and Canada have worked to reduce the use of dental mercury amalgam fillings for pregnant women, children, and patients with kidney problems (Health and Environment Alliance 2007).

In the United States, amalgams are currently being used for 45 % of direct dental restorations (Heintze and Rousson 2012). According to articles published in the *Journal of the American Dental Association*, these fillings are still being used routinely on 53.4 % of Black/African Americans, on 72.9 % of American Indians/Alaska Natives/Asians/Pacific Islanders (Makhija et al. 2011), and on more than 75 % of posterior restorations for new recruits to the US Navy and Marine Corps (Simecek et al. 2009).

In research published in 2011, Richardson and his colleagues reported that more than 67 million Americans aged 2 years and older exceed the intake of mercury vapor considered “safe” by the USEPA due to the presence of dental mercury amalgam fillings, whereas over 122 million Americans exceed the intake of mercury vapor considered “safe” by the California EPA due to their dental mercury amalgam fillings (Richardson et al. 2011).

Employee exposure to mercury is regulated in the United States by the 1970 Occupational Safety and Health Act (Occupational Safety and Health Administration (OSHA) 1970) and Workers’ Rights Handbook (OSHA 2011) that require employers to train employees to avoid or minimize exposure to hazardous chemicals.

Although the US Food and Drug Administration (USFDA) classified dental mercury amalgam in Class II (i.e., does not require any significant special controls) in 2009 (USFDA 2009a, c), the Dental Products Panel met to reevaluate the issue in December 2010. A decision was expected from the USFDA by December 31, 2011 (Moore 2012); however, it took until January 27, 2015, for the USFDA to reaffirm its previously established Class II classification (USFDA 2015a, b).

7.1.2 *Purpose of Review*

Our review was designed to evaluate the scientific evidence of dental mercury amalgam risk and to examine trends in the literature about this issue. We present that evidence in this chapter by providing an overview and highlights of scientific research related to human health risks caused by dental mercury amalgam throughout the life span. We discuss this issue from the perspective of the general population, pregnant women, fetuses, children, and dental professionals. We also present research in relation to genetic predisposition, mercury allergies, and other health conditions potentially caused or exacerbated by dental mercury exposure.

7.2 Methods

In this article, brief background information about global regulations and the use of dental mercury amalgam has been included in the Introduction to demonstrate the wide-scale potential of human health risks throughout the life span.

Our discussion (Sects. 7.3–7.4) centers on providing scientific evidence of these health risks. We present this information by categorizing these health risks and mercury exposures as they occur in the general population, pregnant women, fetuses, children, and dental professionals. For each area discussed, we begin with an overview of risks and proceed to highlight selected studies of interest. We further evaluate these risks by susceptible populations. We define these populations by the parameters of genetic predisposition, mercury allergies, and adverse health conditions in which dental mercury amalgam fillings have been recognized as a potential factor in contributing to or exacerbating the patient's state.

Our review was generated by evaluating medical and scientific literature, environmental and public health publications, government materials, and regulatory and legal documents. These sources were collected and assessed by a search of the IAOMT database and the PubMed database. Both of these searches are explained in more detail below. For the purposes of limiting the number of references in this article, not all of the sources are cited.

7.2.1 IAOMT Literature Search

Founded in 1984, the International Academy of Oral Medicine and Toxicology (IAOMT, www.iaomt.org) is a nonprofit organization with a membership of over 700 dentists, physicians, and research professionals from over 15 countries. The fundamental mission of the IAOMT is to promote the health of the public, and the IAOMT is an accredited member of the United Nations Environment Programme's Global Mercury Partnership. For over three decades, IAOMT members have catalogued materials from around the world about health risks associated with dental mercury amalgam, at times even funding relevant research.

Over 700 documents about this issue are currently filed in the IAOMT's searchable database known as the IAOMT Library. Numerous searches were conducted on this database, which has documents dating from 1926 to present day. Of these hundreds of sources, those used for this article were limited to the scientific and regulatory documents most relevant to dental mercury amalgam health risks.

Even more specifically, sources used for this paper were found by searching the IAOMT Library for evidence of health risks from dental mercury exposures to the general population, pregnant women, fetuses, children, dental professionals, and those individuals who are genetically predisposed, who have an allergy to mercury, or who suffer from health conditions that have been potentially linked to mercury by scientific research. These health conditions include Alzheimer's disease, amyotrophic lateral sclerosis (Lou Gehrig's disease), antibiotic resistance, autism spectrum disorders, autoimmune disorders/immunodeficiency, cardiovascular problems, chronic fatigue syndrome, hearing loss, kidney disease, multiple sclerosis, oral lichenoid reaction and oral lichen planus, Parkinson's disease, periodontal disease, and reproductive dysfunction.

7.2.2 *PubMed Literature Search*

The PubMed literature search was conducted online at the National Institutes of Health of the US Library of Medicine's PubMed database (www.ncbi.nlm.nih.gov/PubMed) from September 16, 2013, to March 6, 2014. The PubMed search term used was "dental mercury risk," and clinical trials and reviews were included in the search. The search was conducted from March 6, 2014, to as far back as PubMed provided results (1972), and the PubMed search resulted in 280 sources. One hundred and twenty-four of these sources were excluded because they were not in English, they were not relevant (not significantly about dental mercury amalgam), they were an erratum, they were a comment on a different article, and/or the abstract and study could not be found. Many of the articles that could not be located were not peer reviewed and appeared in trade journals or journals of localized dental groups.

Fifteen sources (9.6 % of the results) were deemed by the researchers as "ambiguous," meaning that the literature did not appear to suggest more of a safety or of a risk. Some of the 55 sources (35.3 % of results) that did not identify a risk or suggested safety are included in the discussion for this article. Table 7.1 also provides a summarized listing of these articles. Many of the 86 articles (55.1 % of the results) demonstrating risk have been used as sources for this document since they serve as evidence of the known health hazards of dental mercury amalgam. Table 7.2 provides a summarized listing of these articles.

7.2.3 *The Issue of Risk*

It must be noted that although the majority of articles in both of our searches demonstrated a risk, there were a number of articles which suggested that dental amalgam does not pose a health risk, dental amalgam is safe, releases of mercury from dental amalgam are within acceptable exposure levels, and/or there is insignificant data to prove its hazards. Some of these studies are listed in Table 7.1.

Table 7.1 Summarized list of articles suggesting "no risk" from dental mercury amalgam fillings in PubMed database search (articles from 1972 to March 6, 2014, with search term "dental mercury risk"). *Some articles appear in more than one category*

Subject of no-risk finding	No. of no-risk articles (out of 55 total)	% of no-risk articles reporting (55 = 100 %)	Article/s
Alzheimer's disease	1	2	Hock et al. (1998)
Antibiotic resistance	1	2	Roberts et al. (2008)
Autism	3	6	Hertz-Picciotto et al. (2010) Ng et al. (2007) Woods et al. (2010)
Children	4	7	Bellinger et al. (2007) Herrström et al. (1996) Woods et al. (2008) Ye et al. (2009)
Fatigue	1	2	Michel et al. (1989)
Gender	1	2	Kruzikova et al. (2009)
General and/or exposure levels	26	47	Bailer et al. (2001) Consequences of… (1983); Cordier et al. (1998) Dental amalgam… (2008) Eley (1997) Franzblau et al. (2012) García-Godoy (2000) Herrström et al. (1996) Järup (2003) Jones (1998) Jones (1999) Kehe et al. (2001) Kruzikova et al. (2009) Langworth et al. (2002) Levy (1995) MacEntee and Mojon (1991) Mathewson (1984) McParland and Warnakulasuriya (2012) Melchart et al. (2008) Mitchell et al. (2005) Risher and De Rosa (2007) Schweinsberg (1994) Williams (2008) Zimmer et al. (2002)
Immune system	2	4	Langworth et al. (1993) Mitchell et al. (2005)
Improvement of materials/practices and/or safe handling/hygiene	5	9	Arenholt-Bindslev (1998) Bayne et al. (2013) Eley (1997)

(continued)

Table 7.1 (continued)

Subject of no-risk finding	No. of no-risk articles (out of 55 total)	% of no-risk articles reporting (55 = 100 %)	Article/s
Kidney/renal	4	7	Herrström et al. (1995) Mitchell et al. (2005) Thygesen et al. (2011) Woods et al. (2008)
Multiple resonance imaging (MRI)	1	2	Müller-Miny et al. (1996)
Multiple sclerosis	3	6	Aminzadeh and Etminan (2007) Bangsi et al. (1998) Casetta et al. (2001)
Number of fillings	3	6	Casetta et al. (2001) Joshi et al. (2003) Kruzikova et al. (2009)
Occupational	11	20	Akesson et al. (2000) Arenholt-Bindslev (1998) Atesagaoglu et al. (2006) Ericson and Källén (1989) Franzblau et al. (2012) Heggland et al. (2011) Heidam (1984) Joshi et al. (2003) Mandel (1993) Roberts et al. (2001) Thygesen et al. (2011)
Parkinson's disease	1	2	Thygesen et al. (2011)
Pregnancy	5	9	Ericson and Källén (1989) Heggland et al. (2011) Heidam (1984) Hujoel et al. (2005) Mitchell et al. (2005)

It merits consideration that the technology of studying mercury's impact on human health has evolved over the past several decades, and some studies advocating the safety of dental mercury amalgam failed to take into account genetic factors, susceptible populations, metal allergies, synergistic reactions caused by concurrent and/or cumulative chemical exposures, and other variables that are now known to impact each person's response to mercury. Table 7.3 offers an overview of factors that can influence an individual's reaction to dental mercury.

The authors of a 2014 article entitled "Ecogenctics of mercury: from genetic polymorphisms and epigenetics to risk assessment and decision-making" and published in *Environmental Toxicology and Chemistry* expounded on this issue when they stated: "Studies on mammals (wildlife, humans, rodents) are showing Hg exposures to be related to epigenetic marks such as DNA methylation. Such findings are

Table 7.2 Summarized list of articles suggesting "risk" from dental mercury amalgam fillings in PubMed database search (articles from 1972 to March 6, 2014, with search term "dental mercury risk"). *Some articles appear in more than one category*

Subject of risk finding	No. of risk articles (out of 86 total)	% of risk articles reporting (86 = 100 %)	Article/s
Allergy/hypersensitivity	14	16	Athavale et al. (2003) Forte et al. (2008) Hougeir et al. (2006) Kaaber (1990) Kanerva et al. (1999); Lee et al. (2001) Podzimek et al. (2005) Prochazkova et al. (2004) Spencer (2000) Stejskal et al. (2013) Sterzl et al. (1999) Venclikova et al. (2006) Weber et al. (2012) Wong and Freeman (2003)
Alzheimer's disease	1	1	Mutter et al. (2004)
Autism	4	5	Geier et al. (2009, 2010) Holmes et al. (2003) Mutter et al. (2005)
Cardiovascular	1	1	Bergdahl et al. (2013)
Children	3	4	Dunn et al. (2008) Fuks (2002) Watson et al. (2012)
Dermatitis	1	1	Garner (2004)
Gender	3	4	Gundacker et al. (2006) Richardson et al. (2009) Watson et al. (2012)
General/exposure levels	21	24	Barregård (1993) Choi (1996) Clarkson (1992) Ely (2001) Fredin (1994) Fredin and Krabisch (1993) Goodrich et al. (2013a, b) Gundacker et al. (2006) Jensen (1985) Khwaja and Abbasi (2014) Kingman et al. (1998) Koral (2013) Nielsen (1992) Mackey et al. (2014) Mutter (2011) Pleva (1992) Richardon et al. (2009) Richardson et al. (2011) Siblerud et al. (1994) Weiner et al. (1990) Ylinen and Löfroth (2002)

(continued)

Table 7.2 (continued)

Subject of risk finding	No. of risk articles (out of 86 total)	% of risk articles reporting (86 = 100 %)	Article/s
Genetic	3	4	Godfrey et al. (2003) Mutter et al. (2004) Richardson et al. (2009)
Immune system	7	8	Bartova et al. (2003) Cooper et al. (2004) Pleva (1992) Prochazkova et al. (2004) Siblerud (1992) Sterzl et al. (1999) Venclikova et al. (2006)
Kidney/renal	5	6	Barregård et al. (2010) Mortada et al. (2002) Richardson et al. (2011) Spencer (2000) Weiner et al. (1990)
Number of filling/ surfaces	10	12	Akesson et al. (1991) Apostoli et al. (2002) Bergdahl et al. (2013) Geer et al. (2012) Gibičar et al. (2009) Godfrey et al. (2003) Leistevuo et al. (2002) McGrother et al. (1999) Palkovicova et al. (2008) Pirard et al. (2014)
Occupational	21	24	Akesson et al. (1991) Cooper et al. (2004) de Oliveira et al. (2010) Fabrizio et al. (2007) Goodrich et al. (2013a, b) Hilt et al. (2009) Johnson (1978) Kanerva et al. (1999) Karahalil et al. (2005) Lee et al. (2001) Lindbohm et al. (2007) Lönnroth and Shahnavaz (1995) Martin et al. (1995) Parsell et al. (1996) Pérez-Gómez et al. (2005) Rojas et al. (2006) Richardson et al. (2009) Van Zyl (1999) Votaw and Zey (1991) Warwick et al. (2013) Zahir et al. (2005)

(continued)

Table 7.2 (continued)

Subject of risk finding	No. of risk articles (out of 86 total)	% of risk articles reporting (86 = 100 %)	Article/s
Oral carcinoma	2	2	Hougeir et al. (2006) Weber et al. (2012)
Oral lichenoid lesion/oral lichen planus	3	4	Athavale et al. (2003) Forte et al. (2008) Wong and Freeman (2003)
Practices	11	13	Jensen (1985) Johnson (1978) Kanerva et al. (1999) Lönnroth and Shahnavaz (1995) Martin et al. (1995) Olfert (2006) Parsell et al. (1996) Perim and Goldberg (1984) Pleva (1992) Warwick et al. (2013) Votaw and Zey (1991)
Parkinson's disease	2	2	Ngim and Devathasan (1989) Venclikova et al. (2006)
Patient consent	2	2	Chirba-Martin and Welshhans (2004) Edlich et al. (2009)
Pregnancy/fertility	8	9	Geier et al. (2009) Holmes et al. (2003) Palkovicova et al. (2008) Podzimek et al. (2005) Lindbohm et al. (2007) Richardson et al. (2011) Spencer (2000) Vimy et al. (1997)
Removal recovery	6	7	Forte et al. (2008) Prochazkova et al. (2004) Stejskal et al. (2013) Sterzl et al. (1999) Weber et al. (2012) Wong and Freeman (2003)

beginning to increase understanding of the mechanisms of action of Hg, and in doing so they may help identify candidate biomarkers and pinpoint susceptible groups or life stages. Furthermore, they may help refine uncertainty factors and thus lead to more accurate risk assessments and improved decision-making" (Basu et al. 2014, p. 1248).

Table 7.3 Abbreviated list of variables potentially influencing individual reactions to dental mercury exposure

Factors related to mercury vapor release from dental amalgam filling
Age of amalgam filling
Cleaning, polishing, and other dental procedures
Contents of other materials mixed with the mercury, such as tin, copper, silver, etc.
Dental plaque
Deterioration of amalgam filling
Habits such as brushing, bruxism, chewing (including gum chewing, especially nicotine gum), consumption of hot liquids, diet (especially acidic foods), smoking, etc.
Infections in the mouth
Number of amalgam fillings
Other metals in the mouth, such as gold fillings or titanium implants
Root canals and other dental work
Saliva content
Size of amalgam filling
Surface area of amalgam filling
Techniques and safety measures applied when removing amalgam filling
Techniques used when placing amalgam filling

Personal traits and conditions related to mercury exposure response
Alcohol consumption
Allergy or hypersensitivity to mercury
Bacteria, including mercury resistant and antibiotic resistant
Burdens in organs and tissues such as the kidney, pituitary gland, liver, and brain
Diet
Drug use (prescription, recreational, and addiction)
Exercise
Exposure to other forms of mercury (i.e., fish consumption), lead, pollution, and any toxic substances (presently or previously)
Fetal or breast milk exposure to mercury, lead, and any toxic substances
Gender
Genetic traits and variants
Infections
Microbes in the gastrointestinal tract
Milk consumption
Nutrient levels, especially copper, zinc, and selenium
Occupational exposures to toxic substances
Overall health (see below for additional details)
Parasites and helminths
Stress/trauma
Yeast

Health conditions potentially contributed to or exacerbated by dental mercury exposures
Allergies
Alzheimer's disease
Amyotrophic lateral sclerosis, also known as Lou Gehrig's disease

(continued)

Table 7.3 (continued)

Health conditions potentially contributed to or exacerbated by dental mercury exposures
Antibiotic resistance
Autism spectrum disorders
Autoimmune disorders/immunodeficiency
Cardiovascular problems
Chronic fatigue syndrome
Complaints of unclear causation
Fibromyalgia
Hearing loss
Kidney disease
Micromercurialism
Multiple sclerosis
Oral lichenoid reaction and oral lichen planus
Parkinson's disease
Periodontal disease
Psychological conditions such as anxiety, depression, mood disorders, and suicidal ideations
Reproductive dysfunction
Symptoms of chronic mercury poisoning

Another area of consideration in relation to research about dental mercury amalgam is agreeing upon the definition of "risk." To illustrate this point, a number of scientific articles have claimed that dental mercury amalgam is safe for the "general population." Yet, given the current knowledge that sensitivities, biological predispositions, and a gamut of other conditions influence an individual's reaction to mercury exposure, the concept of accurately applying safety to the "general population" becomes highly subjective. This also applies to evaluating dental mercury amalgam risk for individuals manifesting specific health conditions such as Alzheimer's disease, autism, or multiple sclerosis.

Further issues with defining "risk" for dental mercury amalgam arise when considering the impact these restorations might have on an individual for a short amount of time versus long-term exposure, especially since many individuals have these fillings in their mouths for many years of their lives.

Specifically, research has shown that an individual accumulates a chronic dose of mercury ranging from "0.2 to 0.4 [mcg]/day per amalgam-filled tooth surface, or 0.5–1 [mcg]/day/amalgam-filled tooth, depending on age and other factors" (Richardson et al. 2011, p. 4257). Science has also established that once inside the mouth, mercury remains a retained heavy metal until and if the body can excrete the toxin (Heintze et al. 1983; Leistevuo et al. 2001).

As detailed in our Discussion (Sect. 7.3), how each person processes mercury exposure is dependent on a wide range of circumstances. Very few studies have explored this issue with the explicit purpose of an epigenetic perspective, and thus, without the due consideration for specific conditions, populations (including fetuses and children), genetic traits, occupational exposures, and other factors, many of the studies used to defend amalgam fail to accurately assess risk.

However, a question of ethics arises when suggesting the need for more conclusive data exploring these risk factors. Because certain populations are known to be at higher risk for toxicity from dental mercury amalgam, purposely exposing them to mercury would be knowingly endangering their health. For this reason, some researchers have suggested applying the precautionary approach in evaluating risks of materials such as dental mercury amalgam.

The precautionary approach was established in June 1992 for member states when the United Nations Environment Programme ratified the Rio Declaration on Environment and Development (UNEP 1992). Further to the Rio Declaration, in January 1998 at an international conference involving scientists, lawyers, policy makers, and environmentalists from the United States, Canada, and Europe, a formalized statement was signed and became known as the "Wingspread Statement on the Precautionary Principle" (Science and Environmental Health Network 1998). In it, the following advice was given: "When an activity raises threats of harm to human health or the environment, precautionary measures should be taken even if some cause and effect relationships are not fully established scientifically. In this context, the proponent of an activity, rather than the public, should bear the burden of proof" (Science and Environmental Health Network 1998).

Thus, new research on dental mercury amalgam should provide for the inclusion of the latest science and factors in toxicity, and risk should be evaluated by applying the precautionary principle on behalf of the millions of patients being exposed to mercury from their dental amalgam fillings.

7.3 Discussion

7.3.1 General Patient Risk from Dental Mercury Amalgam Overview

Mercury particulate can be discharged from dental mercury amalgam fillings, and mercury vapor is continuously emitted from dental mercury amalgam fillings (Al-Saleh and Al-Sedairi 2011; Barregård 1993; Boyd et al. 1991; Danscher et al. 1990; Fredin 1987; Fredin and Krabisch 1993; Godfrey et al. 2003; Hahn et al. 1989, 1990; Haley 2005; Hanson and Pleva 1991; Kraußet al. 1997; Martin et al. 1995; Molin et al. 1990; Mortada et al. 2002; Redhe and Pleva 1994; Richardson 2003; Richardson et al. 2009; Snapp et al. 1989; Summers et al. 1993; Vimy and Lorscheider 1985a, b). These releases mean that people are directly exposed to mercury from dental amalgam.

The output of mercury is intensified by the number of fillings present and other activities such as chewing, teeth grinding, brushing, dental treatments and procedures, and the consumption of hot liquids. This includes mercury released during hygiene, cleaning, and polishing procedures. This includes mercury released during placement of new restorations and removal of old ones. Ergo, men, women, and children patients, as well as fetuses, are all at risk from the mercury released from dental mercury amalgam fillings. All of these scenarios are described below with supporting scientific evidence.

7.3.2 General Patient Risk from Dental Mercury Amalgam Detail

A series of studies have demonstrated that urinary mercury concentrations consistently increase as the number of amalgam fillings increases (Kingman et al. 1998; Pesch et al. 2002; Woods et al. 2007).

Numerous studies have also demonstrated that the mercury exposure or concentration increases as a result of dental mercury amalgam in the following tissues and situations:

- Due to chewing, brushing, and bruxism (Isacsson et al. 1997; Sällsten et al. 1996; Vimy and Lorscheider 1985a, b)
- In exhaled or intraoral air of persons with amalgam fillings (Jokstad et al. 1992; Patterson et al. 1985; Skare and Engqvist 1994; Vimy and Lorscheider 1985a)
- In saliva of persons with amalgam fillings (Björkman et al. 1997; Fakour et al. 2010; Pizzichini et al. 2000)
- In blood of persons with amalgam fillings (Abraham et al. 1984; Akesson et al. 1991; Jokstad et al. 1992; Lindberg et al. 2004; Molin et al. 1990; Oskarsson et al. 1996; Snapp et al. 1989; Vahter et al. 2000)
- In various organs and tissues of amalgam bearers, including the kidney, pituitary gland, liver, and brain or parts thereof (Barregård et al. 2010; Björkman et al. 2007; Nylander et al. 1989)
- In feces of amalgam bearers (Björkman et al. 1997; Skare and Engqvist 1994)
- In amniotic fluid, cord blood, placenta, and various fetal tissues including the liver, kidney, and brain, in association with maternal amalgam load (Ask et al. 2002; Lutz et al. 1996; Palkovicova et al. 2008; Vahter et al. 2000)
- In colostrum and breast milk in association with maternal amalgam load (da Costa et al. 2005; Norouzi et al. 2012; Oskarsson et al. 1996)

Mercury can adversely influence each individual differently based on a wide range of coexisting factors, including the following:

- The number of amalgam fillings in the mouth (Barregård et al. 2010; Bergdahl et al. 2013; Dye et al. 2005; Geer et al. 2012; Geier et al. 2009; Gibičar et al. 2009; Kraußet al. 1997; McGrother et al. 1999; Pesch et al. 2002; Richardson et al. 2011; Rothwell and Boyd 2008)
- Gender (Gundacker et al. 2006; Haley 2005; Watson et al. 2012; Woods et al. 2012; Richardson et al. 2011)
- Genetic predisposition (Echeverria et al. 2006; Godfrey et al. 2003; Haley 2005; Homme et al. 2014; Kern et al. 2014; Mutter et al. 2004; Richardson et al. 2009; Weiner et al. 1990; Wojcik et al. 2006; Zamm 1991)

- Dental plaque (Lyttle and Bowden 1993)
- Selenium levels (Raymond and Ralston 2004)
- Exposure to lead (Pb) (Haley 2005, 2007; Ingalls 1983; Schubert et al. 1978)
- Consumption of milk (Kostial et al. 1979; Mata et al. 1997)
- Consumption of alcohol (Hursh et al. 1980)

Obviously, other circumstances can also play a role in each person's unique response to mercury.

Whereas individual response varies, evidence supports the potential for a decrease of symptoms related to mercury exposure and chronic mercury toxicity when dental mercury amalgam fillings are safely removed (Hanson 2004; Huggins and Levy 1998; Laine et al. 1992; Prochazkova et al. 2004; Redhe and Pleva 1994; Siblerud 1992; Siblerud and Kienholz 1994; Stejskal et al. 1999; Tomka et al. 2011; Wojcik et al. 2006; Zamm 1986). However, if safe protocols are not followed during amalgam replacement with nonmercury fillings, a patient's symptoms could potentially worsen due to mercury released during the removal process.

This is because if adequate safety measures are not applied, an outcome of dental mercury amalgam removal is acute exposure to mercury vapor and particulate for dentists, dental staff, dental students, and dental patients (Health Canada 1996; Karahalil et al. 2005; Lönnroth and Shahnavaz 1995; Martin et al. 1995; Nimmo et al. 1990; Richardson 2003; Warwick et al. 2013). This particularly endangers pregnant women, lactating women, women of childbearing age, fetuses, children being breastfed, and other sensitive populations.

7.3.3 Genetic Predisposition Overview

Mercury exposure from dental mercury amalgam can markedly threaten individuals with genetic traits that impact their response to mercury exposures (Echeverria et al. 2005, 2006; Godfrey et al. 2003; Haley 2005; Homme et al. 2014; Kern et al. 2014; Mutter et al. 2004; Richardson et al. 2011; Weiner et al. 1990; Wojcik et al. 2006; Zamm 1991). This includes individuals with CPOX4, APO-E3/4, and BDNF polymorphisms. Recent research has also identified a genetic predisposition to neurological impacts by mercury exposure from dental amalgam in male children (Homme et al. 2014; Woods et al. 2012, 2013, 2014).

7.3.4 Genetic Predisposition Detail

While the concept that certain individuals possess genetic traits capable of influencing response to mercury is complex, this area of research is essential in understanding one reason why people's reaction to their amalgam fillings is so variable.

In one study, Echeverria et al. (2005) identified polymorphisms in gene encoding for brain-derived neurotrophic factor (BDNF). Various detriments in neurobehavioral performance were associated with the presence of the BDNF polymorphism, which occurred in 25–35 % of study subjects independent of mercury exposure level. However, the combined effects of the polymorphism and the mercury exposure appeared to be additive.

The presence of a polymorphism for coproporphyrinogen oxidase (CPOX4) has also been observed and is associated with detriments in neurobehavioral response independent of mercury exposure. Yet, as with BDNF, the influence of the CPOX4 polymorphism and the mercury exposure together appeared to be additive. The frequency of CPOX4 was 15 % of study subjects in one study (Woods et al. 2012) and 25 % of study subjects in another study (Echeverria et al. 2006).

In the Casa Pia Children's Amalgam Trial, DeRouen et al. (2002) found no association between mercury body burden and the production of porphyria (porphyria is a group of disorders caused by enzyme deficiency). In contrast, when reviewing the same data set, Geier et al. (2011) found the characteristic pattern of porphyria associated with mercury body burden and the pattern significantly correlated with dental amalgam exposure in a dose-dependent fashion.

Using data collected from the over 500 children with dental amalgam fillings that were part of the children's amalgam trials, Dr. James Woods of the University of Washington and his fellow researchers also found further instances of neurobehavioral impairments potentially associated with mercury exposures in male children. A study on mercury exposures to boys with genetic variants of catechol-*O*-methyltransferase (*COMT*) resulted in the researchers reporting that "numerous gene-Hg interactions were observed between individual *COMT* SNPs [single-nucleotide polymorphisms], as well as with common *COMT* haplotype affecting multiple domains of neurobehavioral function" (Woods et al. 2014, p. 293). In another study, the researchers documented that mercury exposures to boys with genetic variants of metallothionein (MT) isoforms MT1M (rs2270837) and MT2A (rs10636), whether alone or combined, correlated with adverse impacts in various aspects of neurobehavioral function (Woods et al. 2013).

A mercury exposure study on a cohort of 515 dentists based on fish consumption and dental amalgam had already come to a similar conclusion. Wang et al. (2012) concluded, "our findings suggest that some MT genetic polymorphisms may influence mercury biomarker concentrations at levels of exposure relevant to the general population" (p. 530).

Research on this cohort of 515 dentists was done to evaluate mercury biomarkers based on fish consumption and dental amalgam mercury exposure. In one study, while associations were discovered between methylmercury from fish consumption and hypomethylation of a specific region of a genome in males, the researchers reported that no significant associations could be made between urinary mercury levels from dental mercury exposure and DNA methylation (Goodrich et al. 2013a). However, in another study, the researchers found important links between dental mercury exposure, certain polymorphisms, and mercury excretion. They stated: "Overall, this study suggests that polymorphisms in selenoproteins and glutathione-related genes may influence elimination of mercury in the urine and hair or mercury retention following exposures to elemental mercury (via dental amalgams) and methylmercury (via fish consumption)" (Goodrich et al. 2011, p. 301). In their more recent work, the researchers explained: "Beyond the influence of genetic polymorphisms on Hg toxicokinetics and biomarker values, growing evidence links SNPs and other types of genetic polymorphisms to variability in Hg-associated adverse health outcomes" (Basu et al. 2014, p. 1251).

Considering the percentages of the total population that possess the genetic traits mentioned here, it can be concluded that millions of people are predisposed to complications associated with adverse reactions to mercury exposure. Woods and his colleagues clearly expressed this concept of a large population being impacted in two of their recent studies, where they qualified that the genetic variants in their research associated with mercury and adverse neurobehavioral performance are "common" and "may have important public health implications for future strategies aimed at protecting children and adolescents from the potential health risks associated with Hg exposure" (Woods et al. 2013, p. 43; Woods et al. 2014, p. 309).

It is also essential to recognize that there are long-term consequences of mercury exposures for dental patients, especially those with specific genetic traits and variants, because mercury from amalgam fillings is known to be collected in organs and tissues (Barregård et al. 2010; Björkman et al. 2007; Nylander et al. 1989). In a review of the literature, Clarkson elaborated on this issue and established: "Mouth breathing carries the vapor to the lung, where it is absorbed and distributed to tissues, as discussed above. Mercury levels in autopsy tissue samples, including the brain, have been shown to correlate with the total number of surfaces of amalgam restorations" (Clarkson 2002, p. 18).

With knowledge of this accumulated body burden of dental mercury, genetic conditions that result in a variety of responses to mercury exposures become even more significant. For example, an additional examination of links through the APO-E3/4 gene family to mercury-related illness is described below in the Alzheimer's disease subsection.

7.3.5 *Susceptible Populations Overview*

The mercury in dental mercury amalgam fillings can exacerbate and/or contribute to all of the conditions stated below, as well as a myriad of other health problems:

- Allergies (Prochazkova et al. 2004; Stejskal et al. 1999; Zamm 1991)
- Alzheimer's disease (Godfrey et al. 2003; Mutter et al. 2004)
- Amyotrophic lateral sclerosis, also known as Lou Gehrig's disease (Redhe and Pleva 1994)
- Antibiotic resistance (Summers et al. 1993)
- Autism spectrum disorders (Geier et al. 2009, 2010; Mutter et al. 2005)
- Autoimmune disorders/immunodeficiency (Bartova et al. 2003; Cooper et al. 2004; Holmes et al. 2003; Hultman et al. 1994; Lindqvist and Mörnstad 1996; Mutter et al. 2005; Prochazkova et al. 2004; Sterzl et al. 1999; Venclikova et al. 2006; Weiner et al. 1990)

- Cardiovascular problems (Bergdahl et al. 2013; Houston 2014; Siblerud 1990)
- Chronic fatigue syndrome (Stejskal et al. 1999; Sterzl et al. 1999; Wojcik et al. 2006)
- Complaints of unclear causation (Hanson 2004; Hanson and Pleva 1991; Pleva 1992; Siblerud et al. 1994; Sjursen et al. 2011; Zamm 1991)
- Fibromyalgia (Kern et al. 2014)
- Hearing loss (Rothwell and Boyd 2008)
- Kidney disease (Barregård et al. 2010; Boyd et al. 1991; Fredin 1987; Mortada et al. 2002; Nylander et al. 1987; Richardson et al. 2011; Spencer 2000; Weiner et al. 1990)
- Micromercurialism (Ely et al. 1999)
- Multiple sclerosis (Huggins and Levy 1998; Prochazkova et al. 2004; Siblerud 1992; Siblerud and Kienholz 1994)
- Oral lichenoid reaction (Camisa et al. 1999; Dunsche et al. 2003; Henriksson et al. 1995; Ibbotson et al. 1996; Laine et al. 1992; Lind et al. 1986) and oral lichen planus (Athavale et al. 2003; Finne et al. 1982)
- Parkinson's disease (Ngim and Devathasan 1989; Venclikova et al. 2006)
- Periodontal disease (Goldschmidt et al. 1976; Ziff 1992)
- Psychological conditions such as anxiety, depression, and mood disorders, as well as suicide (Kern et al. 2014)
- Reproductive dysfunction (Podzimek et al. 2005; Rowland et al. 1994)
- Symptoms of chronic mercury poisoning (Wojcik et al. 2006)

Allergies to mercury are a completely separate health issue from toxicity. Most dentists do not test their patients for mercury allergy, but millions of patients are unknowingly allergic or sensitive to the dental mercury amalgam fillings in their mouths from the mercury or the other components. A number of articles have discussed this form of metal allergy (Athavale et al. 2003; Djerassi and Berova 1969; Finne et al. 1982; Hougeir et al. 2006; Kaaber 1990; Laine et al. 1992; Lee et al. 2001; Lind et al. 1986; Lundstrom 1984; Miller et al. 1987; Pang and Freeman 1995; Prochazkova et al. 2004; Stejskal and Stejskal 1999; Sterzl et al. 1999; Tomka et al. 2011; Venclikova et al. 2006; Weber et al. 2012). Studies have also established that exposure to dental mercury amalgam fillings correlates with higher prevalence of mercury allergies (Finne et al. 1982; White and Brandt 1976).

7.3.6 Susceptible Populations Detail: Alzheimer's Disease

There are a number of very serious neurological disorders for which the cause is not yet known. The clinical pictures of several of these are most interesting when considered in light of the documented neurotoxicity of mercury and the potential for neurotoxicity from amalgam fillings. Additionally, the synergistic effects of mercury (Schubert et al. 1978) with many of the toxicants commonly found in our environment make the danger of mercury unpredictable and possibly quite severe, especially when there is any mixture of elemental mercury, organic mercury, and other heavy metals such as lead and aluminum.

That being said, mercury has been linked to Alzheimer's disease (AD) by a number of different studies over the last two decades. Ehmann et al. (1986) reported that samples of AD brain analyzed by neutron activation had significantly elevated amounts of mercury in every area. In some parts of the brain such as the cerebellar hemisphere, mercury levels were tenfold greater in the AD brain than in controls.

The elevated mercury imbalance in the AD brain was confirmed in follow-up studies by Thompson et al. (1987) and others (Cornett et al. 1998; Vance et al. 1988). As an example, through cell fractionation, Wenstrup et al. (1990) were able to trace the accumulation of mercury into the cell organelle called the mitochondria. Mitochondria are tiny organelles contained within cells that produce protein. The follow-up studies mentioned above were all published in high-quality scientific journals that were expert in reviewing such analytical data.

Later, a paper by Saxe et al. (1995) featuring a cohort of nuns was published in the *Journal of the American Dental Association* that supposedly refuted these findings. Problems with the design of this study included a lack of controls and the fact that many of the subjects who were reported not to have amalgams actually had no teeth. The latter was a major oversight given that these nuns might have had amalgam fillings in their teeth before they lost them. However, even in Saxe's research, mercury levels in the brains of Catholic nuns showed that many of them had levels of mercury considered toxic by any scientific standard. Because certain nuns living in the same quarters and eating the same food had such elevated levels of mercury, it appears to be most likely that an inability to excrete mercury placed individuals at danger for retaining high mercury levels in the brain.

In 2001, University of Calgary researchers produced a short video visually representing how mercury, and only mercury, can cause synaptic neurodegeneration by destroying neuron growth cones (Lorscheider et al. 2001). The cultured neurons exposed to low levels of mercury degenerated in a manner indicative of lesions observed in Alzheimer's diseased brain. It is important to note that the level of mercury added to the cell culture in this video was one hundred times lower than is typically detected in the cerebral spinal fluid of those with mercury/silver amalgam tooth fillings. A related paper by the researchers is also important because it demonstrated that mercury, and only mercury, produces the neurofibrillary tangles (NFTs) recognized as the major diagnostic hallmark of AD (Leong et al. 2001).

Dr. Boyd Haley concluded that mercury and other blood–brain permeable toxicants with enhanced specificity for thiol-sensitive enzymes are the etiological source of AD (Haley 2007). Included in this category are other heavy metals such as lead and cadmium that act synergistically to enhance the toxicity of mercury and organic-mercury compounds (Haley 2007). The involvement of these other metals explains why mercury levels alone and severity of AD-like brain damage has not been exhibited in other studies.

At the same time, research done on about five hundred sets of identical twins from veterans of World War II showed that AD is definitely not a directly inherited disease, as it requires a toxic insult (Breitner et al. 1995).

Haley (2007) has also detailed why the apolipoprotein-4 (APO-E4) genetic variant represents a susceptibility to mercury toxicity as a pathogenetic factor and host for AD. Considering that Mutter et al. (2004) demonstrated that persons of African descent have a much higher level of the susceptible APO-E4 gene, this may explain why AD is more prevalent in those with an African heritage.

Moreover, Godfrey et al. (2003) found that there was a statistically significant increase in adverse effects in those patients having APO-E4/4 and APO-E3/4 when those patients were chronically exposed to mercury. Godfrey et al. went on to explain: "Mercury is very destructive at the mitochondrial level where catalase can demethylate organic mercury species into highly reactive inorganic mercury. Inorganic mercury is also an extremely potent enzyme inactivator. Furthermore, chronic micro-mercurial toxicity specifically from dental amalgam has been documented and successfully treated by removal of amalgam and medical detoxification in 796 patients" (2003, p. 193).

Additionally, Carocci and fellow colleagues noted: "The most important mechanism by which mercury causes toxicity appears to be mitochondrial damage via depletion of GSH (Nicole et al. 1998), coupled with binding to thiol groups (–SH), which generates free radicals. Mercury has a high affinity for thiol groups (–SH) and seleno groups (–SeH) that are present in amino acids as cysteine and *N*-acetyl cysteine, lipoic acid, proteins, and enzymes. *N*-acetyl cysteine and cysteine are precursors for the biosynthesis of GSH, which is among the most powerful intracellular antioxidants available to protect against oxidative stress and inflammation" (Carocci et al. 2014, p. 8).

Wojcik et al. (2006) supported a correlation between an inability to eliminate mercury and the inheritance of the APO-E4 allele. The study associated the APO-E4 allele with an increased incidence of common symptoms and signs of chronic mercury toxicity.

Ely (2001) applied information about dental mercury and AD by confirming the substantial release of mercury from amalgam fillings and by citing research estimating that the AD population could grow from its 2001 level of four million to fourteen million by 2050 based upon population age alone. This enormous increase could devastate any healthcare system.

7.3.7 Susceptible Populations Detail: Multiple Sclerosis

Multiple sclerosis ("MS") was first commonly identified in the nineteenth century during the time frame in which amalgam fillings came into common use. Unpublished anecdotal evidence has indicated that a significant number of, but certainly not all, MS victims who have their mercury/silver fillings removed resolve (spontaneous remission) or improve gradually. This anecdotal evidence has been supported by published studies.

For example, Baasch (1966) concluded that multiple sclerosis was an adult form of acrodynia (pink disease) and a neuroallergic reaction caused, in most cases, by mercury from amalgam fillings. Baasch reported several specific cases and cited ongoing studies that showed cessation of progression and improvement of resolution of MS after removal of amalgam fillings.

In a detailed study, Craelius (1978) showed a strong correlation ($P<0.001$) between MS death rates and dental caries. The data demonstrated the improbability that this correlation was due to chance. Numerous dietary factors were ruled out as contributing causes.

A hypothesis presented by T. H. Ingalls, MD, proposed that slow, retrograde seepage of mercury from root canals or amalgam fillings might lead to multiple sclerosis in middle age (Ingalls 1983). He also reexamined the extensive epidemiological data that show a linear correlation between death rates from MS and numbers of decayed, missing, and filled teeth. Ingalls (1986) suggested that investigators studying the causes of MS should carefully examine the patients' dental histories.

Another study by Ahlrot-Westerlund (1987) found that multiple sclerosis patients had eight times the normal level of mercury in their cerebral spinal fluid as compared to neurologically healthy controls.

Siblerud and Kienholz (1994) of the Rocky Mountain Research Institute, Inc., investigated the hypothesis that mercury from silver dental fillings (amalgam) may be related to multiple sclerosis (MS). It compared blood findings between MS subjects who had their amalgams removed and MS subjects with amalgams. MS subjects with amalgams were found to have significantly lower levels of red blood cells, hemoglobin, and hematocrit compared to MS subjects with amalgam removal. Thyroxine levels were also significantly lower in the MS amalgam group, and they had significantly lower levels of total T Lymphocytes and T-8 (CD8) suppressor cells. The MS amalgam group had significantly higher blood urea nitrogen and lower serum IgG. Hair mercury was significantly higher in the MS subjects compared to the non-MS control group. A health questionnaire found that MS subjects with amalgams had significantly more (33.7 %) exacerbations during the past 12 months compared to the MS volunteers with amalgam removal.

The role of myelin, a substance which helps the brain send messages to the body, is one of the few facts on which those who study MS are able to agree, and the MELISA Foundation has developed what they believe is a breakthrough in understanding MS by recognizing the link between metal allergy and the erosion of myelin (Stejskal and Stejskal 1999). They noted that hypersensitive reactions are triggered by metal particles entering the body of a person allergic to the metal in question. These particles then bind to the myelin, slightly changing its protein structure. In hypersensitive people, the new structure (myelin plus metal particle) is falsely identified as a foreign invader and is attacked (an autoimmune response). The culprit appears to be the "myelin plaques" in the brain, which are common in patients with MS. Such plaques can be the result of metal allergy. Already, the MELISA Foundation has seen patients with autoimmunity issues make a partial and, in some cases, a full recovery by removing the source of metal—often dental fillings (Stejskal et al. 1999).

Although more research needs to be done in this area, these results that suggest dental mercury exposure from amalgams, as well as from any other chronic low-grade mercury exposure, must be given very serious consideration as possibly playing a role in the etiology of MS in such patients.

In conclusion, the causation of MS is most likely multifactorial. Mercury is a probable factor in this disease, although genetic variability and individual ability to excrete mercury conceivably play a role as well (Ely et al. 1999).

7.3.8 *Susceptible Populations Detail: Amyotrophic Lateral Sclerosis (Lou Gehrig's Disease)*

Amyotrophic lateral sclerosis (ALS) was first identified a few years after the use of amalgam fillings became a routine dental practice. The correlation of ALS, more commonly known as Lou Gehrig's disease, to mercury exposure was first suggested by Brown (1954), and other studies have since followed suit. A study of 11 cases of chronic mercurialism from consumption of bread treated with a mercury-containing fungicide presented neurological symptoms akin to ALS with some more closely resembling progressive muscular atrophy (Kantarjian 1961). The author of the paper concluded: "Chronic mercurialism is a possible etiologic factor in amyotrophic lateral sclerosis" (p. 643).

A report by Barber (1978) is also noteworthy. This involved two employees in a mercury oxide manufacturing plant who developed previously nonexistent neurological symptoms resembling that of ALS. An additional 19 employees developed signs and symptoms which may be regarded as the early onset of a symptom complex of mercury intoxication that would likely have progressed to the ALS-like syndrome if the advancement had not been interrupted by removal of the individuals' exposure to mercury. All symptoms, signs, and laboratory findings returned completely back to normal after approximately 3 months in a mercury-free work environment.

Research published in the *Journal of the American Medical Association* reported on a 54-year-old man with symptoms resembling ALS after a brief but intense exposure to elemental mercury (Adams et al. 1983). The man, who had breathed mercury vapor while "salvaging the liquid mercury from industrial-grade thermometers" (p. 642), developed symptoms so similar to that of ALS that his neurologists gave him a "presumptive diagnosis of ALS" (p. 642). The man's physicians confirmed his exposure to mercury with a urine test weeks after his exposure, which registered 99 μg of mercury per liter of urine, an alarmingly high concentration. Two months later, the man had recovered nearly completely.

A study in the United States also involved neutron-activated analysis (NAA) of the brain, spinal cord, blood cells, serum, and nails of ALS victims compared to controls (Khare et al. 1989). Imbalances were detected in a number of trace elements in the tissue of ALS patients, and more widespread changes were noted in the concentrations of mercury. The authors cautioned that the variation in mercury concentrations need not necessarily indicate active toxicity, as it could merely represent an enlarged pool of detoxified mercury or perhaps a labeling of a specific cellular ligand by mercury in ALS.

Indeed, another study indicated an association between mercury and ALS (Kasarski et al. 1993); a case report described recoveries from ALS after the removal of mercury/silver fillings (Redhe and Pleva 1994), and another case report documented the development of ALS after the accidental injection of mercury (Schwarz et al. 1996).

However, it is very important to note that there are individuals who have ALS and have never had amalgam fillings. So, while dental mercury appears to be a potential cause of ALS, it certainly is not the only one.

7.3.9 Susceptible Populations Detail: Other Adverse Health Effects

One of the major reasons many scientists are concluding that dental amalgam is an unsuitable restorative material is because of its effects on the kidneys. Mercury concentrates in the kidneys, and experimental evidence has shown that it can inhibit kidney function (Boyd et al. 1991). In research by Hahn et al. (1989), the organ that accumulated the greatest amount of mercury following amalgam placement was the kidneys. Another study noted: "From the nephrotoxicity point of view, dental amalgam is an unsuitable filling material, as it may give rise to Hg [mercury] toxicity... In these exposure conditions, renal damage is possible..." (Mortada et al. 2002, p. 171). Animal studies have supported the conclusion that kidney function can be impaired by dental mercury amalgam (Hahn et al. 1990; Hultman et al. 1994; Vimy et al. 1990; Warfvinge et al. 1995).

The issue of mercury allergy has also been explored, and scientific literature has reflected that anywhere between 3.8 and 38.7 % of the population with amalgams is allergic to mercury (Djerassi and Berova 1969; Miller et al. 1987; Rudner et al. 1973; White and Brandt 1976). These statistics present formidable scientific documentation that a very significant amount of our population is at risk for hypersensitive reactions to mercury derived from dental amalgam.

The association between mercury and Parkinson's disease has been examined as well. For example, one epidemiological study correlated systemic mercury levels with increased risk of idiopathic Parkinson's disease (Ngim and Devathasan 1989).

The effects of amalgam fillings on auditory thresholds have been investigated, too. No significant correlation was found between composite (nonamalgam) filling or drilling data and auditory thresholds. However, there was a significant positive linear correlation between amalgam fillings and auditory thresholds at 8, 11.2, 12.5, 14, and 16 kHz. The strongest association was at 14 kHz, where each additional amalgam filling was associated with a 2.4 dB decline in hearing threshold (Rothwell and Boyd 2008).

Meanwhile, mercury has been found to be a potential factor in autism (Adams et al. 2007; Bernard et al. 2001; Bernard et al. 2002; Chen et al. 2007; DeSoto and Hitlan 2007; Geier and Geier 2006, 2007; Hsu et al. 2007; Nataf et al. 2006; Palmer et al. 2009).

As such, maternal dental amalgam fillings have been linked to autism (Ask et al. 2002; Geier et al. 2009; Holmes et al. 2003; Vahter et al. 2000). For example, an epidemiological study strongly associated prenatal mercury exposure from maternal dental amalgams with significantly increased rates of severe autism (Geier et al. 2009). As another example, Holmes et al. (2003) found that mothers with autistic children had higher levels of mercury exposure through Rh_0(D) immunoglobulin injections and amalgam fillings than control mothers.

Furthermore, mercury implanted in teeth can saturate the jawbone, result in bone loss, and produce inflammation and periodontal breakdown (App 1961; Trivedi and Talim 1973; Trott and Sherkat 1964; Turgeon et al. 1972; Zander 1957). Research has also made it apparent that the presence of dental mercury amalgam can result in chronic inflammation and bleeding in the gingival tissue adjacent to it, or in other words, amalgam has been shown to produce chronic gingivitis (Goldschmidt et al. 1976).

In summary, an important review article highlighting some of the scientific documentation concerning dental amalgam was published in the highly prestigious scientific publication, the *Journal of the Federation of American Societies for Experimental Biology (FASEB Journal)*, in 1995. The authors detailed the scientific data and conclusions from scores of peer-reviewed articles documenting the deleterious effects of mercury vapor on the immune, renal, reproductive, and central nervous systems. It was specifically noted that "the recent medical research findings presented herein strongly contradict the unsubstantiated opinions pronounced by various dental associations and related trade organizations, who offer assurances of amalgam safety to dental personnel and their patients without providing hard scientific data, including animal, cellular and molecular evidence, to support their claims" (Lorscheider et al. 1995, p. 507).

7.4 Focus on Maternal and Child Health

7.4.1 Pregnant Women, Fetuses, and Children Overview

Fetal and infant exposure to mercury via maternal dental mercury amalgam may lead to serious health consequences (da Costa et al. 2005; Geier et al. 2009, 2010; Haley 2005; Lindow et al. 2003; Lutz et al. 1996; Mutter et al. 2005; Norouzi et al. 2012; Oskarsson et al. 1996; Palkovicova et al. 2008; Richardson et al. 2011; Vahter et al. 2000; Vimy et al. 1990, 1997).

Mercury is excreted in breast milk of mothers with dental mercury amalgam fillings, and the mercury concentration in breast milk increases as the number of amalgam fillings in the mother increases (da Costa et al. 2005; Norouzi et al. 2012; Oskarsson et al. 1996).

Children are at risk for health impairments caused by dental mercury amalgam fillings (Al-Saleh and Al-Sedairi 2011; Ask et al. 2002; Dunn et al. 2008; Geier et al. 2011; Guzzi and Pigatto 2008; Haley 2005; Homme et al. 2014; Vahter et al. 2000; Woods et al. 2012).

7.4.2 Pregnant Women and Fetuses Detail

Fetuses and infants are at risk from the mercury in maternal amalgam fillings (Al-Saleh and Al-Sedairi 2011; da Costa et al. 2005; Drasch et al. 1998; Drexler and Schaller 1998; Geier et al. 2009; Gordon 1981; Haley 2005; Lindow et al. 2003; Luglie et al. 2005; Lutz et al. 1996; Norouzi et al. 2012; Oskarsson et al. 1996; Palkovicova et al. 2008; Panova and Dimitrov 1974; Richardson et al. 2009, 2011; Rowland et al. 1994; Svare et al. 1980; Ursinyova et al. 2006; Vahter et al. 2000; Vimy et al. 1990, 1997).

Research has repeatedly proven that the number of amalgam fillings in a mother's mouth raises the amount of mercury in breast milk (Björnberg et al. 2003; da Costa et al. 2005; Drasch et al. 1998; Drexler and Schaller 1998; Norouzi et al. 2012; Oskarsson et al. 1996; Ursinyova et al. 2006).

The relationship between miscarriage, stillborn infants, and mercury has also been confirmed (Koos and Longo 1976), and mercury was recognized for its particular ease of crossing the placental membrane in a study that sampled umbilical cord blood at birth (Pitkin et al. 1976).

In 1982, researchers followed 57 prenatal patients with no known exposure to mercury for changes in whole blood from initial prenatal examination to delivery and postpartum hospitalization. The mothers' whole-blood total mercury increased during pregnancy from 0.79 ppb (i.e., parts per billion) at initial examination to 1.16 ppb at delivery. This represents a 46 % increase during pregnancy (Kuntz et al. 1982). After careful analysis of the data, the lead researcher Kuntz stated: "Previous stillbirths, as well as history of birth defects, exhibited significant positive correlation with background mercury levels" (p. 440). He further concluded that patients with large numbers of dental fillings exhibited a tendency to have higher maternal blood levels of mercury (p. 442).

A decade later, Vimy et al. (1990) confirmed the transport of mercury from amalgam fillings to the fetus in experimental animals (sheep) and the additional exposure through mothers' milk. Berlin et al. (1992) showed that the fetal blood content of mercury was raised dramatically at the end of pregnancy. Early miscarriage, premature birth, and low birth weight with death were observed in this animal study.

Research led by Drasch and his colleagues at the University of Munich in Germany evaluated the levels of mercury found in tissues of human babies who had died before or soon after birth (mostly from sudden infant death syndrome (SIDS))

(Drasch et al. 1994). The levels of mercury found in the tissues of the babies significantly correlated with the number of dental amalgam fillings of the mother. These results confirmed previous findings from an animal study (Vimy et al. 1990) and were subsequently confirmed by a human autopsy study conducted in Sweden (Lutz et al. 1996).

Researchers in Germany continued to determine possible relationships between dental mercury and infertility. Gerhard and her coworkers at the University of Heidelberg Gynecological Clinic examined more than 1000 patients for mercury, fertility problems, and symptoms (Gerhard et al. 1998). The high mercury group had more hormonal disturbances, immune disturbances, recurring fungal infections, hair loss, and allergies. A number of different hormonal disturbances were found, including sex hormones. All of these differences were statistically significant, and some were very marked. According to Gerhard, mercury exposure can lead to hormone and immune disturbances that can reduce fertility, and doctors at her clinic have used this information to successfully treat fertility problems with a combination of vitamins/minerals and amalgam removal.

Problems that may develop in the fetus from maternal exposure are not always evident at birth. In fact, prenatal exposure to mercury vapor has been shown to have an effect on brain development (Fredriksson et al. 1996). Delayed problems can include diminished learning capacity, muscle spasms, and altered electroencephalograms. Exposure continues to increase if the infant is nursed, since mercury concentrates eightfold in breast milk (Pierce et al. 1972; Snyder 1971).

7.4.3 Children Detail

Although two studies, commonly referred to as the "New England Children's Amalgam Trial" (Bellinger et al. 2006) and the "Casa Pia Children's Amalgam Trial" (DeRouen et al. 2006) or collectively as the "CAT," have repeatedly been referenced to defend the use of amalgam in children, other researchers have since demonstrated that factors such as long-term effects, genetic predisposition, and measurement errors must also be taken into account (Geier et al. 2011, 2012; Guzzi and Pigatto 2008).

While many have used the CAT studies to purport the safety of dental amalgam in children despite evidence to the contrary, the same 2006 edition of the *Journal of the American Medical Association* that contained the CAT research also contained an editorial by Herbert Needleman, MD. Dr. Needleman, who has received numerous awards for his research on childhood lead poisoning, cautioned: "It is predictable that some outside interests will expand the modest conclusions of these studies to assert that use of mercury amalgam in dentistry is risk free. That conclusion would be unfortunate and unscientific" (Needleman 2006, p. 1836).

Indeed, other research clearly shows that children are at risk for health impairments caused by dental mercury amalgam fillings (Al-Saleh and Al-Sedairi 2011;

Ask et al. 2002; Dunn et al. 2008; Geier et al. 2011, 2012; Guzzi and Pigatto 2008; Haley 2005; Holmes et al. 2003; Vahter et al. 2000). For example, a 2005 WHO report specifically stated that mercury can cause "developmental deficits during fetal development, and attention deficit and developmental delays during childhood" (WHO 2005, p. 1). A second example is research from Al-Saleh and Al-Sedairi (2011), which found that children with amalgam fillings had more mercury in their urine and hair than those without fillings. The researchers also noted: "Changes in dental practices involving amalgam, especially for children, are highly recommended in order to avoid unnecessary exposure to Hg" (Al-Saleh and Al-Sedairi 2011, p. 3003).

When comparing the dental mercury health risks of children to adults, issues to consider include (1) children's bodies have completely different chemistries due to hormones and other factors, (2) their bodies are smaller than adults, and (3) their bodies, and especially their brains, are still developing.

It also means that exposure and tolerable daily intake levels applied to dental mercury amalgam should be adjusted by accounting for the age and weight of the subject. To demonstrate that such an exposure assessment is possible and feasible, the Canadian government, in its risk assessment of dental amalgam (Richardson 1995), was open and transparent about the prevalence of mercury fillings in the Canadian population, with adults having up to 25 filled teeth and children as young as 3 years of age having filled teeth. Richardson (1995) was explicit in incorporating this data into the methods used to estimate exposures, and as such, Health Canada was provided with estimates of mercury vapor exposure per filled tooth for each of five separate age groups (toddlers, children, teens, adults, and seniors).

7.5 Occupational Exposure Overview

Dentists, dental professionals, dental staff, and dental students are occupationally and chronically exposed to mercury released from dental mercury amalgam, and researchers and clinicians have raised concerns about the safety of dental personnel who work with dental mercury amalgam (Cooper et al. 2004; de Oliveira et al. 2010; Duplinsky and Cicchetti 2012; Echeverria et al. 1995; Fabrizio et al. 2007; Goodrich et al. 2013a, b; Herber et al. 1988; Hilt et al. 2009; Kanerva et al. 1999; Karahalil et al. 2005; Lee et al. 2001; Lönnroth and Shahnavaz 1995; Martin et al. 1995; Moen et al. 2008; Ngim et al. 1992; Nimmo et al. 1990; Nylander et al. 1989; Parsell et al. 1996; Pérez-Gómez et al. 2005; Richardson 2003; Richardson et al. 2009; Rowland et al. 1994; Shapiro et al. 1982; Sikorski et al. 1987; Warwick et al. 2013; Zahir et al. 2005). This includes mercury released during hygiene, cleaning, and polishing procedures. This also includes mercury released during removal of old mercury amalgam fillings and replacement with new ones. Scientific data indicates that female dental personnel are particularly impacted by occupational exposure to mercury (Rowland et al. 1994; Sikorski et al. 1987).

7.5.1 Occupational Exposure Detail

One area of dental amalgam research that is constantly expanding deals with evaluating the impact of mercury exposure on dental workers. This is especially important because occupational exposure through dentistry often results in repeated, long-term contact with mercury. To illustrate this point, consider the fact that research showed 85 % of dentists had aberrant porphyrin metabolism characteristic of low-level mercury poisoning (Echeverria et al. 2005).

Furthermore, Richardson used standard exposure assessment methods to estimate that a dentist who removes four amalgams per day will inhale 38 mg of mercury derived from amalgam particulate, which far exceeds any level considered to be safe (2003).

A number of studies demonstrating neurobehavioral deficits in dental personnel have also been published (Echeverria et al. 1995; Gonzalez-Ramirez et al. 1995; Ngim et al. 1992; Shapiro et al. 1982; Uzzell and Oler 1986).

Other studies have likewise demonstrated the neurobehavioral impact of elemental mercury on dentists. One study detected significant central nervous system effects among dentists and dental assistants at very low levels of mercury exposure (Echeverria et al. 1998). Significantly, the authors observed neurobehavioral deficits in dentists and dental assistants at urine mercury levels essentially equivalent to the urine mercury levels of those people in whom amalgam had been placed.

In a series of experiments utilizing neutron activation analysis (NAA) to study the mercury content of brain tissues of amalgam bearers, nonamalgam bearers, and dentists, Nylander and his coresearchers found that the dentist group (seven dentists and one dental nurse) all had a surprisingly high pituitary mercury content totally out of proportion to the content found in other parts of the brain. Values ranged from 135 to 4000 ng of mercury per gram of tissue (Friberg et al. 1986; Nylander 1986). In a related study, Nylander and his colleagues also found that dentists and dental assistants in Sweden had twice the incidence of brain tumors as nondental personnel (Ahlbom et al. 1986).

More recently, Duplinsky and Cicchetti (2012) determined that dentists are not as healthy as matched controls from the general population. Using data from pharmaceutical prescriptions to evaluate the health status of 600 dentists, it was shown that dentists required significantly more medications than the controls. The diminished health status of the dentists was potentially attributed to their occupational exposure to mercury.

Kidney ailments have also been the subject of studies for occupational exposure to dental mercury because the kidney filters the blood, and, as a result, chronic exposure to chemicals can eventually induce kidney damage. Thus, it is troublesome that a dental assistant's death was reported from kidney failure related to mercury exposure (Cook and Yates 1969).

Additionally, a study by Verschoor et al. (1988) evaluated the kidney function of 68 dentists (63 men, 5 women) and 64 female assistants who were apparently healthy, not pregnant, and not addicted to illegal drugs. They compared the results of their kidney function analysis to 250 workers known to be exposed to lead, cadmium, or chromium through the workplace. Their conclusion was that dentists and dental assistants "appear to have a higher potential risk of kidney function disturbances than the workers in these industries. There is a need to assess the renal hazard of the potential nephrotoxic chemicals used in dental practice" (p. 152).

A major turning point in the issue of dental office mercury safety for workers occurred in March of 2012, when the Norwegian Labour and Welfare Service officially categorized mercury injury as an occupational disease. The decision was made because Tordis Stigen Klausen, a Norwegian dental assistant, proved exposure to dental mercury amalgam caused her to become ill. According to a report from Norway, Gerd-Bang Johansen, chair of the League of Dental Assistants, explained, "This is good...for all the other dental assistants who became ill while working." Johansen added, "Many of these women have exactly the same symptoms as Stigen Klausen and have worked with mercury a lot" (Knudsson 2012).

Current scientific data also indicates that female dental personnel are severely impacted by occupational exposure to mercury, and the textbook *Occupational Hazards in the Health Professions* cautioned against comprehensive amalgam work during pregnancy (Brune and Edling 1989). Studies have also associated mercury with menstrual problems and other reproductive disorders (Goering et al. 1992; Mikhailova et al. 1971; Panova and Dimitrov 1974).

Additional studies support the concept that female dental workers are at risk for reproductive health issues. Sikorski et al. (1987) identified a significant positive correlation between mercury levels in the hair of occupationally exposed women and the occurrence of reproductive failures and menstrual cycle disorders. Koos and Longo (1976) stated that women of childbearing age should not be exposed to mercury concentrations exceeding 10 mcg/m^3 and that pregnant women should not be exposed to mercury at all. Eggleston (1989) established that mercury exposures cannot be avoided by women who work in the field of dentistry. However, since that time, a number of safe protocols have been developed by the IAOMT and other researchers to limit these exposures.

Of particular concern are potential birth defects and neurological disorders in children of dental workers (Gordon 1981; Marinova et al. 1973; Noe 1959; Panova and Dimitrov 1974). One published account reported that a young dentist, professionally exposed to mercury for 35 weeks during her pregnancy, delivered a severely brain-damaged mercury-poisoned infant (Gelbier and Ingram 1989).

7.6 Conclusion

In the case of dental amalgam, our review of potential risks from mercury exposure demonstrates that a variety of factors, including genetic traits, allergies, occupational exposures, pregnancy, fetal exposures, childhood exposures, and other health conditions, can impact an individual's response to dental mercury. We have also established that throughout the life span, from fetal development to advanced age, mercury releases from dental amalgam fillings result in serious dangers to human health.

This review makes it apparent that dental mercury amalgam and all dental restorative materials should be assessed for safety and biocompatibility with special consideration for all populations and all known risk factors. This can also be applied to the evaluation of any health risk, especially because science continues to prove that each person has a unique and personalized set of reactions to substances taken into the human body. As such, health risks must be considered from an evolving variety of perspectives in order to diligently protect all citizens from potentially harmful effects.

Acknowledgments Minimal funding for administrative costs of this work was provided by the International Academy of Oral Medicine and Toxicology (IAOMT). However, the views expressed in this publication do not necessarily reflect the views of the IAOMT, its Executive Council, Scientific Advisory Board, administration, membership, employees, contractors, etc.

The authors of this review are affiliated with the IAOMT in the following capacities: Dr. John Kall is the Chairperson of the Board and Accredited Master of the IAOMT, Amanda Just is the Program Director and a writer and researcher for the IAOMT, and Dr. Michael Aschner is a member of the Scientific Advisory Board for the IAOMT. Dr. Aschner was supported in part by grants from the NIH R01 ES07331 and R01 ES020852

Portions of this document were adapted from two documents created by the IAOMT:

Haley, B. E., Love, J. M., Virtue, W. E. (2012, October 10). *Position statement of the International Academy of Oral Medicine and Toxicology in response to the 'Call for Information' extended by the Scientific Committee on Emerging and Newly Identified Health Risks (SCENIHR) European Commission.* ChampionsGate, FL: IAOMT.

Kall, J., Robertson, K., Sukel, P., Just, A. (2013, April 16). *International Academy of Oral Medicine and Toxicology (IAOMT) Position Statement against Dental Mercury Amalgam Fillings for Medical and Dental Practitioners, Dental Students, and Patients.* ChampionsGate, FL: IAOMT.

The authors also wish to thank the Executive Council, Scientific Advisory Board, and the administration and membership of the IAOMT, especially Dr. Boyd Haley; James Love, Esq.; Dr. William Virtue; Dr. Kindal Robertson; and Dr. Phillip Sukel, for their research and writings that led to the publication of this article.

Abbreviations

AD	Alzheimer's disease
ALS	Amyotrophic lateral sclerosis, also known as Lou Gehrig's disease
APO-E3	Apolipoprotein E3 genetic variant
APO-E4	Apolipoprotein E-e4 genetic variant
BDNF	Brain-derived neurotrophic factor
CAT	Children's amalgam trials
COMT	Catechol-*o*-methyltransferase
CPOX4	Coproporphyrinogen oxidase genetic variant
DNA	Deoxyribonucleic acid
USEPA	United States Environmental Protection Agency
USFDA	United States Food and Drug Administration

IAOMT	International Academy of Oral Medicine and Toxicology
IgG	Immunoglobulin G, the most common type of antibodies in the blood
MELISA	Magnetic enzyme-linked immunosorbent assay
MS	Multiple sclerosis
MT	Metallothionein, a protein that is low in molecular weight and cysteine-rich
MT1M	Metallothionein 1M
MT2A	Metallothionein 2A
NAA	Neutron-activated analysis
NFT	Neurofibrillary tangles
$Rh_0(D)$	$Rh_0(D)$ immunoglobulin injection, an injected medicine administered to prevent a particular fetal immune disease
SNP	Single-nucleotide polymorphism
UNEP	United Nations Environment Programme
WHO	World Health Organization

References

Abraham, J. E., Svare, C. W., & Frank, C. W. (1984). The effect of dental amalgam restorations on blood mercury levels. *Journal of Dental Research, 63*(1), 71–73.

Adams, J. B., Romdalvik, J., Ramanujam, V. S., & Legator, M. S. (2007). Mercury, lead, and zinc in baby teeth of children with autism versus controls. *Journal of Toxicology and Environmental Health, Part A, 70*(12), 1046–1051.

Adams, C. R., Ziegler, D. K., & Lin, J. T. (1983). Mercury intoxication simulating amyotrophic lateral sclerosis. *Journal of the American Medical Association, 250*(5), 642–643.

Ahlbom, A., Norell, S., Rodvall, Y., & Nylander, M. (1986). Dentists, dental nurses, and brain tumours. *British Medical Journal (Clinical Research ed.), 292*(6521), 662.

Ahlrot-Westerlund, B. (1987, August 17–21). Multiple sclerosis and mercury in cerebrospinal fluid. In *Second Nordic symposium on trace elements and human health, Odense University, Denmark*.

Akesson, I., Schutz, A., Attewell, R., Skerfving, S., & Glantz, P. O. (1991). Status of mercury and selenium in dental personnel: Impact of amalgam work and own fillings. *Archives of Environmental Health: An International Journal, 46*(2), 102–109.

Akesson, I., Schütz, A., Horstmann, V., Skerfving, S., & Moritz, U. (2000). Musculoskeletal symptoms among dental personnel; - lack of association with mercury and selenium status, overweight and smoking. *Swedish Dental Journal, 24*(1-2), 23–38.

Al-Saleh, I., & Al-Sedairi, A. (2011). Mercury (Hg) burden in children: The impact of dental amalgam. *Science of the Total Environment, 409*(16), 3003–3015.

Aminzadeh, K. K., & Etminan, M. (2007). Dental amalgam and multiple sclerosis: A systematic review and meta-analysis. *Journal of Public Health Dentistry, 67*(1), 64–66.

Apostoli, P., Cortesi, I., Mangili, A., Elia, G., Drago, I., Gagliardi, T., Soleo, L., Valente, T., Sciarra, G. F., Aprea, C., Ronchi, A., & Minoia, C. (2002). Assessment of reference values for mercury in urine: The results of an Italian polycentric study. *Science of the Total Environment, 289*(1), 13–24.

App, G. R. (1961). Effect of silicate, amalgam, and cast gold on the gingiva. *The Journal of Prosthetic Dentistry, 11*(3), 522–532.

Arenholt-Bindslev, D. (1998). Environmental aspects of dental filling materials. *European Journal of Oral Sciences, 106*(2p2), 713–720.

Ask, K., Akesson, A., Berglund, M., & Vahter, M. (2002). Inorganic mercury and methylmercury in placentas of Swedish women. *Environmental Health Perspectives, 110*(5), 523.

Atesagaoglu, A., Omurlu, H., Ozcagli, E., Sardas, S., & Ertas, N. (2006). Mercury exposure in dental practice. *Operative Dentistry, 31*(6), 666–669.

Athavale, P. N., Shum, K. W., Yeoman, C. M., & Gawkrodger, D. J. (2003). Oral lichenoid lesions and contact allergy to dental mercury and gold. *Contact Dermatitis, 49*(5), 264–265.

Baasch, E. (1966). 1. Theoretische Überlegungen zur Ätiologie der Sclerosis multiplex. Die Multiple Sklerose, eine Quecksilberallergie. *Schweizer Arch Neurol Neurochir Psychiat, 98*, 1–19.

Bailer, J., Rist, F., Rudolf, A., Staehle, H. J., Eickholz, P., Triebig, G., Bader, M., & Pfeifer, U. (2001). Adverse health effects related to mercury exposure from dental amalgam fillings: Toxicological or psychological causes? *Psychological Medicine, 31*(02), 255–263.

Bangsi, D., Ghadirian, P., Ducic, S., Morisset, R., Ciccocioppo, S., McMullen, E., & Krewski, D. (1998). Dental amalgam and multiple sclerosis: A case-control study in Montreal, Canada. *International Journal of Epidemiology, 27*(4), 667–671.

Barber, T. E. (1978). Inorganic mercury intoxication reminiscent of amyotrophic lateral sclerosis. *Journal of Occupational and Environmental Medicine, 20*(10), 667–669.

Barregård, L. (1993). Biological monitoring of exposure to mercury vapor. *Scandinavian Journal of Work, Environment & Health, 19*(1), 45–49.

Barregård, L., Fabricius-Lagging, E., Lundh, T., Mölne, J., Wallin, M., Olausson, M., et al. (2010). Cadmium, mercury, and lead in kidney cortex of living kidney donors: Impact of different exposure sources. *Environmental Research, 110*(1), 47–54.

Bartova, J., Prochazkova, J., Kratka, Z., Benetkova, K., Venclikova, Z., & Sterzl, I. (2003). Dental amalgam as one of the risk factors in autoimmune diseases. *Neuroendocrinology Letters, 24*(1/2), 65–67.

Basu, N., Goodrich, J. M., & Head, J. (2014). Ecogenetics of mercury: From genetic polymorphisms and epigenetics to risk assessment and decision-making. *Environmental Toxicology and Chemistry, 33*(6), 1248–1258.

Bayne, S., Petersen, P. E., Piper, D., Schmalz, G., & Meyer, D. (2013). The challenge for innovation in direct restorative materials. *Advances in Dental Research, 25*(1), 8–17.

Bellinger, D. C., Trachtenberg, F., Barregard, L., Tavares, M., Cernichiari, E., Daniel, D., & McKinlay, S. (2006). Neuropsychological and renal effects of dental amalgam in children: A randomized clinical trial. *Journal of the American Medical Association, 295*(15), 1775–1783.

Bellinger, D. C., Trachtenberg, F., Daniel, D., Zhang, A., Tavares, M. A., & McKinlay, S. (2007). A dose-effect analysis of children's exposure to dental amalgam and neuropsychological function: The New England Children's Amalgam Trial. *Journal of the American Dental Association (1939), 138*(9), 1210–1216.

Bergdahl, I. A., Ahlqwist, M., Barregard, L., Björkelund, C., Blomstrand, A., Skerfving, S., Sundh, V., Wennberg, M., & Lissner, L. (2013). Mercury in serum predicts low risk of death and myocardial infarction in Gothenburg women. *International Archives of Occupational and Environmental Health, 86*(1), 71–77.

Berlin, M., Hua, M., Logdberg, J., & Warvinge, K. (1992). Prenatal exposure to mercury vapor: effects on brain development. *Fundamental and Applied Toxicology, 19*(3), 324–326.

Bernard, S., Enayati, A., Redwood, L., Roger, H., & Binstock, T. (2001). Autism: A novel form of mercury poisoning. *Medical Hypotheses, 56*(4), 462–471.

Bernard, S., Enayati, A., Roger, H., Binstock, T., & Redwood, L. (2002). The role of mercury in the pathogenesis of autism. *Molecular Psychiatry, 7*, S42–S43.

BIO Intelligence Service. (2012). Study on the potential for reducing mercury pollution from dental amalgam and batteries. Final Report prepared for the European Commission—DG ENV. Retrieved from the European Commission Web site: http://ec.europa.eu/environment/chemicals/mercury/pdf/final_report_110712.pdf

Björkman, L., Lundekvam, B. F., Lægreid, T., Bertelsen, B. I., Morild, I., Lilleng, P., Lind, B., Palm, B., & Vahter, M. (2007). Mercury in human brain, blood, muscle and toenails in relation to exposure: An autopsy study. *Environmental Health, 6*(30), 13.

Björkman, L., Sandborgh-Englund, G., & Ekstrand, J. (1997). Mercury in saliva and feces after removal of amalgam fillings. *Toxicology and Applied Pharmacology, 144*(1), 156–162.

Björnberg, K. A., Vahter, M., Petersson-Grawe, K., Glynn, A., Cnattingius, S., Darnerud, P. O., Atuma, S., Aune, M., Becker, W., & Berglund, M. (2003). Methyl mercury and inorganic mercury in Swedish pregnant women and in cord blood: Influence of fish consumption. *Environmental Health Perspectives, 111*(4), 637–641.

Boyd, N. D., Benediktsson, H., Vimy, M. J., Hooper, D. E., & Lorscheider, F. L. (1991). Mercury from dental "silver" tooth fillings impairs sheep kidney function. *American Journal of Physiology, 261*(4 Pt 2), R1010–R1014.

Breitner, J. C., Welsh, K. A., Gau, B. A., McDonald, W. M., Steffens, D. C., Saunders, A. M., Magruder, K. M., Helms, M. J., Plassman, B. L., Folstein, M. F., Brandt, J., Robinette, C. D., & Page, W. F. (1995). Alzheimer's disease in the National Academy of Sciences—National Research Council Registry of Aging Twin Veterans: III. Detection of cases, longitudinal results, and observations on twin concordance. *Archives of Neurology, 52*(8), 763–771.

Brown, I. A. (1954). Chronic mercurialism: A cause of the clinical syndrome of amyotrophic lateral sclerosis. *AMA Archives of Neurology & Psychiatry, 72*(6), 674–681.

Brune, D. K., & Edling, C. (1989). *Occupational hazards in the health professions* (pp. 315–316). Boca Raton, FL: CRC Press.

Camisa, C., Taylor, J. S., Bernat, J. R., Jr., & Helm, T. N. (1999). Contact hypersensitivity to mercury in amalgam restorations may mimic oral lichen planus. *Cutis, 63*(3), 189–192.

Carocci, A., Rovito, N., Sinicropi, M. S., & Genchi, G. (2014). Mercury toxicity and neurodegenerative effects. *Reviews of Environmental Contamination and Toxicology, 229*, 1–18 [Whitacre, D.M., ed. Springer International Publishing Switzerland].

Casetta, I., Invernizzi, M., & Granieri, E. (2001). Multiple sclerosis and dental amalgam: Case-control study in Ferrara, Italy. *Neuroepidemiology, 20*(2), 134–137.

Chen, C. Y., Liu, C. Y., Su, W. C., Huang, S. L., & Lin, K. M. (2007). Factors associated with the diagnosis of neurodevelopmental disorders: A population-based longitudinal study. *Pediatrics, 119*(2), e435–e443.

Chirba-Martin, M. A., & Welshhans, C. M. (2004). An uncertain risk and an uncertain future: Assessing the legal implications of mercury amalgam fillings. *Health Matrix: Journal of Law-Medicine, 14*, 293–324.

Choi, B. C. (1996). Risk assessment in dentistry: Health risks of dental amalgam revisited. *Journal (Canadian Dental Association), 62*(4), 328–332.

Clarkson, T. W. (1992). Principles of risk assessment. *Advances in Dental Research, 6*(1), 22–27.

Clarkson, T. W. (2002). The three modern faces of mercury. *Environmental Health Perspectives, 110*(Suppl 1), 11–23.

Consequences of mercury exposure in dentistry: A review of the literature. (1983). *Florida Dental Journal, 54*(4), 17–9, 21.

Cook, T. A., & Yates, P. O. (1969). Fatal mercury intoxication in a dental surgery assistant. *British Dental Journal, 127*(12), 553–555.

Cooper, G. S., Parks, C. G., Treadwell, E. L., St Clair, E. W., Gilkeson, G. S., & Dooley, M. A. (2004). Occupational risk factors for the development of systemic lupus erythematosus. *The Journal of Rheumatology, 31*(10), 1928–1933.

Cordier, S., Grasmick, C., Paquier-Passelaigue, M., Mandereau, L., Weber, J. P., & Jouan, M. (1998). Mercury exposure in French Guiana: Levels and determinants. *Archives of Environmental Health: An International Journal, 53*(4), 299–303.

Cornett, C. R., Markesbery, W. R., & Ehmann, W. D. (1998). Imbalances of trace elements related to oxidative damage in Alzheimer's disease brain. *Neurotoxicology, 19*(3), 339–345.

Craelius, W. (1978). Comparative epidemiology of multiple sclerosis and dental caries. *Journal of Epidemiology and Community Health, 32*(3), 155–165.

da Costa, S. L., Malm, O., & Dorea, J. G. (2005). Breast-mild mercury concentrations and amalgam surface in mothers from Brasilia, Brasil. *Biological Trace Element Research, 106*(2), 145–151.

Danscher, G., Hørsted-Bindslev, P., & Rungby, J. (1990). Traces of mercury in organs from primates with amalgam fillings. *Experimental and Molecular Pathology, 52*(3), 291–299.

de Oliveira, M. T., Pereira, J. R., Ghizoni, J. S., Bittencourt, S. T., & Molina, G. O. (2010). Effects from exposure to dental amalgam on systemic mercury levels in patients and dental school students. *Photomedicine and Laser Surgery, 28*(S2), S-111.

Dental amalgam: Few proven harmful effects but many ongoing concerns. (2008). *Prescrire International, 17*(98), 246–250.

DeRouen, T. A., Leroux, B. G., Martin, M. D., Townes, B. D., Woods, J. S., Leitão, J., Castro-Caldas, A., & Braveman, N. (2002). Issues in design and analysis of a randomized clinical trial to assess the safety of dental amalgam restorations in children. *Controlled Clinical Trials, 23*(3), 301–320.

DeRouen, T. A., Martin, M. D., Leroux, B. G., Townes, B. D., Woods, J. S., Leitão, J., Castro-Caldas, A., Luis, H., Bernardo, M., Rosenbaum, G., & Martins, I. P. (2006). Neurobehavioral effects of dental amalgam in children: A randomized clinical trial. *Journal of the American Medical Association, 295*(15), 1784–1792.

DeSoto, M. C., & Hitlan, R. T. (2007). Blood levels of mercury are related to diagnosis of autism: A reanalysis of an important data set. *Journal of Child Neurology, 22*(11), 1308–1311.

Djerassi, E., & Berova, N. (1969). The possibilities of allergic reactions from silver amalgam restorations. *International Dental Journal, 19*(4), 481–488.

Drasch, G., Aigner, S., Roider, G., Staiger, E., & Lipowsky, G. (1998). Mercury in human colostrum and early breast milk. Its dependence on dental amalgam and other factors. *Journal of Trace Elements in Medicine and Biology, 12*(1), 23–27.

Drasch, G., Schupp, I., Höfl, H., Reinke, R., & Roider, G. (1994). Mercury burden of human fetal and infant tissues. *European Journal of Pediatrics, 153*(8), 607–610.

Drexler, H., & Schaller, K. H. (1998). The mercury concentration in breast milk resulting from amalgam fillings and dietary habits. *Environmental Research, 77*(2), 124–129.

Dunn, J. E., Trachtenberg, F. L., Barregard, L., Bellinger, D., & McKinlay, S. (2008). Scalp hair and urine mercury content of children in the Northeast United States: The New England Children's Amalgam Trial. *Environmental Research, 107*(1), 79–88.

Dunsche, A., Kastel, I., Terheyden, H., Springer, I. N. G., Christopher, E., & Brasch, J. (2003). Oral lichenoid reactions associated with amalgam: Improvement after amalgam removal. *British Journal of Dermatology, 148*(1), 70–76.

Duplinsky, T. G., & Cicchetti, D. V. (2012). The health status of dentists exposed to mercury from silver amalgam tooth restorations. *International Journal of Statistics in Medical Research, 1*(1), 1–15.

Dye, B. A., Schober, S. E., Dillon, C. F., Jones, R. L., Fryar, C., McDowell, M., & Sinks, T. H. (2005). Urinary mercury concentrations associated with dental restorations in adult women aged 16–49 years: United States, 1999–2000. *Occupational and Environmental Medicine, 62*(6), 368–375.

Echeverria, D., Aposhian, H. V., Woods, J. S., Heyer, N. J., Aposhian, M. M., Bittner, A. C., Jr., & Mahurin, R. K. (1998). Neurobehavioral effects from exposure to dental amalgam Hg^o: New distinctions between recent exposure and Hg body burden. *The FASEB Journal (Federation of American Societies for Experimental Biology), 12*, 971–980.

Echeverria, D., Heyer, N. J., Martin, M. D., Naleway, C. A., Woods, J. S., & Bittner, A. C., Jr. (1995). Behavioral effects of low-level exposure to Hg0 among dentists. *Neurotoxicology and Teratology, 17*(2), 161–168.

Echeverria, D., Woods, J. S., Heyer, N. J., Rohlman, D. S., Farin, F. M., Bittner, A. C., Jr., Li, T., & Garabedian, C. (2005). Chronic low-level mercury exposure, BDNF polymorphism, and associations with cognitive and motor function. *Neurotoxicology and Teratology, 27*(6), 781–796.

Echeverria, D., Woods, J. S., Heyer, N. J., Rohlman, D., Farin, F. M., Li, T., & Garabedian, C. E. (2006). The association between a genetic polymorphism of coproporphyrinogen oxidase, dental mercury exposure and neurobehavioral response in humans. *Neurotoxicology and Teratology, 28*(1), 39–48.

Edlich, R., Mason, S. S., Chase, M. E., Fisher, A. L., Gubler, K. D., Long, W. B., III, & Newkirk, A. T. (2009). Revolutionary advances in the prevention of demyelinating diseases. *Journal of Environmental Pathology, Toxicology and Oncology, 28*(2), 143–152.

Eggleston, D. W. (1989). Dental amalgam—To be or not to be. *Pacific Coast Society of Prosthodontists Newsletter, 9*(2), 4–10.

Ehmann, W. D., Markesbery, W. R., Alauddin, M., Hossain, T. I., & Brubaker, E. H. (1986). Brain trace elements in Alzheimer's disease. *Neurotoxicology, 7*(1), 195–206.

Eley, B. M. (1997). The future of dental amalgam: A review of the literature. Part 7: Possible alternative materials to amalgam for the restoration of posterior teeth. *British Dental Journal, 183*(1), 11–14.

Ely, J. T. A. (2001). Mercury induced Alzheimer's disease: Accelerating incidence? *Bulletin of Environmental Contamination and Toxicology, 67*(6), 800–806.

Ely, J. T. A., Fudenberg, H. H., Muirhead, R. J., LaMarche, M. G., Krone, C. A., Buscher, D., & Stern, E. A. (1999). Urine mercury in micromercurialism: Bimodal distribution and diagnostic implications. *Bulletin of Environmental Contamination and Toxicology, 63*(5), 553–559.

Ericson, A., & Källén, B. (1989). Pregnancy outcome in women working as dentists, dental assistants or dental technicians. *International Archives of Occupational and Environmental Health, 61*(5), 329–333.

Fabrizio, E., Vanacore, N., Valente, M., Rubino, A., & Meco, G. (2007). High prevalence of extrapyramidal signs and symptoms in a group of Italian dental technicians. *BMC Neurology, 7*(1), 24.

Fakour, H., Esmaili-Sari, A., & Zayeri, F. (2010). Scalp hair and saliva as biomarkers in determination of mercury levels in Iranian women: Amalgam as a determinant of exposure. *Journal of Hazardous Materials, 177*(1), 109–113.

Finne, K. A. J., Göransson, K., & Winckler, L. (1982). Oral lichen planus and contact allergy to mercury. *International Journal of Oral Surgery, 11*(4), 236–239.

Forte, G., Petrucci, F., & Bocca, B. (2008). Metal allergens of growing significance: Epidemiology, immunotoxicology, strategies for testing and prevention. *Inflammation & Allergy-Drug Targets (Formerly Current Drug Targets-Inflammation & Allergy), 7*(3), 145–162.

Franzblau, A., d'Arcy, H., Ishak, M. B., Werner, R. A., Gillespie, B. W., Albers, J. W., Hamann, C., Gruninger, S. E., Chou, H. N., & Meyer, D. M. (2012). Low-level mercury exposure and peripheral nerve function. *Neurotoxicology, 33*(3), 299–306.

Fredin, B. (1987). The distribution of mercury in various tissues of guinea-pigs after application of dental amalgam fillings (a pilot study). *Science of the Total Environment, 66*, 263–268.

Fredin, B. (1994). Mercury release from dental amalgam fillings. *The International Journal of Risk and Safety in Medicine, 4*(3), 197–208.

Fredin, B., & Krabisch, L. (1993). In vitro investigation of the possible influence of inorganic mercury and hydrogen peroxide on the formation of peroxides in a polyunsaturated fatty acid system (linoleic acid). *The International Journal of Risk and Safety in Medicine, 4*(1), 19–25.

Fredriksson, A., Dencker, L., Archer, T., & Danielsson, B. R. G. (1996). Prenatal coexposure to metallic mercury vapour and methylmercury produce interactive behavioural changes in adult rats. *Neurotoxicology and Teratology, 18*(2), 129–134.

Friberg, L., Kullman, L., Lind, B., & Nylander, M. (1986). Mercury in the central nervous system in relation to amalgam fillings. *Läkartidningen, 83*(7), 519.

Fuks, A. B. (2002). The use of amalgam in pediatric dentistry. *Pediatric Dentistry, 24*(5), 448–455.

García-Godoy, F. (2000). Resin-based composites and compomers in primary molars. *Dental Clinics of North America, 44*(3), 541–570.

Garner, L. A. (2004). Contact dermatitis to metals. *Dermatologic Therapy, 17*(4), 321–327.

Gay, D., Cox, R., & Reinhardt, J. (1979). Chewing releases mercury from fillings. *The Lancet, 313*(8123), 985–986.

Geer, L. A., Persad, M. D., Palmer, C. D., Steuerwald, A. J., Dalloul, M., Abulafia, O., & Parsons, P. J. (2012). Assessment of prenatal mercury exposure in a predominately Caribbean immigrant community in Brooklyn, NY. *Journal of Environmental Monitoring, 14*(3), 1035–1043.

Geier, D. A., Carmody, T., Kern, J. K., King, P. G., & Geier, M. R. (2011). A significant relationship between mercury exposure from dental amalgams and urinary porphyrins: A further assessment of the Casa Pia children's dental amalgam trial. *Biometals, 24*(2), 215–224.

Geier, D. A., Carmody, T., Kern, J. K., King, P. G., & Geier, M. R. (2012). A dose-dependent relationship between mercury exposure from dental amalgams and urinary mercury levels: A further assessment of the Casa Pia Children's Dental Amalgam Trial. *Human & Experimental Toxicology, 31*(1), 11–17.

Geier, D. A., & Geier, M. R. (2006). A prospective assessment of porphyrins in autistic disorders: A potential marker for heavy metal exposure. *Neurotoxicity Research, 10*(1), 57–63.

Geier, D. A., & Geier, M. R. (2007). A prospective study of mercury toxicity biomarkers in autistic spectrum disorders. *Journal of Toxicology and Environmental Health, Part A, 70*(20), 1723–1730.

Geier, D. A., Kern, J. K., & Geier, M. R. (2009). A prospective study of prenatal mercury exposure from maternal dental amalgams and autism severity. *Acta Neurobiologiae Experimentalis, 69*, 189–197.

Geier, D. A., Kern, J. K., & Geier, M. R. (2010). The biological basis of autism spectrum disorders: Understanding causation and treatment by clinical geneticists. *Acta Neurobiologiae Experimentalis (Wars), 70*(2), 209–226.

Gelbier, S., & Ingram, J. (1989). Possible fetotoxic effects of mercury vapor: A case report. *Public Health, 103*(1), 35–40.

Gerhard, I., Monga, B., Waldbrenner, A., & Runnebaum, B. (1998). Heavy metals and fertility. *Journal of Toxicology and Environmental Health Part A, 54*, 593–612.

Gibičar, D., Horvat, M., Logar, M., Fajon, V., Falnoga, I., Ferrara, R., Lanzillotta, E., Ceccarini, C., Mazzolai, B., Denby, B., & Pacyna, J. (2009). Human exposure to mercury in the vicinity of chlor-alkali plant. *Environmental Research, 109*(4), 355–367.

Godfrey, M. E., Wojcik, D. P., & Krone, C. A. (2003). Apolipoprotein E genotyping as a potential biomarker for mercury neurotoxicity. *Journal of Alzheimer's Disease, 5*(3), 189–195.

Goering, P. L., Galloway, W. D., Clarkson, T. W., Lorscheider, F. L., Berlin, M., & Rowland, A. S. (1992). Toxicity assessment of mercury vapor from dental amalgams. *Fundamental and Applied Toxicology, 19*(3), 319–329.

Goldschmidt, P. R., Cogan, R. B., & Taubman, S. B. (1976). Effects of amalgam corrosion products on human cells. *Journal of Periodontal Research, 11*(2), 108–115.

Gonzalez-Ramirez, D., Maiorino, R. M., Zuniga-Charles, M., Xu, Z., Hurlbut, K. M., Junco-Munoz, P., Aposhian, M. M., Dart, R. C., Echeverria, D., Woods, J. S., & Aposhian, H. V. (1995). Sodium 2, 3-dimercaptopropane-1-sulfonate challenge test for mercury in humans: II. Urinary mercury, porphyrins and neurobehavioral changes of dental workers in Monterrey, Mexico. *Journal of Pharmacology and Experimental Therapeutics, 272*(1), 264–274.

Goodrich, J. M., Basu, N., Franzblau, A., & Dolinoy, D. C. (2013a). Mercury biomarkers and DNA methylation among Michigan dental professionals. *Environmental and Molecular Mutagenesis, 54*(3), 195–203.

Goodrich, J. M., Wang, Y., Gillespie, B., Werner, R., Franzblau, A., & Basu, N. (2011). Glutathione enzyme and selenoprotein polymorphisms associate with mercury biomarker levels in Michigan dental professionals. *Toxicology and Applied Pharmacology, 257*(2), 301–308.

Goodrich, J. M., Wang, Y., Gillespie, B., Werner, R., Franzblau, A., & Basu, N. (2013b). Methylmercury and elemental mercury differentially associate with blood pressure among dental professionals. *International Journal of Hygiene and Environmental Health, 216*(2), 195–201.

Gordon, H. (1981, September). Pregnancy in female dentists: A mercury hazard. In *Proceedings of international conference on mercury hazards in dental practice, Glasgow, Scotland* (pp. 2–4).

Gundacker, C., Komarnicki, G., Zödl, B., Forster, C., Schuster, E., & Wittmann, K. (2006). Whole blood mercury and selenium concentrations in a selected Austrian population: Does gender matter? *Science of the Total Environment, 372*(1), 76–86.

Guzzi, G., & Pigatto, P. D. (2008). Urinary mercury levels in children with amalgam fillings. *Environmental Health Perspectives, 116*(7), A286.

Hahn, L. J., Kloiber, R., Leininger, R. W., Vimy, M. J., & Lorscheider, F. L. (1990). Whole-body imaging of the distribution of mercury released from dental fillings into monkey tissues. *The FASEB Journal (Federation of American Societies for Experimental Biology), 4*(14), 3256–3260.

Hahn, L. J., Kloiber, R., Vimy, M. J., Takahashi, Y., & Lorscheider, F. L. (1989). Dental "silver" tooth fillings: A source of mercury exposure revealed by whole-body image scan and tissue analysis. *The FASEB Journal (Federation of American Societies for Experimental Biology), 3*(14), 2641–2646.

Haley, B. E. (2005). Mercury toxicity: Genetic susceptibility and synergistic effects. *Medical Veritas, 2*(2), 535–542.

Haley, B. E. (2007). The relationship of the toxic effects of mercury to exacerbation of the medical condition classified as Alzheimer's disease. *Medical Veritas, 4*(2), 1484–1498.

Hanson, M. (2004). Health and amalgam removal: A meta-analysis of 25 studies. *Tf-bladet Bull of the Swedish Association of Dental Mercury Patients,* (Tf-bladet no. 2 2004 and SOU 2003:53 appendix 10, Sw. Dept. of Health), 204–216.

Hanson, M., & Pleva, J. (1991). The dental amalgam issue. A review. *Experientia, 47*(1), 9–22.

Health and Environment Alliance and Health Care Without Harm. (2007, May). *Mercury and dental amalgams* [Fact sheet]. Retrieved from Health and Environment Alliance Web site http://www.env-health.org/IMG/pdf/HEA_009-07.pdf

Health Canada. (1996). The safety of dental amalgam. Retrieved from Health Canada Web site http://www.hc-sc.gc.ca/dhp-mps/alt_formats/hpfb-dgpsa/pdf/md-im/dent_amalgam-eng.pdf

Heggland, I., Irgens, Å. I., Tollånes, M., Romundstad, P., Syversen, T., Svendsen, K., Melø, I., & Hilt, B. (2011). Pregnancy outcomes among female dental personnel--a registry-based retrospective cohort study. *Scandinavian Journal of Work, Environment & Health, 37*(6), 539–546.

Heidam, L. Z. (1984). Spontaneous abortions among dental assistants, factory workers, painters, and gardening workers: A follow up study. *Journal of Epidemiology and Community Health, 38*(2), 149–155.

Heintze, U., Edwardsson, S., Dérand, T., & Birkhed, D. (1983). Methylation of mercury from dental amalgam and mercuric chloride by oral streptococci in vitro. *European Journal of Oral Sciences, 91*(2), 150–152.

Heintze, S. D., & Rousson, V. (2012). Clinical effectiveness of direct class II restorations—A meta-analysis. *The Journal of Adhesive Dentistry, 14*(5), 407–431.

Henriksson, E., Mattsson, U., & Håkansson, J. (1995). Healing of lichenoid reactions following removal of amalgam. *Journal of Clinical Periodontology, 22*(4), 287–294.

Herber, R. F., de Gee, A. J., & Wibowo, A. A. (1988). Exposure of dentists and assistants to mercury: Mercury levels in urine and hair related to conditions of practice. *Community Dentistry and Oral Epidemiology, 16*(3), 153–158.

Herrström, P., Högstedt, B., Aronson, S., Holmén, A., & Råstam, L. (1996). Acute glomerulonephritis, Henoch-Schönlein purpura and dental amalgam in Swedish children: A case-control study. *Science of the Total Environment, 191*(3), 277–282.

Herrström, P., Schütz, A., Raihle, G., Holthuis, N., Högstedt, B., & Råstam, L. (1995). Dental amalgam, low-dose exposure to mercury, and urinary proteins in young Swedish men. *Archives of Environmental Health: An International Journal, 50*(2), 103–107.

Hertz-Picciotto, I., Green, P. G., Delwiche, L., Hansen, R., Walker, C., & Pessah, I. N. (2010). Blood mercury concentrations in CHARGE Study children with and without autism. *Environmental Health Perspectives, 118*(1), 161–166.

Hilt, B., Svendsen, K., Syversen, T., Aas, O., Qvenild, T., Sletvold, H., & Melø, I. (2009). Occurrence of cognitive symptoms in dental assistants with previous occupational exposure to metallic mercury. *Neurotoxicology, 30*(6), 1202–1206.

Hock, C., Drasch, G., Golombowski, S., Müller-Spahn, F., Willershausen-Zönnchen, B., Schwarz, P., Hock, U., Growdon, J. H., & Nitsch, R. M. (1998). Increased blood mercury levels in patients with Alzheimer's disease. *Journal of Neural Transmission, 105*(1), 59–68.

Holmes, A. S., Blaxill, M. F., & Haley, B. E. (2003). Reduced levels of mercury in first baby haircuts of autistic children. *International Journal of Toxicology, 22*(4), 277–285.

Homme, K. G., Kern, J. K., Haley, B. E., Geier, D. A., King, P. G., Sykes, L. K., & Geier, M. R. (2014). New science challenges old notion that mercury dental amalgam is safe. *BioMetals, 27*(1), 19–24.

Hougeir, F. G., Yiannias, J. A., Hinni, M. L., Hentz, J. G., & El-Azhary, R. A. (2006). Oral metal contact allergy: A pilot study on the cause of oral squamous cell carcinoma. *International Journal of Dermatology, 45*(3), 265–271.

Houston, M. C. (2014). The role of mercury in cardiovascular disease. *Journal of Cardiovascular Diseases and Diagnosis, 2*(170), 2.

Hsu, C. S., Liu, P. L., Chien, L. C., Chou, S. Y., & Han, B. C. (2007). Mercury concentration and fish consumption in Taiwanese pregnant women. *BJOG: An International Journal of Obstetrics & Gynaecology, 114*(1), 81–85.

Huggins, H. A., & Levy, T. E. (1998). Cerebrospinal fluid protein changes in multiple sclerosis after dental amalgam removal. *Alternative Medicine Review, 3*, 295–300.

Hujoel, P. P., Lydon-Rochelle, M., Bollen, A. M., Woods, J. S., Geurtsen, W., & Del Aguila, M. A. (2005). Mercury exposure from dental filling placement during pregnancy and low birth weight risk. *American Journal of Epidemiology, 161*(8), 734–740.

Hultman, P., Johansson, U., Turley, S. J., Lindh, U., Eneström, S., & Pollard, K. M. (1994). Adverse immunological effects and autoimmunity induced by dental amalgam and alloy in mice. *The FASEB Journal (Federation of American Societies for Experimental Biology), 8*(14), 1183–1190.

Hursh, J. B., Greenwood, M. R., Clarkson, T. W., Allen, J., & Demuth, S. (1980). The effect of ethanol on the fate of mercury vapor inhaled by man. *Journal of Pharmacology and Experimental Therapeutics, 214*(3), 520–527.

Hyson, J. M., Jr. (2006). Amalgam: Its history and perils. *CDA Journal, 34*(3), 215–229.

Ibbotson, S. H., Speight, E. L., Macleod, R. I., Smart, E. R., & Lawrence, C. M. (1996). The relevance and effect of amalgam replacement in subjects with oral lichenoid reactions. *British Journal of Dermatology, 134*(3), 420–423.

Ingalls, T. H. (1983). Epidemiology, etiology, and prevention of multiple sclerosis: Hypothesis and fact. *The American Journal of Forensic Medicine and Pathology, 4*(1), 55–62.

Ingalls, T. (1986). Triggers for multiple sclerosis. *The Lancet, 328*(8499), 160.

Isacsson, G., Barregård, L., Seldén, A., & Bodin, L. (1997). Impact of nocturnal bruxism on mercury uptake from dental amalgams. *European Journal of Oral Sciences, 105*(3), 251–257.

Järup, L. (2003). Hazards of heavy metal contamination. *British Medical Bulletin, 68*(1), 167–182.

Jensen, S. J. (1985). Maximum contents of mercury in dental silver amalgams. *European Journal of Oral Sciences, 93*(1), 84–88.

Johnson, K. F. (1978). Mercury hygiene. *Dental Clinics of North America, 22*(3), 477.

Jokstad, A., Thomassen, Y., Bye, E., Clench-Aas, J., & Aaseth, J. (1992). Dental amalgam and mercury. *Pharmacology & Toxicology, 70*(4), 308–313.

Jones, D. W. (1998). A Canadian perspective on the dental amalgam issue. *British Dental Journal, 184*(12), 581–586.

Jones, D. W. (1999). Exposure or absorption and the crucial question of limits for mercury. *Journal of the Canadian Dental Association, 65*(1), 42–46.

Joshi, A., Douglass, C. W., Kim, H. D., Joshipura, K. J., Park, M. C., Rimm, E. B., Carino, M. J., Garcia, R. I., Morris, J. S., & Willett, W. C. (2003). The relationship between amalgam restorations and mercury levels in male dentists and nondental health professionals. *Journal of Public Health Dentistry, 63*(1), 52–60.

Kaaber, S. (1990). Allergy to dental materials with special reference to the use of amalgam and polymethylmethacrylate. *International Dental Journal, 40*(6), 359–365.

Kanerva, L., Lahtinen, A., Toikkanen, J., Forss, H., Estlander, T., Susitaival, P., & Jolanki, R. (1999). Increase in occupational skin diseases of dental personnel. *Contact Dermatitis, 40*(2), 104–108.

Kantarjian, A. D. (1961). A syndrome clinically resembling amyotrophic lateral sclerosis following chronic mercurialism. *Neurology, 11*(7), 639–644.

Karahalil, B., Rahravi, H., & Ertas, N. (2005). Examination of urinary mercury levels in dentists in Turkey. *Human & Experimental Toxicology, 24*(8), 383–388.

Kasarski, E. J., Haley, B. E., & Gunnerson, D. (1993). GTP-binding proteins in amyotrophic lateral sclerosis cerebrospinal fluid. *Annals of Neurology, 2*(34), 297.

Kehe, K., Reichl, F. X., Durner, J., Walther, U., Hickel, R., & Forth, W. (2001). Cytotoxicity of dental composite components and mercury compounds in pulmonary cells. *Biomaterials, 22*(4), 317–322.

Kern, J. K., Geier, D. A., Björklund, G., King, P. G., Homme, K. G., Haley, B. E., Sykes, L. K., & Geier, M. R. (2014). Evidence supporting a link between dental amalgams and chronic illness, fatigue, depression, anxiety, and suicide. *Neuroendocrinology Letters, 35*(7), 535–552.

Khare, S. S., Ehmann, W. D., Kasarskis, E. J., & Markesbery, W. R. (1989). Trace element imbalances in amyotrophic lateral sclerosis. *Neurotoxicology, 11*(3), 521–532.

Khwaja, M. A., & Abbasi, M. S. (2014). Mercury poisoning dentistry: High-level indoor air mercury contamination at selected dental sites. *Reviews on Environmental Health, 29*(1-2), 29–31.

Kingman, A., Albertini, T., & Brown, L. J. (1998). Mercury concentrations in urine and whole blood associated with amalgam exposure in a US military population. *Journal of Dental Research, 77*(3), 461–471.

Knudsson, K. (2012, April 19). NAV finally believes mercury victim. In *Tannhelsesekretærenes Forbund, Oslo, Norway.*

Koos, B. J., & Longo, L. D. (1976). Mercury toxicity in the pregnant woman, fetus, and newborn infant. A review. *American Journal of Obstetrics and Gynecology, 126*(3), 390–409.

Koral, S. M. (2013). Mercury from dental amalgam: Exposure and risk assessment. *Compendium of Continuing Education in Dentistry, 34*(2), 138–140.

Kostial, K., Rabar, I., Ciganovic, M., & Simonovic, I. (1979). Effect of milk on mercury absorption and gut retention in rats. *Bulletin of Environmental Contamination and Toxicology, 23*(1), 566–571.

Kraußß, P., Deyhle, M., Maier, K. H., Roller, E., Weiß, H. D., & Clédon, P. (1997). Field study on the mercury content of saliva. *Toxicological & Environmental Chemistry, 63*(1-4), 29–46.

Kruzikova, K., Kensova, R., Blahova, J., Harustiakova, D., & Svobodova, Z. (2009). Using human hair as an indicator for exposure to mercury. *Neuroendocrinology Letters, 30*, 177–181.

Kuntz, W. D., Pitkin, R. M., Bostrom, A. W., & Hughes, M. S. (1982). Maternal and cord blood background mercury levels: A longitudinal surveillance. *American Journal of Obstetrics and Gynecology, 143*(4), 440–443.

Laine, J., Kalimo, K., Forssell, H., & Happonen, R. (1992). Resolution of oral lichenoid lesions after replacement of amalgam restorations in patients allergic to mercury compounds. *British Journal of Dermatology, 126*(1), 10–15.

Langworth, S., Björkman, L., Elinder, C. G., Järup, L., & Savlin, P. (2002). Multidisciplinary examination of patients with illness attributed to dental fillings. *Journal of Oral Rehabilitation, 29*(8), 705–713.

Langworth, S., Elinder, C. G., & Sundqvist, K. G. (1993). Minor effects of low exposure to inorganic mercury on the human immune system. *Scandinavian Journal of Work, Environment & Health, 19*(6), 405–413.

Lee, J. Y., Yoo, J. M., Cho, B. K., & Kim, H. O. (2001). Contact dermatitis in Korean dental technicians. *Contact Dermatitis, 45*(1), 13–16.

Leistevuo, J., Leistevuo, T., Helenius, H., Pyy, L., Huovinen, P., & Tenovuo, J. (2002). Mercury in saliva and the risk of exceeding limits for sewage in relation to exposure to amalgam fillings. *Archives of Environmental Health: An International Journal, 57*(4), 366–370.

Leistevuo, J., Leistevuo, T., Helenius, H., Pyy, L., Österblad, M., Huovinen, P., & Tenovuo, J. (2001). Dental amalgam fillings and the amount of organic mercury in human saliva. *Caries Research, 35*(3), 163–166.

Leong, C. C., Syed, N. I., & Lorscheider, F. L. (2001). Retrograde degeneration of neurite membrane structural integrity of nerve growth cones following in vitro exposure to mercury. *NeuroReport, 12*(4), 733–737.

Levy, M. (1995). Dental amalgam: Toxicological evaluation and health risk assessment. *Journal (Canadian Dental Association), 61*(8), 667–668, 671–674.

Lind, P. O., Hurlen, B., Lyberg, T., & Aas, E. (1986). Amalgam-related oral lichenoid reaction. *European Journal of Oral Sciences, 94*(5), 448–451.

Lindberg, A., Ask Björnberg, K., Vahter, M., & Berglund, M. (2004). Exposure to methylmercury in non-fish-eating people in Sweden. *Environmental Research, 96*(1), 28–33.

Lindbohm, M. L., Ylöstalo, P., Sallmén, M., Henriks-Eckerman, M. L., Nurminen, T., Forss, H., & Taskinen, H. (2007). Occupational exposure in dentistry and miscarriage. *Occupational and Environmental Medicine, 64*(2), 127–133.

Lindow, S. W., Knight, R., Batty, J., & Haswell, S. J. (2003). Maternal and neonatal hair mercury concentrations-the effect of dental amalgam. *Journal of Obstetrics & Gynecology, 23*(S1), S48–S49.

Lindqvist, B., & Mörnstad, H. (1996). Effects of removing amalgam fillings from patients with diseases affecting the immune system. *Medical Science Research, 24*(5), 355–356.

Lönnroth, E. C., & Shahnavaz, H. (1995). Dental clinics--a burden to environment? *Swedish Dental Journal, 20*(5), 173–181.

Lorscheider, F. L., Leong, C. C. W., & Syed, N. I. (2001). *How mercury causes brain neuron degeneration* [Video file]. University of Calgary, Canada.

Lorscheider, F. L., Vimy, M. J., & Summers, A. O. (1995). Mercury exposure from "silver" tooth fillings: Emerging evidence questions a traditional dental paradigm. *The FASEB Journal (Federation of American Societies for Experimental Biology), 9*(7), 504–508.

Luglie, P. F., Campus, G., Chessa, G., Spano, G., Capobianco, G., Fadda, G. M., & Dessole, S. (2005). Effect of amalgam fillings on the mercury concentration in human amniotic fluid. *Archives of Gynecology and Obstetrics, 271*(2), 138–142.

Lundstrom, I. M. (1984). Allergy and corrosion of dental materials in patients with oral lichen planus. *International Journal of Oral Surgery, 13*(1), 16.

Lutz, E., Lind, B., Herin, P., Krakau, I., Bui, T. H., & Vahter, M. (1996). Concentrations of mercury, cadmium and lead in brain and kidney of second trimester fetuses and infants. *Journal of Trace Elements in Medicine and Biology, 10*(2), 61–67.

Lyttle, H. A., & Bowden, G. H. (1993). The level of mercury in human dental plaque and interaction in vitro between biofilms of Streptococcus mutans and dental amalgam. *Journal of Dental Research, 72*(9), 1320–1324.

MacEntee, M. I., & Mojon, P. (1991). Issues in the amalgam debate. *Journal (Canadian Dental Association), 57*(12), 931–936.

Mackey, T. K., Contreras, J. T., & Liang, B. A. (2014). The Minamata Convention on Mercury: Attempting to address the global controversy of dental amalgam use and mercury waste disposal. *Science of the Total Environment, 472*, 125–129.

Mahler, D. B., Adey, J. D., & Fleming, M. A. (1994). Hg emission from dental amalgam as related to the amount of Sn in the Ag-Hg (γ1) phase. *Journal of Dental Research, 73*(10), 1663–1668.

Makhija, S. K., Gordan, V. V., Gilbert, G. H., Litaker, M. S., Rindal, D. B., Pihlstrom, D. J., & Gvist, V. (2011). Dental practice-based research network restorative material: Findings from the characteristics associated with type of practitioner, patient and carious lesion. *Journal of the American Dental Association, 142*, 622–632.

Mandel, I. D. (1993). Occupational risks in dentistry: Comforts and concerns. *Journal of the American Dental Association (1939), 124*(10), 40–49.

Marinova, G., Chakarova, O., & Kaneva, Y. A. (1973). A study of reproductive function of women working with mercury. *Probl Akush Ginekol, 1*, 75–77.

Martin, M. D., Naleway, C., & Chou, H. N. (1995). Factors contributing to mercury exposure in dentists. *Journal of the American Dental Association, 126*(11), 1502–1511.

Mata, L., Sanchez, L., & Calvo, M. (1997). Interaction of mercury with human and bovine milk proteins. *Bioscience, Biotechnology, and Biochemistry, 61*(10), 1641–1644.

Mathewson, R. J. (1984). Restoration of primary teeth with amalgam. *Dental Clinics of North America, 28*(1), 137–143.

McGrother, C. W., Dugmore, C., Phillips, M. J., Raymond, N. T., Garrick, P., & Baird, W. O. (1999). Epidemiology: Multiple sclerosis, dental caries and fillings: A case-control study. *British Dental Journal, 187*(5), 261–264.

McParland, H., & Warnakulasuriya, S. (2012). Oral lichenoid contact lesions to mercury and dental amalgam—A review. *BioMed Research International, 2012*, 589569.

Melchart, D., Köhler, W., Linde, K., Zilker, T., Kremers, L., Saller, R., & Halbach, S. (2008). Biomonitoring of mercury in patients with complaints attributed to dental amalgam, healthy amalgam bearers, and amalgam-free subjects: A diagnostic study. *Clinical Toxicology, 46*(2), 133–140.

Michel, I., Norbäck, D., & Edling, C. (1989). An epidemiologic study of the relation between symptoms of fatigue, dental amalgam and other factors. *Swedish Dental Journal, 13*(1–2), 33–38.

Mikhailova, L. M., Kobyets, G. P., Lyubomudrov, V. E., & Braga, G. F. (1971). The influence of occupational factors on diseases of the female reproductive organs. *Pediatr Akush Ginekol, 33*, 56–58.

Miller, E. G., Perry, W. L., & Wagner, M. J. (1987). Prevalence of mercury hypersensitivity in dental students. *The Journal of Prosthetic Dentistry, 58*(2), 235–237.

Ministry of the Environment, Norway. (2007, December 21). *Minister of the Environment and International Development Erik Solheim: Bans mercury in products* [Press release]. Retrieved from Government of Norway Web site https://www.regjeringen.no/en/aktuelt/Bans-mercury-in-products/id495138/

Mitchell, R. J., Osborne, P. B., & Haubenreich, J. E. (2005). Dental amalgam restorations: Daily mercury dose and biocompatibility. *Journal of Long-Term Effects of Medical Implants, 15*(6), 709–721.

Moen, B., Hollund, B. E., & Riise, T. (2008). Neurological symptoms among dental assistants: A cross-sectional study. *Journal of Occupational Medicine and Toxicology, 3*, 10.

Molin, M., Bergman, B., Marklund, S. L., Schütz, A., & Skerfving, S. (1990). Mercury, selenium, and glutathione peroxidase before and after amalgam removal in man. *Acta Odontologica Scandinavica, 48*(3), 189–202.

Moore, R. [Mercury Exposure.] (2012, January 8). *Jeffrey Shuren, director of FDA's CDRH, will make end year (2011) announcement on dental amalgam* [Video File from FDA Townhall Meeting in California]. Retrieved from http://www.youtube.com/watch?v=H2t0J2_1yr0

Mortada, W. L., Sobh, M. A., El-Defrawy, M. M., & Farahat, S. E. (2002). Mercury in dental restoration: Is there a risk of nephrotoxicity? *Journal of Nephrology, 15*(2), 171–176.

Müller-Miny, H., Erber, D., Möller, H., Müller-Miny, B., & Bongartz, G. (1996). Is there a hazard to health by mercury exposure from amalgam due to MRI? *Journal of Magnetic Resonance Imaging, 6*(1), 258–260.

Mutter, J. (2011). Is dental amalgam safe for humans? The opinion of the scientific committee of the European Commission. *Journal of Occupational Medicine and Toxicology, 6*(1), 2.

Mutter, J., Naumann, J., Sadaghiani, C., Schneider, R., & Walach, H. (2004). Alzheimer disease: Mercury as pathogenetic factor and apolipoprotein E as a moderator. *Neuroendocrinology Letters, 25*(5), 331–339.

Mutter, J., Naumann, J., Schneider, R., Walach, H., & Haley, B. (2005). Mercury and autism: Accelerating evidence. *Neuroendocrinology Letters, 26*(5), 439–446.

Nataf, R., Skorupka, C., Amet, L., Lam, A., Springbett, A., & Lathe, R. (2006). Porphyrinuria in childhood autistic disorder: Implications for environmental toxicity. *Toxicology and Applied Pharmacology, 214*(2), 99–108.

Needleman, H. L. (2006). Mercury in dental amalgam—A neurotoxic risk? *Journal of the American Medical Association, 295*(15), 1835–1836.

Ng, D. K. K., Chan, C. H., Soo, M. T., & Lee, R. S. Y. (2007). Low-level chronic mercury exposure in children and adolescents: Meta-analysis. *Pediatrics International, 49*(1), 80–87.

Ngim, C., & Devathasan, G. (1989). Epidemiologic study on the association between body burden mercury level and idiopathic Parkinson's disease. *Neuroepidemiology, 8*(3), 128–141.

Ngim, C. H., Foo, S. C., Boey, K. W., & Jeyaratnam, J. (1992). Chronic neurobehavioural effects of elemental mercury in dentists. *British Journal of Industrial Medicine, 49*(11), 782–790.

Nicole, A., Santiard-Baron, D., & Ceballos-Picot, I. (1998). Direct evidence for glutathione as mediator of apoptosis in neuronal cells. *Biomedicine & Pharmacotherapy, 52*(9), 349–355.

Nielsen, J. B. (1992). Toxicokinetics of mercuric chloride and methylmercuric chloride in mice. *Journal of Toxicology and Environmental Health, Part A Current Issues, 37*(1), 85–122.

Nimmo, A., Werley, M. S., Martin, J. S., & Tansy, M. F. (1990). Particulate inhalation during the removal of amalgam restorations. *The Journal of Prosthetic Dentistry, 63*(2), 228–233.

Noe, F. E. (1959). Mercury as a potential hazard in medical laboratories. *The New England Journal of Medicine, 261*, 1002–1006.

Norouzi, E., Bahramifar, N., & Ghasempouri, S. M. (2012). Effect of teeth amalgam on mercury levels in the colostrums human milk in Lenjan. *Environmental Monitoring and Assessment, 184*(1), 375–380.

Nylandor, M. (1986). Mercury in pituitary glands of dentists. *The Lancet, 327*(8478), 442.

Nylander, M., Friberg, L., Eggleston, D., & Björkman, L. (1989). Mercury accumulation in tissues from dental staff and controls in relation to exposure. *Swedish Dental Journal, 13*(6), 235–243.

Nylander, M., Friberg, L., & Lind, B. (1987). Mercury concentrations in the human brain and kidneys in relation to exposure from dental amalgam fillings. *Swedish Dental Journal, 11*(5), 179–187.

Occupational Safety and Health Administration. (1970). *OSHA Act of 1970.* Retrieved from OSHA web site http://www.osha.gov/pls/oshaweb/owasrch.search_form?p_doc_type=OSHACT

Occupational Safety and Health Administration. (2011). *Workers' rights* (OSHA 3021-09R 2011). Retrieved from OSHA web site http://www.osha.gov/Publications/osha3021.pdf

Olfert, S. M. (2006). Reproductive outcomes among dental personnel: A review of selected exposures. *Journal (Canadian Dental Association), 72*(9), 821.

Oskarsson, A., Schütz, A., Skerfving, S., Hallén, I. P., Ohlin, B., & Lagerkvist, B. J. (1996). Total and inorganic mercury in breast milk and blood in relation to fish consumption and amalgam fillings in lactating women. *Archives of Environmental Health: An International Journal, 51*(3), 234–241.

Palkovicova, L., Ursinyova, M., Masanova, V., Yu, Z., & Hertz-Picciotto, I. (2008). Maternal amalgam dental fillings as the source of mercury exposure in developing fetus and newborn. *Journal of Exposure Science and Environmental Epidemiology, 18*(3), 326–331.

Palmer, R. F., Blanchard, S., & Wood, R. (2009). Proximity to point sources of environmental mercury release as a predictor of autism prevalence. *Health & Place, 15*(1), 18–24.

Pang, B. K., & Freeman, S. (1995). Oral lichenoid lesions caused by allergy to mercury in amalgam fillings. *Contact Dermatitis, 33*(6), 423–427.

Panova, Z., & Dimitrov, G. (1974). Ovarian function in women having professional contact with metallic mercury. *Akusherstvoi Ginekologiya, 13*(1), 29–34.

Parsell, D. E., Karns, L., Buchanan, W. T., & Johnson, R. B. (1996). Mercury release during autoclave sterilization of amalgam. *Journal of Dental Education, 60*(5), 453–458.

Patterson, J. E., Weissberg, B. G., & Dennison, P. J. (1985). Mercury in human breath from dental amalgams. *Bulletin of Environmental Contamination and Toxicology, 34*(1), 459–468.

Pérez-Gómez, B., Aragonés, N., Gustavsson, P., Plato, N., López-Abente, G., & Pollán, M. (2005). Cutaneous melanoma in Swedish women: Occupational risks by anatomic site. *American Journal of Industrial Medicine, 48*(4), 270–281.

Perim, S. I., & Goldberg, A. F. (1984). Mercury in hospital dentistry. *Special Care in Dentistry, 4*(2), 54–55.

Pesch, A., Wilhelm, M., Rostek, U., Schmitz, N., Weishoff-Houben, M., Ranft, U., & Idel, H. (2002). Mercury concentrations in urine, scalp hair, and saliva in children from Germany. *Journal of Exposure Analysis and Environmental Epidemiology, 12*(4), 252–258.

Pierce, P. E., Thompson, J. F., Likosky, W. H., Nickey, L. N., Barthel, W. F., & Hinman, A. R. (1972). Alkyl mercury poisoning in humans: Report of an outbreak. *Journal of the American Medical Association, 220*(11), 1439–1442.

Pirard, C., Koppen, G., De Cremer, K., Van Overmeire, I., Govarts, E., Dewolf, M. C., et al. (2014). Hair mercury and urinary cadmium levels in Belgian children and their mothers within the framework of the COPHES/DEMOCOPHES projects. *Science of the Total Environment, 472*, 730–740.

Pitkin, R. M., Bahns, J. A., Filer, L. J., & Reynolds, W. A. (1976). Mercury in human maternal and cord blood, placenta, and milk. *Experimental Biology and Medicine, 151*(3), 565–567.

Pizzichini, M., Fonzi, M., Gasparoni, A., & Fonzi, L. (2000). Salivary mercury levels in healthy donors with and without amalgam fillings. *Bulletin du Groupement International pour la Recherche Scientifique en Stomatologie et Odontologie, 42*(2), 88–93.

Pleva, J. (1992). Mercury from dental amalgams: Exposure and effects. *The International Journal of Risk and Safety in Medicine, 3*(1), 1–22.

Podzimek, S., Prochazkova, J., Buitasova, L., Bartova, J., Ulcova-Gallova, Z., Mrklas, L., & Stejskal, V. D. (2005). Sensitization to inorganic mercury could be a risk factor for infertility. *Neuroendocrinology Letters, 26*(4), 277–282.

Prochazkova, J., Sterzl, I., Kucerova, H., Bartova, J., & Stejskal, V. D. (2004). The beneficial effect of amalgam replacement on health in patients with autoimmunity. *Neuroendocrinology Letters, 25*(3), 211–218.

Raymond, L. J., & Ralston, N. V. (2004). Mercury: Selenium interactions and health implications. *Seychelles Medical and Dental Journal, 7*(1), 72–77.

Redhe, O., & Pleva, J. (1994). Recovery from amyotrophic lateral sclerosis and from allergy after removal of dental amalgam fillings. *The International Journal of Risk and Safety in Medicine, 4*(3), 229–236.

Richardson, G. M. (1995). *Assessment of mercury exposure risks from dental amalgam: Final report*. Health Canada.

Richardson, G. M. (2003). Inhalation of mercury-contaminated participate matter by dentists: An overlooked occupational risk. *Human and Ecological Risk Assessment: An International Journal, 9*(6), 1519–1531.

Richardson, G. M., Brecher, R. W., Scobie, H., Hamblen, J., Samuelian, J., & Smith, C. (2009). Mercury vapour (Hg0): Continuing toxicological uncertainties, and establishing a Canadian reference exposure level. *Regulatory Toxicology and Pharmacology, 53*(1), 32–38.

Richardson, G. M., Wilson, R., Allard, D., Purtill, C., Douma, S., & Graviere, J. (2011). Mercury exposure and risks from dental amalgam in the US population, post-2000. *Science of the Total Environment, 409*(20), 4257–4268.

Risher, J. F. (2003). Elemental mercury and inorganic mercury compounds: Human health aspects (Concise International Chemical Assessment Document 50. Published under the joint sponsorship of the United Nations Environment Programme, the International Labour Organization, and the World Health Organization). Retrieved from the International Programme on Chemical Safety Web site http://www.inchem.org/documents/cicads/cicads/cicad50.htm

Risher, J. F., & De Rosa, C. T. (2007). Inorganic: The other mercury. *Journal of Environmental Health, 70*(4), 9–16.

Roberts, H. W., Leonard, D., & Osborne, J. (2001). Potential health and environmental issues of mercury-contaminated amalgamators. *Journal of the American Dental Association, 132*(1), 58–64.

Roberts, M. C., Leroux, B. G., Sampson, J., Luis, H. S., Bernardo, M., & Leitão, J. (2008). Dental amalgam and antibiotic-and/or mercury-resistant bacteria. *Journal of Dental Research, 87*(5), 475–479.

Rojas, M., Seijas, D., Agreda, O., & Rodríguez, M. (2006). Biological monitoring of mercury exposure in individuals referred to a toxicological center in Venezuela. *Science of the Total Environment, 354*(2), 278–285.

Rothwell, J. A., & Boyd, P. J. (2008). Amalgam dental fillings and hearing loss. *International Journal of Audiology, 47*(12), 770–776.

Rowland, A. S., Baird, D. D., Weinberg, C. R., Shore, D. L., Shy, C. M., & Wilcox, A. J. (1994). The effect of occupational exposure to mercury vapour on the fertility of female dental assistants. *Occupational and Environmental Medicine, 51*(1), 28–34.

Rudner, E. J., Clendenning, W. E., Epstein, E., Fisher, A. A., Jillson, O. F., Jordan, W. P., Kanof, N., Larsen, W., Maibach, H., Mitchell, J. C., O'Quinn, S. E., Schorr, W. F., & Sulzberg, M. B. (1973). Epidemiology of contact dermatitis in North America. *Archives of Dermatology, 108*(4), 537–540.

Sällsten, G., Thoren, J., Barregård, L., Schütz, A., & Skarping, G. (1996). Long-term use of nicotine chewing gum and mercury exposure from dental amalgam fillings. *Journal of Dental Research, 75*(1), 594–598.

Saxe, S. R., Snowdon, D. A., Wekstein, M. W., Henry, R. G., Grant, F. T., Donegan, S. J., & Wekstein, D. R. (1995). Dental amalgam and cognitive function in older women: Findings from the Nun Study. *Journal of the American Dental Association, 126*(11), 1495–1501.

Schubert, J., Riley, E. J., & Tyler, S. A. (1978). Combined effects in toxicology-a rapid systematic testing procedure: Cadmium, mercury, and lead. *Journal of Toxicology and Environmental Health, Part A Current Issues, 4*(5-6), 763–776.

Schwarz, S., Husstedt, I., Bertram, H. P., & Kuchelmeister, K. (1996). Amyotrophic lateral sclerosis after accidental injection of mercury. *Journal of Neurology, Neurosurgery, and Psychiatry, 60*(6), 698.

Schweinsberg, F. (1994). Risk estimation of mercury intake from different sources. *Toxicology Letters, 72*(1), 345–351.

Science and Environmental Health Network. (1998). *Wingspread conference on the precautionary principle.* Retrieved from Science and Environmental Health Network Web site http://www.sehn.org/wing.html

Shapiro, I. M., Cornblath, D. R., Sumner, A. J., Sptiz, L. K., Uzzell, B., Ship, I. I., & Bloch, P. (1982). Neurophysiological and neuro-psychological function in mercury-exposed dentists. *Lancet, 319*(8282), 1147–1150.

Siblerud, R. L. (1990). The relationship between mercury from dental amalgam and the cardiovascular system. *Science of the Total Environment, 99*(1), 23–35.

Siblerud, R. L. (1992). A comparison of mental health of multiple sclerosis patients with silver/mercury dental fillings and those with fillings removed. *Psychological Reports, 70*(3c), 1139–1151.

Siblerud, R. L., & Kienholz, E. (1994). Evidence that mercury from silver dental fillings may be an etiological factor in multiple sclerosis. *Science of the Total Environment, 142*(3), 191–205.

Siblerud, R. L., Motl, J., & Kienholz, E. (1994). Psychometric evidence that mercury from silver dental fillings may be an etiological factor in depression, excessive anger, and anxiety. *Psychological Reports, 74*(1), 67–80.

Sikorski, R., Juszkiewicz, T., Paszkowski, T., & Szprengier-Juszkiewicz, T. (1987). Women in dental surgeries: Reproductive hazards in occupational exposure to metallic mercury. *International Archives of Occupational and Environmental Health, 59*(6), 551–557.

Simecek, J. W., Diefenderfer, K. E., & Cohen, M. E. (2009). An evaluation of replacement rates for posterior resin-based composite and amalgam restorations in US Navy and Marine Corps recruits. *Journal of the American Dental Association, 140*(2), 200–209.

Sjursen, T. T., Lygre, G. B., Dalen, K., Helland, V., Laegreid, T., Svahn, J., Lundekvam, B. F., & Björkman, L. (2011). Changes in health complaints after removal of amalgam fillings. *Journal of Oral Rehabilitation, 38*(11), 835–848.

Skare, I., & Engqvist, A. (1994). Human exposure to mercury and silver released from dental amalgam restorations. *Archives of Environmental Health: An International Journal, 49*(5), 384–394.

Snapp, K. R., Boyer, D. B., Peterson, L. C., & Svare, C. W. (1989). The contribution of dental amalgam to mercury in blood. *Journal of Dental Research, 68*(5), 780–785.

Snyder, R. D. (1971). Congenital mercury poisoning. *New England Journal of Medicine, 284*(18), 1014–1016.

Spencer, A. J. (2000). Dental amalgam and mercury in dentistry. *Australian Dental Journal, 45*(4), 224–234.

Stejskal, V. D., Danersund, A., Lindvall, A., Hudecek, R., Nordman, V., Yaqob, A., Mayer, W., Bieger, W., & Lindh, U. (1999). Metal-specific lymphocytes: Biomarkers of sensitivity in man. *Neuroendocrinology Letters, 20*, 289–298.

Stejskal, V., Ocker, K., & Bjørklund, G. (2013). Metal-induced inflammation triggers fibromyalgia in metal-allergic patients. *Neuroendocrinology Letters, 34*(6), 559–565.

Stejskal, J., & Stejskal, V. D. (1999). The role of metals in autoimmunity and the link to neuroendocrinology. *Neuroendocrinology Letters, 20*(6), 351–366.

Sterzl, I., Procházková, J., Hrdá, P., Bártová, J., Matucha, P., & Stejskal, V. D. (1999). Mercury and nickel allergy: Risk factors in fatigue and autoimmunity. *Neuroendocrinology Letters, 20*, 221–228.

Summers, A. O., Wireman, J., Vimy, M. J., Lorscheider, F. L., Marshall, B., Levy, S. B., Bennet, S., & Billard, L. (1993). Mercury released from dental "silver" fillings provokes an increase in mercury-and antibiotic-resistant bacteria in oral and intestinal floras of primates. *Antimicrobial Agents and Chemotherapy, 37*(4), 825–834.

Svare, C. W., Peterson, L. C., Reinhardt, J. W., Frank, C. W., & Boyer, D. B. (1980). Dental amalgam: A potential source of mercury vapor exposure. *Journal of Dental Research, 59*(special issue A), 341. Abstract #293.

Swedish Chemicals Agency. (2008). The Swedish Chemicals Agency's chemical products and biotechnical organisms regulations (KIFS 2008: 2 in English, consolidated up to KIFS 2012: 3).

Thompson, C. M., Markesbery, W. R., Ehmann, W. D., Mao, Y. X., & Vance, D. E. (1987). Regional brain trace-element studies in Alzheimer's disease. *Neurotoxicology, 9*(1), 1–7.

Thygesen, L. C., Flachs, E. M., Hanehøj, K., Kjuus, H., & Juel, K. (2011). Hospital admissions for neurological and renal diseases among dentists and dental assistants occupationally exposed to mercury. *Occupational and Environmental Medicine, 68*(12), 895–901.

Tomka, M., Machovcová, A., Pelclová, D., Petanová, J., Arenbergerová, M., & Procházková, J. (2011). Orofacial granulomatosis associated with hypersensitivity to dental amalgam. *Oral Surgery, Oral Medicine, Oral Pathology, Oral Radiology, and Endodontology, 112*(3), 335–341.

Trivedi, S. C., & Talim, S. T. (1973). The response of human gingiva to restorative materials. *The Journal of Prosthetic Dentistry, 29*(1), 73–80.

Trott, J. R., & Sherkat, A. (1964). Effect of class II amalgam restorations on health of the gingiva: A clinical survey. *Journal of the Canadian Dental Association, 30*(12), 766–770.

Turgeon, J., LeMay, L. P., & Cleroux, R. (1972). Periodontal effects of restoring proximal tooth surfaces with amalgam: A clinical evaluation in children. *Journal of the Canadian Dental Association, 38*(7), 255–256.

United Nations Environment Programme. (1992). *Rio declaration on environment and development*. Retrieved from UNEP Web site http://www.unep.org/documents.multilingual/default.asp?documentid=78&articleid=1163

United Nations Environment Programme. (2013). *Minamata convention on mercury*. (UNEP[DTIE]/Hg/INC.5/3). Retrieved from UNEP Mercury Convention Web site http://www.mercuryconvention.org/Portals/11/documents/conventionText/Minamata%20Convention%20on%20Mercury_e.pdf

United States Food and Drug Administration. (2009a). *Dental devices: Classification of dental amalgam, reclassification of dental mercury, designation of special controls for dental amalgam, mercury, and amalgam alloy. Final rule* [Fed Regist. 2009; 74(148):38685-714. 21 CFR Part 872. (Docket No. FDA-2008-N-0163) {formerly Docket No. 2001N-0067} RIN 0910-AG21]. Retrieved from FDA Web site http://www.fda.gov/downloads/MedicalDevices/ProductsandMedicalProcedures/DentalProducts/DentalAmalgam/UCM174024.pdf

United States Food and Drug Administration. (2009b). Medical devices: About dental amalgam fillings. Retrieved from FDA web site http://www.fda.gov/medicaldevices/productsandmedicalprocedures/dentalproducts/dentalamalgam/ucm171094.htm

United States Food and Drug Administration. (2009c, July 28). *FDA issues final regulation on dental amalgam* [Press release]. Retrieved from FDA web site http://www.fda.gov/NewsEvents/Newsroom/Pressannouncements/ucm173992.htm

United States Food and Drug Administration. (2015a, January 27). Letter to James Love and Robert Reeves re: Citizen Petition Citizen Position Docket No. FDA-2009-P-0357.

United States Food and Drug Administration. (2015b, January 27). Letter to James Love and Robert Reeves re: Citizen Petition Citizen Position Docket No. FDA-2009-P-0610 (formerly FDA 2008-N-0163).

Ursinyova, M., Masanova, V., Palkovicova, L., & Wsolova, L. (2006). The influence of mother's dental amalgam fillings on prenatal and postnatal exposure of children to mercury. *Epidemiology, 17*(6), S494–S495.

Uzzell, B. P., & Oler, J. (1986). Chronic low-level mercury exposure and neuropsychological functioning. *Journal of Clinical and Experimental Neuropsychology, 8*(5), 581–593.

Vahter, M., Åkesson, A., Lind, B., Björs, U., Schütz, A., & Berglund, M. (2000). Longitudinal study of methylmercury and inorganic mercury in blood and urine of pregnant and lactating women, as well as in umbilical cord blood. *Environmental Research, 84*(2), 186–194.

Van Zyl, I. (1999). Mercury amalgam safety: A review. *The Journal of the Michigan Dental Association, 81*(1), 40–48.

Vance, D. E., Ehmann, W. D., & Markesbery, W. R. (1988). Trace element imbalances in hair and nails of Alzheimer's disease patients. *Neurotoxicology, 9*(2), 197–208.

Venclikova, Z., Benada, O., Bartova, J., Joska, L., Mrklas, L., Prochazkova, J., Stejskal, V., & Podzimek, S. (2006). In vivo effects of dental casting alloys. *Neuro Endocrinology Letters, 27*, 61–68.

Verschoor, M. A., Herber, R. F., & Zielhuis, R. L. (1988). Urinary mercury levels and early changes in kidney function in dentists and dental assistants. *Community Dentistry and Oral Epidemiology, 16*(3), 148–152.

Vimy, M. J., Hooper, D. E., King, W. W., & Lorscheider, F. L. (1997). Mercury from maternal "silver" tooth fillings in sheep and human breast milk. *Biological Trace Element Research, 56*(2), 143–152.

Vimy, M. J., & Lorscheider, F. L. (1985a). Intra-oral air mercury released from dental amalgam. *Journal of Dental Research, 64*(8), 1069–1071.

Vimy, M. J., & Lorscheider, F. L. (1985b). Serial measurements of intra-oral air mercury: Estimation of daily dose from dental amalgam. *Journal of Dental Research, 64*(8), 1072–1075.

Vimy, M. J., Luft, A. J., & Lorscheider, F. L. (1986). Estimation of mercury body burden from dental amalgam: Computer simulation of a metabolic compartmental model. *Journal of Dental Research, 65*(12), 1415–1419.

Vimy, M. J., Takahashi, Y., & Lorscheider, F. L. (1990). Maternal-fetal distribution of mercury (203 Hg) released from dental amalgam fillings. *American Physiology Society-Regulatory, Integrative and Comparative Physiology, 258*(4), R939–R945.

Votaw, A. L., & Zey, J. (1991). Vacuuming a mercury-contaminated dental office may be hazardous to your health. *The Dental Assistant, 60*(1), 27–29.

Wang, Y., Goodrich, J. M., Gillespie, B., Werner, R., Basu, N., & Franzblau, A. (2012). An investigation of modifying effects of metallothionein single-nucleotide polymorphisms on the association between mercury exposure and biomarker levels. *Environmental Health Perspectives, 120*(4), 530–534.

Warfvinge, K., Hansson, H., & Hultman, P. (1995). Systemic autoimmunity due to mercury vapor exposure in genetically susceptible mice: Dose-response studies. *Toxicology and Applied Pharmacology, 132*(2), 299–309.

Warwick, R., O'Connor, A., & Lamey, B. (2013). Mercury vapour exposure during dental student training in amalgam removal. *Journal of Occupational Medicine and Toxicology, 8*(1), 27.

Watson, G. E., Evans, K., Thurston, S. W., van Wijngaarden, E., Wallace, J. M., McSorley, E. M., Bonham, M. P., Mulhern, M. S., McAfee, A. J., Davidson, P. W., Shamlaye, C. F., Strain, J. J., Love, T., Zareba, G., & Myers, G. J. (2012). Prenatal exposure to dental amalgam in the Seychelles Child Development Nutrition Study: Associations with neurodevelopmental outcomes at 9 and 30 months. *Neurotoxicology, 33*(6), 1511–1517.

Weber, M. E., Yiannias, J. A., Hougeir, F. G., Kyle, A., Noble, B. N., Landry, A. M., & Hinni, M. L. (2012). Intraoral metal contact allergy as a possible risk factor for oral squamous cell carcinoma. *Annals of Otology, Rhinology & Laryngology, 121*(6), 389–394.

Weiner, J. A., Nylander, M., & Berglund, F. (1990). Does mercury from amalgam restorations constitute a health hazard? *Science of the Total Environment, 99*(1), 1–22.

Wenstrup, D., Ehman, W. D., & Markesbery, W. R. (1990). Trace element imbalances in isolated subcellular fractions of Alzheimer's disease brains. *Brain Research, 533*(1), 125–131.

White, R. R., & Brandt, R. L. (1976). Development of mercury hypersensitivity among dental students. *Journal of the American Dental Association, 92*(6), 1204–1207.

Williams, D. (2008). Caution and causation: Lessons from the delicate story of dental amalgam. *Medical Device Technology, 19*(5), 8–10.

Wojcik, D. P., Godfrey, M. E., Christie, D., & Haley, B. E. (2006). Mercury toxicity presenting as chronic fatigue, memory impairment and depression: Diagnosis, treatment, susceptibility, and outcomes in a New Zealand general practice setting (1994-2006). *Neuro Endocrinology Letters, 27*(4), 415–423.

Wong, L., & Freeman, S. (2003). Oral lichenoid lesions (OLL) and mercury in amalgam fillings. *Contact Dermatitis, 48*(2), 74–79.

Woods, J. S., Armel, S. E., Fulton, D. I., Allen, J., Wessels, K., Simmonds, P. L., Granpeesheh, D., Mumper, E., Bradstreet, J. J., Echeverria, D., Heyer, N. J., & Rooney, J. P. (2010). Urinary porphyrin excretion in neurotypical and autistic children. *Environmental Health Perspectives, 118*(10), 1450–1457.

Woods, J. S., Heyer, N. J., Echeverria, D., Russo, J. E., Martin, M. D., Bernardo, M. F., Luis, H. S., Vaz, L., & Farin, F. M. (2012). Modification of neurobehavioral effects of mercury by a genetic polymorphism of coproporphyrinogen oxidase in children. *Neurotoxicology and Teratology, 34*(5), 513–521.

Woods, J. S., Heyer, N. J., Russo, J. E., Martin, M. D., Pillai, P. B., Bammler, T. K., & Farin, F. M. (2014). Genetic polymorphisms of catechol-*O*-methyltransferase modify the neurobehavioral effects of mercury in children. *Journal of Toxicology and Environmental Health, Part A, 77*(6), 293–312.

Woods, J. S., Heyer, N. J., Russo, J. E., Martin, M. D., Pillai, P. B., & Farin, F. M. (2013). Modification of neurobehavioral effects of mercury by genetic polymorphisms of metallothionein in children. *Neurotoxicology and Teratology, 39*, 36–44.

Woods, J. S., Martin, M. D., Leroux, B. G., DeRouen, T. A., Bernardo, M. F., Luis, H. S., Leitao, J. G., Kushleika, J. V., Rue, T. C., & Korpak, A. M. (2008). Biomarkers of kidney integrity in children and adolescents with dental amalgam mercury exposure: Findings from the Casa Pia children's amalgam trial. *Environmental Research, 108*(3), 393–399.

Woods, J. S., Martin, M. D., Leroux, B. G., DeRouen, T. A., Leitão, J. G., Bernardo, M. F., Luis, H. S., Simmonds, P. L., Kushleika, J. V., & Huang, Y. (2007). The contribution of dental amalgam to urinary mercury excretion in children. *Environmental Health Perspectives, 115*(10), 1527–1531.

World Health Organization. (1991). *Environmental health criteria 118: Inorganic mercury* (Draft by L. Friberg published jointly by the United Nations Environment Programme, the International Labour Organisation, and the World Health Organization). Retrieved from International Programme on Chemical Safety Web site http://www.inchem.org/documents/ehc/ehc/ehc118.htm

World Health Organization. (2005). *Mercury in health care: Policy paper*. Retrieved from World Health Organization Web site http://www.who.int/water_sanitation_health/medicalwaste/mercurypolpaper.pdf

Ye, X., Qian, H., Xu, P., Zhu, L., Longnecker, M. P., & Fu, H. (2009). Nephrotoxicity, neurotoxicity, and mercury exposure among children with and without dental amalgam fillings. *International Journal of Hygiene and Environmental Health, 212*(4), 378–386.

Ylinen, K., & Löfroth, G. (2002). Nordic dentists' knowledge and attitudes on dental amalgam from health and environmental perspectives. *Acta Odontologica, 60*(5), 315–320.

Zahir, F., Rizwi, S. J., Haq, S. K., & Khan, R. H. (2005). Low dose mercury toxicity and human health. *Environmental Toxicology and Pharmacology, 20*(2), 351–360.

Zamm, A. V. (1986). Candida albicans therapy. *Journal of Orthomolecular Medicine, 1*(4), 261–265.

Zamm, A. (1991). Dental mercury: A factor that aggravates and induces xenobiotic intolerance. *Journal of Orthomolecular Medicine, 6*(2), 67–77.

Zander, H. A. (1957). Effect of silicate cement and amalgam on the gingiva. *Journal of the American Dental Association, 55*(1), 11.

Ziff, M. F. (1992). Documented clinical side-effects to dental amalgam. *Advances in Dental Research, 6*(1), 131–134.

Zimmer, H., Ludwig, H., Bader, M., Bailer, J., Eickholz, P., Staehle, H. J., & Triebig, G. (2002). Determination of mercury in blood, urine and saliva for the biological monitoring of an exposure from amalgam fillings in a group with self-reported adverse health effects. *International Journal of Hygiene and Environmental Health, 205*(3), 205–211.

Chapter 8
Translational Animal Models for the Study of Epigenetics and the Environment

Chris Murgatroyd and Steven Bradburn

8.1 Introduction

Cells of a multicellular organism are genetically identical but structurally and functionally distinct, owing to the differential expression of genes. Many of these differences in gene expression arise during development and are subsequently retained through mitosis. To represent the way that these developmental decisions are made, Conrad Hal Waddington in the 1940s and 1950s formulated the concept of the epigenetic landscape and the term epigenetics to refer to mechanisms that act on genes to govern a resulting phenotype. The term epigenetics is now commonly used to describe the study of stable alterations in gene expression potential that arise during development and differentiation and under the influence of the environment (for review, see Murgatroyd and Spengler 2011a). Therefore, in contrast to DNA sequence that is identical in all tissues, the patterns of epigenetic marks are tissue specific. Hence, a genome can be considered to contain two layers of information: the DNA sequence inherited from our parents which is conserved throughout life and mostly identical in all cells and tissues of our body and epigenetic marks (i.e., chromatin and DNA methylation patterns) which are cell and tissue specific. As such, an epigenome denotes the ensemble of coordinated epigenetic marks that govern accessibility of the DNA to the machinery driving gene expression; inaccessible genes become silenced, whereas accessible genes are actively transcribed. In this way, epigenetics could be thought of as conferring additional plasticity to the hard-coded genome (Fig. 8.1).

Although the understanding of the interplay between epigenetic modifications is still evolving, DNA methylation at CpG dinucleotide sequences and modification

C. Murgatroyd, Ph.D. (✉) • S. Bradburn, Ph.D. Candidate
School of Healthcare Science, Manchester Metropolitan University,
Chester Street, Manchester M1 5GD, UK
e-mail: c.murgatroyd@mmu.ac.uk; s.bradburn@mmu.ac.uk

D. Hollar (ed.), *Epigenetics, the Environment, and Children's Health Across Lifespans*, DOI 10.1007/978-3-319-25325-1_8

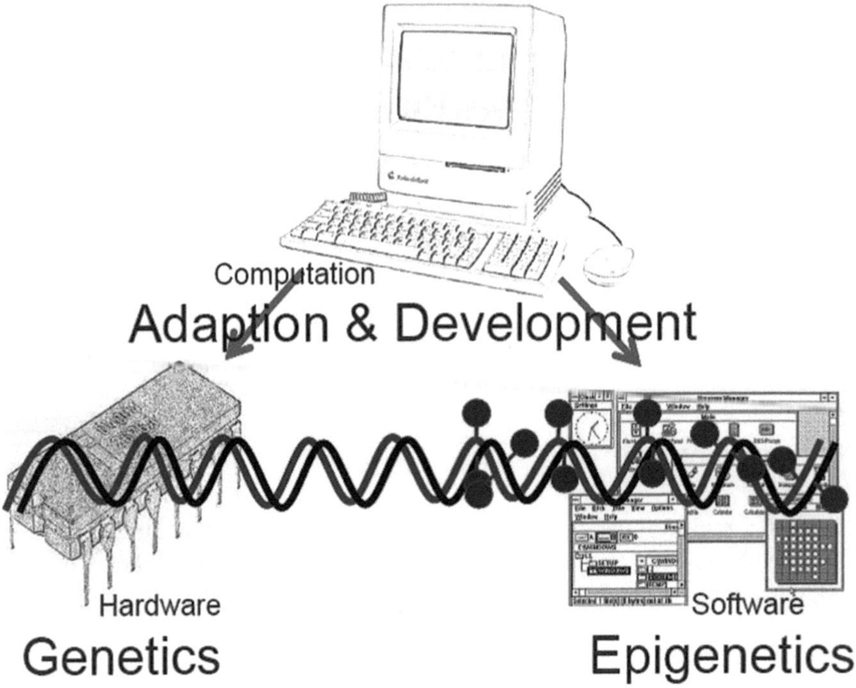

Fig. 8.1 Using an analogy of a computer, DNA sequence could be likened to the hardware, while the operating system, i.e., the epigenetic programming, enables the processing of decisions regarding which functions the DNA hardware does and does not perform. Environmental conditions could be thought of as inputted data that allow the epigenetic software to meet nature's goals of adaptation

of core histones (e.g., histone deacetylation and methylation of histones such as H3 at lysine 9) that package DNA into chromatin are the best-characterized covalent modifications associated with a repressed chromatin state and gene silencing.

Epigenetic regulation of gene expression therefore allows the integration of intrinsic and environmental signals in the genome, thus facilitating the adaptation of an organism to changing environment through alterations in gene activity. In the context of the early-life environment, epigenetic changes offer a plausible mechanism by which experiences during prenatal, postnatal and adulthood could be integrated into the genome to program long-term physiological, hormonal and behavioural responses (Fig. 8.2).

The field of epigenetics, particularly in relation to environmental exposures, has rapidly expanded with animal models, particularly rodents, providing a valuable resource in this field. Numerous studies have been able to shed light on epigenetic mechanisms regulating various environmental exposures such as diet, toxicants and stress, either prenatally, postnatally, in adulthood or transgenerationally. Here we briefly discuss some of the rodent models that have been employed to understand how these epigenetic changes underpin diverse physiological phenomena and particularly how these translate to the clinic and human disease.

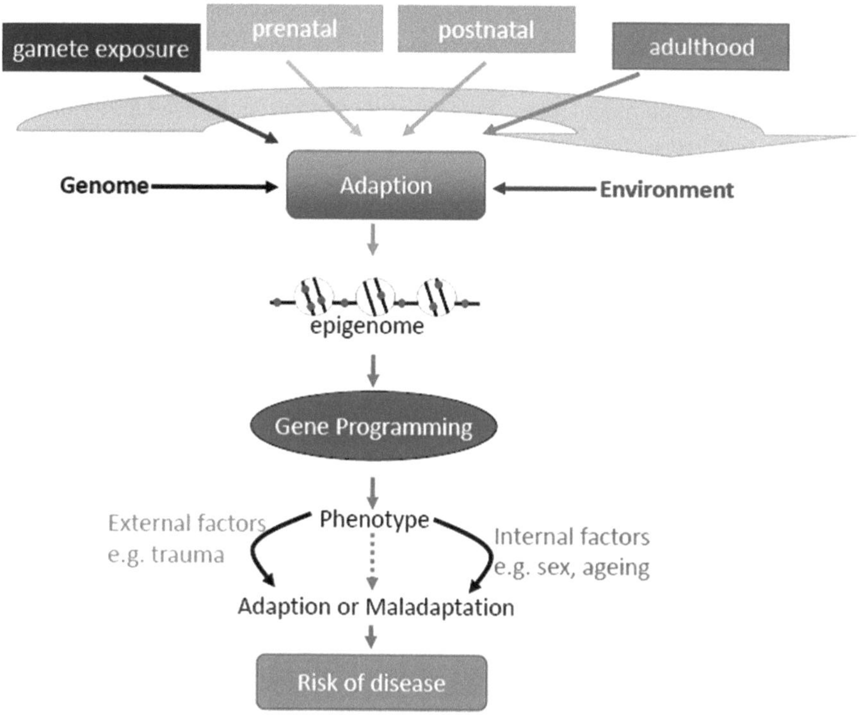

Fig. 8.2 Long-term programming of the stress system takes place during critical developmental time windows. The environment from gamete, prenatal, postnatal and adulthood can persistently alter expression levels of key genes through epigenetic marking, thus initiating adjustments in biology throughout later life. The nature of the environment and experiences throughout later life, in addition to the impact of biological processes associated with ageing and genetic sex, may exacerbate the programming established during early life, resulting in increased vulnerability to and manifestation of diseases

8.2 Animal Models for Epigenetic Studies

For epigenetic-based studies, rodent models have proved to be particularly useful for numerous reasons. They have a relatively short life span (mice >2.5, rats >3.5 years) that can allow rapid testing, establishment or development of experimental models and paradigms covering early life to ageing. They reach sexual maturity early (6–8 weeks) and demonstrate an abbreviated gestational period (rats, 19–22; mice, 18–21 days), further allowing several generations to be analysed within a relatively short time frame. That rats and mice are polytocous (litter bearing) is also an important feature when biological replicates, i.e., studying epigenetic alterations in multiple animals under the same environmental cues (e.g., exposed to the same dam), are considered and when tissue availability is a limiting factor, especially if several methods are used to address different types of epigenetic changes, i.e., either

DNA methylation or chromatin modifications. Using inbred mice, it is also possible to control for genetic background, avoiding genetic influences on epigenetics while there is also a range of genetically engineered mouse lines. While the mouse model, as opposed to the rat model, benefits from better support in the availability of various assay platforms (especially arrays), rats are considered by some better suited to some behavioural and cognitive tests.

Although rodent models are highly versatile for studying epigenetics, human and rodent epigenetic programs cannot be considered as 100 % identically conserved, for obvious reasons. So there are important limitations. For example, patterns of DNA methylation during development, particularly in the late blastocyst stage where cells of the inner cell mass and trophectoderm are differentially methylated, are different in the mouse and human embryo (inverse relationship between trophectoderal cells and inner cell mass) (Fulka et al. 2004). Of further importance is the role genetic background plays in regulating epigenetic marks. One study using seven mouse strains exposed to 1,3-butadiene identified major epigenetic differences in global DNA methylations and histone modifications between the lines (Koturbash et al. 2011), while another study revealed strain- or genotype-specific epigenetic effects as a consequence of stress (Uchida et al. 2011).

8.3 Animal Model Mutations in the Epigenetic Machinery

Deregulation of epigenetic pathways can alter regulation of specific sets of genes manifesting with diseases. As such, there are a large number of genes encoding epigenetic regulators that, when mutated, can give rise to various diseases, particularly mental retardation (for review, see Murgatroyd and Spengler 2012). Animal models are particularly useful for studying diseases with a genetic basis as technological advances in genetic engineering of mice allow systems for functional genomic investigations such as knockout and knockin strategies in which either an existing gene is inactivated by replacing it or disrupting it with an artificial piece of DNA or where a desired gene is inserted into a specific locus. Conditional and inducible methods further permit cell-specific regulation of gene expression within select tissues at specific stages of development or periods of time.

One of the most common causes of mental retardation in females is Rett syndrome, a progressive neurodevelopmental disorder frequently caused by a mutation to the gene for methyl-binding protein MECP2 (Amir et al. 1999). The onset of symptoms at 6–18 months of age coincides with a time in early development when sensory experience is driving dendritic pruning and shaping connections in the brain (Samaco and Neul 2011). Although MECP2 is expressed by the majority of cells, it is particularly important for normal neuronal function. In mice, a conditional knockout of Mecp2 specifically in neurons is sufficient to recapitulate the majority of Rett symptoms (e.g., Guy et al. 2001). The phenotype of Mecp2 mutant mice can be reversed by restoration of the Mecp2 gene in postmitotic neurons (e.g., Guy et al. 2007). Mecp2 protein is regulated by neuronal activity (e.g., Murgatroyd et al. 2009) and in turn regulates the expression of Bdnf, which has enhanced expression following depolarization (Martinowich et al. 2003). The overexpression of Bdnf in postmitotic neurons of Mecp2 mutant mice ameliorates their phenotype, suggesting that Mecp2 is critical for regulating the expression of genes like Bdnf that are regulated by neuronal activity and essential for normal cognitive function (Chang et al. 2006).

In another example, patients with hereditary sensory and autonomic neuropathy type 1 (HSAN1) develop dementia in adulthood that results from a mutation in the N-terminal regulatory domain of DNA methyltransferase 1 (DNMT1) (Klein et al. 2011). Complete knockout of Dnmt1 gene in mice results in embryonic lethality between embryonic day (E) 8 and E10.5 (Li et al. 1992), whereas a conditional knockout of Dnmt1 between E8.5 and E13.5, a time period that coincides with neurogenesis, leads to hypomethylation of differentiating neurons and demethylation of the Gfap promoter in neural precursor cells, thus accelerating astrogliosis (e.g., Fan et al. 2001).

8.4 Agouti and Axin 1 Fused Mice Models as "Epigenetic Biosensors"

A number of select markers have been identified in animals that have a well-characterized locus whose methylation pattern governs dramatic phenotypic outcomes, providing distinct visual markers (e.g., coat colour with the agouti viable yellow (Avy) mice and tail morphology in the axin 1 fused (Axin1Fu) mice). These are examples of metastable epialleles; identical alleles (often associated with retro-elements and transgenesis) that are variably expressed due to epigenetic modifications established early in development such as in response to maternal nutrition and early environmental exposures. From this standpoint, these mice are a valuable animal model for studying how maternal diet can alter epigenetic and accompanying phenotypic responses in her offspring. Approximately, a thousand copies of IAP retrotransposons are present in the mouse genome, and about 40 % of the human genome is comprised of transposable elements, of which approximately 9 % are retrotransposons. So, though neither of Avy or Axin1Fu genes are present in the human genome, changes to retrotransposons in general might have important implications to humans. These models as sensors of epigenetic marks established early in development have been used to investigate the impacts of nutritional and environmental influences on the foetal epigenome. In the agouti mouse model, DNA methylation and histone modification status determines the relative extent to which pheomelanin and eumelanin pigment proteins are expressed in hair follicles to establish the characteristic yellow coat colour (Fig. 8.3). As such, this model has proven to be an exceptional model in which to interrogate environmental and nutritional factors that are capable of altering critical outcomes (for review, see Dolinoy 2008).

Interestingly, expression of the agouti gene at sites other than the skin and follicle can have further significant systemic effects. For example, yellow Avy mice are

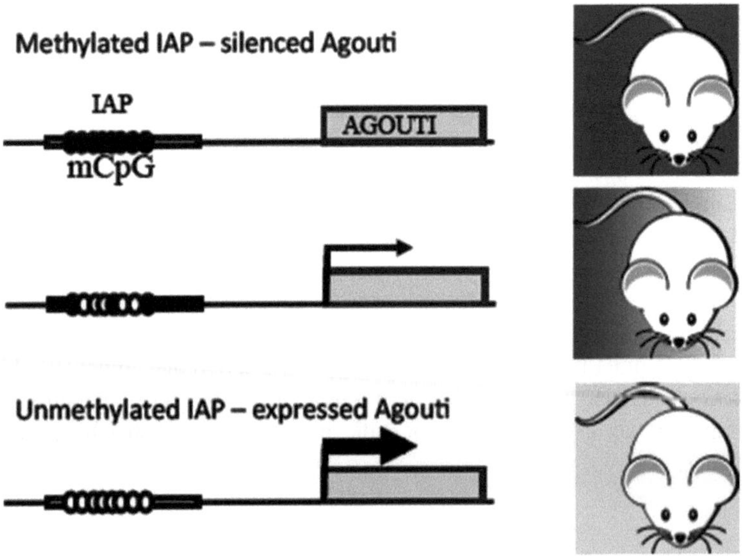

Fig. 8.3 In the agouti wild-type mouse, transcription of the agouti gene is controlled by a follicle-specific promoter. Sensitivity to DNA methylation (mCpG) status arises from the insertion of an IAP element that has master regulation of agouti expression. Hypomethylation of the IAP results in constitutive ectopic agouti expression and is manifested as the characteristic yellow coat colour. Increasing degrees of DNA methylation in the IAP result in progressive mottling until the mice eventually appear completely brown, an indication that complete hypermethylation of IAP has occurred to create a pseudoagouti mouse

characteristically obese because of the selective agouti-mediated antagonism of the melanocortin-4 receptor in the hypothalamus (Krashes et al. 2011). The impacts of other environmental impacts on this model, such as maternal stress, would also be interesting to test.

8.5 Animal Models of Epigenetics and Nutrition

Studies have applied various nutrients or diets to rodents, either before and during gestation or postnatally, to determine impact upon the epigenetic status. These include studies of methyl donors (e.g., folate, choline, methionine, betaine) and related cofactors (e.g., vitamins B_{12} and B_6) that have been described to induce epigenetic modifications at different exposure times and ages. For example, mice fed a folate-deficient diet for 32 weeks exhibited global DNA hypomethylation (Linhart et al. 2009), while choline has been reported to alter both global and gene-specific DNA methylation (Mehedint et al. 2010). The epigenetic role of methyl donors has also been studied in mouse cancer models (Min mice) as reported by Sibani and colleagues (2002). Specifically, folate supplementation has been directly linked with Nat2 promoter hypermethylation and subsequent decrease in gene expression (Wakefield et al. 2010). A number of (but not all!) epidemiological and clinical studies support some of the dietary findings above, demonstrating protection against some (particularly neurodevelopmental) diseases associated with altered DNA methylation.

8.6 Animal Models of Epigenetics and Diet

Epigenetic involvement in obesity and metabolic syndrome is a major interest in supporting the role of early-life nutrition in the outset of chronic disease later in life and in ageing (developmental origins of health and disease theory) (Gluckman et al. 2011). Epidemiological studies show that a poor intrauterine environment induced by unbalanced maternal diet or body composition, placental insufficiency or endocrine factors induces an offspring phenotype that is characterized by an increased risk of developing chronic non-communicable diseases such as cardiovascular disease and metabolic syndrome in later life. Some of these findings have been replicated in animal models using restricted nutrition during pregnancy, as described later. This association between poor intrauterine growth and increased risk of disease in later life may result from a predictive adaptive response where the foetus responds to environmental cues during development with permanent adjustments in its development and homeostatic systems to aid later survival and reproductive fitness. However, if these adaptations are inappropriate for the postnatal environment, they may ultimately lead to an increased risk of disease because its homeostatic capacity is mismatched to that environment.

Animal models have been used to investigate the role of maternal consumption of a high-fat diet (HFD) during certain vulnerable periods, such as pregnancy and early postnatal life on the potential to cause long-term effects on metabolic disorders, obesity and responsivity to stress and immune challenges in offspring (Ashino et al. 2012; Spencer 2013). This may be, in part, due to the dysregulation of the early-life epigenome. Dams from mice fed an HFD during the pregnancy and lactation period display global DNA hypomethylation in their brains along with hypomethylation at promoters in genes involved in the food reward circuitry (Dat, Mor and Penk) with corresponding increases in gene expression (Vucetic et al. 2010). Longitudinal analysis in humans has also suggested that the prenatal maternal, rather than paternal, diet is particularly influential in determining the dietary habits, and possibly the future metabolic profile, of their offspring (Brion et al. 2010).

Leptin, the major satiety adipokine, maintains food intake and energy homeostasis through a negative feedback loop. Milagro and colleagues (2009) assigned Wistar rats to either a conventional diet or HFD over an 11-week period. Coinciding with an obese phenotype by the end of the dietary intervention, the latter group also displayed hypermethylation of the leptin promoter from retroperitoneal adipocytes. Additionally, this same pattern has also been observed in the oocytes and liver of offspring from HFD-induced mice, indicating that this nutritional effect can be transgenerational (Ge et al. 2014).

8.7 Animal Models to Study Translational Epigenetic Effects

Jean-Baptiste Lamarck (1744–1829) is chiefly remembered as a proponent of the theory of inheritance of acquired characteristics (though he was a highly respected naturalist in his own right), which suggests that individuals can pass on certain features that they acquired during their lifetime to their offspring. Some of the examples he chose, including the gradual lengthening of the neck of the giraffe as a result of its foraging lifestyle, fell into disfavour, first as a result of Darwin's theory of natural selection and later by the implications of Mendelian inheritance and the notion of the gene.

However, inheritance of acquired characteristics has gradually acquired new currency; the idea that the transfer of epigenetic information across generations might confer "memory" of environmental stresses experienced in earlier generations to subsequent generations. The existence of a definite molecular mechanism(s) underlying such a process largely still eludes us. It was previously thought that a complete erasure of epigenetic variation occurred within the genome during the early stages of embryogenesis. However, the discovery of imprinted genes (genes that are expressed dependent on the parent from whom they are inherited) has led to increasing speculation that some form of an epigenetic "memory" of the previous generation might be maintained and transmitted on the DNA through the germ line.

Studies on epigenetic inheritance in the aforementioned metastable epialleles have indicated some levels of transgenerational inheritance of phenotype (Anway et al. 2005). In the Avy mouse, yellow coat colour females are significantly more likely to birth yellow coat colour offspring than genetically identical mice with agouti markings, and these effects persist into the F(2) generation. Importantly, no paternal germ line effects on coat colour have been observed in these mice (Blewitt et al. 2006). In other words, the epigenetic mark is preserved through the oocyte but not through the sperm. As discussed earlier, Avy mice are also prone to obesity, and this obesity phenotype, like coat colour, is transmitted through the female germ line. However, early exposure of these obese, founder dams to a methyl-supplemented diet reduces the obesity in their offspring and tends to lead to darker coats, suggesting that maternal diet can offset the transgenerational effects (Waterland et al. 2006). In contrast with Axin1Fu mice, both the paternal and maternal germ lines are capable of transmitting the kinked tails to the next generation (Rakyan et al. 2003).

Protein restriction during gestation can likewise result in transgenerational methylation changes in some candidate genes (Burdge et al. 2007). In utero exposure of rats to a protein-restricted diet, for example, results in altered methylation of the peroxisome proliferator-activated receptor alpha (Ppara) and glucocorticoid receptor (Nr3c1) promoters in F1 male offspring, a change that can be transmitted to the F2 generation. Exposure to a range of toxicants, including the endocrine disruptors (e.g., vinclozolin (Nilsson et al. 2008) and diethylstilbestrol (Walker and Haven 1997)), can result in transgenerational disease states, including testis defects, prostate disease, kidney disease, reproductive and breast cancer development and immune abnormalities, which in many cases are due to epigenetic alterations within the male germ line.

In rodents, there is evidence that maternal care, such as maternal stress, abusive caregiving and variation in maternal licking and grooming (LG) behaviour can shift the development of female offspring such that these maternal traits are also observed in both offspring and grand-offspring (Murgatroyd and Nephew 2013; Champagne et al. 2007). Rat models of maternal LG behaviour and chronic social stress have been reported to induce, in the female offspring respectively, the DNA methylation and expression of Esr1 in the medial preoptic area of the hypothalamus (Champagne et al. 2006) and Nr3c1 methylation and expression in the paraventricular nucleus of the hypothalamus (PVN) (Murgatroyd et al. in submission).

Pathways through which perinatal epigenetic and behavioural effects may persist across generations may either occur through indirect mechanisms via the mother's behaviour and hormones (as described above) or through the direct inheritance of epigenetic modifications. This can be specifically tested, by determining if the transgenerational-inherited phenotype is explicitly carried only through the maternal line or if paternal germ line effects occur. For example, a study of maternal separation revealed that the offspring of separated male mice (F2) and the grand-offspring of separated males (F3) showed depressive-like behaviours together with specific DNA methylation patterns at the Mecp2 and Crfr2 genes present in the sperm and brains of F1 and the brains of F2 males (Franklin et al. 2010). Thus, it would appear that epigenetic variation that is induced through the quality of the early-life environment can become encoded into the germ cells, leading to a transgenerational inheritance of the effects of stress and social experiences.

In humans, the best evidence for the occurrence of transgenerational effects induced by environmental factors comes from studies investigating nutrition and food availability. A well-known example is the Dutch famine of 1944–1945, during which a food embargo led to a devastating 5-month famine. Mothers who were pregnant during this time gave birth to children who had reduced birth weight and developed a host of clinical disorders during adulthood, ranging from obesity to glucose intolerance and coronary heart disease (e.g., Kyle and Pichard 2006). Surprisingly, the grandchildren of mothers exposed to the famine during gestation showed similar effects suggestive of epigenetic changes in the germ line of the exposed embryos that were then passed on to their own children (e.g., Painter et al. 2008). Indeed a study investigating the long-term effects of prenatal exposure to famine on DNA methylation found hypomethylation at the imprinted IGF2 gene when analysed six decades later. Interestingly, no differences in DNA methylation were observed in individuals exposed to famine late in gestation (Heijmans et al. 2008). A paternal translational effect has been identified in a large Swedish epidemiological study reporting that early smoking in fathers associated with a greater body mass index in their sons. Additionally, they found a correlation between mortality risk ratio of grandsons and paternal grandfather's food supply in mid-childhood (Pembrey et al. 2006). Importantly, though it is possible to explain these observations based on transgenerational epigenetic inheritance, other equally plausible explanations may exist such as various cultural confounders.

8.8 Modelling Early Adverse Experiences

Comprehensive clinical studies show that adverse conditions in early life can severely impact the developing brain and increase vulnerability to mood disorders later in life (for review, see Murgatroyd and Spengler 2011a). During early postnatal life, the brain exhibits high plasticity, which allows environmental signals to alter the trajectories of rapidly developing circuits. Adversity in early life is able to shape the experience-dependent maturation of stress-regulating pathways underlying emotional functions and endocrine responses to stress, such as the hypothalamic–pituitary–adrenal (HPA) system, leading to long-lasting altered stress responsivity during adulthood. Evidently there are ethical limitations to conducting prospective early-life stress experiments in human participants.

Therefore, animal models are invaluable in gaining insight into the behavioural and physiological mechanisms underlying the long-term effects of early experiences on emotional reactivity and the stress response. Rodents are particularly well suited to studies of the effects of early-life experience (for review see Nephew & Murgatroyd 2013). Even quite subtle alterations in the experience of rats during the early postnatal or prenatal periods can provoke long enduring consequences in behavioural and endocrine phenotypes (for review, see Holmes et al. 2005).

8.8.1 The Epigenome and Postnatal Adversity

Animal models of neglect, abuse and variation in maternal care are increasingly incorporating analyses of epigenetic mechanisms to account for the persistent effects of these experiences. For example, early-life adversity through the use of maternal separation in mice (involving separating pups from their mothers for 3 h each day during the first 10 days of life) has been found to induce sustained expression of hypothalamic Avp mRNA and peptide specific to the parvocellular subpopulation of the PVN. This underlined elevated corticosterone secretion, heightened endocrine responsiveness to subsequent stressors and altered HPA axis feedback inhibition. Importantly, this altered expression associated with reduced levels of DNA methylation in the PVN at particular CpG dinucleotides within an enhancer region important for Avp gene activity. Further investigations revealed that hypomethylation at this region reduced the ability of Mecp2 to bind and recruit further epigenetic machinery such as HDACs and DNMTs (Murgatroyd and Spengler 2014), supporting previous evidences for a role of Mecp2 as an epigenetic platform upon which histone and DNA modifications are carried out to confer transcriptional repression. Signals controlling Mecp2 occupancy at this early step were then explored, revealing that neuronal depolarization is able to trigger Ca2+-dependent phosphorylation of Mecp2. This causes Mecp2 to dissociate from the Avp enhancer, resulting in a loss of both its repressing activity and its organization of DNA methylation and histone marks (for review, see Murgatroyd and Spengler 2011b) (Fig. 8.4).

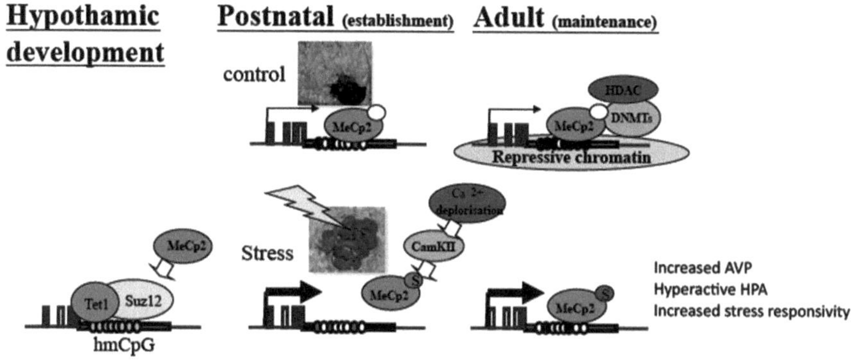

Fig. 8.4 A stepwise pathway in the epigenetic programming of *Avp* in the hypothalamic neurons in response to early-life stress in a time-dependent manner. Mecp2 is recruited to *Avp* enhancer through polycomb (suz12) binding during early hypothalamus development and recruits HDACs and DNMTs to target DNA methylation and repressive chromatin. Stress, through neuronal stimulation and CamKII activation, phosphorylates Mecp2, restricting binding at the *Avp* enhancer, relieving its repression and leading to increased *Avp* production and HPA axis activity

By means of a similar maternal separation paradigm, male offspring exposed to postnatal separation were found to have elevated levels of DNA methylation within the Mecp2 gene and decreased methylation within the Crh receptor (Crhr2) gene (Franklin et al. 2010). A further study in rats revealed that daily exposure to 30 min of aggressive caregiving on postnatal days 1 through 7 is associated with increased DNA methylation of the Bdnf promoter at postnatal days 8, 30 and 90 (Roth et al. 2009). Although a limited range of targets has been explored, these initial studies suggest that epigenetic modifications, particularly DNA methylation, are associated with early-life manipulation of the quality and frequency of contact between dams and pups.

One well-characterized model established to study the effects of early postnatal maternal care, as measured by levels of LG behaviour, found that rats that received high levels of maternal care during early life developed sustained elevations in GR expression within the hippocampus and reduced HPA axis responses to stress. Epigenetic analyses revealed that the enhanced GR expression associated with a persistent DNA hypomethylation at specific CpG dinucleotides within the hippocampal GR (Nr3c1) exon 17 promoter and increased histone acetylation. The lower CpG methylation facilitated binding of the transcriptional activator nerve growth factor-inducible protein A (Ngf1a) to this region (Weaver et al. 2004). Cross-fostering of pups between high and low LG dams has indicated that these epigenetic effects are related to the quality of postnatal care rather than prenatal or genetic factors. Histone acetylation was also increased within the differentially methylated region of the Nr3c1 promoter, such that among offspring reared by high LG dams, there are elevated levels of histone acetylation.

8.8.2 The Epigenome and Prenatal Adversity

Chronic variable stress experienced by pregnant females has been demonstrated to induce a long-term impact on HPA pathways, including altered gene expression within the hypothalamus. For example, in mice, stress during the 1st week of pregnancy has been found to induce significant impairments in male offspring. This associated with increased CRF and decreased GR gene expression together with reduced DNA methylation within the promoter region of the Crh gene and increased methylation within the promoter region of the Nr3c1 gene (encoding GR) in hypothalamic tissue of prenatally stressed offspring (Mueller and Bale 2008).

8.8.3 The Epigenome in Adolescence and Adulthood

Although plasticity in epigenetic pathways was initially thought to be limited to the early stages of development and postnatal life, it is becoming increasingly evident that experiences occurring across the life span are capable of inducing epigenetic variation. Moreover, the capacity to modify DNA methylation and histone tails may be a critical aspect of learning and memory from infancy to adulthood (Miller and Sweatt 2007). In the context of studies on the influence of social interactions and stressors, epigenetic variation, likewise, has been associated with changes in gene expression and phenotype during the later stages of development. Among adult mice that have a genetically induced memory impairment, 4 weeks of exposure to complex housing environments was found to be associated with increased histone acetylation in the hippocampus and cortex and improvements in memory (Fischer et al. 2007). Interestingly, these enrichment-induced effects on both histones and learning and memory could also be achieved in non-memory-impaired mice with pharmacological treatments that promote histone acetylation.

In contrast with environmental enrichment, which has been demonstrated to increased expression of BDNF together with associated chromatin changes (particularly histone methylation) at the promoter (Kuzumaki et al. 2011), chronic stress is associated with epigenetically reduced expression of this gene. In rats, immobilization stress combined with exposure to predator odour has been found to induce significant increases in hippocampal DNA methylation within the Bdnf gene (Roth et al. 2011). Within the social defeat model, reductions in Bdnf are observed a month after exposure to the social stressor, and hippocampal histone demethylation at the Bdnf promoter may account for this effect (Tsankova et al. 2006). Histone acetylation is transiently decreased and then exhibits prolonged increases among socially defeated mice, and this effect may be associated with long-term, stress-induced reductions in histone deacetylase levels (Covington et al. 2009).

Similarly, in rats, increased histone acetylation has been observed for up to 24 h after the experience of repeated social defeat (Hollis et al. 2010). Moreover, the behavioural consequences of social defeat such as decreased social behaviour can be reversed pharmacologically with a drug that inhibits histone deacetylases

(Covington et al. 2009). Interestingly, although social defeat has been found to induce long-term effects, a percentage of mice display resilience to this stressor. In a recent study, stress-susceptible mice were found to have increased levels of Crh mRNA in the PVN and decreased DNA methylation within the Crh gene (Elliott et al. 2010). In contrast, stress-resilient mice were found to have no changes in Crh mRNA or DNA methylation of this gene.

Many of the above animal studies highlight the importance of the temporal nature of brain developmental processes in relation to the effects that environmental stressors can have on epigenetic programming of long-term changes in neurodevelopment and behaviour. Developing brain regions typically pass through critical "windows" of sensitivity that stretch over different perinatal periods. Therefore, it stands to reason that the impact of stressors at different time points will confer more pronounced and long-lasting effects within brain regions that are actively developing at that particular time. For example, late prenatal and early postnatal life is the critical period for hippocampal development, possibly explaining why environmental exposures during this time strongly associate with cellular, morphological and epigenetic changes within these structures (for review, see McCrory et al. 2010). In contrast, environmental exposures during later life tend to confer their phenotypic effects by altering other areas of the brain. For example, repeated restraint stress in adult rats fails to cause long-term hippocampal-related effects such as those observed following similar stress exposure during prenatal and early postnatal life (Conrad et al. 1999). On the other hand, substantial cortical development and differentiation is known to continue well into adolescence (e.g., Wang and Gao 2009), possibly explaining why these regions are more susceptible to epigenetic changes in response to later-life environmental factors. In addition to the temporal nature of brain development, we should also consider the temporal and gene-specific manner in which epigenetic marks are established. This will be of crucial importance in understanding how experience-dependent epigenetic marks may undergo the transition from a preliminary labile state to a hard-coded stable print. For example, those studies on epigenetic marking of the Avp enhancer in response to early-life stress persisted under vasopressin receptor blockade in adult (3-month) mice consistent with the concept that the early-life stress had already engraved a lasting cellular memory (Murgatroyd et al. 2011a). The question however remains of whether critical "windows" for timely psychotherapeutic and pharmacological interventions following exposure to severe trauma might exist at periods prior to the establishment of stable epigenetic marks.

8.9 Translating Epigenetic Programming in Animal Models to Clinical Studies

Translational approaches seek to bridge the gaps between basic animal research and medical practice. Though not a new concept in medicine, it is becoming popular with its introduction in the National Institutes of Health Roadmap initiative

(Zerhouni 2005; McArthur and Borsini 2008) particularly in the field of psychiatry, and especially to affective disorders, including depression and anxiety-related disorders.

Such translational approaches obviously require rodent models, such as some of those described above that are suitable to study the clinical condition. Numerous studies in animals and humans demonstrate that early postnatal experiences also have long-term effects in adulthood and prenatal stress can lead to sustained effects on stress reactivity and behaviour in the offspring (for review, see Murgatroyd 2011a). Meaney and colleague's studies linking LG in rats with enhanced HPA axis feedback regulation, reduced anxiety-related behaviour and GR methylation (Weaver et al. 2004) have translated hypothesis-led research in clinical studies. For example, one study reported elevated NR3C1 1-F promoter methylation and reduced GR expression in post-mortem hippocampal tissue of suicide completers who were abused during childhood when compared to non-abused (McGowan et al. 2009). Other studies using peripheral DNA, from blood or saliva of infants, adolescents or adults, have shown increased levels of NR3C1 methylation in response to perinatal stress (Oberlander et al. 2008; Radtke et al. 2011; Mulligan et al. 2012) and abuse or neglect during childhood (Perroud et al. 2011; Tyrka et al. 2012). Most studies thus far reported on DNA methylation after enduring stress or traumatic events such as abuse or neglect (McGowan et al. 2009; Perroud et al. 2011, 2013; Tyrka et al. 2012). Interestingly, many of these studies have looked at different regions and CpG residues of the NR3C1 1-F promoter regions, though several recent clinical studies examining leukocytes have reported elevated methylation of the homologous human NR3C1 1-F promoter (homologous to the rat 1–7 promoter) at a specific CpG (CpG unit 22,23, Fig. 8.5) associated with prenatal maternal depression (Oberlander et al. 2008; Conradt et al. 2013; Hompes et al. 2013) and childhood stress (Tyrka et al. 2012; Melas et al. 2013). This suggests perhaps that different CpG residues might be sensitive to specific environmental stressors across this region (Fig. 8.5).

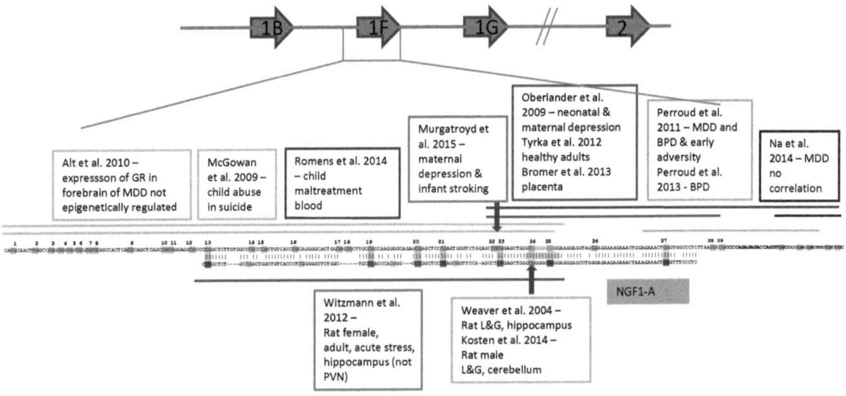

Fig. 8.5 Schematic of human (*upper*) and rat (*lower*) *Nr3c1* gene promoter exon 1F/7 detailing conserved sequences and areas investigated for DNA methylation in human and rat studies reported in the literature

Very few studies have identified candidates for maternal behaviours in humans that may be the equivalent of rat licking and grooming. Strikingly, in rats, the effects of the LG can be mimicked by stroking the pups with a brush (Mulligan et al. 2012). A recent study has reported moderation of prenatal stress effects by early human tactile stimulation (assessed by how often mothers reported stroking their babies) over the first weeks of life on behavioural and physiological stress reactivity in infancy (Sharp et al. 2012), demonstrating that the effect of maternal prenatal depression on infant reactivity would be modified by tactile. This was followed up by a new study showing a reduction of NR3C1 gene methylation associated with maternal stroking on these children, supporting the role of epigenetic mechanisms linking early-life stress with long-term effects of maternal care (Murgatroyd et al. 2015). Interestingly, the same study further found interactive effects of prenatal and postnatal maternal depression on NR3C1 1-F promoter methylation in young children; specifically, infants of mothers with low prenatal depression were vulnerable to the effects of postnatal depression, at the level of DNA methylation—consistent with an interplay between prenatal and postnatal environments. In general terms, this is consistent with the foetal origin hypothesis of human disease that proposes that in utero environmental exposures lead to modifications in foetal development which are adaptive where the subsequent postnatal environment is similar. Together, this study highlights the importance of translational research in linking studies in animals to humans and in understanding the earliest origins of neurobiological and behavioural development and psychiatric disorders.

8.9.1 From Clinical to Basic Research; from Bedside to Nest

Most efforts have been devoted to the "bench to the bedside" (from animals to humans or from basic to clinical research) approaches focusing on the design of animal models (particularly using rodents) that would be relevant to study the human disorders and to predict the therapeutic outcomes. Unfortunately, less research follows the opposite direction, using the reverse-translational approach and thus going from the bedside to the bench (from humans to animals) (Belzung and Lemoine 2011; Malkesman et al. 2009) (Fig. 8.6). However, when trying to assess the function of mechanisms discovered in animal models in the pathophysiology of human disorders and developing new treatments for these conditions, such reverse-translational approaches are crucial. For example, the role of hippocampal neurogenesis in the pathophysiology of depression and in the therapeutic efficacy of pharmacological treatments has been shown using rodent models (Surget et al. 2008; Snyder et al. 2011) and then confirmed in human studies (Boldrini et al. 2009; Lucassen et al. 2010). BDNF knockout and polymorphism research in animals and humans represents a powerful tool for understanding the involvement of BDNF on the pathogenesis of anxiety disorders (Domingos da Silveira da Luz et al. 2013).

Concerning such translational studies, some may show scepticism—that animal models have limited relevance in the case of largely specific features of the human

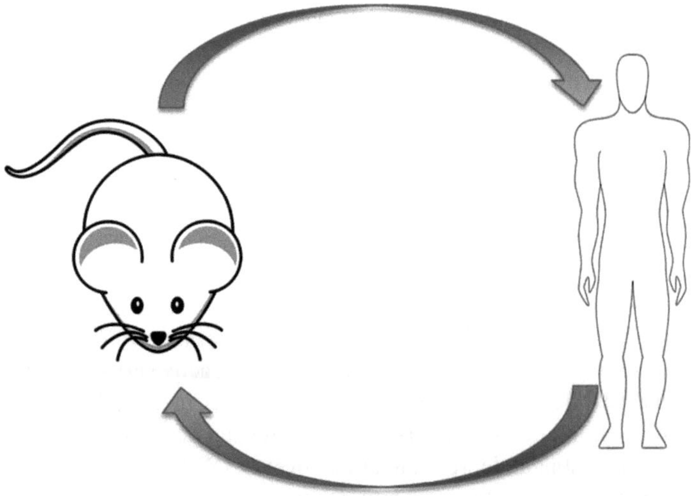

Fig. 8.6 Crosstalk between basic research experiments and the clinic incorporating translational and back-translational studies, from "nest" to "bedside" and back again

species, such as maternal behaviours. For instance, in relation to other mammals, humans have vastly greater and more complex social competences (Bullinger et al. 2011). Variations in well-established dimensions of parenting such as sensitivity, intrusiveness or hostility with effects on attachment security, emotionality and social attributions may be expected to be the most salient for long-term effects (McElwain and Booth-Laforce 2006). Equally, however, long-term parental effects on multiple physiological systems are common in biology and likely to be mediated via epigenetic modifications (Weaver et al. 2004; Murgatroyd et al. 2009).

Importantly, tactile stimulation derived from parental care has immediate effects on endocrine systems that regulate somatic growth in rodents and humans (Levine 1994; Schanberg and Field 1987), suggesting that the ability of the infant to respond to specific forms of parental care is conserved at least among mammals. In the context of the above findings, the genetic mechanisms in glucocorticoid regulation are highly conserved across species (Suderman et al. 2012; Labonté et al. 2012), and so an effect of tactile stimulation on GR expression may have been conserved across rodents and humans. The examination of dynamic gene expression and epigenetic mechanisms in infants is currently limited to peripheral tissue; however, against the background of extensive animal work, demonstrating differences in GR expression and epigenetic regulation, it would be important to investigate links between peripheral and CNS gene regulation. Importantly, it also needs to be tested whether stroking through skin-to-skin contact is the direct mechanism or only a proxy for another causal aspect of parenting. To address these questions, it would appear imperative to further back-translate findings from the human studies back to animal model. Such translational crosstalk between basic research experiments and the clinic would be important to allow fine-tuning of animal models to address such pertinent questions.

8.10 Conclusion

Ultimately, to succeed in identifying the role of epigenetic mechanisms leading to complex phenotype and disease, researchers must integrate the various animal models, human clinical approaches and human population approaches, paying attention to the times of sensitivity and model system of evaluation. As highlighted above, it is increasingly recognized that chemical, nutritional, behavioural, social and physical factors alter gene expression and affect health and disease by not only mutating promoter and coding regions of genes but also modifying the epigenome. The use of animal models in these investigations has informed the fields of molecular biology and toxicology by elucidating the mechanisms underlying developmental exposure and adult disease. Candidate gene approaches have recently been enhanced by concomitant whole epigenome technologies. Thus, the evaluation of epigenetic mechanisms in health and disease is now poised for enhanced investigation in animal models as well as expansion into clinical and population health approaches. Animal models will continue to help inform the evaluation of vulnerable time periods and multigenerational studies that are not feasible in human populations. Additionally, the epigenome, in contrast with the genome, is particularly affected by cell-type specificity.

Thus, animal model studies, in which cell-type specificity is more readily evaluated than in humans, can serve as important proof-of-principle approaches to evaluate the use of peripheral tissue (e.g., blood, saliva) in human epigenetic epidemiology studies. Ultimately, to fully succeed in elucidating epigenetic mechanisms underlying disease susceptibility, researchers must integrate animal models and human approaches to generate the best prescriptions for human health evaluation and disease prevention.

Abbreviations

AVP	Arginine vasopressin
Avy	Agouti viable yellow
Axin1Fu	Axin 1 fused
BDNF	Brain-derived neurotrophic factor
CRH	Corticotropin-releasing hormone
DNMT	DNA methyltransferase
Esr1	Estrogen receptor alpha
GFAP	Glial fibrillary acidic protein
GR	Glucocorticoid receptor
HFD	High-fat diet
HPA	Hypothalamic–pituitary–adrenal
LG	Licking and grooming
MECP2	Methyl-binding protein MECP2
Nr3c1	Nuclear receptor subfamily 3, group c1
Ppara	Peroxisome proliferator-activated receptor alpha

References

Amir, R. E., Van den Veyver, I. B., Wan, M., Tran, C. Q., Francke, U., & Zoghti, H. Y. (1999). Rett syndrome is caused by mutations in X-linked MECP2, encoding methyl-CpG-binding protein 2. *Nature Genetics, 23*, 185–188.

Anway, M. D., Cupp, A. S., Uzumcu, M., & Skinner, M. K. (2005). Epigenetic transgenerational actions of endocrine disruptors and male fertility. *Science, 308*, 1466–1469.

Ashino, N. G., Saito, K. N., Souza, F. D., Nakutz, F. S., Roman, E. A., Velloso, L. A., Torsoni, A. S., & Torsoni, M. A. (2012). Maternal high-fat feeding through pregnancy and lactation predisposes mouse offspring to molecular insulin resistance and fatty liver. *Journal of Nutritional Biochemistry, 23*(4), 341–348.

Belzung, C., & Lemoine, M. (2011). Criteria of validity for animal models of psychiatric disorders: focus on anxiety disorders and depression. *Biology of Mood and Anxiety Disorders, 1*(1), 9.

Blewitt, M. E., Vickaryous, N. K., Paldi, A., Koseki, H., & Whitelaw, E. (2006). Dynamic reprogramming of DNA methylation at an epigenetically sensitive allele in mice. *PLoS Genetics, 2*, e49. http://dx.doi.org/10.1371/journal.pgen.0020049.

Boldrini, M., Underwood, M. D., Hen, R., Rosoklija, G. B., Dwork, A. J., John Mann, J., & Arango, V. (2009). Antidepressants increase neural progenitor cells in the human hippocampus. *Neuropsychopharmacology, 34*(11), 2376–2389.

Brion, M. J., Ness, A. R., Rogers, I., Emmett, P., Cribb, V., Davey Smith, G., & Lawlor, D. A. (2010). Maternal macronutrient and energy intake in pregnancy and offspring intake at 10 y: exploring paternal comparisons and prenatal effects. *American Journal of Clinical Nutrition, 91*(3), 748–756.

Bullinger, A. F., Zimmermann, F., Kaminski, J., & Tomasello, M. (2011). Different social motives in the gestural communication of chimpanzees and human children. *Developmental Science, 14*, 58–68.

Burdge, G. C., Slater-Jefferies, J., Torrens, C., Phillips, E. S., Hanson, M. A., & Lillycrop, K. A. (2007). Dietary protein restriction of pregnant rats in the F0 generation induces altered methylation of hepatic gene promoters in the adult male offspring in the F1 and F2 generations. *British Journal of Nutrition, 97*(435), 439.

Champagne, F. A., & Meaney, M. J. (2007). Transgenerational effects of social environment on variations in maternal care and behavioral response to novelty. *Behavioural Neuroscience, 121*(6), 1353–1363.

Champagne, F., Weaver, I., Diorio, J., Dymov, S., Szyf, M., & Meaney, M. J. (2006). Maternal care associated with methylation of the estrogen receptor-alpha1b promoter and estrogen receptor-alpha expression in the medial preoptic area of female offspring. *Endocrinology, 147*, 2909–2915.

Chang, Q., Khare, G., Dani, V., Nelson, S., & Jaenisch, R. (2006). The disease progression of Mecp2 mutant mice is affected by the level of BDNF expression. *Neuron, 49*, 341–348.

Conrad, C. D., LeDoux, J. E., Magariños, A. M., & McEwen, B. S. (1999). Repeated restraint stress facilitates fear conditioning independently of causing hippocampal CA3 dendritic atrophy. *Behavioral Neuroscience, 113*, 902–913.

Conradt, E., Lester, B. M., Appleton, A. A., Armstrong, D. A., & Marsit, C. J. (2013). The roles of DNA methylation of NR3C1 and 11β-HSD2 and exposure to maternal mood disorder in utero on newborn neurobehavior. *Epigenetics, 8*, 1321–1329.

Covington, H. E., III, Maze, I., LaPlant, Q. C., Vialou, V. F., Ohnishi, Y. N., Berton, O., Fass, D. M., Renthal, W., Rush, A. J., III, Wu, E. Y., Ghose, S., Krishnan, V., Russo, S. J., Tamminga, C., Haggarty, S. J., & Nestler, E. J. (2009). Antidepressant actions of histone deacetylase inhibitors. *Journal of Neuroscience, 29*(37), 11451–11460.

Dolinoy, D. C. (2008). The agouti mouse model: an epigenetic biosensor for nutritional and environmental alterations on the fetal epigenome. *Nutrition Reviews, 66*(Suppl 1), S7–S11.

Domingos da Silveira da Luz, A. C., Pereira Dias, G., do Nascimento Bevilaqua, M. C., Cocks, G., Gardino, P. F., Thuret, S., & Nardi, A. E. (2013). Translational findings on brain-derived neurotrophic factor and anxiety: contributions from basic research to clinical practice. *Neuropsychobiology, 68*(3), 129–138.

Elliott, E., Ezra-Nevo, G., Regev, L., Neufeld-Cohen, A., & Chen, A. (2010). Resilience to social stress coincides with functional DNA methylation of the Crf gene in adult mice. *Nature Neuroscience, 13*(11), 1351–1353.

Fan, G., Beard, C., Chen, R. Z., Csankovszki, G., Sun, Y., Siniaia, M., et al. (2001). DNA hypomethylation perturbs the function and survival of CNS neurons in postnatal animals. *Journal of Neuroscience, 21*, 788–797.

Fischer, A., Sananbenesi, F., Wang, X., Dobbin, M., & Tsai, L. H. (2007). Recovery of learning and memory is associated with chromatin remodelling. *Nature, 447*(7141), 178–182.

Franklin, T. B., Russig, H., Weiss, I. C., Graff, J., Linder, N., Michalon, A., Vizi, S., & Mansuy, I. M. (2010). Epigenetic transmission of the impact of early stress across generations. *Biological Psychiatry, 68*, 408–415.

Fulka, H., Mrazek, M., Tepla, O., & Fulka, J., Jr. (2004). DNA methylation pattern in human zygotes and developing embryos. *Reproduction, 128*, 703–708.

Ge, Z. J., Luo, S. M., Lin, F., Liang, Q. X., Huang, L., Wei, Y. C., Hou, Y., Han, Z. M., Schatten, H., & Sun, Q. Y. (2014). DNA methylation in oocytes and liver of female mice and their offspring: effects of high-fat-diet-induced obesity. *Environmental Health Perspectives, 122*, 159–164.

Gluckman, P. D., Hanson, M. A., & Low, F. M. (2011). The role of developmental plasticity and epigenetics in human health. *Birth Defects Research. Part C, Embryo Today, 93*, 12–18.

Guy, J., Gan, J., Selfridge, J., Cobb, S., & Bird, A. (2007). Reversal of neurological defects in a mouse model of Rett syndrome. *Science, 315*, 1143–1147.

Guy, J., Hendrich, B., Holmes, M., Martin, J. E., & Bird, A. (2001). A mouse Mecp2-null mutation causes neurological symptoms that mimic Rett syndrome. *Nature Genetics, 27*, 322–326.

Heijmans, B. T., Tobi, E. W., Stein, A. D., Putter, H., Blauw, G. J., Susser, E. S., Slagboom, P. E., & Luney, L. H. (2008). Persistent epigenetic differences associated with prenatal exposures to famine in humans. *Proceedings of the National Academy of Sciences of the United States of America, 105*(44), 17046–17049.

Hollis, F., Wang, H., Dietz, D., Gunjan, A., & Kabbaj, M. (2010). The effects of repeated social defeat on long-term depressive-like behavior and short-term histone modifications in the hippocampus in male Sprague-Dawley rats. *Psychopharmacology, 211*(1), 69–77.

Holmes, A., le Guisquet, A. M., Vogel, E., Millstein, R. A., Leman, S., & Belzung, C. (2005). Early life genetic, epigenetic and environmental factors shaping emotionality in rodents. *Neuroscience and Biobehavioral Review, 29*, 1335–1346.

Hompes, T., Izzi, B., Gellens, E., Morreels, M., Fieuws, S., Pexsters, A., Schops, G., Dom, M., Van Bree, R., Freson, K., Verhaeghe, J., Spitz, B., Demyttenaere, K., Glover, V., Van den Bergh, B., Allegaert, K., & Claes, S. (2013). Investigating the influence of maternal cortisol and emotional state during pregnancy on the DNA methylation status of the glucocorticoid receptor gene (NR3C1) promoter region in cord blood. *Journal of Psychiatric Research, 47*, 880–891.

Klein, C. J., Botuyan, M. V., Wu, Y., Ward, C. J., Nicholson, G. A., Hammans, S., et al. (2011). Mutations in DNMT1 cause hereditary sensory neuropathy with dementia and hearing loss. *Nature Genetics, 43*, 595–600.

Koturbash, I., Scherhag, A., Sorrentino, J., Sexton, K., Bodnar, W., Swenberg, J. A., Beland, F. A., Pardo-Manuel Devillena, F., Rusyn, I., & Pogribny, I. P. (2011). Epigenetic mechanisms of mouse inter-strain variability in genotoxicity of the environmental toxicant 1,3-butadiene. *Toxicological Sciences, 122*, 448–456.

Krashes, M. J., Koda, S., Ye, C., et al. (2011). Rapid, reversible activation of AgRP neurons drives feeding behavior in mice. *Journal of Clinical Investigation, 121*, 1424–1428.

Kuzumaki, N., Ikegami, D., Tamura, R., Hareyama, N., Imai, S., Narita, M., Torigoe, K., Niikura, K., Takeshima, H., Ando, T., Igarashi, K., Kanno, J., Ushijima, T., Suzuki, T., & Narita, M. (2011). Hippocampal epigenetic modification at the brain-derived neurotrophic factor gene induced by an enriched environment. *Hippocampus, 21*(2), 127–132.

Kyle, U. G., & Pichard, C. (2006). The Dutch Famine of 1944–1945: a pathophysiological model of long-term consequences of wasting disease. *Current Opinion in Clinical Nutrition and Metabolic Care, 9*, 388–394.

Labonté, B., Suderman, M., Maussion, G., Navaro, L., Yerko, V., Mahar, I., Bureau, A., Mechawar, N., Szyf, M., Meaney, M. J., & Turecki, G. (2012). Genome-wide epigenetic regulation by early-life trauma. *Archives of General Psychiatry, 69*(7), 722–731.

Levine, S. (1994). The ontogeny of the hypothalamic pituitary adrenal axis: the influence of maternal factors. *Annals of the New York Academy of Sciences, 746*, 275–288.

Li, E., Bestor, T. H., & Jaenisch, R. (1992). Targeted mutation of the DNA methyltransferase gene results in embryonic lethality. *Cell, 69*, 915–926.

Linhart, H. G., Troen, A., Bell, G. W., Cantu, E., Chao, W. H., Moran, E., Steine, E., He, T., & Jaenisch, R. (2009). Folate deficiency induces genomic uracil misincorporation and hypomethylation but does not increase DNA point mutations. *Gastroenterology, 136*, 227–235.

Lucassen, P. J., Stumpel, M. W., Wang, Q., & Aronica, E. (2010). Decreased numbers of progenitor cells but no response to antidepressant drugs in the hippocampus of elderly depressed patients. *Neuropharmacology, 58*(6), 940–949.

Malkesman, O., Austin, D. R., Chen, G., & Manji, H. K. (2009). Reverse translational strategies for developing animal models of bipolar disorder. *Disease Models and Mechanisms, 2*(5–6), 238–245.

Martinowich, K., Hattori, D., Wu, H., Fouse, S., He, F., Hu, Y., et al. (2003). DNA methylation-related chromatin remodeling in activity-dependent BDNF gene regulation. *Science, 302*, 890–893.

McArthur, R. A., & Borsini, F. (2008). What do you mean by "Translational research"? An enquiry through animal and translational models for CNS drug discovery: Psychiatric disorders. In R. A. McArthur & F. Borsini (Eds.), *Animal and translational models for CNS drug discovery* (Psychiatric disorders, Vol. 1, pp. xvii–xxxviii). New York: Academic.

McCrory, E., De Brito, S. A., & Viding, E. (2010). Research review: The neurobiology and genetics of maltreatment and adversity. *Journal of Child Psychology and Psychiatry, 51*, 1079–1095.

McElwain, N. L., & Booth-Laforce, C. (2006). Maternal sensitivity to infant distress and nondistress as predictors of infant-mother attachment security. *Journal of Family Psychology, 20*, 247–255.

McGowan, P. O., Sasaki, A., D'Alessio, A. C., Dymov, S., Labonte, B., Szyf, M., et al. (2009). Epigenetic regulation of the glucocorticoid receptor in human brain associates with childhood abuse. *Nature Neuroscience, 12*, 342–348.

Mehedint, M. G., Niculescu, M. D., Craciunescu, C. N., & Zeisel, S. H. (2010). Choline deficiency alters global histone methylation and epigenetic marking at the Re1 site of the calbindin 1 gene. *FASEB Journal, 24*, 184–195.

Melas, P. A., Wei, Y., Wong, C. C., Sjöholm, L. K., Aberg, E., Mill, J., Schalling, M., Forsell, Y., & Lavebratt, C. (2013). Genetic and epigenetic associations of MAOA and NR3C1 with depression and childhood adversities. *International Journal of Neuropsychopharmacology, 1*, 1–16.

Milagro, F. I., Campion, J., Garcia-Diaz, D. F., Goyenechea, L., Paternain, L., & Martinez, J. A. (2009). High fat diet-induced obesity modifies the methylation pattern of leptin promoter in rats. *Journal of Physiology and Biochemistry, 65*(1), 1–10.

Miller, C. A., & Sweatt, J. D. (2007). Covalent modification of DNA regulates memory formation. *Neuron, 53*(6), 857–869.

Mueller, B. R., & Bale, T. L. (2008). Sex-specific programming of offspring emotionality after stress early in pregnancy. *Journal of Neuroscience, 28*(36), 9055–9065.

Mulligan, C. J., D'Errico, N. C., Stees, J., & Hughes, D. A. (2012). Methylation changes at NR3C1 in newborns associate with maternal prenatal stress exposure and newborn birth weight. *Epigenetics, 7*, 853–857.

Murgatroyd, C. (2014). Epigenetic programming of neuroendocrine systems during early life. *Experimental Physiology, 99*(1), 62–65.

Murgatroyd, C. A., & Nephew, B. C. (2013). Effects of early life social stress on maternal behavior and neuroendocrinology. *Psychoneuroendocrinology, 38*(2), 219–228.

Murgatroyd, C. A., Taliefar, M., Bradburn, S., Carini, L. M., Babb, J. A., & Nephew, B. C. (2015). Social stress during lactation, depressed maternal care, and neuropeptidergic gene expression. *Behavioural Pharmacology, 26*(7), 642–653. doi:10.1097/FBP.0000000000000147.

Murgatroyd, C., Patchev, A. V., Wu, Y., Micale, V., Bockmühl, Y., Fischer, D., Holsboer, F., Wotjak, C. T., Almeida, O. F., & Spengler, D. (2009). Dynamic DNA methylation programs persistent adverse effects of early-life stress. *Nature Neuroscience, 12*(12), 1559–1566.

Murgatroyd, C., Quinn, J., Sharp, H., Pickles, A., & Hill, J. (2015). Effects of prenatal and postnatal depression, and maternal stroking, at the glucocorticoid receptor gene. *Translational Psychiatry, 5*, e560. http://dx.doi.org/10.1038/tp.2014.140.

Murgatroyd, C., & Spengler, D. (2011a). Epigenetics of early child development. *Frontiers in Psychiatry, 2*, 16. http://dx.doi.org/10.3389/fpsyt.2011.00016.

Murgatroyd, C., & Spengler, D. (2011b). Epigenetic programming of the HPA axis: Early life decides. *Stress, 14*(6), 581–589.

Murgatroyd, C., & Spengler, D. (2012). Genetic variation in the epigenetic machinery and mental health. *Current Psychiatry Reports, 14*(2), 138–149.

Murgatroyd, C., & Spengler, D. (2014). Polycomb binding precedes early-life stress responsive DNA methylation at the Avp enhancer. *PLoS One, 9*(3), e90277. http://dx.doi.org/10.1371/journal.pone.0090277.

Nephew, B., & Murgatroyd, C. (2013). The role of maternal care in shaping CNS function. *Neuropeptides, 47*(6), 371–378.

Nilsson, E. E., Anway, M. D., Stanfield, J., & Skinner, M. K. (2008). Transgenerational epigenetic effects of the endocrine disruptor vinclozolin on pregnancies and female adult onset disease. *Reproduction, 135*, 713–721.

Oberlander, T. F., Weinberg, J., Papsdorf, M., Grunau, R., Misri, S., & Devlin, A. M. (2008). Prenatal exposure to maternal depression, neonatal methylation of human glucocorticoid receptor gene (NR3C1) and infant cortisol stress responses. *Epigenetics, 3*, 97–106.

Painter, R. C., Osmond, C., Gluckman, P., Hanson, M., Phillips, D. I., & Roseboom, T. J. (2008). Transgenerational effects of prenatal exposure to the Dutch famine on neonatal adiposity and health in later life. *BJOG, 115*, 1243–1249.

Pembrey, M. E., Bygren, L. O., Kaati, G., Edvinsson, S., Northstone, K., Sjostrom, M., Golding, J., & the ALSPAC Study Team. (2006). Sex-specific, male-line transgenerational responses in humans. *European Journal of Human Genetics, 14*, 159–166.

Perroud, N., Dayer, A., Piguet, C., Nallet, A., Favre, S., Malafosse, A., et al. (2013). Childhood maltreatment and methylation of the glucocorticoid receptor gene NR3C1 in bipolar disorder. *British Journal of Psychiatry, 204*(1), 30–35.

Perroud, N., Paoloni-Giacobino, A., Prada, P., Olie, E., Salzmann, A., Nicastro, R., et al. (2011). Increased methylation of glucocorticoid receptor gene (NR3C1) in adults with a history of childhood maltreatment: a link with the severity and type of trauma. *Translational Psychiatry, 1*, e59. http://dx.doi.org/10.1038/tp.2011.60.

Radtke, K. M., Ruf, M., Gunter, H. M., Dohrmann, K., Schauer, M., Meyer, A., et al. (2011). Transgenerational impact of intimate partner violence on methylation in the promoter of the glucocorticoid receptor. *Translational Psychiatry, 1*, e21. http://dx.doi.org/10.1038/tp.2011.21.
Rakyan, V. K., Chong, S., Champ, M. E., Cuthbert, P. C., Morgan, H. D., Luu, K. V., & Whitelaw, E. (2003). Transgenerational inheritance of epigenetic states at the murine Axin(Fu) allele occurs after maternal and paternal transmission. *Proceedings of the National Academy of Sciences of the United States of America, 100*, 2538–2543.
Roth, T. L., Lubin, F. D., Funk, A. J., & Sweatt, J. D. (2009). Lasting epigenetic influence of early-life adversity on the BDNF gene. *Biological Psychiatry, 65*, 760–769.
Roth, T. L., Zoladz, P. R., Sweatt, J. D., & Diamond, D. M. (2011). Epigenetic modification of hippocampal Bdnf DNA in adult rats in an animal model of post-traumatic stress disorder. *Journal of Psychiatry Research, 245*(7), 919–926.
Samaco, R. C., & Neul, J. L. (2011). Complexities of Rett syndrome and MeCP2. *Journal of Neuroscience, 31*(22), 7951–7959. doi:10.1523/JNEUROSCI.0169-11.2011.
Schanberg, S. M., & Field, T. M. (1987). Sensory deprivation stress and supplemental stimulation in the rat pup and preterm human neonate. *Child Development, 58*, 1431–1447.
Sharp, H., Pickles, A., Meaney, M., Marshall, K., Tibu, F., & Hill, J. (2012). Frequency of infant stroking reported by mothers moderates the effect of prenatal depression on infant behavioural and physiological outcomes. *PLoS One, 7*, e45446. http://dx.doi.org/10.1371/journal.pone.0045446.
Sibani, S., Melnyk, S., Pogribny, I. P., Wang, W., Hiou-Time, F., Deng, L., Trasler, J., James, S. J., & Rozen, R. (2002). Studies of methionine cycle intermediates (SAM, SAH), DNA methylation and the impact of folate deficiency on tumor numbers in Min mice. *Carcinogenesis, 23*, 61–65.
Snyder, J. S., Soumier, A., Brewer, M., Pickel, J., & Cameron, H. A. (2011). Adult hippocampal neurogenesis buffers stress responses and depressive behaviour. *Nature, 476*(7361), 458–461.
Spencer, S. J. (2013). Perinatal nutrition programs neuroimmune function long-term: mechanisms and implications. *Frontiers in Neuroscience, 7*, 144. http://dx.doi.org/10.3389/fnins.2013.00109.
Suderman, M., McGowan, P. O., Sasaki, A., Huang, T. C., Hallett, M. T., Meaney, M. J., Turecki, G., & Szyf, M. (2012). Conserved epigenetic sensitivity to early life experience in the rat and human hippocampus. *Proceedings of the National Academy of Sciences of the United States of America, 109*(Suppl 2), 17266–17272.
Surget, A., Saxe, M., Leman, S., Ibarguen-Vargas, Y., Chalon, S., Griebel, G., Hen, R., & Belzung, C. (2008). Drug-dependent requirement of hippocampal neurogenesis in a model of depression and of antidepressant reversal. *Biological Psychiatry, 64*, 293–301.
Tsankova, N. M., Berton, O., Renthal, W., Kumar, A., Neve, R. L., & Nestler, E. J. (2006). Sustained hippocampal chromatin regulation in a mouse model of depression and antidepressant action. *Nature Neuroscience, 9*(4), 519–525.
Tyrka, A. R., Price, L. H., Marsit, C., Walters, O. C., & Carpenter, L. L. (2012). Childhood adversity and epigenetic modulation of the leukocyte glucocorticoid receptor: preliminary findings in healthy adults. *PLoS One, 7*, e30148. http://dx.doi.org/10.1371/journal.pone.0030148.
Uchida, S., Hara, K., Kobayashi, A., Otsuki, K., Yamagata, H., Hobara, T., Suzuki, T., Miyata, N., & Watanabe, Y. (2011). Epigenetic status of Gdnf in the ventral striatum determines susceptibility and adaptation to daily stressful events. *Neuron, 69*, 359–372.
Vucetic, Z., Kimmel, J., Totoki, K., Hollenbeck, E. A., & Reyes, T. M. (2010). Maternal high-fat diet alters methylation and gene expression of dopamine and opiod-related genes. *Endocrinology, 151*(10), 4756–4764.
Wakefield, L., Boukouvala, S., & Sim, E. (2010). Characterisation of CpG methylation in the upstream control region of mouse Nat2: Evidence for a gene-environment interaction in a polymorphic gene implicated in folate metabolism. *Gene, 452*, 16–21.
Walker, B. E., & Haven, M. I. (1997). Intensity of multigenerational carcinogenesis from diethylstilbestrol in mice. *Carcinogenesis, 18*(791), 793.

Wang, H.-X., & Gao, W.-J. (2009). Cell type-specific development of NMDA receptors in the interneurons of rat prefrontal cortex. *Neuropsychopharmacology, 34*, 2028–2040.

Waterland, R. A., Dolinoy, D. C., Lin, J. R., Smith, C. A., Shi, X., & Tahiliani, K. G. (2006). Maternal methyl supplements increase offspring DNA methylation at Axin fused. *Genesis, 44*(401), 406.

Weaver, I. C., Cervoni, N., Champagne, F. A., D'Alessio, A. C., Sharma, S., Seckl, J. R., Dymov, S., Szyf, M., & Meaney, M. J. (2004). Epigenetic programming by maternal behavior. *Nature Neuroscience, 7*(8), 847–854.

Zerhouni, E. A. (2005). US biomedical research: basic, translational, and clinical sciences. *JAMA, 294*, 1352–1358.

Chapter 9
Epigenetic Regulation of ES Cell Pluripotency Maintenance

Ping Hu

9.1 Introduction

Stem cells can be divided into four categories according to their plasticity: totipotent stem cells, pluripotent stem cells, multipotent stem cells, and unipotent stem cells. Totipotent stem cells are represented by fertilized eggs, which can differentiate to any type of cells in mammals and the whole organism. ES cells are pluripotent stem cells. They are derived from the inner cell mass (ICM) of blastocyst and capable of differentiating to all cell types from the three germ layers, namely, ectoderm, mesoderm, and endoderm. But ES cells cannot differentiate to extraembryonic cells and do not have the abilities to form the whole organism. Multipotent stem cells can be differentiated to limited types of cells. The adult stem cells such as hematopoietic stem cells, which can only differentiate to blood cells, and neuronal stem cells, which can only differentiate to neural cells are the examples of multipotent stem cells. Unipotent stem cells have only one differentiation potential. For example, spermatogonial stem cells have single differentiation potential to become sperm cells. Due to their abilities to differentiate to functional somatic cells, stem cells have great clinical application potentials. Especially, ES cells can be self-renewed unlimitedly and differentiate to a variety of cell types. Therefore, these cells are important cell sources for regenerative medicine. Here we will focus on the epigenetic regulation of ES pluripotency maintenance.

P. Hu, Ph.D. (✉)
Shanghai Institutes for Biological Sciences, Institute of Biochemistry and Cell Biology, Chinese Academy of Sciences, 320 Yueyang Road, Room 1211, Shanghai 200031, China
e-mail: hup@sibcb.ac.cn

D. Hollar (ed.), *Epigenetics, the Environment, and Children's Health Across Lifespans*, DOI 10.1007/978-3-319-25325-1_9

9.2 Chromatin Remodeling

Genomic DNA is packed in a highly organized chromosome structure. DNA and its associated proteins (histones) are collectively named chromatin. The basic structural unit of chromatin is the nucleosome. Each nucleosome contains 146 base pairs of DNA wrapped around a histone octamer comprised of two copies of each of the histones H2A, H2B, H3, and H4. Histone H1 promotes higher-order chromatin structures by wrapping another 20 base pairs of DNA and forming a structure called chromatosome. Numerous nucleosomes can form a "beads on a string" structure and be further packed into a 30 nm chromatin fiber. The 30 nm fiber is further looped and packed into 250 nm chromatin fibers. The 250 nm chromatin fibers are further coiled into a higher-order chromosome structure (Luger et al. 1997; Woodcock 2005). The chromosome structure packs billions of base pairs of DNA in the microminiature space of nuclei to allow rich genetic information being passed from generation to generation.

At the same time, the highly compacted chromosome structure prevents transcription factors from accessing DNA and inhibits transcription. In order to activate transcription, cells evolve a complicated process to remove the chromosome barrier named chromatin remodeling. Chromatin remodeling is a dynamic process of chromatin architecture modifications to allow the access of transcription factors to the condensed DNA to regulate transcription thereby. Chromatin remodeling is carried by ATP-dependent chromatin remodeling complexes, including BAF/PBAF (SWI/SNF) complexes, NuRD complexes, ISWI complexes, and CHD1complexes (Hargreaves and Crabtree 2011).

9.3 Chromatin Architecture in ES Cells

Chromatin architecture changes play essential roles in pluripotency maintenance in ES cells. In ES cells, chromatin assumes a more flexible structure than in differentiated cells. ES cells have more diffused heterochromatin and less heterochromatin foci (Lessard and Crabtree 2010). The distribution pattern of heterochromatin protein 1α (HP1α) and heterochromatin histone mark H3K9me3 in ES cells is different from that in somatic cells. In somatic cells, heterochromatin proteins HP1α and H3K9me3 are distributed in punctuated foci, while these proteins are more evenly distributed in ES cells, suggesting changes in chromatin structure in ES cells. HP1 exchange rate at the heterochromatin foci is significantly faster in ES cells, and less HP1 binds chromatin in ES cells, further supporting the point that less heterochromatin is present in ES cells. Idenfication of several ES cell specific chromatin recomdeling complexes further support the notion that chromatin structure is a key player in ES cell pluripotency maintenance.

A sharp increase in euchromatin markers, such as H3Ac, H3K9Ac, H4Ac, H4K5Ac, H3K4me3, and H3K36me2, can be observed when somatic cells were induced to iPSCs. The increase of these euchromatin markers is accompanied with the expression of Nanog, suggesting that Nanog is important for converting the heterochromatin to euchromatin during reprogramming (Mattout et al. 2011). Coherently, less heterochromatin and more open chromatin architecture are also observed in iPSCs (induced pluripotent stem cells) (Mattout et al. 2011). The highly compacted chromocenter domains are turned to loosely packed 10 nm chromatin fiber domains when somatic cells are induced to iPSCs (Fussner et al. 2011). In reprogramming, HP1 impedes reprogramming by repressing Nanog reactivation (Sridharan et al. 2013), suggesting that heterochromatin is one of the obstacles for iPSC induction.

9.4 BAF Chromatin Remodeling Complexes

ES cells require a unique chromatin remodeling complex termed esBAF for self-renewal and pluripotency maintenance. esBAF has distinctive subunit components from BAF complexes in somatic cells. esBAF contains Baf250, Baf180, Brg1, Baf155, Baf60a, Baf60b, Baf57, Baf53a, actin, Baf47, Baf45a, Baf45b, Brd7, and Brd9. It is marked by the presence of Baf155 and the absence of Baf170, Baf60c, and Brm (Ho et al. 2009b). Consistently, Baf155 is highly expressed in ES cells and iPSCs, while Baf170 is upregulated upon ES cell differentiation. The expression of Baf155 is increased, while Baf170 is downregulated during reprogramming (Singhal et al. 2010). Baf155 knockout ES cells cannot be derived from blastocysts (Kim et al. 2001). Overexpression of Baf170 causes reduction of the ES cell self-renewal ability (Ho et al. 2009b). The ATPase Brg1 in esBAF plays critical roles to support ES cell self-renewal. Brg1 knockdown by shRNA leads to spontaneous differentiation of ES cells. Furthermore, ES cells cannot survive while being derived from Brg1 knockout blastocysts (Ho et al. 2009a, b; Kidder et al. 2009). Baf250a knockout ES cells display impaired abilities to differentiate to mesoderm (Gao et al. 2008). Baf47 is also required for the derivation of ES cells from blastocysts (Guidi et al. 2001; Klochendler-Yeivin et al. 2000). All these observations support the notion that BAF chromatin remodeling complexes are essential for ES cell self-renewal and pluripotency maintenance. Indeed, the esBAF plays a direct role in mediating the regulatory functions of ES cell-specific transcription factors Oct4, Sox2, and Nanog. BAF complex interacts with Nanog and Oct4 to repress the expression of developmental genes (Liang et al. 2008). Upon ES cell differentiation, the functions of BAF complex change. It helps turn off the expression of Nanog and promotes the expression of developmental genes. The alternation of BAF functions may be due to the changes of complex components and their interacting proteins (Fazzio and Panning 2010).

Reprogramming from somatic cells to iPS cells is considered a process to regain the ES cell self-renewal abilities. Consistent with the functions of BAF in ES cells, Baf155 and Brg1 can synergistically increase reprogramming efficiency by enhancing Oct4 binding on promoters. They could also facilitate the demethylation of Oct4 and Nanog promoter to activate the expression of these two key pluripotency genes (Singhal et al. 2010). Klf4 can enhance reprogramming through direct interaction with the SWI/SNF catalytic subunit BRN and promote active chromatin remodeling (Mak et al. 2010).

9.5 NuRD Chromatin Remodeling Complexes

The nucleosome remodeling and deacetylase (NuRD) complex is another key chromatin remodeling complex. It is one of the four major ATP-dependent chromatin remodeling complexes. NuRD complex possesses both ATP-dependent chromatin remodeling activity and HDAC (histone deacetylases) activity. It can play both repressive and active roles in target gene transcription. NuRD complex contains multiple subunits encoded by gene families, namely, Mi2a/2b, methyl-CpG-binding protein Mbd2/3, metastasis-associated protein Mta1/2/3, WD40 domain containing proteins Rbbp7 and Rbbp8, Hdac1, Hdac2, and zinc figure protein p66a/66b. Hdac1 and Hdac2 in the complex have histone deacetylase activities. Mi2a/2b has ATPase activities (Wade et al. 1999). NuRD complex can directly interact with histone and bind DNA to regulate gene expression and genome integrity (Basta and Rauchman 2015). NuRD subunit Mbd3 has been shown to be important for ES cell pluripotency maintenance. *Mbd3* knockout disrupts the formation of NuRD complex. *Mbd3−/−* ES cells can be maintained in the absence of LIF and displayed persistent self-renewal ability. These cells display persistent expression of ES cell marker *Oct4* and *Nanog* upon differentiation induction and fail to commit to more differentiated cell lineages (Kaji et al. 2006, 2007). Consistently, in *Mbd3* knockout mice, the ICM of blastocysts fails to develop into mature epiblasts. The ICM of *Mbd3* knock out mice also has defects. Pluripotent ES cells cannot be derived from *Mbd3−/−* ICM; instead, robust endoderm outgrowth can be detected from *Mbd3−/−* ICM (Kaji et al. 2007). These results suggest that Mbd3 is essential for cell fate determination. And its function is highly context-dependent. The molecular mechanism needs further investigation.

Consistent with its role in ES cells, knockdown of *Mbd3* significantly reduced the time to convert somatic cells to iPSCs and improved the induction efficiency to close to 100 % (dos Santos et al. 2014). These results suggest that chromatin structure is the major obstacle for the conversion of somatic cells to iPSCs.

An isoform of NuRD complex termed NODE (Nanog and Oct4-associated deacetylase) exists specifically in ES cells. NODE contains Nanog, Oct4, Hdac1/2, and Mta1/2 but lacks Mbd3 and Rbbp7. Mta1 in NODE complex is required for repression of the expression of lineage determination genes such as Foxa2 and Gata6 (Liang et al. 2008). In summary, NODE complex directly participates in the pluripotency factor-mediated circuitry to inhibit the expression of development-related genes and promote pluripotency maintenance.

9.6 INO80

The INO80 chromatin remodeling complex is a member of the SWI/SNF family of chromatin remodelers. The INO80 subfamily is distinguished from other SWI/SNF family members by the presence of a spacer region that splits the conserved ATPase domain. Knocking down of INO80 subunits in ES cells leads to the loss of normal ES cell morphology. The ATPase subunit of INO80 chromatin remodeling complex is Ino80. Knocking down of *Ino80* leads to the loss of alkaline phosphatase staining and decreases of the expression level of pluripotency factors such as *Oct4*, *Nanog*, *Sox2*, *Klf4*, and *Esrrb*. The expression levels of differentiation-related genes such as *Cdx4*, *Fgf5*, *Nestin*, and *Pax3* are increased upon Ino80 knockdown (Wang et al. 2014a). INO80 complex is required for ES cell self-renewal and pluripotency gene expression. It binds all the currently known pluripotency genes together with Oct4 and Klf4 to help maintain open chromatin structure and promote mediator and RNA Pol II recruitment. Consistent with its critical roles in ES cell self-renewal and pluripotency maintenance, Ino80 is also required for reprogramming. Knockdown of Ino80 greatly reduced the efficiency of iPSC formation (Wang et al. 2014a).

9.7 Histone Modifications

Histones are organized to form octamer to wrap genomic DNA. Each of the core histones bears a 20–35-amino-acid-long highly dynamic tails. These tails protrude from the nucleosome surface and are subjected to varieties of posttranslational modifications, including phosphorylation, acetylation, methylation, ubiquitination, succinylation, ADP-ribosylation, and *O*-GlcNAC glycosylation (Rothbart and Strahl 2014). The modifications mainly occur on amino acids lysine, arginine, serine, threonine, tyrosine, aspartic acid, glutamic acid, and histidine in the tails (Table 9.1). The combination of modifications on variant amino acids at different positions constitutes the "histone codes" to regulate transcription.

Functions of only a few histone modifications in cellular process have been extensively studied, including histone methylation and acetylation. Histone modifications are added and removed by many enzymes. These enzymes also play essential roles in nearly every aspect of cellular activities. Here we are focusing on the functions of histone acetylation and methylation and the enzymes involved in these processes in ES cell pluripotency maintenance.

The histone modification landscape of ES cells is dramatically different from that of somatic cells. The ratio of active (H3K9Ac, H3K4me3) to repressive (H3K9me3, H3K27me3) histone modification marks is relatively high (Gaspar-Maia et al. 2011). In ES cells, the pluripotency genes are marked by active histone marks H3K4me3 and acetylation, while the key lineage commitment-related genes, whose expression is repressed in ES stage, are marked by bivalent histone marks. They are both labeled by the active histone H3K4me3 and the repressive H3K27me3

Table 9.1 Summary of main histone tail modifications

markers (Azuara et al. 2006; Bernstein et al. 2006). The genes labeled by bivalent histone marks have the potential to assume more active or repressive chromatin conformations depending on the ratio of trithorax group (TrxG) protein and polycomb group (PcG) complexes. TrxG and PcG complexes are all histone methyltransferase complexes. TrxG complex executes the active H3K4 trimethylation; PcG complexes PRC2 catalyze the repressive H3K27 trimethylation. Both complexes have been shown to play critical roles in ES cell self-renewal and pluripotency maintenance.

Members of TrxG proteins were initially identified in *Drosophila* as mutations mimicking Hox mutations or suppression of Pc(Su(Pc)) mutations. Sixteen proteins have been identified as the members of TrxG proteins. The human homolog of TrxG complex is named Mll (myeloid/lymphoid or mixed lineage leukemia). Mll complex consists of multiple subunits including Mll1/2/3/4, Wdr5, Ash2l, Rbbp5, Mof, Menin, Ledgf, and Dpy-30 (Slany 2009). Knockdown of TrxG subunit Wdr5 in ES cells leads to global downregulation of the H3K4me3 levels and reduced expression levels of the key pluripotency genes. The ES cells lack of Wdr5 display defects in pluripotency maintenance. Ash2l knockout mice are embryonic lethal. ES cells cannot be derived from the Ash2l knockout blastocyst (Stoller et al. 2010). Knockdown of Dpy-30 causes global downregulation of H3K4me3, and the development-related genes cannot be properly activated during differentiation (Vastenhouw and Schier 2012). *Mll2* knockout mice are embryonic lethal with global downregulation of H3K4me3. *Mll2−/−* ES cells displayed defects in proliferation and differentiation (Glaser et al. 2006). These results show the importance of H3K4me3 and TrxG proteins in ES cell derivation and proliferation.

PcG complexes have two members: PRC1 and PRC2. PRC1 brings the repressive mark H2AK119Ub (Schuettengruber et al. 2011). PRC2 executes the H3K27 trimethylation. PRC2 has four subunits: zinc finger protein Suz12, histone methyltransferases Ezh1 and Ezh2, histone binding domain containing protein RbAp48, and Eed. Ezh2 knockout mice are embryonic lethal due to gastrulation defects. Depletion of Ezh2 in ES cells leads to decreased levels of H3K27me2 and H3K27me3. The ES cells fail to differentiate to mesendoderm (Shen et al. 2008). Simultaneous depletion of *Ezh1* and *Ezh2* abolishes the global H3K27 methylation and activates the lineage commitment-related genes prematurely (Shen et al. 2008). *Eed* knockout mice are embryonic lethal due to gastrulation failure. The global level of H3K27 methylation is significantly decreased in *Eed−/−* ES cells, and the expression levels of development-related genes are upregulated (Chamberlain et al. 2008). *Suz12* knockout mice are embryonic lethal due to gastrulation failure and proliferation defects. Depletion of *Suz12* in ES cells leads to global loss of H3K27me2 and H3K27me3. The expression levels of lineage commitment-related genes are abnormally upregulated in ES cells (Pasini et al. 2007).

To echo their important functions in ES cell pluripotency maintenance and self-renewal, both TrxG and PcG complexes are required for the process of reprogramming. TrxG complexes are required for the activation of pluripotency genes. Wdr5 is activated during reprogramming, and knocking down Wdr5 during reprogramming causes a significant reduction in the reprogramming efficiency (Ang et al. 2011). PcG complexes are required for early silencing of lineage-specific genes. Knockdown of PRC2 subunit *EED* and *SUZ12* can significantly reduce the reprogramming efficiency, while consistently, overexpression of *Ezh2* improves reprogramming efficiency (Pereira et al. 2010). An ES cell-specific PRC2 subunit esPRC2p48 (ES cell-specific PRC2 subunit p48) has been identified. Together with Ezh1, Jarid2, and Mtf2, esPRC2p48 can interact with Oct4, Sox2, and Klf4 to mediate H3K27 methylation on MEF-specific genes to increase reprogramming efficiency (Zhang et al. 2011). It is proposed that reprogramming is initiated by the recruitment of transcription factors to a nucleosome-depleted region generated by TrxG binding and removal of PcG, and the cohesin complex is subsequently recruited to facilitate enhancer–promoter interaction by DNA looping (Taberlay et al. 2011).

During ES cell self-renewal, the bivalent histone marks can be inherited by the daughter cells. It has been shown in *Drosophila* embryos that H3K4me3 and H3K27me3 are all replaced with unmodified histones at S phase. TrxG and PcG proteins remain associated with their target genes during DNA replication and put the methyl group back to histones right after the passage of DNA polymerase. These results emphasize the importance of TrxG and PcG proteins in histone mark inheritance. Whether the same mechanism is applied to mammalian cells remains to be proved (Petruk et al. 2012).

Histone modifications can also interplay with chromatin remodeling factors to regulate the ES cell self-renewal and pluripotency maintenance. NuRD chromatin remodeling complex is enriched on genes with bivalent marks. NuRD binding at these genes is required for the recruitment of PRC2 complex. NuRD works cooperatively with PRC2 to maintain differentiation-related genes in a silent but poised state. This state allows rapid activation of the cell lineage commitment-related genes in response to developmental cues (Basta and Rauchman 2015). Besides its ATPase-dependent chromatin remodeling activities, NuRD can also deacetylate histone H3K9. Deacetylation of H3K9 enables Lsd1 to bind its target genes and demethylation of histone H3K4me1. Loss of H3K4me1silences the enhancers of the pluripotency-related genes and allows cells to differentiate (Kaji et al. 2006; Shimbo et al. 2013; Whyte et al. 2012).

Histone modification enzymes also play important roles in reprogramming. Histone acetylation is predominant in ES cells; decreased levels of H3 and H4 acetylations are observed after lineage commitment (Mattout and Meshorer 2010). Along with this line of evidences, increasing histone acetylation levels by HDAC inhibitor treatment can increase the efficiency of reprogramming (Huangfu et al. 2008b; Mali et al. 2010; Stadtfeld et al. 2010). Indeed, the HDAC inhibitor can substitute for c-Myc and Klf4 to induce reprogramming (Huangfu et al. 2008a, b; Mali et al. 2010).

Histone modifications are important targets to manipulate reprogramming efficiency. Repressive histone mark H3K9me is increased after ES cell differentiation (Azuara et al. 2006). H3K9 lysine methyltransferase Kmt1c increases the levels of H3K9me and inhibits the re-expression of Oct4 during reprogramming. Knockdown of Kmt1c accelerates reprogramming. Overexpression of H3K9 lysine demethylase Kdm3a has the similar effects (Ma et al. 2008). SUV39H1/2 demethylates H3K9me3 to reduce H3K9 level. It enhances reprogramming by facilitating Oct4/Sox2 binding on the target genes (Onder et al. 2012). G9a is another H3K9 methyltransferase. Reducing the global H3K9me2/3 levels by knocking down G9a improves reprogramming efficiency. Chemical inhibition of Kmt1c enables reprogramming of MEFs to iPSCs by activating Oct4 and Klf4 (Shi et al. 2008).

H3K79 methyltransferase Kmt4 catalyzes H3K79 methylation on lineage-specific genes. Knockout of Kmt4 significantly reduced H3K79me2 levels on MEF-specific genes and increases reprogramming (Onder et al. 2012). Dotl1 (Dot1-like histone H3K79 methyltransferase) is a histone methyltransferase that methylates H3 lysine 79. Dotl1 enhances reprogramming by reducing H3K79me3 level to promote the silence of lineage-specific genes (Nguyen and Zhang 2011; Venkatesh et al. 2012).

H3K36me2/3 demethylase Kdm2b can suppress Ink4/Arf expression and activate mir302/367 expression to improve the reprogramming efficiency (Wang et al. 2011).

Loss of H3K27me3 is observed at the earliest stage of reprogramming. Jmjd3 is a histone H3K27 demethylase. Depletion of Jmjd3 enhances iPS cell induction efficiency, while overexpression of Jmjd3 inhibits iPS cell formation. Overexpression of Jmjd3 demethylates H3K27me3 at the Ink4/Arf gene, thereby removing the repression of this gene during reprogramming. The activation of Ink4/Arf dramatically inhibits the formation of iPS cells. Utx is the enzyme that specifically catalyzes H3K27me2/3 demethylation. Depletion of *Utx* completely blocks the iPS cell formation (Mansour et al. 2012). Utx can interact with Oct4, Sox2, and Klf4 to remove the repressive mark from pluripotency-related genes such as *Fgf4*, *Sall4*, and *Sall1* during reprogramming (Mansour et al. 2012). However, complete loss of H3K27me3 blocks reprogramming, suggesting that H3K27me3 needs to be redistributed from pluripotency genes to lineage-specific genes during reprogramming.

9.8 Histone Variants

Histone variants can replace canonical histones to integrate into chromatin, regulating transcription, chromatin structure, DNA damage and repair, and epigenetic silencing (Talbert and Henikoff 2010). More and more evidences are accumulated to show histone variants as emerging important regulators for ES cell self-renewal and pluripotency maintenance. The incorporation of histone variants is tightly regulated by a group of proteins named histone chaperons.

Eleven H1 variants are identified in mammals, namely, H1.0– H1.5, germ cell-specific H1 (H1τ, H1T2, HILS1, and H1oo), and H1 (Happel and Doenecke 2009). H1.2 and H1.3 are highly enriched in ES cells, while H1.0 is enriched in differentiated cells (Turinetto and Giachino 2015). H1.2 and H1.3 are located at the major satellite repeat sequence at the pericentric heterochromatin region. Depletion of *H1.3*, *H1.4*, and *H1.5* induces significant increase in transcription from the major satellite repeats, but does not affect repressive histone marks such as H3K9me3, H4K20me3, and K3K27me3 (Cao et al. 2013). ES cells without H1.3, H1.4, and H1.5 are more resistant to spontaneous differentiation induced by LIF removal (Christophorou et al. 2014). Similarly, H1.3 and H1.5 are highly expressed in human ES cells. These two H1 variants may contribute to pluripotency maintenance and self-renewal of human ES cells (Terme et al. 2011). Overexpression of oocyte-specific H1, H1foo keeps the global methylation level low and maintains the pluripotency gene expression in ES cells (Hayakawa et al. 2012).

H2A.X is an important form of H2A variant. It is one of the highly expressed histone variants in ES cells and preimplantation embryos (Kafer et al. 2010). After fertilization, H2A.X stored in oocyte will quickly incorporate to the chromatin, suggesting that H2A.X plays a critical role in ES cell pluripotency establishment and maintenance. *H2A.X* knockout ES cells have reduced self-renewal capacity (Turinetto et al. 2012). Knockdown of *H2A.X* completely inhibits iPSC formation (Wu et al. 2014). H2A.X is phosphorylated at Serine 139 by DNA damage-dependent protein kinases ATR/ATM upon DNA damages to form γH2A.X foci. The formation of γH2A.X foci recruits DNA damage repair machinery and chromatin remodeling complex INO80 and SWR1 (Stummvoll et al. 2009; van Attikum et al. 2007). High γH2A.X level is detected in ES cells, and it is associated with global chromatin decondensation (Banath et al. 2009). High basal level of γH2A.X contributes to sustained self-renewal and pluripotency maintenance in ES cells and iPSCs (Turinetto et al. 2012). It seems that the high γH2A.X level in ES cells is not the consequence of DNA damage. How the high γH2A.X level is achieved in ES cells needs more investigation.

H2A.Z is one of the most conserved histone variants and universally expressed across species (Biterge and Schneider 2014). It is incorporated into chromatin region flanking transcription start sites (Barski et al. 2007; Creyghton et al. 2008). H2A.Z is specifically enriched in the trophectoderm (Rangasamy et al. 2003). The incorporation of H2A.Z can influence nucleosome positioning, H1 linker binding, and chromatin remodeling enzyme activity (Li et al. 2005). H2A.Z is enriched at the promoters of differentiation-related genes to recruit repressive histone modifications. It can also facilitate the expression of pluripotency-related genes; depletion of H2A.Z leads to premature differentiation of ES cells (Creyghton et al. 2008).

MacroH2A is the largest histone variant and contains a macro domain at the C-terminal globular domain. There are three macroH2A variants characterized in mammals: macroH2A1.1, macroH2A1.2, and macroH2A.2 (Buschbeck and Di Croce 2010). All three isoforms can be detected in ES cells. MacroH2A is enriched at H3K27me3 sites. It serves as an epigenetic barrier during iPSC induction. Knockdown of *macroH2A* significantly enhances the iPS cell induction efficiency (Barrero et al. 2013; Gaspar-Maia et al. 2013; Pasque et al. 2012). In ES cells, the exchange rate of macroH2A is higher than in differentiated cells, indicating that macroH2A is important for the maintenance of the dynamic chromatin status. In ES cells with macroH2A.1 depletion, the activation of cell lineage commitment-related genes is significantly delayed when the cells were induced to differentiate by EB (embryonic body) formation (Creppe et al. 2012).

There are five H3 variants in mammals: H3.1, H3.2, H3.3, the centromere-specific variant Cen H3, and the testis-specific histone H3t (Turinetto and Giachino 2015). After fertilization, H3.3 stored in the oocyte will quickly replace the H3 (Nashun et al. 2011), suggesting that the change of histone variants could have a big impact on early development of fertilized oocytes. Histone H3.3, which associates with transcriptionally active regions, incorporates more stably in the chromatin in ES cells. HirA is a histone chaperone that preferentially places histone H3.3 into nucleosomes. Consistent with the notion that more H3.3 is incorporated in chromatin in ES cells, HirA knockout ES cells displayed accelerated differentiation due to the decreasing level of H3.3 incorporation in the open chromatin region. HirA–/– ES cells have more heterochromatin regions. These results suggest that H3.3 is important for maintenance of an open chromatin region in ES cells and the pluripotency maintenance of ES cells (Meshorer et al. 2006).

9.9 DNA Methylation

DNA methylation has been emerging as an important epigenetic regulator. DNA methylation occurs primarily at the 5′ position of cytosine residue (5-mC). It is a heritable DNA modification mainly occurring at the CpG region. The functions of DNA methylation start to be elucidated.

DNA methylation levels are low in the naive pluripotent cells both in vivo and in vitro (Guo et al. 2014; Habibi et al. 2013; Leitch et al. 2013; Smith et al. 2012, 2014; Takashima et al. 2014; Wang et al. 2014b). Consistently, a gradual change in DNA methylation patterns takes place during reprogramming, shifting from hypermethylation to hypomethylation in pluripotency genes (Mikkelsen et al. 2008; Simonsson and Gurdon 2004). During reprogramming, DNA methylome needs to be reprogrammed. The incomplete reprogramming of the methylome affects the differentiation capacity of the cells (Bar-Nur et al. 2011; Kim et al. 2010; Lister et al. 2011; Ohi et al. 2011; Polo et al. 2010). Incomplete DNA methylome is correlated with incomplete transcriptional reprogramming.

Inhibition of Dnmt1, a DNA methyltransferase, to add methyl group to DNA by 5-aza-2’ deoxycytidine greatly improves the reprogramming process (Kim et al. 2010; Mikkelsen et al. 2008), suggesting that hypomethylation can improve reprogramming. Activation-induced deaminase (AID) can deaminate 5-mC to thymidine, which leads to a T:G mismatch and is repaired by the based excision DNA repair pathway. It is required for the expression of pluripotency genes during reprogramming (Bhutani et al. 2010). The 10–11 translocation proteins (TET) can catalyze the oxidation of 5-methylcytosine (5-mC) to 5-hydroxymethylcytosine (5-hmC), 5-formylcytosine (5-fC), and 5-carboxycytosine (5-caC) (He et al. 2011; Ito et al. 2011; Tahiliani et al. 2009) and lead to demethylation of cysteine. Tet is required for reprogramming of somatic cells to iPSCs. Triple knockout of Tet1, Tet2, and Tet3 severely abolishes the iPSC induction. Tet proteins are only required for mesenchymal-to-epithelial transition during reprogramming by oxidizing the MET regulator mir200 cluster to elevate their expression levels. When the MET step is rescued, Tet triple knockout cells can be reprogrammed efficiently (Hu et al. 2014). Tet3 plays important roles in demethylation and reactivation of the somatic Oct4 promoter in iPSCs generated by nuclei transfer (Gu et al. 2011). More work needs to be carried out to further elucidate the function of DNA methylation in ES cells.

9.10 Noncoding RNAs

Noncoding RNAs (ncRNAs) have long been considered to be the noise of transcription. However, recent studies have shown that these RNAs play important roles in almost every aspect of cellular behaviors. NcRNA makes up the majority of the transcription products from eukaryotic genome. They can be divided into structural and regulatory RNAs. Ribosomal RNA (rRNA), small nuclear RNA (snRNA), small nucleolar RNA (snoRNA), and tRNA are grouped as structural RNAs whose functions have been thoroughly studied for decades. Small regulatory RNA comprises microRNAs (miRNAs), piwi-interacting RNAs (piRNAs), and small interfering RNAs (siRNAs). Regulatory RNA longer than 200 nucleotides are characterized as long noncoding RNAs (lncRNAs) (Erriquez et al. 2013).

9.11 LncRNAs

Over 100 lncRNAs have been shown to be enriched in ES cells and iPS cells and 104 lncRNAs specifically expressed in ES cells and iPS cells (Guttman et al. 2010; Loewer et al. 2010). Several of these lncRNAs have been shown to be able to regulate ES cell pluripotency maintenance and differentiation through interactions with epigenetic regulators. In a genome-wide study, 12 lncRNAs were identified to correlate with the expression of pluripotency genes, such as *Oct4*, *Nanog*, and *Sox2* (Dinger et al. 2008). Among these lncRNAs, *Gomafu* (Miat, AK028326) was identified to be the direct target of Oct4. Knockdown of *Oct4* and *Nanog* can decrease the expression levels of these two lncRNAs (Sheik Mohamed et al. 2010). During differentiation, *Gomafu* was downregulated. Knockdown of

Gomafu in ES cells downregulates *Oct4* expression and promotes differentiation. *Gomafu* and Oct4 may work in a potential synergistic feedback mechanism (Sheik Mohamed et al. 2010). *AK141205* was identified as direct target repressed by Nanog. *Nanog* RNAi can upregulate *AK141205*. Consistently, the expression level of *AK141205* is upregulated by RA-induced differentiation. Knockdown of *AK141205* significantly downregulates *Oct4*, but not *Nanog*, suggesting that *AK141205* does not form an auto feedback loop with Nanog, but can modulate the expression of *Oct4* with an unknown mechanism (Sheik Mohamed et al. 2010). *HOTAIRM1* interacts with PRC1, PRC2, and CBX1 in ES cells to regulate *HoxA* cluster genes (Lin et al. 2011). LncRNA *Rian* binds a number of chromatin remodeling proteins such as PRC1, PRC2, JARID1B, JARID1C, and Cbx3 in ES cells (Guttman et al. 2011), but the functions of the lncRNA binding to chromatin remodelers remain to be uncovered. LncRNA *RoR* is enriched in ES cells and iPS cells, and plays important roles in iPS cell formation. Knockdown of *RoR* reduced colony formation in iPS cell induction, while overexpression increases the number of colonies (Ohnuki et al. 2014).

9.12 MicroRNAs

MicroRNAs (miRs) represent another important group of regulators in ES cells. The members of the miR290–295 cluster are highly enriched in ES cells (Houbaviy et al. 2003; Marson et al. 2008). MicroRNAs from *miR290–295* cluster share the seed sequence of AAGUGCU. They can target multiple proteins in the cyclin E-cdk2 pathway to regulate G1-S transition and promote rapid proliferation (Wang et al. 2008). Consistently, these microRNAs can also enhance iPS cell induction (Judson et al. 2009). MiR290–295 can repress the expression of *Rbl2*. Rbl2 directly represses the expression of *Dnmt1*. Therefore, *miR290–295* specifically expressed in ES cell keep the *Rbl2* expression level low and allow the DNA methyltransferase Dnmt1 to have relatively high activities to maintain the normal DNA methylation in ES cells (Benetti et al. 2008).

Multiple components of the BAF chromatin remodeling can be targeted by microRNAs in ES cells. miR294 directly targets Baf170, which is absent from the esBAF and expressed in differentiated cells. *MiR294* is highly expressed in ES cells and keeps the expression level of Baf170 low. Another microRNA let-7 can antagonize the effects of the miR290–295 cluster of miRNAs in early differentiation. Let-7 is widely expressed in differentiated somatic cells, but not in ES cells. In ES cells, the processing of let-7 is inhibited by the pluripotency-related protein Lin28 (Heo et al. 2008; Rybak et al. 2008; Viswanathan et al. 2008). Let-7 targets several pluripotency-related factors like Sall4, c-Myc, and esBAF component Baf155 (Kumar et al. 2007; Melton et al. 2010). The interplay between miR290–295 cluster and let-7 is important for stem cell pluripotency maintenance, but the mechanism remains to be elucidated.

Abbreviations

BAF	Brahma-associated factor
CpG	Cytosine-guanine nucleotide-rich sequence
DNMT	DNA methyltransferase
ES Cells	Embryonic stem cell
HP1α	Heterochromatin protein 1α
iPS Cells	Induced pluripotent stem cells
ICM	Inner cell mass
INO80 chromatin remodeling complex	A chromatin remodeling complex
lncRNA	Long noncoding RNA
NODE	Nanog and Oct4-associated deacetylase
ncRNA	Noncoding RNA
NuRD	Nucleosome remodeling and deacetylase
c-Myc	Oncogene master regulator
PcG	Polycomb group protein
TrxG	Trithorax group protein

References

Ang, Y. S., Tsai, S. Y., Lee, D. F., Monk, J., Su, J., Ratnakumar, K., Ding, J., Ge, Y., Darr, H., Chang, B., et al. (2011). Wdr5 mediates self-renewal and reprogramming via the embryonic stem cell core transcriptional network. *Cell, 145*, 183–197.

Azuara, V., Perry, P., Sauer, S., Spivakov, M., Jorgensen, H. F., John, R. M., Gouti, M., Casanova, M., Warnes, G., Merkenschlager, M., et al. (2006). Chromatin signatures of pluripotent cell lines. *Nature Cell Biology, 8*, 532–538.

Banath, J. P., Banuelos, C. A., Klokov, D., MacPhail, S. M., Lansdorp, P. M., & Olive, P. L. (2009). Explanation for excessive DNA single-strand breaks and endogenous repair foci in pluripotent mouse embryonic stem cells. *Experimental Cell Research, 315*, 1505–1520.

Bar-Nur, O., Russ, H. A., Efrat, S., & Benvenisty, N. (2011). Epigenetic memory and preferential lineage-specific differentiation in induced pluripotent stem cells derived from human pancreatic islet beta cells. *Cell Stem Cell, 9*, 17–23.

Barrero, M. J., Sese, B., Kuebler, B., Bilic, J., Boue, S., Marti, M., & Izpisua Belmonte, J. C. (2013). Macrohistone variants preserve cell identity by preventing the gain of H3K4me2 during reprogramming to pluripotency. *Cell Reports, 3*, 1005–1011.

Barski, A., Cuddapah, S., Cui, K., Roh, T. Y., Schones, D. E., Wang, Z., Wei, G., Chepelev, I., & Zhao, K. (2007). High-resolution profiling of histone methylations in the human genome. *Cell, 129*, 823–837.

Basta, J., & Rauchman, M. (2015). The nucleosome remodeling and deacetylase complex in development and disease. *Translational Research, 165*, 36–47.

Benetti, R., Gonzalo, S., Jaco, I., Munoz, P., Gonzalez, S., Schoeftner, S., Murchison, E., Andl, T., Chen, T., Klatt, P., et al. (2008). A mammalian microRNA cluster controls DNA methylation and telomere recombination via Rbl2-dependent regulation of DNA methyltransferases. *Nature Structural and Molecular Biology, 15*, 998.

Bernstein, B. E., Mikkelsen, T. S., Xie, X., Kamal, M., Huebert, D. J., Cuff, J., Fry, B., Meissner, A., Wernig, M., Plath, K., et al. (2006). A bivalent chromatin structure marks key developmental genes in embryonic stem cells. *Cell, 125*, 315–326.

Bhutani, N., Brady, J. J., Damian, M., Sacco, A., Corbel, S. Y., & Blau, H. M. (2010). Reprogramming towards pluripotency requires AID-dependent DNA demethylation. *Nature, 463*, 1042–1047.

Biterge, B., & Schneider, R. (2014). Histone variants: Key players of chromatin. *Cell and Tissue Research, 356*, 457–466.

Buschbeck, M., & Di Croce, L. (2010). Approaching the molecular and physiological function of macroH2A variants. *Epigenetics, 5*, 118–123.

Cao, K., Lailler, N., Zhang, Y., Kumar, A., Uppal, K., Liu, Z., Lee, E. K., Wu, H., Medrzycki, M., Pan, C., et al. (2013). High-resolution mapping of h1 linker histone variants in embryonic stem cells. *PLoS Genetics, 9*, e1003417.

Chamberlain, S. J., Yee, D., & Magnuson, T. (2008). Polycomb repressive complex 2 is dispensable for maintenance of embryonic stem cell pluripotency. *Stem Cells, 26*, 1496–1505.

Christophorou, M. A., Castelo-Branco, G., Halley-Stott, R. P., Oliveira, C. S., Loos, R., Radzisheuskaya, A., Mowen, K. A., Bertone, P., Silva, J. C., Zernicka-Goetz, M., et al. (2014). Citrullination regulates pluripotency and histone H1 binding to chromatin. *Nature, 507*, 104–108.

Creppe, C., Janich, P., Cantarino, N., Noguera, M., Valero, V., Musulen, E., Douet, J., Posavec, M., Martin-Caballero, J., Sumoy, L., et al. (2012). MacroH2A1 regulates the balance between self-renewal and differentiation commitment in embryonic and adult stem cells. *Molecular and Cellular Biology, 32*, 1442–1452.

Creyghton, M. P., Markoulaki, S., Levine, S. S., Hanna, J., Lodato, M. A., Sha, K., Young, R. A., Jaenisch, R., & Boyer, L. A. (2008). H2AZ is enriched at polycomb complex target genes in ES cells and is necessary for lineage commitment. *Cell, 135*, 649–661.

Dinger, M. E., Amaral, P. P., Mercer, T. R., Pang, K. C., Bruce, S. J., Gardiner, B. B., Askarian-Amiri, M. E., Ru, K., Solda, G., Simons, C., et al. (2008). Long noncoding RNAs in mouse embryonic stem cell pluripotency and differentiation. *Genome Research, 18*, 1433–1445.

dos Santos, R. L., Tosti, L., Radzisheuskaya, A., Caballero, I. M., Kaji, K., Hendrich, B., & Silva, J. C. (2014). MBD3/NuRD facilitates induction of pluripotency in a context-dependent manner. *Cell Stem Cell, 15*, 102–110.

Erriquez, D., Perini, G., & Ferlini, A. (2013). Non-coding RNAs in muscle dystrophies. *International Journal of Molecular Sciences, 14*, 19681–19704.

Fazzio, T. G., & Panning, B. (2010). Control of embryonic stem cell identity by nucleosome remodeling enzymes. *Current Opinion in Genetics and Development, 20*, 500–504.

Fussner, E., Djuric, U., Strauss, M., Hotta, A., Perez-Iratxeta, C., Lanner, F., Dilworth, F. J., Ellis, J., & Bazett-Jones, D. P. (2011). Constitutive heterochromatin reorganization during somatic cell reprogramming. *Embo Journal, 30*, 1778–1789.

Gao, X., Tate, P., Hu, P., Tjian, R., Skarnes, W. C., & Wang, Z. (2008). ES cell pluripotency and germ-layer formation require the SWI/SNF chromatin remodeling component BAF250a. *Proceedings of the National Academy of Sciences of the United States of America, 105*, 6656–6661.

Gaspar-Maia, A., Alajem, A., Meshorer, E., & Ramalho-Santos, M. (2011). Open chromatin in pluripotency and reprogramming. *Nature Reviews Molecular Cell Biology, 12*, 36–47.

Gaspar-Maia, A., Qadeer, Z. A., Hasson, D., Ratnakumar, K., Leu, N. A., Leroy, G., Liu, S., Costanzi, C., Valle-Garcia, D., Schaniel, C., et al. (2013). MacroH2A histone variants act as a barrier upon reprogramming towards pluripotency. *Nature Communications, 4*, 1565.

Glaser, S., Schaft, J., Lubitz, S., Vintersten, K., van der Hoeven, F., Tufteland, K. R., Aasland, R., Anastassiadis, K., Ang, S. L., & Stewart, A. F. (2006). Multiple epigenetic maintenance factors implicated by the loss of Mll2 in mouse development. *Development, 133*, 1423–1432.

Gu, T. P., Guo, F., Yang, H., Wu, H. P., Xu, G. F., Liu, W., Xie, Z. G., Shi, L., He, X., Jin, S. G., et al. (2011). The role of Tet3 DNA dioxygenase in epigenetic reprogramming by oocytes. *Nature, 477*, 606–610.

Guidi, C. J., Sands, A. T., Zambrowicz, B. P., Turner, T. K., Demers, D. A., Webster, W., Smith, T. W., Imbalzano, A. N., & Jones, S. N. (2001). Disruption of Ini1 leads to peri-implantation lethality and tumorigenesis in mice. *Molecular and Cellular Biology, 21*, 3598–3603.

Guo, H., Zhu, P., Yan, L., Li, R., Hu, B., Lian, Y., Yan, J., Ren, X., Lin, S., Li, J., et al. (2014). The DNA methylation landscape of human early embryos. *Nature, 511*, 606–610.

Guttman, M., Donaghey, J., Carey, B. W., Garber, M., Grenier, J. K., Munson, G., Young, G., Lucas, A. B., Ach, R., Bruhn, L., et al. (2011). lincRNAs act in the circuitry controlling pluripotency and differentiation. *Nature, 477*, 295–300.

Guttman, M., Garber, M., Levin, J. Z., Donaghey, J., Robinson, J., Adiconis, X., Fan, L., Koziol, M. J., Gnirke, A., Nusbaum, C., et al. (2010). Ab initio reconstruction of cell type-specific transcriptomes in mouse reveals the conserved multi-exonic structure of lincRNAs. *Nature Biotechnology, 28*, 503–510.

Habibi, E., Brinkman, A. B., Arand, J., Kroeze, L. I., Kerstens, H. H., Matarese, F., Lepikhov, K., Gut, M., Brun-Heath, I., Hubner, N. C., et al. (2013). Whole-genome bisulfite sequencing of two distinct interconvertible DNA methylomes of mouse embryonic stem cells. *Cell Stem Cell, 13*, 360–369.

Happel, N., & Doenecke, D. (2009). Histone H1 and its isoforms: contribution to chromatin structure and function. *Gene, 431*, 1–12.

Hargreaves, D. C., & Crabtree, G. R. (2011). ATP-dependent chromatin remodeling: Genetics, genomics and mechanisms. *Cell Research, 21*, 396–420.

Hayakawa, K., Ohgane, J., Tanaka, S., Yagi, S., & Shiota, K. (2012). Oocyte-specific linker histone H1foo is an epigenomic modulator that decondenses chromatin and impairs pluripotency. *Epigenetics, 7*, 1029–1036.

He, Y. F., Li, B. Z., Li, Z., Liu, P., Wang, Y., Tang, Q., Ding, J., Jia, Y., Chen, Z., Li, L., et al. (2011). Tet-mediated formation of 5-carboxylcytosine and its excision by TDG in mammalian DNA. *Science, 333*, 1303–1307.

Heo, I., Joo, C., Cho, J., Ha, M., Han, J., & Kim, V. N. (2008). Lin28 mediates the terminal uridylation of let-7 precursor MicroRNA. *Molecular Cell, 32*, 276–284.

Ho, L., Jothi, R., Ronan, J. L., Cui, K., Zhao, K., & Crabtree, G. R. (2009a). An embryonic stem cell chromatin remodeling complex, esBAF, is an essential component of the core pluripotency transcriptional network. *Proceedings of the National Academy of Sciences of the United States of America, 106*, 5187–5191.

Ho, L., Ronan, J. L., Wu, J., Staahl, B. T., Chen, L., Kuo, A., Lessard, J., Nesvizhskii, A. I., Ranish, J., & Crabtree, G. R. (2009b). An embryonic stem cell chromatin remodeling complex, esBAF, is essential for embryonic stem cell self-renewal and pluripotency. *Proceedings of the National Academy of Sciences of the United States of America, 106*, 5181–5186.

Houbaviy, H. B., Murray, M. F., & Sharp, P. A. (2003). Embryonic stem cell-specific MicroRNAs. *Developmental Cell, 5*, 351–358.

Hu, X., Zhang, L., Mao, S. Q., Li, Z., Chen, J., Zhang, R. R., Wu, H. P., Gao, J., Guo, F., Liu, W., et al. (2014). Tet and TDG mediate DNA demethylation essential for mesenchymal-to-epithelial transition in somatic cell reprogramming. *Cell Stem Cell, 14*, 512–522.

Huangfu, D., Maehr, R., Guo, W., Eijkelenboom, A., Snitow, M., Chen, A. E., & Melton, D. A. (2008a). Induction of pluripotent stem cells by defined factors is greatly improved by small-molecule compounds. *Nature Biotechnology, 26*, 795–797.

Huangfu, D., Osafune, K., Maehr, R., Guo, W., Eijkelenboom, A., Chen, S., Muhlestein, W., & Melton, D. A. (2008b). Induction of pluripotent stem cells from primary human fibroblasts with only Oct4 and Sox2. *Nature Biotechnology, 26*, 1269–1275.

Ito, S., Shen, L., Dai, Q., Wu, S. C., Collins, L. B., Swenberg, J. A., He, C., & Zhang, Y. (2011). Tet proteins can convert 5-methylcytosine to 5-formylcytosine and 5-carboxylcytosine. *Science, 333*, 1300–1303.

Judson, R. L., Babiarz, J. E., Venere, M., & Blelloch, R. (2009). Embryonic stem cell-specific microRNAs promote induced pluripotency. *Nature Biotechnology, 27*, 459–461.

Kafer, G. R., Lehnert, S. A., Pantaleon, M., Kaye, P. L., & Moser, R. J. (2010). Expression of genes coding for histone variants and histone-associated proteins in pluripotent stem cells and mouse preimplantation embryos. *Gene Expression Patterns, 10*, 299–305.

Kaji, K., Caballero, I. M., MacLeod, R., Nichols, J., Wilson, V. A., & Hendrich, B. (2006). The NuRD component Mbd3 is required for pluripotency of embryonic stem cells. *Nature Cell Biology, 8*, 285–292.

Kaji, K., Nichols, J., & Hendrich, B. (2007). Mbd3, a component of the NuRD co-repressor complex, is required for development of pluripotent cells. *Development, 134*, 1123–1132.
Kidder, B. L., Palmer, S., & Knott, J. G. (2009). SWI/SNF-Brg1 regulates self-renewal and occupies core pluripotency-related genes in embryonic stem cells. *Stem Cells, 27*, 317–328.
Kim, K., Doi, A., Wen, B., Ng, K., Zhao, R., Cahan, P., Kim, J., Aryee, M. J., Ji, H., Ehrlich, L. I., et al. (2010). Epigenetic memory in induced pluripotent stem cells. *Nature, 467*, 285–290.
Kim, J. K., Huh, S. O., Choi, H., Lee, K. S., Shin, D., Lee, C., Nam, J. S., Kim, H., Chung, H., Lee, H. W., et al. (2001). Srg3, a mouse homolog of yeast SWI3, is essential for early embryogenesis and involved in brain development. *Molecular and Cellular Biology, 21*, 7787–7795.
Klochendler-Yeivin, A., Fiette, L., Barra, J., Muchardt, C., Babinet, C., & Yaniv, M. (2000). The murine SNF5/INI1 chromatin remodeling factor is essential for embryonic development and tumor suppression. *EMBO Reports, 1*, 500–506.
Kumar, M. S., Lu, J., Mercer, K. L., Golub, T. R., & Jacks, T. (2007). Impaired microRNA processing enhances cellular transformation and tumorigenesis. *Nature Genetics, 39*, 673–677.
Leitch, H. G., McEwen, K. R., Turp, A., Encheva, V., Carroll, T., Grabole, N., Mansfield, W., Nashun, B., Knezovich, J. G., Smith, A., et al. (2013). Naive pluripotency is associated with global DNA hypomethylation. *Nature Structural and Molecular Biology, 20*, 311–316.
Lessard, J. A., & Crabtree, G. R. (2010). Chromatin regulatory mechanisms in pluripotency. *Annual Review of Cell and Developmental Biology, 26*, 503–532.
Li, B., Pattenden, S. G., Lee, D., Gutierrez, J., Chen, J., Seidel, C., Gerton, J., & Workman, J. L. (2005). Preferential occupancy of histone variant H2AZ at inactive promoters influences local histone modifications and chromatin remodeling. *Proceedings of the National Academy of Sciences of the United States of America, 102*, 18385–18390.
Liang, J., Wan, M., Zhang, Y., Gu, P., Xin, H., Jung, S. Y., Qin, J., Wong, J., Cooney, A. J., Liu, D., et al. (2008). Nanog and Oct4 associate with unique transcriptional repression complexes in embryonic stem cells. *Nature Cell Biology, 10*, 731–739.
Lin, M., Pedrosa, E., Shah, A., Hrabovsky, A., Maqbool, S., Zheng, D., & Lachman, H. M. (2011). RNA-Seq of human neurons derived from iPS cells reveals candidate long non-coding RNAs involved in neurogenesis and neuropsychiatric disorders. *PLoS One, 6*, e23356.
Lister, R., Pelizzola, M., Kida, Y. S., Hawkins, R. D., Nery, J. R., Hon, G., Antosiewicz-Bourget, J., O'Malley, R., Castanon, R., Klugman, S., et al. (2011). Hotspots of aberrant epigenomic reprogramming in human induced pluripotent stem cells. *Nature, 471*, 68–73.
Loewer, S., Cabili, M. N., Guttman, M., Loh, Y. H., Thomas, K., Park, I. H., Garber, M., Curran, M., Onder, T., Agarwal, S., et al. (2010). Large intergenic non-coding RNA-RoR modulates reprogramming of human induced pluripotent stem cells. *Nature Genetics, 42*, 1113–1117.
Luger, K., Mader, A. W., Richmond, R. K., Sargent, D. F., & Richmond, T. J. (1997). Crystal structure of the nucleosome core particle at 2.8 A resolution. *Nature, 389*, 251–260.
Ma, D. K., Chiang, C. H., Ponnusamy, K., Ming, G. L., & Song, H. (2008). G9a and Jhdm2a regulate embryonic stem cell fusion-induced reprogramming of adult neural stem cells. *Stem Cells, 26*, 2131–2141.
Mak, A. B., Ni, Z., Hewel, J. A., Chen, G. I., Zhong, G., Karamboulas, K., Blakely, K., Smiley, S., Marcon, E., Roudeva, D., et al. (2010). A lentiviral functional proteomics approach identifies chromatin remodeling complexes important for the induction of pluripotency. *Molecular and Cellular Proteomics, 9*, 811–823.
Mali, P., Chou, B. K., Yen, J., Ye, Z., Zou, J., Dowey, S., Brodsky, R. A., Ohm, J. E., Yu, W., Baylin, S. B., et al. (2010). Butyrate greatly enhances derivation of human induced pluripotent stem cells by promoting epigenetic remodeling and the expression of pluripotency-associated genes. *Stem Cells, 28*, 713–720.
Mansour, A. A., Gafni, O., Weinberger, L., Zviran, A., Ayyash, M., Rais, Y., Krupalnik, V., Zerbib, M., Amann-Zalcenstein, D., Maza, I., et al. (2012). The H3K27 demethylase Utx regulates somatic and germ cell epigenetic reprogramming. *Nature, 488*, 409–413.
Marson, A., Levine, S. S., Cole, M. F., Frampton, G. M., Brambrink, T., Johnstone, S., Guenther, M. G., Johnston, W. K., Wernig, M., Newman, J., et al. (2008). Connecting microRNA genes to the core transcriptional regulatory circuitry of embryonic stem cells. *Cell, 134*, 521–533.

Mattout, A., Biran, A., & Meshorer, E. (2011). Global epigenetic changes during somatic cell reprogramming to iPS cells. *Journal of Molecular Cell Biology, 3*, 341–350.
Mattout, A., & Meshorer, E. (2010). Chromatin plasticity and genome organization in pluripotent embryonic stem cells. *Current Opinion in Cell Biology, 22*, 334–341.
Melton, C., Judson, R. L., & Blelloch, R. (2010). Opposing microRNA families regulate self-renewal in mouse embryonic stem cells. *Nature, 463*, 621–626.
Meshorer, E., Yellajoshula, D., George, E., Scambler, P. J., Brown, D. T., & Misteli, T. (2006). Hyperdynamic plasticity of chromatin proteins in pluripotent embryonic stem cells. *Developmental Cell, 10*, 105–116.
Mikkelsen, T. S., Hanna, J., Zhang, X., Ku, M., Wernig, M., Schorderet, P., Bernstein, B. E., Jaenisch, R., Lander, E. S., & Meissner, A. (2008). Dissecting direct reprogramming through integrative genomic analysis. *Nature, 454*, 49–55.
Nashun, B., Akiyama, T., Suzuki, M. G., & Aoki, F. (2011). Dramatic replacement of histone variants during genome remodeling in nuclear-transferred embryos. *Epigenetics, 6*, 1489–1497.
Nguyen, A. T., & Zhang, Y. (2011). The diverse functions of Dot1 and H3K79 methylation. *Genes and Development, 25*, 1345–1358.
Ohi, Y., Qin, H., Hong, C., Blouin, L., Polo, J. M., Guo, T., Qi, Z., Downey, S. L., Manos, P. D., Rossi, D. J., et al. (2011). Incomplete DNA methylation underlies a transcriptional memory of somatic cells in human iPS cells. *Nature Cell Biology, 13*, 541–549.
Ohnuki, M., Tanabe, K., Sutou, K., Teramoto, I., Sawamura, Y., Narita, M., Nakamura, M., Tokunaga, Y., Watanabe, A., Yamanaka, S., et al. (2014). Dynamic regulation of human endogenous retroviruses mediates factor-induced reprogramming and differentiation potential. *Proceedings of the National Academy of Sciences of the United States of America, 111*, 12426–12431.
Onder, T. T., Kara, N., Cherry, A., Sinha, A. U., Zhu, N., Bernt, K. M., Cahan, P., Marcarci, B. O., Unternaehrer, J., Gupta, P. B., et al. (2012). Chromatin-modifying enzymes as modulators of reprogramming. *Nature, 483*, 598–602.
Pasini, D., Bracken, A. P., Hansen, J. B., Capillo, M., & Helin, K. (2007). The polycomb group protein Suz12 is required for embryonic stem cell differentiation. *Molecular and Cellular Biology, 27*, 3769–3779.
Pasque, V., Radzisheuskaya, A., Gillich, A., Halley-Stott, R. P., Panamarova, M., Zernicka-Goetz, M., Surani, M. A., & Silva, J. C. (2012). Histone variant macroH2A marks embryonic differentiation in vivo and acts as an epigenetic barrier to induced pluripotency. *Journal of Cell Science, 125*, 6094–6104.
Pereira, C. F., Piccolo, F. M., Tsubouchi, T., Sauer, S., Ryan, N. K., Bruno, L., Landeira, D., Santos, J., Banito, A., Gil, J., et al. (2010). ESCs require PRC2 to direct the successful reprogramming of differentiated cells toward pluripotency. *Cell Stem Cell, 6*, 547–556.
Petruk, S., Sedkov, Y., Johnston, D. M., Hodgson, J. W., Black, K. L., Kovermann, S. K., Beck, S., Canaani, E., Brock, H. W., & Mazo, A. (2012). TrxG and PcG proteins but not methylated histones remain associated with DNA through replication. *Cell, 150*, 922–933.
Polo, J. M., Liu, S., Figueroa, M. E., Kulalert, W., Eminli, S., Tan, K. Y., Apostolou, E., Stadtfeld, M., Li, Y., Shioda, T., et al. (2010). Cell type of origin influences the molecular and functional properties of mouse induced pluripotent stem cells. *Nature Biotechnology, 28*, 848–855.
Rangasamy, D., Berven, L., Ridgway, P., & Tremethick, D. J. (2003). Pericentric heterochromatin becomes enriched with H2A.Z during early mammalian development. *EMBO Journal, 22*, 1599–1607.
Rothbart, S. B., & Strahl, B. D. (2014). Interpreting the language of histone and DNA modifications. *Biochimica et Biophysica Acta, 1839*, 627–643.
Rybak, A., Fuchs, H., Smirnova, L., Brandt, C., Pohl, E. E., Nitsch, R., & Wulczyn, F. G. (2008). A feedback loop comprising lin-28 and let-7 controls pre-let-7 maturation during neural stem-cell commitment. *Nature Cell Biology, 10*, 987–993.
Schuettengruber, B., Martinez, A. M., Iovino, N., & Cavalli, G. (2011). Trithorax group proteins: Switching genes on and keeping them active. *Nature Reviews Molecular Cell Biology, 12*, 799–814.

Sheik Mohamed, J., Gaughwin, P. M., Lim, B., Robson, P., & Lipovich, L. (2010). Conserved long noncoding RNAs transcriptionally regulated by Oct4 and Nanog modulate pluripotency in mouse embryonic stem cells. *RNA, 16*, 324–337.

Shen, X., Liu, Y., Hsu, Y. J., Fujiwara, Y., Kim, J., Mao, X., Yuan, G. C., & Orkin, S. H. (2008). EZH1 mediates methylation on histone H3 lysine 27 and complements EZH2 in maintaining stem cell identity and executing pluripotency. *Molecular Cell, 32*, 491–502.

Shi, Y., Desponts, C., Do, J. T., Hahm, H. S., Scholer, H. R., & Ding, S. (2008). Induction of pluripotent stem cells from mouse embryonic fibroblasts by Oct4 and Klf4 with small-molecule compounds. *Cell Stem Cell, 3*, 568–574.

Shimbo, T., Du, Y., Grimm, S. A., Dhasarathy, A., Mav, D., Shah, R. R., Shi, H., & Wade, P. A. (2013). MBD3 localizes at promoters, gene bodies and enhancers of active genes. *PLoS Genetics, 9*, e1004028.

Simonsson, S., & Gurdon, J. (2004). DNA demethylation is necessary for the epigenetic reprogramming of somatic cell nuclei. *Nature Cell Biology, 6*, 984–990.

Singhal, N., Graumann, J., Wu, G., Arauzo-Bravo, M. J., Han, D. W., Greber, B., Gentile, L., Mann, M., & Scholer, H. R. (2010). Chromatin-remodeling components of the BAF complex facilitate reprogramming. *Cell, 141*, 943–955.

Slany, R. K. (2009). The molecular biology of mixed lineage leukemia. *Haematologica, 94*, 984–993.

Smith, Z. D., Chan, M. M., Humm, K. C., Karnik, R., Mekhoubad, S., Regev, A., Eggan, K., & Meissner, A. (2014). DNA methylation dynamics of the human preimplantation embryo. *Nature, 511*, 611–615.

Smith, Z. D., Chan, M. M., Mikkelsen, T. S., Gu, H., Gnirke, A., Regev, A., & Meissner, A. (2012). A unique regulatory phase of DNA methylation in the early mammalian embryo. *Nature, 484*, 339–344.

Sridharan, R., Gonzales-Cope, M., Chronis, C., Bonora, G., McKee, R., Huang, C., Patel, S., Lopez, D., Mishra, N., Pellegrini, M., et al. (2013). Proteomic and genomic approaches reveal critical functions of H3K9 methylation and heterochromatin protein-1gamma in reprogramming to pluripotency. *Nature Cell Biology, 15*, 872–882.

Stadtfeld, M., Apostolou, E., Akutsu, H., Fukuda, A., Follett, P., Natesan, S., Kono, T., Shioda, T., & Hochedlinger, K. (2010). Aberrant silencing of imprinted genes on chromosome 12qF1 in mouse induced pluripotent stem cells. *Nature, 465*, 175–181.

Stoller, J. Z., Huang, L., Tan, C. C., Huang, F., Zhou, D. D., Yang, J., Gelb, B. D., & Epstein, J. A. (2010). Ash2l interacts with Tbx1 and is required during early embryogenesis. *Experimental Biology and Medicine (Maywood, N.J.), 235*, 569–576.

Stummvoll, G. H., Fritsch, R. D., Meyer, B., Hoefler, E., Aringer, M., Smolen, J. S., & Steiner, G. (2009). Characterisation of cellular and humoral autoimmune responses to histone H1 and core histones in human systemic lupus erythaematosus. *Annals of the Rheumatic Diseases, 68*, 110–116.

Taberlay, P. C., Kelly, T. K., Liu, C. C., You, J. S., De Carvalho, D. D., Miranda, T. B., Zhou, X. J., Liang, G., & Jones, P. A. (2011). Polycomb-repressed genes have permissive enhancers that initiate reprogramming. *Cell, 147*, 1283–1294.

Tahiliani, M., Koh, K. P., Shen, Y., Pastor, W. A., Bandukwala, H., Brudno, Y., Agarwal, S., Iyer, L. M., Liu, D. R., Aravind, L., et al. (2009). Conversion of 5-methylcytosine to 5-hydroxymethylcytosine in mammalian DNA by MLL partner TET1. *Science, 324*, 930–935.

Takashima, Y., Guo, G., Loos, R., Nichols, J., Ficz, G., Krueger, F., Oxley, D., Santos, F., Clarke, J., Mansfield, W., et al. (2014). Resetting transcription factor control circuitry toward ground-state pluripotency in human. *Cell, 158*, 1254–1269.

Talbert, P. B., & Henikoff, S. (2010). Histone variants--ancient wrap artists of the epigenome. *Nature Reviews Molecular Cell Biology, 11*, 264–275.

Terme, J. M., Sese, B., Millan-Arino, L., Mayor, R., Izpisua Belmonte, J. C., Barrero, M. J., & Jordan, A. (2011). Histone H1 variants are differentially expressed and incorporated into chromatin during differentiation and reprogramming to pluripotency. *The Journal of Biological Chemistry, 286*, 35347–35357.

Turinetto, V., & Giachino, C. (2015). Histone variants as emerging regulators of embryonic stem cell identity. *Epigenetics, 10*, 563–573.

Turinetto, V., Orlando, L., Sanchez-Ripoll, Y., Kumpfmueller, B., Storm, M. P., Porcedda, P., Minieri, V., Saviozzi, S., Accomasso, L., Cibrario Rocchietti, E., et al. (2012). High basal gammaH2AX levels sustain self-renewal of mouse embryonic and induced pluripotent stem cells. *Stem Cells, 30*, 1414–1423.

van Attikum, H., Fritsch, O., & Gasser, S. M. (2007). Distinct roles for SWR1 and INO80 chromatin remodeling complexes at chromosomal double-strand breaks. *EMBO Journal, 26*, 4113–4125.

Vastenhouw, N. L., & Schier, A. F. (2012). Bivalent histone modifications in early embryogenesis. *Current Opinion in Cell Biology, 24*, 374–386.

Venkatesh, S., Smolle, M., Li, H., Gogol, M. M., Saint, M., Kumar, S., Natarajan, K., & Workman, J. L. (2012). Set2 methylation of histone H3 lysine 36 suppresses histone exchange on transcribed genes. *Nature, 489*, 452–455.

Viswanathan, S. R., Daley, G. Q., & Gregory, R. I. (2008). Selective blockade of microRNA processing by Lin28. *Science, 320*, 97–100.

Wade, P. A., Gegonne, A., Jones, P. L., Ballestar, E., Aubry, F., & Wolffe, A. P. (1999). Mi-2 complex couples DNA methylation to chromatin remodelling and histone deacetylation. *Nature Genetics, 23*, 62–66.

Wang, Y., Baskerville, S., Shenoy, A., Babiarz, J. E., Baehner, L., & Blelloch, R. (2008). Embryonic stem cell-specific microRNAs regulate the G1-S transition and promote rapid proliferation. *Nature Genetics, 40*, 1478–1483.

Wang, T., Chen, K., Zeng, X., Yang, J., Wu, Y., Shi, X., Qin, B., Zeng, L., Esteban, M. A., Pan, G., et al. (2011). The histone demethylases Jhdm1a/1b enhance somatic cell reprogramming in a vitamin-C-dependent manner. *Cell Stem Cell, 9*, 575–587.

Wang, L., Du, Y., Ward, J. M., Shimbo, T., Lackford, B., Zheng, X., Miao, Y. L., Zhou, B., Han, L., Fargo, D. C., et al. (2014a). INO80 facilitates pluripotency gene activation in embryonic stem cell self-renewal, reprogramming, and blastocyst development. *Cell Stem Cell, 14*, 575–591.

Wang, L., Zhang, J., Duan, J., Gao, X., Zhu, W., Lu, X., Yang, L., Li, G., Ci, W., Li, W., et al. (2014b). Programming and inheritance of parental DNA methylomes in mammals. *Cell, 157*, 979–991.

Whyte, W. A., Bilodeau, S., Orlando, D. A., Hoke, H. A., Frampton, G. M., Foster, C. T., Cowley, S. M., & Young, R. A. (2012). Enhancer decommissioning by LSD1 during embryonic stem cell differentiation. *Nature, 482*, 221–225.

Woodcock, C. L. (2005). A milestone in the odyssey of higher-order chromatin structure. *Nature Structural and Molecular Biology, 12*, 639–640.

Wu, T., Liu, Y., Wen, D., Tseng, Z., Tahmasian, M., Zhong, M., Rafii, S., Stadtfeld, M., Hochedlinger, K., & Xiao, A. (2014). Histone variant H2A.X deposition pattern serves as a functional epigenetic mark for distinguishing the developmental potentials of iPSCs. *Cell Stem Cell, 15*, 281–294.

Zhang, Z., Jones, A., Sun, C. W., Li, C., Chang, C. W., Joo, H. Y., Dai, Q., Mysliwiec, M. R., Wu, L. C., Guo, Y., et al. (2011). PRC2 complexes with JARID2, MTF2, and esPRC2p48 in ES cells to modulate ES cell pluripotency and somatic cell reprogramming. *Stem Cells, 29*, 229–240.

Chapter 10
Epigenetic Mechanisms of Adverse Neurodevelopment in Response to Lead Exposure and Prenatal Stress and the Combination: The Road Ahead

Jay S. Schneider and Deborah A. Cory-Slechta

10.1 Introduction: The Vulnerability of Early Neurodevelopmental Periods

Early development constitutes a well-known period of vulnerability of the brain as a time in which brain systems continue to undergo substantial development that must proceed in an orchestrated and precisely timed fashion. Increasingly understood is that environmental conditions during such periods have the potential to reprogram brain actions in a dynamic and/or permanent capacity. Such reprogramming can have significant consequences for subsequent behavioral and brain function, a phenomenon often referred to as fetal basis of adult disease or developmental origins of adult health and disease.

Numerous adverse environmental conditions have now been shown to have the potential to produce brain reprogramming during early development via epigenetic alterations. Epigenetics refers to molecular modifications of DNA that switch genes on and off based on molecular modifications to DNA without a change in the DNA sequence (Jirtle and Skinner 2007). These include modifications in DNA methylation at the carbon-5 position of cytosine in CpG dinucleotides and changes to the chromatin packaging of DNA by posttranslational histone modifications (Jirtle and Skinner 2007). Such epigenetic changes can incorrectly silence or activate a gene

J.S. Schneider, PhD (✉)
Department of Pathology, Anatomy and Cell Biology, Thomas Jefferson University, Philadelphia, PA 19107, USA
e-mail: Jay.Schneider@jefferson.edu

D.A. Cory-Slechta, PhD
Department of Environmental Medicine, University of Rochester School of Medicine, Box EHSC, Rochester, NY 14642, USA
e-mail: deborah_cory-slechta@urmc.rochester.edu

D. Hollar (ed.), *Epigenetics, the Environment, and Children's Health Across Lifespans*, DOI 10.1007/978-3-319-25325-1_10

(or genes), influencing cell health and functionality. Epigenetic changes conferred prenatally in response to certain environmental factors can even be inherited transgenerationally, thereby potentially affecting the health of future generations (Jirtle and Skinner 2007).

In the central nervous system (CNS), known environmental conditions that have been demonstrated to alter epigenetic profiles include nutritional alterations/deficiencies, maternal stress, and/or enhanced fetal glucocorticoids, reduced birth weight, prematurity, and maternal infections (Auger and Auger 2011; Bale 2015; Matthews and Phillips 2010). A clear elaboration of the mechanisms by which such epigenetic reprogramming occurs is of critical importance to understanding the diseases and disorders that have been shown to result from such changes.

It is becoming increasingly clear that there is a significant complexity to epigenetic processes in the central nervous system (CNS), where numerous factors influence the development of epigenetic changes and reversals and their consequent impact on brain development and function. For example, epigenetic outcomes resulting from environmental perturbations are frequently found to be sex dependent, particularly in relation to neuroendocrine-acting systems. These differences may reflect sex-dependent alterations in such aspects of epigenetic programming as levels of the DNA methyltransferases (DNMTs). Higher levels of DNMT3a mRNA have been reported in the amygdala of females as compared to males early in development (Kolodkin and Auger 2011), while higher levels of DNMT1 activity were observed in the male brain (McCarthy et al. 2009). Sex differences have also been noted in DNA methylation patterns and expression of methyl-binding proteins and corepressor proteins (McCarthy et al. 2009). The existence of such sex differences early in development has been suggested to be critical to brain differentiation (Auger and Auger 2011), but it correspondingly provides a different biological substrate for each sex on which environmental stressors, including chemical exposures, then act.

Another factor that can introduce differences in the epigenetic outcomes produced by environmental factors is the developmental period during which these environmental factors act. For example, the higher levels of DNMT3a mRNA expression in female amygdala were seen at postnatal day 1 but had disappeared by postnatal day 10, with these changes related to sex differences in steroid exposure levels (Kolodkin and Auger 2011). It is then possible that different environmental events occurring prior to birth, around the time of birth, or later in postnatal life could interact differently with the epigenome and have different physiological and behavioral consequences.

With respect to the CNS in particular, a critical feature of epigenetic function is that both permanent and plastic changes can occur in the epigenome, and, moreover, these may be brain region dependent (Auger and Auger 2011). For example, in a time course study (postnatal day 7, 6 weeks, and 1 year) of acquisition of 5-hydroxymethylation in mice, dynamic increases were seen at each successive time point in the cerebellum, whereas in the hippocampus, increases were seen between postnatal day 7 and 6 weeks but not between 6 weeks and 1 year, underscoring the additional complexity of regional brain differences in epigenetic modifications (Szulwach et al. 2011). It is important to point out that this potential for plasticity of epigenetic changes, coupled with their regional specificity, is likely to significantly complicate the ability to find peripheral biomarkers that reflect brain function, a clear need for human studies.

Also increasingly appreciated with respect to the epigenome is that behavioral experience itself is a factor that can modify epigenetic profiles. For example, differences in the level/quality of maternal care were found to affect the levels of DNA methylation of both estrogen and glucocorticoid receptor gene promoter regions in the mouse brain (Weaver et al. 2004). A study comparing suicide victims with or without early childhood abuse revealed significant differences in epigenetic regulation of hippocampal (McGowan et al. 2009) but not anterior cingulate cortex (Labonte et al. 2012) glucocorticoid receptor expression. Several reports now document hippocampal epigenetic changes associated with the imposition of environmental enrichment procedures (Morse et al. 2015; Szulwach et al. 2011; Kuzumaki et al. 2011). The ability to produce dynamic and plastic epigenetic alterations seems entirely consistent with the need for dynamic changes in behavior with rapid alterations in environmental conditions. Differences in behavioral experiences seem a likely explanation for the significant differences in epigenetic profiles that can be found in monozygotic twins (Fraga et al. 2005), who, although even while living in the same environment, will inevitably have different trajectories of behavioral experience over the lifetime.

This impact of behavioral experience on the epigenome has several notable implications. The dynamic modulation of epigenetic profiles in relation to behavioral experience, an ever-changing component of the human environment, could modulate the epigenetic response of the brain to various environmental factors, e.g., nutrition and stress, or interact with such factors to change outcomes. Further, it raises the possibility that specific behavioral interventions may be effective in reversing or mitigating reprogramming changes and thus have therapeutic efficacy. Another concern, however, arises with respect to experimental designs, in that measurement of epigenetic changes following behavioral testing will potentially reveal epigenetic profiles that differ from what may have been present before or in the absence of behavioral testing, i.e., that have been altered by behavioral testing. Thus, it will be difficult in animal models to interpret epigenetic changes that are only examined in animals that also have a behavioral testing history. Further, such possibilities of earlier behavioral experiences must be considered in human studies.

In addition to the risk factors resulting in epigenetic reprogramming cited above, there is increasing concern and recognition for the potential of environmental chemical exposures, including those known to produce adverse neurodevelopmental effects, to produce epigenetic reprogramming. Elevated levels of lead exposure represent one such environmental concern, as lead is a well-documented neurotoxicant (National Toxicology Program 2012). Exposures to environmental chemicals always effectively occur in multiplicity and not in isolation (Kortenkamp et al. 2007; Mumtaz et al. 2007) and inevitably in conjunction with risk factors such as prenatal stress that have already shown to produce epigenetic changes (McEwen and Tucker 2011; Gee and Payne-Sturges 2004), underscoring the potential for interactions and for cumulative neurotoxicity. Based on the current understanding, it seems likely that epigenetic patterns produced by combined exposures to such risk factors will differ from those produced by each risk factor alone. Such combined exposures will also occur in an environment that includes behavioral experiences with the potential to further modify epigenetic profiles.

In the following sections, we review the current understanding of the effects of lead and prenatal stress on children's neurodevelopment and the effects of lead and prenatal stress on the epigenome and discuss the rationale for studying and understanding the combined effects of these (and other) environmental factors.

10.2 Lead and Children's Neurodevelopment

The adverse neurodevelopmental consequences of exposures to elevated levels of lead in the environment are well established (National Toxicology Program 2012). At high levels of exposure, lead is associated with acute encephalopathy that can result in childhood fatalities (Chisolm and Harrison 1957), an outcome that eventually resulted in the removal of lead from its two primary human exposure sources, i.e., paint and gasoline. But even children who have survived an episode of acute encephalopathy frequently sustained permanent neurological consequences (Byers and Lord 1943; Cecil et al. 2008).

The declines in population blood lead levels (Pirkle et al. 1994) that followed the removal of lead from paint and gasoline also provided the opportunity to assess the "dose–response" nature of neurotoxicity of lead in children with cross-sectional studies, followed by prospective longitudinal cohort studies, examining the neurodevelopmental consequences at ever-decreasing blood lead concentrations. Those studies have now established that it is not possible to identify a level of lead, in terms of measurable blood lead values in field instruments that are without adverse effects in children (Canfield et al. 2003; Lanphear et al. 2005; Lucchini et al. 2012). Notably, these impacts have included significant reductions in IQ scores that occur even in the range of blood lead values of 1–5 μg/dl and deficits in attention-related behaviors. Further, there is a supralinearity of effects of elevated lead exposure on IQ, with greater reductions at blood lead values below 10 μg/dl than above.

Though the removal of lead from paint and gasoline, in countries where it occurred, certainly reduced the numbers/levels of lead-affected children, lead exposure nevertheless remains a significant environmental health problem, particularly in low-socioeconomic-status communities where the greatest residual contamination from prior uses of leaded paint and gasoline remains (CDC 2013). Moreover, in many developing countries where exposure regulations are lax or nonexistent, exposure levels are far higher and many fatalities have resulted, as a consequence of battery recycling activities, housing near smelting plants (e.g., China), and unregulated lead smelting and gold mining activities. As evidence of the latter, recent reports document the deaths of over 400 children as of 2013 from artisanal gold mining in Nigeria (Plumlee et al. 2013).

Studies reporting association of lead exposure with reduced IQ, executive function, and attention deficits in children are supported by a substantive body of experimental studies that have included both rodents and nonhuman primates that likewise demonstrate changes in learning and executive function, as well as in attention-related behaviors (Cory-Slechta 1997; Cory-Slechta et al. 1997, 1998, 1999; Gilbert et al. 1999; Guilarte et al. 1994; Rice 1988, 1990; Rice and Karpinski 1988). Such effects, moreover, have been seen at even the lowest blood lead concentrations that have been studied in animal models (e.g., 6–9 μg/dl). The fact that toxicokinetics and toxicodynamics of lead exposure are similar in rodents and humans has also supported the utility of rodent models for advancing the understanding of mechanisms by which such effects occur (Cory-Slechta et al. 1989; Gulson et al. 1997; Leggett 1993; O'Flaherty 1991). Although epidemiological studies have typically controlled for sex and not examined differences in outcome in relation to this factor, and earlier animal studies typically focused primarily on males, increasing evidence from animal studies now makes clear that the neurodevelopmental consequences of lead exposure are sex specific, in addition to reflecting differences in timing and duration of exposures (Cory-Slechta et al. 2010, 2012). The sex specificity does not identify one sex as more vulnerable than the other, but highlights different profiles of effects by sex, as both sexes are impacted.

10.3 Prenatal Stress and Children's Neurodevelopment

A substantial literature including both human and animal studies has documented the adverse neurodevelopmental consequences of prenatal stress on offspring. Human studies have demonstrated outcomes of maternal psychosocial stress that are lifelong and that can be seen independently of changes in postnatal depression and anxiety (Babenko et al. 2015; Brunton 2015; Bock et al. 2015; Richetto and Riva 2014; Schuurmans and Kurrasch 2013; Talge et al. 2007). In both human and animal studies, maternal stress can be associated with premature delivery and with reduced birth weight and in the case of humans, with lower neonatal neurobehavioral assessment scores (Babenko et al. 2015; Talge et al. 2007), outcomes which themselves are associated with impaired neurodevelopment. Maternal stressors have been associated with impaired cognitive function in offspring, detected early in development, and with impaired attention and impulsivity and disrupted development of brain laterality, as evidenced by increased numbers of mixed-handedness children (Talge et al. 2007). A significant literature also documents the relationship of maternal stress to subsequent psychopathology, including increased odds of later development of schizophrenia and affective disorders.

Moreover, maternal stress can result in brain morphological changes. For example, Qiu et al. (2013) reported that maternal anxiety measured at 26 weeks of pregnancy was associated with reduced growth rates of both left and right hippocampus over the first 6 weeks of life. Correspondingly, adult offspring of mothers who experienced major negative life events during pregnancy were found to exhibit altered learning strategies, including a more rigid and less flexible strategy for spatial learning (Schwabe et al. 2012).

Similar consequences of prenatal stress are replicated in experimental animal models and likewise display sex-dependent effects. For example, deficits in learning were more pronounced in males, while anxiety-type responses predominated in female rats subjected to maternal stress (Weinstock 2011), and these behavioral alterations were accompanied by reductions in neurogenesis and in complexity of dendritic morphology in regions critical to executive functioning, specifically the hippocampus and prefrontal cortex. Attention-related behaviors including reductions in sustained attention and inhibitory response control are also reported to be impaired in offspring of prenatally stressed rats (Wilson et al. 2012).

As with elevated lead exposure, effects of maternal stress on offspring neurodevelopment appear to be time (e.g., trimester), duration, and sex specific, and these factors are likely contributing to differences in outcomes of prenatal stress that are reported across some studies. Furthermore, the time of assessment in offspring is also likely to modify the outcomes (e.g., Lupien et al. 2009). Studies, for example, have shown differences in the profile of consequences of prenatal stress dependent upon the period of gestation during which stress is imposed (Davis and Sandman 2010) and sex of the offspring (e.g., Wang et al. 2015a, b).

An additional important consideration in regard to understanding the influences of prenatal stress relates to the paradigm used to induce stress itself. It is important to note that different stress paradigms can actually yield quite distinct outcomes, a phenomenon that has generated confusion in defining biomarkers of resilience vs. pathology. Stress does not invariably lead to behavioral/mental pathology. Instead, outcome (stress vs. development of resilience) appears to depend critically upon "dose" (magnitude of negative experience) and in particular the presence or absence of associated protective factors such as self-control, problem-solving skills, and family/community relationships (Masten and Narayan 2012). One hypothesis, paralleled in animal models (see below), is that more moderate negative experiences, through learned adaptations, can produce resilience, thereby preparing the organism for future challenges, whereas experiencing no challenges or extreme negative challenges (Rutter 2006; Seery et al. 2010), where no such adaptations are learned, may lead to a "learned helplessness" phenotype (DiCorcia and Tronick 2011). In other words and as has been elegantly demonstrated in experimental animal models, controllable and predictable negative experiences lead to future resiliency, whereas uncontrollable and unpredictable experiences produce subsequent behavioral pathology (Koolhaas et al. 2011).

10.4 A Biological Framework for Cumulative Toxicity of Lead and Prenatal Stress

As with all environmental chemicals, elevated lead exposure does not occur in isolation, but in conjunction with numerous other environmental conditions, including other risk factors for adverse child neurodevelopment. One other such risk factor is prenatal stress. Indeed, elevated lead exposure and prenatal stress can be co-occurring and/or sequentially occurring risk factors in a population, and, like lead exposure, prenatal stress appears to be a significant risk factor in lower-socioeconomic-status communities (e.g., Keenan et al. 2007; Thayer and Kuzawa 2014).

Importantly from a neurodevelopmental perspective, in addition to their co-occurrence, lead exposure and prenatal stress also share biological substrates and can produce common adverse outcomes. For example, both elevated lead exposure and prenatal stress impact the hypothalamic–pituitary–adrenal (HPA) axis and associated glucocorticoid receptors, and both also target brain mesocorticolimbic dopamine–glutamate systems (Cory-Slechta et al. 1998, 1999; Rossi-George et al. 2011; Virgolini et al. 2008a; Martinez-Tellez et al. 2009; Barros et al. 2004; Berger et al. 2002). Human studies have reported associations of concurrent blood lead levels in pregnant women with cortisol awakening response (Braun et al. 2014).

Both lead exposure and prenatal stress also impact brain hippocampal systems and associated neurocircuitry as well. Elevated lead exposure-related effects, while clearly differing by exposure and time of measurement conditions, have long been recognized and can include alterations in synaptogenesis, IGF1 and IGF2 expressions, altered gene expression profiles, neuronal numbers, oxidative stress levels, and glutamatergic, particularly NMDA-related, function (Hu et al. 2014; Li et al. 2014, 2015; Baranowska-Bosiacka et al. 2013; Stansfield et al. 2012; Neal et al. 2012). Similarly, long-term changes in the hippocampus arise from maternal stress that include inhibited neurogenesis, altered dendritic branching, hypomyelination, altered levels of NDMA receptors and glutamate release, decreased synaptic plasticity, and altered BDNF signaling (Focking et al. 2014; Belnoue et al. 2013; Xu et al. 2013).

These overlapping biological targets of lead and prenatal stress, i.e., the HPA axis and mesocorticolimbic/hippocampal dopamine and glutamate function, may be considered to establish a biological framework for the enhanced effects that can be observed when these two risk factors occur in combination. This makes assessment of their combined or cumulative toxicity of particular interest, particularly in light of the well-documented interactions of the HPA axis with brain mesocorticolimbic systems (Deroche et al. 1995; Tidey and Miczek 1996; Sullivan and Dufresne 2006) that are critical to mediation of cognitive behavioral functions, including those adversely impacted by elevated lead exposure and by prenatal stress. Perhaps as a result of these conjoint biological substrates, elevated lead exposure and prenatal can also produce common adverse outcomes such as changes in executive functions and attention-related behaviors (Anderson and Armstead 1995; Bradley and Corwyn 2002; Cory-Slechta 1995; Dietrich et al. 2001; Dohrenwend 1973; Needleman et al. 1996; Schwartz 1994).

10.5 Epigenetic Mechanisms May Contribute to Persistence of Neurodevelopmental Consequences of Lead Exposure, Prenatal Stress, and Enhanced Toxicity Produced by Their Combination

The hypothesis suggested above that combined exposures to lead and prenatal stress, by virtue of their overlapping profile of biological targets, could result in enhanced neurotoxicity profiles has been shown in several of our prior studies examining the effects of combined developmental lead exposure and prenatal and/or offspring stress relative to either factor alone (Cory-Slechta et al. 2008, 2010, 2012, 2013; Virgolini et al. 2008a). For example, enhanced learning deficits are seen with combined developmental lead exposure and prenatal stress in female rats only (Cory-Slechta et al. 2010). Increased rates of responding on a fixed interval schedule of food reinforcement were also observed in females that had been subjected to maternal lead exposure combined with prenatal and offspring stress (Virgolini et al. 2008b). In both instances, these deficits were greater than those observed in response to exposure to either lead or prenatal stress alone.

Heightened neurotoxicity was not restricted to behavioral consequences, however. Enhanced levels of brain mesocorticolimbic system monoamines were found in response to combined lead and prenatal stress, where neither lead alone nor prenatal stress alone produced any changes. These monoaminergic changes included increased frontal cortex dopamine and norepinephrine levels and increased nucleus accumbens and dorsal striatal dopamine, DOPAC and HVA levels in males, whereas in females, such increases occurred in nucleus accumbens norepinephrine levels and in serotonergic neurotransmission in the frontal cortex, nucleus accumbens, and striatum (Virgolini et al. 2008b). In sum, combined lead exposure and prenatal stress increased these neurotransmitter levels to those associated with a higher exposure level of lead alone.

The breadth of biological targets of elevated lead exposure and prenatal stress suggests that multiple potential mechanisms could account for the overlap in the adverse neurodevelopmental consequences that arise from these combined exposures and/or their enhanced toxicity in combination, including alterations in HPA axis and glucocorticoid function, altered hippocampal/glutamatergic structure and function, and/or altered mesocorticolimbic dopamine/serotonin functioning. An emergent consideration for both risk factors is the impact of epigenetic changes and reprogramming of function. As noted above, epigenetic reprogramming is already seen in response to prenatal stress, but the elaboration of effects in response to elevated lead exposure is still in its infancy. Moreover, it is important to remember that because such risk factors occur in combination in human environments, any full mechanistic understanding of the role of epigenetic changes in the long-term consequences of combined elevated lead and prenatal stress will involve the assessment of their interactive epigenetic consequences.

Notably, both elevated lead exposure and prenatal stress are also often thought of as multigenerational stressors. A growing literature has begun to document such effects of prenatal stress (Bale 2015). A recent study reported progressive changes up to the F2 generation (F0–F2) in gestational length, maternal weight gain, behavioral activity, blood glucose levels, and delayed behavioral development in response to transgenerational maternal stress (Yao et al. 2014). In a mouse model, maternal stress-induced increases in offspring in stress sensitivity and HPA axis function were passed in a sex-specific capacity to the subsequent generation of males (Morgan and Bale 2011).

In the case of lead exposure, one potential mechanism of multigenerational transmission may relate to its toxicokinetic and/or toxicodynamic properties. As a result of its calcium-mimetic properties, more than 90 % of ingested lead resides in the bone where it maintains a dynamic equilibrium with blood; during physiological conditions that impose significant calcium requirements, e.g., pregnancy and lactation, calcium/lead is mobilized from the bone back into the bloodstream and into breast milk where it can then pass into fetal circulation. Thus, the higher the levels of lead prior to pregnancy, the greater the exposure of the fetus/infant. In the case of low SES communities, in which successive generations will also likely reside in a higher lead-contaminated environment, additional exposures will occur and additional amounts of lead will be passed on during pregnancy. Currently unknown, however, is the actual potential for epigenetically induced effects of lead exposure on central nervous system function and its potential for transgenerational consequences.

10.5.1 Epigenetics of Lead Exposure: Human Studies

Persistent effects from early-life lead exposure are consistent with a model of an early-life basis of adult disability in which insults that occur early in life produce changes that arise from physiological reprogramming (Cottrell and Seckl 2009). One way in which this might occur is through epigenetic reprogramming. Epigenetic modifications such as DNA methylation and posttranslational histone modifications are key regulators of gene expression and gene–environment interactions. However, there have been few studies that have examined specific epigenetic alterations as a consequence of elevated lead exposure in humans. In the first study to examine the effect of maternal lead (Pb) burden on offspring DNA methylation levels, Pilsner et al. (2009) assessed the relationship between lead exposure and DNA methylation by examining LINE-1 (long interspersed nuclear elements-1) and Alu methylation by pyrosequencing on umbilical cord blood samples from participants in the ELEMENT study in Mexico City. LINE-1 and Alu methylation levels were assessed in relation to maternal bone lead levels (maternal patella and tibia lead levels, reflective of cumulative exposure) and cord blood lead levels. An inverse dose–response relationship was found for patella lead measures and cord LINE-1 methylation and between tibia lead measures and Alu methylation. No associations were found between cord blood lead levels and cord genomic DNA methylation (Pilsner et al. 2009). These results suggest that these particular epigenetic marks are influenced more by maternal cumulative lead burden and chronic exposure (assessed by maternal bone lead levels) than by recent exposure (assessed by cord blood lead levels). The authors speculated that lead-induced oxidative stress may have played a role in the effect of lead on DNA methylation, with the developing fetus being particularly prone to epigenetic errors due to high rates of DNA synthesis and epigenetic reprogramming that occurs shortly after implantation (Pilsner et al. 2009).

Another study suggesting an influence of elevated lead exposure on DNA methylation assessed LINE-1 and Alu methylation as a function of patella, tibia, or blood lead levels in elderly men participating in the Normative Aging Study (Wright et al. 2010). In this study, patella lead levels were inversely associated with LINE-1 methylation but not with Alu, and neither tibia lead nor blood lead levels associated with global methylation for either Alu or LINE-1 (Wright et al. 2010). Again, these findings appear consistent with a marker of cumulative lead exposure and not a marker of recent exposure being associated with DNA methylation.

Although these novel findings are interesting and suggest lead-induced changes in epigenetic programming, the interpretations of these findings are limited by some important factors. Although changes in methylation of LINE-1 and Alu could have significant health implications, Alu and LINE-1 are surrogate markers for estimating global DNA methylation levels (Weisenberger et al. 2005; Yang et al. 2004). At least one-third of DNA methylation occurs in repetitive elements that represent a large portion of the human genome (Lander et al. 2001). Among these repetitive sequences, Alu and LINE-1 represent about 30 % of the human genome (Zhu et al. 2012). Because of this high representation throughout the genome, Alu and LINE-1 have been used as surrogate markers for estimating global DNA methylation levels, particularly in epidemiological studies (Zhu et al. 2012). However, methylation levels of these repetitive elements may not be equivalent to global DNA methylation content; moreover, the mechanisms that control methylation of these repetitive elements may differ from the rest of the genome, and thus, measuring Alu and LINE-1 methylation levels may not necessarily provide an accurate representation of total genomic methylation (Yang et al. 2004). Also, these estimates of global methylation that represent a pool of multiple unique Alu or LINE-1 repeats in the entire genome may not be predictive of results from measures of gene-specific or genome-wide array-based methylation analyses (Zhu et al. 2012). Thus, the extent to which conclusions from these studies using this surrogate measure of DNA methylation may be relevant to site-specific methylation measures, particularly those related to the central nervous system and associated function, is uncertain at this time.

Additionally, leukocyte DNA methylation levels have been used as a proxy for lead-induced changes in epigenetic patterns in target tissues of lead toxicity. Unfortunately, there are no data currently available to confirm that the extent of methylation levels observed within circulating DNA is representative of DNA methylation changes in other tissues, particularly the brain. Studies have shown the existence of tissue-specific methylation profiles (Eckhardt et al. 2006). Further, care must be taken in the interpretation of results from studies using a peripheral biomarker for estimating effects on a different target tissue, as recent data have shown that the proportions of white blood cell subsets affect LINE-1 methylation levels measured in blood DNA, with differences in the percent of neutrophils and lymphocytes, the major types of white blood cells, exhibiting different and opposite effects on LINE-1 methylation levels (Zhu et al. 2012). This could be particularly important in toxicology studies as leukocyte DNA is derived from a mix of numerous cell types and various toxicants, including lead, may affect the composition of white blood cell populations (Sarasua et al. 2000).

Other studies, also using a peripheral marker in a surrogate tissue (blood), have examined relationships between lead exposure and methylation status of particular genes or sets of genes. In one such study, it was shown that lead exposure may affect promoter methylation in the delta-aminolevulinic acid dehydratase gene *ALAD*. Lead exposure has well-documented effects on heme biosynthesis and increases levels of delta-aminolevulinic acid; correspondingly, *ALAD1* and *ALAD2* have been suggested to modify lead toxicokinetics, impacting individual susceptibility to lead poisoning (Wetmur et al. 1991). Genomic DNA was extracted from blood of lead-exposed battery plant workers and nonexposed controls, and significantly increased ALAD methylation was found in exposed factory workers compared to controls (Li et al. 2011). The authors concluded that increases in the level of *ALAD* CpG methylation might contribute to the risk of lead poisoning.

Kovatsi et al. (2010) examined promoter methylation of the tumor suppressor gene *p16* in a small number of adult men ($N=9$) with occupational lead exposure. Methylation of *p16*, a cyclin-dependent kinase inhibitor, was examined because of its role in carcinogenesis and its overexpression in neurodegenerative disorders, such as Alzheimer's disease (Kovatsi et al. 2010). Men with higher blood lead levels (51–100 μg/dl) had high levels of *p16* methylation, whereas men with lower blood lead levels (6–11 μg/dl) showed partial methylation.

In a more recent study, DNA was extracted from whole blood collected from adult women undergoing in vitro fertilization; this study used the Illumina GoldenGate Cancer Panel I bead array (Illumina, Inc., San Diego, CA, USA) that assays 1505 CpG sites throughout the genome that are associated with cell replication, cell differentiation, or oxidative stress (Hanna et al. 2012). Significantly reduced methylation in the *COLIA2* promoter was correlated with higher blood lead levels (Hanna et al. 2012). These results suggested that increased lead exposure may be associated with increased *COLIA2* expression, which could affect reproductive outcomes, as increased expression of this gene may be associated with preterm rupture of membranes and preterm birth in humans (Hanna et al. 2012), and lead exposure has been associated with a variety of impaired reproductive outcomes (e.g., Jelliffe-Pawlowski et al. 2006; Hernandez-Avila et al. 2002; Taylor et al. 2015).

Using a somewhat different approach, Senut et al. (2014) examined DNA methylation profiles of human embryonic stem cells (hESCs) and neural progenitor cells (NPCs) acutely or chronically exposed (throughout the neural differentiation process) to lead concentrations ranging from the equivalents of 8 to 40 μg/dl, as assessed using the Infinium Human Methylation 450K BeadChip system. Lead exposure appeared to have different effects depending on the differentiation stage at which hESCs were exposed. Although it was not entirely clear at what lead exposure levels significant changes in methylation occurred, lead exposure rapidly modified the methylation profile of hESCs, and in cells engaged in the differentiation process, lead-associated hypomethylation predominated (Senut et al. 2014). These data suggest the potential for lead exposure to affect the methylation status of the developing nervous system and produce aberrant gene expression profiles that could negatively impact proper brain development. More recently, Sen et al. (2015) examined the relationship between lead exposure and 5-methylcytosine (5mC) and 5-hydroxymethylcytosine (5hmC) modifications in hESCs and in DNA extracted from umbilical cord blood and suggest that a mechanism through which lead exposure might contribute to altering gene expression is by changing the 5hmC and 5mC profiles in the genome. Although the precise biological role of 5hmC is still unclear, it is proposed that 5hmC is an intermediate base in an active or passive DNA demethylation process that operates during important reprogramming phases of brain development (Pfeifer et al. 2013). Analysis of lead-associated changes in cord blood DNA identified several 5hmC and 5mC clusters as potential candidates for sex-independent or sex-specific epigenetic biomarkers for prenatal lead exposure. These data provide additional support for an epigenetic basis of sex-dependent outcomes from developmental lead exposure.

In a preliminary study (Schneider et al. 2010), we attempted to identify differentially methylated DNA regions across the genome that might be associated with low-level childhood lead exposures. We randomly sampled 12 white, non-Hispanic males from the HOME Study, a prospective, longitudinal cohort study at Cincinnati Children's Hospital Medical Center that examines the effect of low-level exposures (prenatal and postnatal) of prevalent toxicants (e.g., lead, mercury, tobacco, PCBs, bisphenol A, phthalates, organophosphate, and pyrethroid insecticides) on neurobehavioral outcomes and developmental disorders and tests the efficacy of interventions for the primary prevention of lead toxicity. Subjects had mean maternal prenatal blood lead levels of 0.53 μg/dl, mean cord blood lead levels of 0.38 μg/dl, and mean blood lead levels in the highest (3.7 μg/dl, group 1) or lowest (0.82 μg/dl, group 2) terciles at 2 years of age. We extracted genomic DNA from blood samples taken at 2 years of age and analyzed methylated DNA on Human NimbleGen Methylation Promoter Arrays. Ninety-nine genes were identified with differentially methylated promoters between the two groups ($t=p<0.05$). Knowledge-based and pathway analyses suggested significant differences in methylation status of genes related to nervous system development and function, cardiovascular disease, renal and reproductive system functioning and key biological functions including cellular growth and proliferation, carbohydrate/sphingolipid metabolism, and cell signaling. These data, although preliminary and from a very small sample, suggest the possibility of widespread changes in methylation status associated with developmental exposure to very low levels of lead that may have far-reaching effects on brain development and function and the risk for developing systemic diseases later in life. These intriguing data now need to be validated in a much larger sample.

While the epidemiological data on epigenetic effects of lead are still limited and have been confined to examining effects on methylation, the picture that is emerging is that exposure to lead, either during development, childhood, or adulthood, can effect DNA methylation profiles and has the potential to exert wide-ranging effects on a variety of genes and processes involved in diverse physiological activities and represents an important pathway through which lead exposure can influence a variety of outcomes including disease risk across the life span. That these effects can be produced even at relatively low exposure levels enforces the need to eliminate lead from our environments and to reinforce primary prevention. Research is now needed to expand the limited findings from the existing epidemiological work to both global genome-wide and candidate gene DNA methylation, expand studies to include assessment of posttranslational histone modifications, and to better understand which pathways and mechanisms may be particularly at risk following lead exposure.

Although the data discussed thus far suggests that lead may have effects on the epigenome throughout life, there are no direct data available at this time to draw definitive links between lead exposures, epigenetics, and specific human diseases or conditions. Yet, associations between lead exposure, epigenetics, and disease have been proposed based on some intriguing indirect evidence. For example, lead exposure has been positively correlated with homocysteine (HCY) levels (which modulate cellular methylation reactions), and elevated HCY (linked to DNA hypomethylation) has been associated with increased risk of developing Alzheimer's disease (AD) and increased rate of disease progression in individuals with AD (Bakulski et al. 2012). Continued work on several fronts including epidemiological studies, genome-wide methyl-seq analyses, postmortem studies on tissues of interest such as the brain, and well-designed animal experimental studies across both sexes and multiple species/strains will provide an integrated approach to understanding the influence of lead exposure on epigenetics and the role of lead-induced epigenetic changes in neurological function and various neurological and non-neurological diseases.

10.5.2 Epigenetics of Lead Exposure: Animal Studies

As with the study of lead and epigenetics in humans, there have been relatively few studies published examining epigenetic effects of lead exposure in animals. In 2013, we were the first to describe an epigenetic effect of developmental lead exposure on DNA methyltransferases and DNA-binding proteins in the rat hippocampus. We showed that these effects were different in males and females and are further differentiated by the period during development when the exposure to lead occurred and the level of lead exposure (Schneider et al. 2013). Developmental lead exposure particularly affected DNMT1, DNMT3a, and MeCP2 expression that could influence the modulation of DNA methylation that occurs during brain development as well as during a variety of neuronal processes including learning and memory (Feng et al. 2007). Faulk et al. (2013) also reported exposure concentration- and sex-specific DNA methylation responses to perinatal lead exposures in the agouti (A^{vy}) mouse model, used as a biosensor to explore the epigenetic response to exposure to lead in utero and in early life, with significant epigenetic effects observable even at relatively low exposure levels (approximate blood lead levels of 4.1 μg/dl). Sanchez-Martin et al. (2015) analyzed whole-genome DNA methylation (global methyl-seq analysis) in the cortex (region(s) not specified) and presumably the whole hippocampus of 2-month-old mice exposed (or unexposed) to low levels of lead (producing 0.9–1.3 μg/dl blood lead levels in dams) during gestation and lactation. Lead exposure resulted in no significant hyper- or hypomethylated sites in the male hippocampus, whereas hypermethylation of three differentially methylated regions in the hippocampus was found in females. No DNA hypomethylation was observed in females, and the hippocampus was reported to have significantly higher levels of differentially methylated CpG sites than the cortex (Sanchez-Martin et al. 2015). A sample of differentially methylated genes was tested for mRNA expression and showed a trend toward negative correlation between mRNA expression and methylation in exposed females but not in males. These results are in general agreement with the hypothesis that lead exposure acts in a locus-specific way on the epigenome, depending on the genomic features in which affected CpG sites are located (Faulk et al. 2013, 2014) and also highlights the sex dependence of these effects.

Together, the results from these studies suggest that exposure to lead during embryonic life and during the early postnatal period may have complex effects on the epigenome that at the very least appear to be sex, exposure concentration, and tissue dependent. While these effects relate to effects on DNA methylation, effects on posttranslational histone modifications are currently unknown. However, effects of lead exposure on DNA methylation, likely in concert with effects on posttranslational histone modifications, are expected to contribute to a variety of epigenetically induced adverse outcomes that may persist into adulthood.

There have also been reports that early life exposure to lead can result in epigenetic modifications (such as changes in levels of DNMT1, DNMT3a, and MeCP2) that result in changes in the brain (e.g., overexpression of amyloid-β protein precursor (AβPP) and amyloid-β (Aβ)) that may lead to the existence of Alzheimer's-like changes in later life (Bihaqi et al. 2011). These findings were based on results from four female monkeys, examined at approximately 23 years of age, following exposure to 1.5 mg/kg/day of lead acetate from birth to 400 days of age (Bihaqi et al. 2011). Blood lead levels at 400 days of age ranged from 19 to 26 μg/dl. Previous work from this group had also shown that exposure of rats to lead from birth to postnatal day 20 resulted in a delayed overexpression of APP and elevation of its product Aβ later in life (Basha et al. 2005). Possible epigenetic mechanisms underlying these effects in these animals were not explored. While the nonhuman primate data are interesting, they need to be interpreted cautiously. First, only females were studied and there are well-documented sex-specific effects of lead on epigenetics and gene/protein expression, particularly when exposures encompass periods of brain differentiation. Second, it is not possible to tell which of the monkeys originally prepared for this developmental lead exposure were analyzed by Bihaqi et al. (2011) as nine females were reported to be prepared for the experimental group described (Rice and Gilbert 1990) and all of the animals in this group underwent extensive behavioral training and testing beginning at approximately 5–6 years of age. As the behavioral experience of the animals could result in dynamic and potentially long-lasting impact on epigenetic modifications, it is difficult to interpret the findings from these animals without knowing the precise behavioral history of all experimental and control (non-lead exposed) animals. Also, the fact that there might be reduced DNMT1 activity in the brains of these animals (also reported by the same group in 2008 (Wu et al. 2008) does not in and of itself argue in favor of infant lead exposure epigenetically modifying Alzheimer's-related genes such as *APP* or *BACE1*. A change in the methylation status of CpG dinucleotides in the specific Alzheimer's disease-related genes of interest would need to be demonstrated in these animals.

10.5.3 Epigenetics of Prenatal Stress: Human Studies

Although there have been several animal studies (see below) that have examined the effects of prenatal stress on epigenetic programming in offspring, there is little known at this time about the epigenetic effects of prenatal stress on offspring in humans. It is now generally accepted that early life experience and environmental factors occurring early in life can have persisting effects on a variety of systems including the central nervous system (Welberg et al. 2001; Oberlander et al. 2008). These early occurring factors can result in persistent organization effects, referred to as "programming" that can affect the functioning of various physiological systems throughout the life span, with the brain particularly sensitive to prenatal programming (Welberg et al. 2001). Prenatal stress is one environmental factor that can have long-lasting effects on the hypothalamic–pituitary–adrenal (HPA) axis, resulting in an altered capacity to regulate responses to stressful events in offspring. Although there are a number of potential prenatal stressors that could impact the HPA axis stress system, prenatal exposure to maternal depressed and anxious mood confers risk for behavioral/emotional disturbances in childhood and beyond (Oberlander et al. 2008; O'Connor et al. 2003). Oberlander et al. (2008) investigated the extent to which prenatal exposure to maternal depressed/anxious mood may be associated with epigenetic modification (i.e., methylation status) of the promoter and exon1F of the human *NR3C1* (glucocorticoid receptor) gene in newborns and whether methylation status of this specific region of *NR3C1* at birth is associated with altered HPA function and stress responsivity during infancy. Using DNA extracted from cord blood, increased methylation of *NR3C1* at a predicted NGFI-A binding site (in particular, methylation of CpG3, the 5′CpG in the potential NGFI-A consensus binding site 5′ to exon 1F of the human *NR3C1* gene) was associated to increased third-trimester maternal depressed/anxious mood and increased salivary cortisol responses at 3 months of age (Oberlander et al. 2008). These findings suggest a potential epigenetic link between prenatal stress, HPA function, and stress reactivity.

In another study, Mulligan et al. (2012) examined the relationship between a variety of maternal psychosocial stressors in women in the Democratic Republic of the Congo on birth outcome (birth weight) and methylation status of the *NR3C1* gene. There was a strong inverse correlation between maternal stress and newborn birth weight and a significant effect of maternal stress on newborn *NR3C1* methylation status. It is anticipated that these epigenetic effects would influence *NR3C1* gene expression, stress responses, and the risk for developing stress-related adult-onset diseases (Mulligan et al. 2012).

10.5.4 Epigenetics of Prenatal Stress: Animal Studies

Stress during pregnancy may not only adversely affect maternal health but may also alter the brain and behavior of offspring, potentially with lifelong consequences (Owen et al. 2005; Zucchi et al. 2013). As early life (postnatal) stress effects on the brain and epigenome are likely quite different from those effects that occur during the prenatal period, we will restrict this discussion to effects of prenatal stress. Gestational stress may directly influence fetal brain development and programming of HPA axis function (Cottrell and Seckl 2009), resulting in persisting changes in stress responsiveness in offspring (Meaney et al. 1996) and possibly enhanced vulnerability to behavioral/emotional disorders (Ellenbogen et al. 2011; Markham and Koenig 2011). However, there is considerable variation in the behavioral and epigenetic outcomes of studies utilizing prenatal stress with some data even contradictory. As reviewed by Weinstock (2005, 2008), results from different studies need to be reviewed carefully as many parameters that may affect outcomes, including the type and timing of the stress exposure during pregnancy, the age of testing of the offspring, the genetic strain of animal used, the sex of the offspring, and the time of day of testing the offspring (Glover et al. 2010). In studies examining effects on the brain, the brain region(s) analyzed is also a critical variable as effects (and particularly epigenetic effects) may differ dramatically by brain region. A study by Pena et al. (2012) that examined the epigenetic mechanisms contributing to the impact of prenatal stress (chronic restraint stress during gestational days 14–20) in Long Evans rats on *HSD11B2* mRNA in the placenta and fetal brain (E20) highlighted the issue of tissue specificity of epigenetic effects of maternal stress and raised the intriguing possibility of using the epigenetic status of the placenta to predict corresponding changes in the brain. Further, the intensity of the prenatal stress appears to be an important factor influencing behavioral and epigenetic outcomes (Mychasiuk et al. 2011). For example, "mild" prenatal stress (10-min sessions on an elevated platform twice daily on gestational days 12–16) and "extreme" stress (30-min sessions on an elevated platform twice daily on gestational days 12–16) had sex-specific effects on brain weight and several behavioral outcomes. Mild prenatal stress increased global DNA methylation levels in the frontal cortex of males only, while extreme stress decreased methylation levels in both sexes (Mychasiuk et al. 2011). In the hippocampus, mild stress increased methylation in both sexes and extreme stress decreased methylation in both sexes. These results suggest that the intensity of prenatal stress is an important factor that needs to be taken into consideration in studies of the effects of prenatal stress on both behavior and the epigenome.

Despite variability reported in several studies due to methodological differences, sex-specific effects are reported consistently. Sex-specific programming appears to begin very early in pregnancy and may contribute significantly to the timing and vulnerability of the developing fetus to maternal environmental influences (Auger and Auger 2011; Mueller and Bale 2008). For example, Mueller and Bale (2008) examined the effects of chronic, variable stress during different gestational periods and found significant sex differences in stress sensitivity and responsivity in male offspring, particularly in males exposed to stress early in gestation. Changes in corticotropin-releasing factor (CRF) and glucocorticoid receptor (GR) gene methylation in brain regions associated with the HPA axis correlated with altered gene expression, providing evidence of sex-specific epigenetic programming during early prenatal stress (Mueller and Bale 2008).

In addition to studying prenatal stress effects on the epigenetic programming of the GR, other studies have examined other epigenetic targets related to prenatal stressors and risk for developing cognitive and behavioral abnormalities. For example, Matrisciano et al. (2013) reported a number of changes in expression levels of DNA methyltransferases (DNMT1, DNMT3a) in the frontal cortex and hippocampus, particularly in GABAergic interneurons in mice exposed to prenatal restraint stress. These investigators also found altered expression of schizophrenia-related genes (i.e., *GAD67* and *reelin*) in frontal cortex of mice exposed to prenatal restraint stress that was epigenetically regulated and consistent with a schizophrenia-like behavioral phenotype expressed in these animals (Matrisciano et al. 2013).

Also, while DNA methylation has been predominantly investigated in regard to the epigenetic effects of prenatal stress, clearly changes in posttranslational histone modifications must also play a role. Recently, it has been shown that maternal stress can also affect the fetal transcriptome through microRNA (miRNA) regulation (Zucchi et al. 2013). Putative gene targets for miRNAs differentially affected by prenatal stress included genes related to miRNA biogenesis, apoptosis, brain pathologies, neurotransmission, neurodevelopment, hormonal regulation, neurotrophic factors, brain angiogenesis, cell signaling, stress response, and metabolism (Zucchi et al. 2013). Unfortunately, only the whole brain from male offspring was analyzed so it is not possible to know the extent to which prenatal stress affected different brain regions or differentially affected males and females.

10.5.5 Epigenetic Alterations in the Glucocorticoid Receptor (GR) Gene as a Potential Mechanism of Toxicity of Developmental Lead Exposure and Prenatal Stress

Potential changes in expression of the glucocorticoid receptor gene *NR3C1* via alterations in epigenetic markers is an intriguing potential mechanism that could account for the permanent consequences of developmental lead exposure and prenatal stress alone or in combination, based on the fact that both of these risk factors target the HPA axis. Essential for development, DNA methylation is regulated by developmental cues and environmental stimuli. It is possible that exposure to environmental toxicants (including lead) and other developmental insults (early stress) may be able to alter gene expression and phenotype, in part by modifying the epigenome, and inducing (at crucial stages of prenatal or postnatal development) epigenetic adaptations that can potentially change behavior as well as disease susceptibility. If so, such findings will have significant public health risk implications. Moreover, when more fully clarified, they will undoubtedly yield critical new insights into brain–behavior relationships that set the stage for opportunities for prevention and/or intervention. Indeed, as noted above, both lead exposure and early stress alone have been reported to produce epigenetic alterations; however, the potential interactive effects of Pb and prenatal stress are only now beginning to be examined (Anderson et al. 2015). How these interactive effects relate to subsequent behavioral consequences will also be a subject of much additional research.

Unlike genetic changes, epigenetic alterations are potentially reversible (Auger and Auger 2011; Dolinoy et al. 2007), establishing the possibility for therapies for conditions, such as the cognitive/behavioral sequelae associated with developmental lead exposure and early stress, that have been considered permanent. Certain drugs and dietary manipulations can remove aberrant hypermethylation to reactivate genes that have been silenced. Alternatively, hypermethylating dietary supplements can potentially reduce the DNA hypomethylation effects of environmental toxicants, reversing associated epigenetic effects (Jirtle and Skinner 2007). Likewise, pharmacological manipulation of posttranslational histone modifications can potentially reverse environmentally altered histone effects and associated behavioral phenotypes (Weaver et al. 2005). Thus, although there is significant potential for interactive effects of lead and prenatal stress to adversely affect a variety of cognitive/behavioral/emotional outcomes via epigenetic programming, accumulating evidence suggests that there might be a variety of strategies that could be employed to manipulate these environmentally induced epigenetic alterations and improve outcomes from environmentally related disorders.

10.6 Conclusions

Understanding of the role of epigenetic alterations in mediating the adverse neurodevelopmental impacts of environmental, chemical, and nonchemical stressors will ultimately require assessment of these factors in combination, consistent with the realities of the human environment and experience. Moreover, as the human environment inevitably includes behavioral experience, the extent to which behavioral experience modifies the epigenetic consequences of these environmental risk factors will need to be discerned. We schematically diagram this integrative trajectory in Fig. 10.1. Assessment of epigenetic functions under conditions of combined exposure will assist in understanding cumulative toxicity.

While such an integrated assessment is a long way off from this early stage of our current understanding, the need to reach this goal poses challenging but interesting questions to pursue. These include defining pathological vs. healthy epigenetic

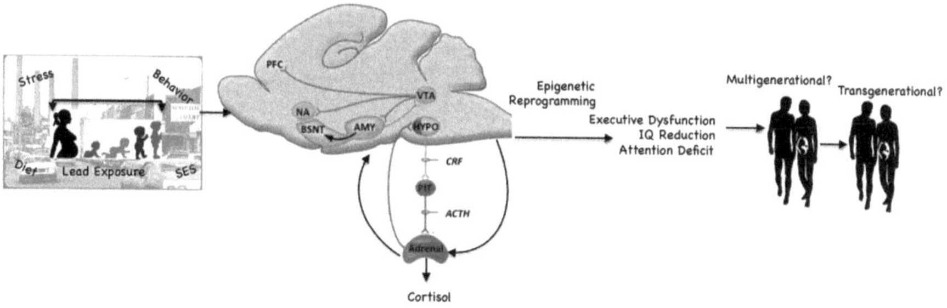

Fig. 10.1 Multiple environmental risk factors known or suspected to produce adverse effects on CNS development via epigenetic reprogramming are co-occurring or sequentially occurring in the human environment. Importantly, these can include plastic alterations, in accord with the need for rapid and dynamic changes in behavior in response to environmental conditions. In the human environment, these can include stress, diet, behavioral experience, and environmental chemical exposures. When these act on joint biological substrates, it raises the potential for interactive effects, especially enhanced adverse effects on subsequent brain and/or behavioral function. A clear understanding of how epigenetic profiles are modified by these human environmental conditions will therefore necessitate understanding the interactive effects of such risk factors. The importance of the observation that behavioral experiences can modify epigenetic profiles raises the possibility that behavioral experiences can be identified that could be used in a therapeutic capacity to mitigate adverse condition-generated epigenetic changes

alterations. Such definitions could then be used to assist in defining appropriate intervention strategies. What are healthy vs. toxic behavioral experiences, and do these experiences further modify epigenetic profiles, and if so, in what way? Can specific behavioral experiences be used in interventions that would reverse or mitigate damaging epigenetic changes? Can they be used to define resilience to environmental stressors?

Abbreviations

5hmC	5-Hydroxymethylcytosine
5mC	5-Methylcytosin
Aβ	Amyloid-β
AβPP	Amyloid-β protein precursor
AD	Alzheimer's disease
ALAD	Aminolevulinic acid dehydratase
BACE1	Beta-secretase 1
BDNF	Brain-derived neurotrophic factor
CNS	Central nervous system
CpG	–C–phosphate–G–
CRF	Corticotropin-releasing factor
DNA	Deoxyribonucleic acid

DNMTs	DNA methyltransferases
DOPAC	3,4-Dihydroxyphenylacetic acid
GR	Glucocorticoid receptor
HCY	Homocysteine
hESCs	Human embryonic stem cells
HPA axis	Hypothalamic–pituitary–adrenal axis
IQ	Intelligence quotient
NMDA	*n*-methyl-D-aspartate
SES	Socioeconomic status

References

Anderson, N. B., & Armstead, C. A. (1995). Toward understanding the association of socioeconomic status and health: A new challenge for the biopsychosocial approach. *Psychosomatic Medicine, 57*, 213–225.

Anderson, D. W., Schneider, J. S., Sobolewski, M., & Cory-Slechta, D. (2015). Sex-dependent effects of lead and prenatal stress on adult glucocorticoid receptor expression and its epigenetic control. *Toxicological Sciences, 144*(1), 207.

Auger, A. P., & Auger, C. J. (2011). Epigenetic turn ons and turn offs: Chromatin reorganization and brain differentiation. *Endocrinology, 152*(2), 349–353.

Babenko, O., Kovalchuk, I., & Metz, G. A. (2015). Stress-induced perinatal and transgenerational epigenetic programming of brain development and mental health. *Neuroscience & Biobehavioral Reviews, 48*, 70–91.

Bakulski, K. M., Rozek, L. S., Dolinoy, D. C., Paulson, H. L., & Hu, H. (2012). Alzheimer's disease and environmental exposure to lead: The epidemiologic evidence and potential role of epigenetics. *Current Alzheimer Research, 9*(5), 563–573.

Bale, T. L. (2015). Epigenetic and transgenerational reprogramming of brain development. *Nature Reviews Neuroscience, 16*(6), 332–344.

Baranowska-Bosiacka, I., Struzynska, L., Gutowska, I., Machalinska, A., Kolasa, A., Klos, P., et al. (2013). Perinatal exposure to lead induces morphological, ultrastructural and molecular alterations in the hippocampus. *Toxicology, 303*, 187–200.

Barros, V. G., Berger, M. A., Martijena, I. D., Sarchi, M. I., Perez, A. A., Molina, V. A., et al. (2004). Early adoption modifies the effects of prenatal stress on dopamine and glutamate receptors in adult rat brain. *Journal of Neuroscience Research, 76*(4), 488–496.

Basha, M. R., Wei, W., Bakheet, S. A., Benitez, N., Siddiqi, H. K., Ge, Y. W., et al. (2005). The fetal basis of amyloidogenesis: Exposure to lead and latent overexpression of amyloid precursor protein and beta-amyloid in the aging brain. *Journal of Neuroscience, 25*(4), 823–829.

Belnoue, L., Grosjean, N., Ladeveze, E., Abrous, D. N., & Koehl, M. (2013). Prenatal stress inhibits hippocampal neurogenesis but spares olfactory bulb neurogenesis. *PLoS One, 8*(8), e72972.

Berger, M. A., Barros, V. G., Sarchi, M. I., Tarazi, F. I., & Antonelli, M. C. (2002). Long-term effects of prenatal stress on dopamine and glutamate receptors in adult rat brain. *Neurochemical Research, 27*(11), 1525–1533.

Bihaqi, S. W., Huang, H., Wu, J., & Zawia, N. H. (2011). Infant exposure to lead (Pb) and epigenetic modifications in the aging primate brain: Implications for Alzheimer's disease. *Journal of Alzheimer's Disease, 27*(4), 819–833.

Bock, J., Wainstock, T., Braun, K., & Segal, M. (2015). Stress in utero: Prenatal programming of brain plasticity and cognition. *Biol Psychiatry, 78*(5), 315–326.

Bradley, R. H., & Corwyn, R. F. (2002). Socioeconomic status and child development. *Annual Review of Psychology, 53*, 371–399.

Braun, J. M., Wright, R. J., Just, A. C., Power, M. C., Tamayo, Y. O. M., Schnaas, L., et al. (2014). Relationships between lead biomarkers and diurnal salivary cortisol indices in pregnant women from Mexico City: A cross-sectional study. *Environmental Health, 13*(1), 50.

Brunton, P. J. (2015). Programming the brain and behaviour by early-life stress: A focus on neuroactive steroids. *Journal of Neuroendocrinology, 27*(6), 468–480.

Byers, R., & Lord, E. (1943). Late effects of lead poisoning on mental development. *American Journal of Diseases of Children, 66*, 471–494.

Canfield, R. L., Henderson, C. R., Jr., Cory-Slechta, D. A., Cox, C., Jusko, T. A., & Lanphear, B. P. (2003). Intellectual impairment in children with blood lead concentrations below 10 microg per deciliter. *The New England Journal of Medicine, 348*(16), 1517–1526.

Cecil, K. M., Brubaker, C. J., Adler, C. M., Dietrich, K. N., Altaye, M., Egelhoff, J. C., et al. (2008). Decreased brain volume in adults with childhood lead exposure. *PLoS Medicine, 5*(5), e112.

Centers for Disease Control and Prevention (CDC). (2013). Blood lead levels in children aged 1-5 years - United States, 1999-2010. *Morbidity and Mortality Weekly Report, 62*, 245–248.

Chisolm, J. J., Jr., & Harrison, H. E. (1957). The treatment of acute lead encephalopathy in children. *Pediatrics, 19*(1), 2–20.

Cory-Slechta, D. A. (1995). Relationships between lead-induced learning impairments and changes in dopaminergic, cholinergic, and glutamatergic neurotransmitter system functions. *Annual Review of Pharmacology and Toxicology, 35*, 391–415.

Cory-Slechta, D. A. (1997). Relationships between Pb-induced changes in neurotransmitter system function and behavioral toxicity. *Neurotoxicology, 18*, 673–688.

Cory-Slechta, D. A., Garcia-Osuna, M., & Greenamyre, J. T. (1997). Lead-induced changes in NMDA receptor complex binding: Correlations with learning accuracy and with sensitivity to learning impairments caused by MK-801 and NMDA administration. *Behavioural Brain Research, 85*, 161–174.

Cory-Slechta, D. A., Merchant-Borna, K., Allen, J., Liu, S., Weston, D., & Conrad, K. (2013). Variations in the nature of behavioral experience can differentially alter the consequences of developmental exposures to lead, prenatal stress and the combination. *Toxicological Sciences, 131*, 194–205.

Cory-Slechta, D. A., O'Mara, D. J., & Brockel, B. J. (1998). Nucleus accumbens dopaminergic mediation of fixed interval schedule-controlled behavior and its modulation by low-level lead exposure. *Journal of Pharmacology and Experimental Therapeutics, 286*(3), 794–805.

Cory-Slechta, D. A., O'Mara, D. J., & Brockel, B. J. (1999). Learning versus performance impairments following regional administration of MK-801 into nucleus accumbens and dorsomedial striatum. *Behavioural Brain Research, 102*, 181–194.

Cory-Slechta, D. A., Stern, S., Weston, D., Allen, J. L., & Liu, S. (2010). Enhanced learning deficits in female rats following lifetime Pb exposure combined with prenatal stress. *Toxicological Sciences, 117*(2), 427–438.

Cory-Slechta, D. A., Virgolini, M. B., Liu, S., & Weston, D. (2012). Enhanced stimulus sequence-dependent repeated learning in male offspring after prenatal stress alone or in conjunction with lead exposure. *Neurotoxicology, 33*(5), 1188–1202.

Cory-Slechta, D. A., Virgolini, M. B., Rossi-George, A., Thiruchelvam, M., Lisek, R., & Weston, D. (2008). Lifetime consequences of combined maternal lead and stress. *Basic and Clinical Pharmacology and Toxicology, 102*(2), 218–227.

Cory-Slechta, D. A., Weiss, B., & Cox, C. (1989). Tissue distribution of Pb in adult vs. old rats: A pilot study. *Toxicology, 59*(2), 139–150.

Cottrell, E. C., & Seckl, J. R. (2009). Prenatal stress, glucocorticoids and the programming of adult disease. *Frontiers in Behavioral Neuroscience, 3*, 19.

Davis, E. P., & Sandman, C. A. (2010). The timing of prenatal exposure to maternal cortisol and psychosocial stress is associated with human infant cognitive development. *Child Development, 81*(1), 131–148.

Deroche, V., Marinelli, M., Maccari, S., Le Moal, M., Simon, H., & Piazza, P. V. (1995). Stress-induced sensitization and glucocorticoids. I. Sensitization of dopamine-dependent locomotor

effects of amphetamine and morphine depends on stress-induced corticosterone secretion. *Journal of Neuroscience, 15*(11), 7181–7188.
DiCorcia, J. A., & Tronick, E. (2011). Quotidian resilience: Exploring mechanisms that drive resilience from a perspective of everyday stress and coping. *Neuroscience & Biobehavioral Reviews, 35*(7), 1593–1602.
Dietrich, K. N., Ris, M. D., Succop, P. A., Berger, O. G., & Bornschein, R. L. (2001). Early exposure to lead and juvenile delinquency. *Neurotoxicology and Teratology, 23*(6), 511–518.
Dohrenwend, B. P. (1973). Social status and stressful life events. *Journal of Personal and Social Psychology, 28*, 225–235.
Dolinoy, D. C., Das, R., Weidman, J. R., & Jirtle, R. L. (2007). Metastable epialleles, imprinting, and the fetal origins of adult diseases. *Pediatric Research, 61*(5 Pt 2), 30R–37R.
Eckhardt, F., Lewin, J., Cortese, R., Rakyan, V. K., Attwood, J., Burger, M., et al. (2006). DNA methylation profiling of human chromosomes 6, 20 and 22. *Nature Genetics, 38*(12), 1378–1385.
Ellenbogen, M. A., Hodgins, S., Linnen, A. M., & Ostiguy, C. S. (2011). Elevated daytime cortisol levels: A biomarker of subsequent major affective disorder? *Journal of Affective Disorders, 132*(1-2), 265–269.
Faulk, C., Barks, A., Liu, K., Goodrich, J. M., & Dolinoy, D. C. (2013). Early-life lead exposure results in dose- and sex-specific effects on weight and epigenetic gene regulation in weanling mice. *Epigenomics, 5*(5), 487–500.
Faulk, C., Liu, K., Barks, A., Goodrich, J. M., & Dolinoy, D. C. (2014). Longitudinal epigenetic drift in mice perinatally exposed to lead. *Epigenetics, 9*(7), 934–941.
Feng, J., Fouse, S., & Fan, G. (2007). Epigenetic regulation of neural gene expression and neuronal function. *Pediatric Research, 61*(5 Pt 2), 58R–63R.
Focking, M., Opstelten, R., Prickaerts, J., Steinbusch, H. W., Dunn, M. J., van den Hove, D. L., et al. (2014). Proteomic investigation of the hippocampus in prenatally stressed mice implicates changes in membrane trafficking, cytoskeletal, and metabolic function. *Developmental Neurosciences, 36*(5), 432–442.
Fraga, M. F., Ballestar, E., Paz, M. F., Ropero, S., Setien, F., Ballestar, M. L., et al. (2005). Epigenetic differences arise during the lifetime of monozygotic twins. *Proceedings of the National Academy of Sciences of the United States of America, 102*(30), 10604–10609.
Gee, G. C., & Payne-Sturges, D. C. (2004). Environmental health disparities: A framework integrating psychosocial and environmental concepts. *Environmental Health Perspectives, 112*(17), 1645–1653.
Gilbert, M. E., Mack, C. M., & Lasley, S. M. (1999). Chronic developmental lead exposure and hippocampal long-term potentiation: Biphasic dose-response relationship. *Neurotoxicology, 20*(1), 71–82.
Glover, V., O'Connor, T. G., & O'Donnell, K. (2010). Prenatal stress and the programming of the HPA axis. *Neuroscience & Biobehavioral Reviews, 35*(1), 17–22.
Guilarte, R. T., Miceli, R. C., & Jett, D. A. (1994). Neurochemical aspects of hippocampal and cortical Pb2+ neurotoxicity. *Neurotoxicology, 15*(3), 459–466.
Gulson, B. L., Jameson, C. W., Mahaffey, K. R., Mizon, K. J., Korsch, M. J., & Vimpani, G. (1997). Pregnancy increases mobilization of lead from maternal skeleton. *Journal of Laboratory and Clinical Medicine, 130*, 51–62.
Hanna, C. W., Bloom, M. S., Robinson, W. P., Kim, D., Parsons, P. J., vom Saal, F. S., et al. (2012). DNA methylation changes in whole blood is associated with exposure to the environmental contaminants, mercury, lead, cadmium and bisphenol A, in women undergoing ovarian stimulation for IVF. *Human Reproduction, 27*(5), 1401–1410.
Hernandez-Avila, M., Peterson, K. E., Gonzalez-Cossio, T., Sanin, L. H., Aro, A., Schnaas, L., et al. (2002). Effect of maternal bone lead on length and head circumference of newborns and 1-month-old infants. *Archives of Environmental Health, 57*(5), 482–488.
Hu, F., Xu, L., Liu, Z. H., Ge, M. M., Ruan, D. Y., & Wang, H. L. (2014). Developmental lead exposure alters synaptogenesis through inhibiting canonical Wnt pathway in vivo and in vitro. *PLoS One, 9*(7), e101894.

Jelliffe-Pawlowski, L. L., Miles, S. Q., Courtney, J. G., Materna, B., & Charlton, V. (2006). Effect of magnitude and timing of maternal pregnancy blood lead (Pb) levels on birth outcomes. *Journal of Perinatology, 26*(3), 154–162.

Jirtle, R. L., & Skinner, M. K. (2007). Environmental epigenomics and disease susceptibility. *Nature Reviews Genetics, 8*(4), 253–262.

Keenan, K., Gunthorpe, D., & Grace, D. (2007). Parsing the relations between SES and stress reactivity: Examining individual differences in neonatal stress response. *Infant Behavior & Development, 30*(1), 134–145.

Kolodkin, M. H., & Auger, A. P. (2011). Sex difference in the expression of DNA methyltransferase 3a in the rat amygdala during development. *Journal of Neuroendocrinology, 23*(7), 577–583.

Koolhaas, J. M., Bartolomucci, A., Buwalda, B., de Boer, S. F., Flugge, G., Korte, S. M., et al. (2011). Stress revisited: A critical evaluation of the stress concept. *Neuroscience & Biobehavioral Reviews, 35*(5), 1291–1301.

Kortenkamp, A., Faust, M., Scholze, M., & Backhaus, T. (2007). Low-level exposure to multiple chemicals: Reason for human health concerns? *Environmental Health Perspectives, 115*(Suppl 1), 106–114.

Kovatsi, L., Georgiou, E., Ioannou, A., Haitoglou, C., Tzimagiorgis, G., Tsoukali, H., et al. (2010). p16 promoter methylation in Pb2+ -exposed individuals. *Clinical Toxicology, 48*(2), 124–128.

Kuzumaki, N., Ikegami, D., Tamura, R., Hareyama, N., Imai, S., Narita, M., et al. (2011). Hippocampal epigenetic modification at the brain-derived neurotrophic factor gene induced by an enriched environment. *Hippocampus, 21*(2), 127–132.

Labonte, B., Yerko, V., Gross, J., Mechawar, N., Meaney, M. J., Szyf, M., et al. (2012). Differential glucocorticoid receptor exon 1(B), 1(C), and 1(H) expression and methylation in suicide completers with a history of childhood abuse. *Biological Psychiatry, 72*(1), 41–48.

Lander, E. S., Linton, L. M., Birren, B., Nusbaum, C., Zody, M. C., Baldwin, J., et al. (2001). Initial sequencing and analysis of the human genome. *Nature, 409*(6822), 860–921.

Lanphear, B. P., Hornung, R., Khoury, J., Yolton, K., Baghurst, P., Bellinger, D. C., et al. (2005). Low-level environmental lead exposure and children's intellectual function: An international pooled analysis. *Environmental Health Perspectives, 113*(7), 894–899.

Leggett, R. W. (1993). An age-specific kinetic model of lead metabolism in humans. *Environmental Health Perspectives, 101*(7), 598–616.

Li, W., Song, X., Zhang, H., Yang, Y., Jiang, C., Xiao, B., et al. (2011). Association study of RELN polymorphisms with schizophrenia in Han Chinese population. *Progress in Neuropsychopharmacology and Biological Psychiatry, 35*(6), 1505–1511.

Li, N., Zhang, P., Qiao, M., Shao, J., Li, H., & Xie, W. (2015). The effects of early life lead exposure on the expression of P2X7 receptor and synaptophysin in the hippocampus of mouse pups. *Journal of Trace Elements in Medicine and Biology, 30*, 124–128.

Li, N., Zhao, G., Qiao, M., Shao, J., Liu, X., Li, H., et al. (2014). The effects of early life lead exposure on the expression of insulin-like growth factor 1 and 2 (IGF1, IGF2) in the hippocampus of mouse pups. *Food and Chemical Toxicology, 63*, 48–52.

Lucchini, R. G., Zoni, S., Guazzetti, S., Bontempi, E., Micheletti, S., Broberg, K., et al. (2012). Inverse association of intellectual function with very low blood lead but not with manganese exposure in Italian adolescents. *Environmental Research, 118*, 65–71.

Lupien, S. J., McEwen, B. S., Gunnar, M. R., & Heim, C. (2009). Effects of stress throughout the lifespan on the brain, behaviour and cognition. *Nature Reviews Neuroscience, 10*, 434–445.

Markham, J. A., & Koenig, J. I. (2011). Prenatal stress: Role in psychotic and depressive diseases. *Psychopharmacology, 214*(1), 89–106.

Martinez-Tellez, R. I., Hernandez-Torres, E., Gamboa, C., & Flores, G. (2009). Prenatal stress alters spine density and dendritic length of nucleus accumbens and hippocampus neurons in rat offspring. *Synapse, 63*(9), 794–804.

Masten, A. S., & Narayan, A. J. (2012). Child development in the context of disaster, war, and terrorism: Pathways of risk and resilience. *Annual Review of Psychology, 63*, 227–257.

Matrisciano, F., Tueting, P., Dalal, I., Kadriu, B., Grayson, D. R., Davis, J. M., et al. (2013). Epigenetic modifications of GABAergic interneurons are associated with the schizophrenia-like phenotype induced by prenatal stress in mice. *Neuropharmacology, 68*, 184–194.

Matthews, S. G., & Phillips, D. I. (2010). Minireview: Transgenerational inheritance of the stress response: A new frontier in stress research. *Endocrinology, 151*(1), 7–13.

McCarthy, M. M., Auger, A. P., Bale, T. L., De Vries, G. J., Dunn, G. A., Forger, N. G., et al. (2009). The epigenetics of sex differences in the brain. *Journal of Neuroscience, 29*(41), 12815–12823.

McEwen, B., & Tucker, P. (2011). Critical biological pathways for chronic psychosocial stress and research opportunities to advance the consideration of stress in chemical risk assessment. *American Journal of Public Health, 101*(S1), S131–S139.

McGowan, P. O., Sasaki, A., D'Alessio, A. C., Dymov, S., Labonte, B., Szyf, M., et al. (2009). Epigenetic regulation of the glucocorticoid receptor in human brain associates with childhood abuse. *Nature Neuroscience, 12*(3), 342–348.

Meaney, M. J., Diorio, J., Francis, D., Widdowson, J., LaPlante, P., Caldji, C., et al. (1996). Early environmental regulation of forebrain glucocorticoid receptor gene expression: Implications for adrenocortical responses to stress. *Developmental Neurosciences, 18*(1-2), 49–72.

Morgan, C. P., & Bale, T. L. (2011). Early prenatal stress epigenetically programs dysmasculinization in second-generation offspring via the paternal lineage. *Journal of Neuroscience, 31*(33), 11748–11755.

Morse, S. J., Butler, A. A., Davis, R. L., Soller, I. J., & Lubin, F. D. (2015). Environmental enrichment reverses histone methylation changes in the aged hippocampus and restores age-related memory deficits. *Biology, 4*(2), 298–313.

Mueller, B. R., & Bale, T. L. (2008). Sex-specific programming of offspring emotionality after stress early in pregnancy. *The Journal of Neuroscience, 28*(36), 9055–9065.

Mulligan, C. J., D'Errico, N. C., Stees, J., & Hughes, D. A. (2012). Methylation changes at NR3C1 in newborns associate with maternal prenatal stress exposure and newborn birth weight. *Epigenetics, 7*(8), 853–857.

Mumtaz, M. M., Ruiz, P., & De Rosa, C. T. (2007). Toxicity assessment of unintentional exposure to multiple chemicals. *Toxicology and Applied Pharmacology, 223*(2), 104–113.

Mychasiuk, R., Ilnytskyy, S., Kovalchuk, O., Kolb, B., & Gibb, R. (2011). Intensity matters: Brain, behaviour and the epigenome of prenatally stressed rats. *Neuroscience, 180*, 105–110.

National Toxicology Program. (2012). *Health effects of low-level lead.* Office of Health Assessment and Translation.

Neal, A. P., Stansfield, K. H., & Guilarte, T. R. (2012). Enhanced nitric oxide production during lead (Pb(2)(+)) exposure recovers protein expression but not presynaptic localization of synaptic proteins in developing hippocampal neurons. *Brain Research, 1439*, 88–95.

Needleman, H. L., Riess, J. A., Tobin, M. J., Biesecker, G. E., & Greenhouse, J. B. (1996). Bone lead levels and delinquent behavior. *Journal of the American Medical Association, 275*(5), 363–369.

Oberlander, T. F., Weinberg, J., Papsdorf, M., Grunau, R., Misri, S., & Devlin, A. M. (2008). Prenatal exposure to maternal depression, neonatal methylation of human glucocorticoid receptor gene (NR3C1) and infant cortisol stress responses. *Epigenetics, 3*(2), 97–106.

O'Connor, T. G., Heron, J., Golding, J., & Glover, V. (2003). Maternal antenatal anxiety and behavioural/emotional problems in children: A test of a programming hypothesis. *Journal of Child Psychology and Psychiatry, and Allied Disciplines, 44*(7), 1025–1036.

O'Flaherty, E. J. (1991). Physiologically based models for bone-seeking elements. II. Kinetics of lead disposition in rats. *Toxicology and Applied Pharmacology, 111*(2), 313–331.

Owen, D., Andrews, M. H., & Matthews, S. G. (2005). Maternal adversity, glucocorticoids and programming of neuroendocrine function and behaviour. *Neuroscience & Biobehavioral Reviews, 29*(2), 209–226.

Pena, C. J., Monk, C., & Champagne, F. A. (2012). Epigenetic effects of prenatal stress on 11β-hydroxysteroid dehydrogenase-2 in the placenta and fetal brain. *PLoS ONE, 7*(6): e39791. doi:10.1371/journal.pone.0039791.

Pfeifer, G. P., Kadam, S., & Jin, S. G. (2013). 5-Hydroxymethylcytosine and its potential roles in development and cancer. *Epigenetics & Chromatin, 6*(1), 10.

Pilsner, J. R., Hu, H., Ettinger, A., Sanchez, B. N., Wright, R. O., Cantonwine, D., et al. (2009). Influence of prenatal lead exposure on genomic methylation of cord blood DNA. *Environmental Health Perspectives, 117*(9), 1466–1471.

Pirkle, J. L., Brody, D. J., Gunter, E. W., Kramer, R. A., Paschal, D. C., Flegal, K. M., et al. (1994). The decline in blood lead levels in the United States. The National Health and Nutrition Examination Surveys (NHANES). *Journal of the American Medical Association, 272*(4), 284–291.

Plumlee, G. S., Durant, J. T., Morman, S. A., Neri, A., Wolf, R. E., Dooyema, C. A., et al. (2013). Linking geological and health sciences to assess childhood lead poisoning from artisanal gold mining in Nigeria. *Environmental Health Perspectives, 121*(6), 744–750.

Qiu, A., Rifkin-Graboi, A., Chen, H., Chong, Y. S., Kwek, K., Gluckman, P. D., et al. (2013). Maternal anxiety and infants' hippocampal development: Timing matters. *Translational Psychiatry, 3*, e306.

Rice, D. C. (1988). Schedule-controlled behavior in infant and juvenile monkeys exposed to lead from birth. *Neurotoxicology, 9*, 75–88.

Rice, D. C. (1990). Lead-induced behavioral impairment on a spatial discrimination reversal task in monkeys during different periods of development. *Toxicology and Applied Pharmacology, 106*, 327–333.

Rice, D. C., & Gilbert, S. G. (1990). Sensitive periods for lead-induced behavioral impairment (nonspatial discrimination reversal) in monkeys. *Toxicology and Applied Pharmacology, 102*, 101–109.

Rice, D. C., & Karpinski, K. F. (1988). Lifetime low-level lead exposure produces deficits in delayed alternation in adult monkeys. *Neurotoxicology and Teratology, 10*, 207–214.

Richetto, J., & Riva, M. A. (2014). Prenatal maternal factors in the development of cognitive impairments in the offspring. *Journal of Reproductive Immunology, 104–105*, 20–25.

Rossi-George, A., Virgolini, M. B., Weston, D., Thiruchelvam, M., & Cory-Slechta, D. A. (2011). Interactions of lifetime lead exposure and stress: Behavioral; neurochemical and HPA axis effects. *Neurotoxicology, 32*(1), 83–99.

Rutter, M. (2006). Implications of resilience concepts for scientific understanding. *Annals of the New York Academy of Sciences, 1094*, 1–12.

Sanchez-Martin, F. J., Lindquist, D. M., Landero-Figueroa, J., Zhang, X., Chen, J., Cecil, K. M., et al. (2015). Sex- and tissue-specific methylome changes in brains of mice perinatally exposed to lead. *Neurotoxicology, 46*, 92–100.

Sarasua, S. M., Vogt, R. F., Henderson, L. O., Jones, P. A., & Lybarger, J. A. (2000). Serum immunoglobulins and lymphocyte subset distributions in children and adults living in communities assessed for lead and cadmium exposure. *Journal of Toxicology and Environmental Health, Part A, 60*(1), 1–15.

Schneider, J. S., Chen, J., Setzo, C., Anderson, D. W., Lanphear, B. P., & Ho, S.-M. (2010). *Methylation status of genes related to nervous system development and function as well as other key biological pathways is significantly affected by low level childhood lead exposure* (Program No. 360.5). 2010 Neuroscience Meeting Planner.

Schneider, J. S., Kidd, S. K., & Anderson, D. W. (2013). Influence of developmental lead exposure on expression of DNA methyltransferases and methyl cytosine-binding proteins in hippocampus. *Toxicology Letters, 217*(1), 75–81.

Schuurmans, C., & Kurrasch, D. M. (2013). Neurodevelopmental consequences of maternal distress: What do we really know? *Clinical Genetics, 83*(2), 108–117.

Schwabe, L., Bohbot, V. D., & Wolf, O. T. (2012). Prenatal stress changes learning strategies in adulthood. *Hippocampus, 22*(11), 2136–2143.

Schwartz, J. (1994). Low-level lead exposure and children's IQ: A meta-analysis and search for a threshold. *Environmental Research, 65*, 42–55.

Seery, M. D., Holman, E. A., & Silver, R. C. (2010). Whatever does not kill us: Cumulative lifetime adversity, vulnerability, and resilience. *Journal of Personal and Social Psychology, 99*(6), 1025–1041.

Sen, A., Heredia, N., Senut, M. C., Hess, M., Land, S., Qu, W., et al. (2015). Early life lead exposure causes gender-specific changes in the DNA methylation profile of DNA extracted from dried blood spots. *Epigenomics, 7*(3), 379–393.

Senut, M. C., Sen, A., Cingolani, P., Shaik, A., Land, S. J., & Ruden, D. M. (2014). Lead exposure disrupts global DNA methylation in human embryonic stem cells and alters their neuronal differentiation. *Toxicological Sciences, 139*(1), 142–161.

Stansfield, K. H., Pilsner, J. R., Lu, Q., Wright, R. O., & Guilarte, T. R. (2012). Dysregulation of BDNF-TrkB signaling in developing hippocampal neurons by Pb(2+): Implications for an environmental basis of neurodevelopmental disorders. *Toxicological Sciences, 127*(1), 277–295.

Sullivan, R. M., & Dufresne, M. M. (2006). Mesocortical dopamine and HPA axis regulation: Role of laterality and early environment. *Brain Research, 1076*(1), 49–59.

Szulwach, K. E., Li, X., Li, Y., Song, C. X., Wu, H., Dai, Q., et al. (2011). 5-hmC-mediated epigenetic dynamics during postnatal neurodevelopment and aging. *Nature Neuroscience, 14*(12), 1607–1616.

Talge, N. M., Neal, C., & Glover, V. (2007). Antenatal maternal stress and long-term effects on child neurodevelopment: How and why? *Journal of Child Psychology and Psychiatry, 48*(3–4), 245–261.

Taylor, C. M., Golding, J., & Emond, A. M. (2015). Adverse effects of maternal lead levels on birth outcomes in the ALSPAC study: A prospective birth cohort study. *BJOG, 122*(3), 322–328.

Thayer, Z. M., & Kuzawa, C. W. (2014). Early origins of health disparities: Material deprivation predicts maternal evening cortisol in pregnancy and offspring cortisol reactivity in the first few weeks of life. *American Journal of Human Biology, 26*(6), 723–730.

Tidey, J. W., & Miczek, K. A. (1996). Social defeat stress selectively alters mesocorticolimbic dopamine release: An in vivo microdialysis study. *Brain Research, 721*(1-2), 140–149.

Virgolini, M. B., Rossi-George, A., Lisek, R., Weston, D. D., Thiruchelvam, M., & Cory-Slechta, D. A. (2008a). CNS effects of developmental Pb exposure are enhanced by combined maternal and offspring stress. *Neurotoxicology, 29*(5), 812–827.

Virgolini, M. B., Rossi-George, A., Weston, D., & Cory-Slechta, D. A. (2008b). Influence of low level maternal Pb exposure and prenatal stress on offspring stress challenge responsivity. *Neurotoxicology, 29*(6), 928–939.

Wang, Y., Ma, Y., Cheng, W., Jiang, H., Zhang, X., Li, M., et al. (2015a). Sexual differences in long-term effects of prenatal chronic mild stress on anxiety-like behavior and stress-induced regional glutamate receptor expression in rat offspring. *International Journal of Developmental Neuroscience, 41*, 80–91.

Wang, Y., Ma, Y., Hu, J., Cheng, W., Jiang, H., Zhang, X., et al. (2015b). Prenatal chronic mild stress induces depression-like behavior and sex-specific changes in regional glutamate receptor expression patterns in adult rats. *Neuroscience, 301*, 363–374.

Weaver, I. C., Cervoni, N., Champagne, F. A., D'Alessio, A. C., Sharma, S., Seckl, J. R., et al. (2004). Epigenetic programming by maternal behavior. *Nature Neuroscience, 7*(8), 847–854.

Weaver, I. C., Champagne, F. A., Brown, S. E., Dymov, S., Sharma, S., Meaney, M. J., et al. (2005). Reversal of maternal programming of stress responses in adult offspring through methyl supplementation: Altering epigenetic marking later in life. *Journal of Neuroscience, 25*(47), 11045–11054.

Weinstock, M. (2005). The potential influence of maternal stress hormones on development and mental health of the offspring. *Brain, Behavior, and Immunity, 19*(4), 296–308.

Weinstock, M. (2008). The long-term behavioural consequences of prenatal stress. *Neuroscience & Biobehavioral Reviews, 32*(6), 1073–1086.

Weinstock, M. (2011). Sex-dependent changes induced by prenatal stress in cortical and hippocampal morphology and behaviour in rats: An update. *Stress, 14*(6), 604–613.

Weisenberger, D. J., Campan, M., Long, T. I., Kim, M., Woods, C., Fiala, E., et al. (2005). Analysis of repetitive element DNA methylation by MethyLight. *Nucleic Acids Research, 33*(21), 6823–6836.

Welberg, L. A., Seckl, J. R., & Holmes, M. C. (2001). Prenatal glucocorticoid programming of brain corticosteroid receptors and corticotrophin-releasing hormone: Possible implications for behaviour. *Neuroscience, 104*(1), 71–79.

Wetmur, J. G., Kaya, A. H., Plewinska, M., & Desnick, R. J. (1991). Molecular characterization of the human delta-aminolevulinate dehydratase 2 (ALAD2) allele: Implications for molecular screening of individuals for genetic susceptibility to lead poisoning. *American Journal of Human Genetics, 49*(4), 757–763.

Wilson, C. A., Schade, R., & Terry, A. V., Jr. (2012). Variable prenatal stress results in impairments of sustained attention and inhibitory response control in a 5-choice serial reaction time task in rats. *Neuroscience, 218*, 126–137.

Wright, R. O., Schwartz, J., Wright, R. J., Bollati, V., Tarantini, L., Park, S. K., et al. (2010). Biomarkers of lead exposure and DNA methylation within retrotransposons. *Environmental Health Perspectives, 118*(6), 790–795.

Wu, J., Basha, M. R., Brock, B., Cox, D. P., Cardozo-Pelaez, F., McPherson, C. A., et al. (2008). Alzheimer's disease (AD)-like pathology in aged monkeys after infantile exposure to environmental metal lead (Pb): Evidence for a developmental origin and environmental link for AD. *Journal of Neuroscience, 28*(1), 3–9.

Xu, J., Yang, B., Yan, C., Hu, H., Cai, S., Liu, J., et al. (2013). Effects of duration and timing of prenatal stress on hippocampal myelination and synaptophysin expression. *Brain Research, 1527*, 57–66.

Yang, A. S., Estecio, M. R., Doshi, K., Kondo, Y., Tajara, E. H., & Issa, J. P. (2004). A simple method for estimating global DNA methylation using bisulfite PCR of repetitive DNA elements. *Nucleic Acids Research, 32*(3), e38.

Yao, Y., Robinson, A. M., Zucchi, F. C., Robbins, J. C., Babenko, O., Kovalchuk, O., et al. (2014). Ancestral exposure to stress epigenetically programs preterm birth risk and adverse maternal and newborn outcomes. *BMC Medicine, 12*, 121.

Zhu, Z. Z., Hou, L., Bollati, V., Tarantini, L., Marinelli, B., Cantone, L., et al. (2012). Predictors of global methylation levels in blood DNA of healthy subjects: A combined analysis. *International Journal of Epidemiology, 41*(1), 126–139.

Zucchi, F. C., Yao, Y., Ward, I. D., Ilnytskyy, Y., Olson, D. M., Benzies, K., et al. (2013). Maternal stress induces epigenetic signatures of psychiatric and neurological diseases in the offspring. *PLoS One, 8*(2), e56967.

Chapter 11
Posttraumatic Stress Disorder: Neurological, Genetic, and Epigenetic Bases

Jennifer S. Lewis

11.1 Evolving Definition of Posttraumatic Stress Disorder

11.1.1 Earliest References

Even before the condition was named, references in early Greek literature such as the Iliad described symptoms associated with posttraumatic stress disorder (PTSD; Turnbull 1998). Moreover, Shakespeare described symptoms of PTSD in a Midsummer Night's Dream (Bennet 2011). When describing large-scale fires such as the 1666 Great Fire of London, references to the fear and mental anguish related to PTSD appear in literature of that time (Turnbull 1998). As these references indicate, the symptoms of what ultimately became known as PTSD existed long before a term was coined to clearly reference these symptoms.

11.1.2 Type of Incidents

Notably, the condition continued to increase in incidence particularly following periods of war (Iribarren et al. 2005). During the Civil War, the term DaCosta's Syndrome described a disorder whose symptoms mirror those of PTSD (DaCosta 1871). As the incidence of PTSD conditioned to increased following the Civil War, the environment was set for coining a name for the term (Iribarren et al. 2005). In addition to war, the descriptions related to what would ultimately be coined PTSD seemed to also be used when describing such traumatic experiences such as railway

J.S. Lewis, M.H.A., M.Ed. (✉)
University of Southern Mississippi, 118 College Drive, Hattiesburg, MS 39406, USA
e-mail: Jennifer.Lewis@usm.edu

D. Hollar (ed.), *Epigenetics, the Environment, and Children's Health Across Lifespans*, DOI 10.1007/978-3-319-25325-1_11

accidents, which in addition to war constituted the typical traumatic experiences occurring within the late nineteenth as well as the early twentieth century (Turnbull 1998). Currently, the following represent examples of situations that give rise to the level of trauma that culminates in the development of PTSD: military combat, sexual assault, natural and human-created disasters, and accidents of a serious nature (Ouimette et al. 1998).

11.1.3 An Evolving Definition

In 1870, ABR Myers labeled the condition posttraumatic stress disorder (PTSD) as fear. In detailing the symptoms of the disorder, Myers described it as an emotional state that manifested in persons who surrendered to heightened emotions such as fear (Iribarren et al. 2005). By the time of the First World War, the public understood that trauma potentially led to symptoms that resulted from physical forces exerting force on the central nervous system resulting in a temporary disconnection from the normal behavior patterns (Turnbull 1998). Upon the release of the International Classification of Diseases ninth edition in 1978, the edition included a listing of the symptoms of PTSD (Turnbull 1998). Turnbull contends that these references to the symptoms of PTSD within the 1978 edition marked the precursors to the changes to the 1980 edition of the Diagnostic and Statistical Manual of Mental Disorders-III-R [Table 11.1; DSM-III-R (Turnbull 1998)].

Within DSM-III, the American Psychiatric Association (APA) defined PTSD as an anxiety disorder that occurs after an event that produces psychological distress of level that exceeds the range typically experienced (American Psychiatric Association 1987). Notably, the 1994 APA definition of PTSD characterizes the condition as one that involves experiencing intrusive thoughts, often manifests itself in flashbacks or night terrors, and results in avoidance as well as hyperarousal all of which continue for over a month following a traumatic event (Domschke 2012). Moreover, the traumatic event involves an extreme event in which death occurs, a belief exists that death will occur, the

Table 11.1 Evolution of PTSD

1870	ABR Meyers labels condition posttraumatic stress disorder and characterizes it as an emotional state manifesting in those who surrender to heightened emotions such as fear.
1978	ICD 9th ed. References symptoms of posttraumatic stress disorder.
1987	DSM-III defines PTSD as an anxiety disorder occurring after an event producing psychological distress exceeding range typically experienced.
1994 DSM	Edition notes that PTSD results from intrusive thoughts, often manifested with flashbacks or night terrors and resulting in avoidance as well as hyperarousal extending for more than 1 month following a traumatic event that involves death, perceived risk of death, serious injury, or threats to an individual's physical integrity.

occurrence of a serious injury, or a physical threat to the individual's physical integrity (Domschke 2012). Ultimately, the 1994 definition also expands the definition of PTSD to include those who witnessed rather than directly experienced a traumatic event (Turnbull 1998). Such an expansion reflects research confirming an increased incidence of PTSD among rescue workers (Alexander and Wells 1991).

Turnbull notes that each subsequent edition of the DMS reflects an updated definition based on incorporating research findings (Turnbull 1998). Ultimately, such findings and the subsequent updates in the definition reflect the findings of research done on those involved in wars such as World War II and Vietnam (Turnbull 1998). Only as of the 1980s did research begin to focus on other types of trauma such as rape (Turnbull 1998). Following the expansion of the scope of PTSD research, the data from the research supported considering PTSD as an appropriate category applicable to a range of group suffering from trauma including children victimized by abuse, individuals seeking refuge from their home country as well as those exploited by violence within their relationships (Turnbull 1998). Examples of PTSD include terrorist attack, violent crime, military combat, abuse, dire accidents, natural disasters, or personal assaults of a violent nature (Iribarren et al. 2005).

11.1.4 Overview of PTSD

With an understanding of the definition of PTSD, one can move toward examining how the condition actually manifests itself as well as why. When an event rises to the level of being considered traumatic, it is by definition significantly stressful (Iribarren et al. 2005). Typically, individuals are innately resilient to stress thereby resulting in their body's automatic ability to restrict increases in corticotrophin-releasing hormones and corticosterone (CORT), which are induced by stress (Talilaz et al. 2011). Although a stressful event instigates an array of outcomes of a psychologically emotional nature only in the gravest form does the response to stress rise to the level of being diagnosed as a psychiatric disorder occurring because of traumatic events (Iribarren et al. 2005). Before examining what distinguishes whether an individual is able to cope with stress that rises to the level of trauma or not, one needs an understanding of the specific components involved and an understanding of how the response to stress occurs.

11.2 The Role of the Brain and Its Constituent Limbic System in Stress Response

When an individual perceives a disturbance either internal or external that moves them away from homeostasis, the brain processes the disturbance as stress (Joëls 2007). Upon the perception of stress, both the brain stem and the areas of the limbic system become active (Joëls 2007). As a result of the activation, the information relaying the occurrence of stress moves through the hypothalamus to activate two systems: the sympathoadrenomedullary system as well as the hypothalamo pituitary system that increases levels of adrenaline and cortisol with adrenalin precipitating noradrenalin release (Joëls 2007). Researchers agree that cortisol represents a means of providing energy for the body during the stress response as well as stopping the production of macrophages along with other white cells typically triggered in response to inflammation thereby allowing the body to cope with stress (Greenstein and Greenstein 2007). Meanwhile, adrenaline or epinephrine represents one of the medullary catecholamines, which include norepinephrine (Everly and Lating 2013). The medullary catecholamines result in the overall increase of nerve activity and extend the sympathetic response of nerves (Everly and Lating 2013).

11.2.1 The Brain: Structures and Functions

In terms of overall structure, the brain consists of two hemispheres (Glick 2011). Each hemisphere controls the opposite side of the body (Glick 2011). Located around each hemisphere is cortex and contains within it neurons or nerve cells (Glick 2011). In the posterior of the brain, the cerebellum contains more neurons than other parts of the brain and facilitates motor as well as mental proficiency (Glick 2011). Within the middle of the brain, the brain stem connects the spinal cord to the brain and controls automatic functioning of mechanisms such as breathing and blood pressure (Glick 2011).

11.2.2 Frontal Lobes

In terms of the lobes of the brain, several lobes exist including the frontal lobe, the parietal lobe, the occipital lobe, and the temporal lobe (Glick 2011). The frontal lobes are located at the front of the head immediately behind the forehead and represent about a third of the total size of the cortex (Glick 2011). Frontal lobes control the characteristics that together create an individual's personality (Miller and Cummings 2007). Notably, distinct areas of the frontal lobe control specific aspects of behavior (Miller and Cummings 2007). The medial frontal area controls motivation (Miller and Cummings 2007). Additionally, the orbitofrontal area manages the rule of social convention (Miller and Cummings 2007). Moreover, the dorsolateral prefrontal area facilitates the planning of which actions to take and represents an ability akin to executive functioning (Miller and Cummings 2007). Within specific regions of the frontal lobes, the ability to speak, learn languages, executive abilities, basic neurological functions, and the ability to get along with others on a social level are all handled by the frontal lobes (Miller and Cummings 2007).

11.2.3 Prefrontal Cortex

Additionally, the frontal lobes include within them the prefrontal cortex (Glick 2011). In terms of a more specific location, the prefrontal cortex represents the part of the cerebral cortex receiving projections coming from the thalamus' mediodorsal nucleus (Fuster 2015). In terms of functioning, the prefrontal cortex manages executive functions including solving problems, critical thinking, and overall creativity (Glick 2011). Moreover, the prefrontal cortex serves to store short-term memory (Kane and Engle 2002).

11.2.4 Medial Prefrontal Cortex

References to the medial prefrontal cortex indicate the middle wall of the anterior frontal lobes (Koenigs and Grafman 2009). Composed of the anterior cingulate cortex, subcallosal cortex, and medial frontal gyrus, the medial prefrontal cortex functions in extinguishing the conditioning of fear and retaining the conditioning as extinguished (Milad and Quirk 2002; Morgan et al. 1993). Moreover, the medial prefrontal cortex communicates with the amygdala through densely packed white matter (Koenigs and Grafman 2009).

11.2.5 Parietal Lobes

Located across the top portion of the forehead in the middle lobe of the cerebrum, the parietal lobe is comprised of the postcentral gyrus and superior and inferior parietal lobes (Jacobson and Marcus 2008). In terms of functionality, the parietal lobe furthers integration of sensory information about the environment including processing information visually and spatially; functionally, this allows individuals to understand their physical space in terms of things surrounding them (Glick 2011). Functionally, what function the parietal lobe controls depends on which side the lobe is located with each controlling distinct functions (Jacobson and Marcus 2008). Hence, the left parietal lobe controls spatial language that allows an individual to communicate about spatial concepts while the right parietal lobe controls body imagery (Jacobson and Marcus 2008).

11.2.6 Occipital Lobes

Located at the back of the head, the occipital lobes represents the area of the brain responsible for processing what an individual sees including the ability to distinguish between shapes and colors (Glick 2011). Within the occipital lobe, several distinct areas exist including the striate or primary visual cortex and the extrastriate, which is also known as the association visual cortex (Hale and Fiorello 2004).

Essentially, the primary visual cortex obtains the information by way of the visual pathways and then transmits that information to the association visual cortex to have it begin interpretation (Hale and Fiorello 2004).

11.2.7 Temporal Lobes

Located near the temples, the temporal lobes manage language, process what we hear, and operationalizes memory (Glick 2011). Overall, the temporal lobe facilitates memory movement from short-term memory to working memory (Glick 2011). Moreover, the temporal lobes control the ability to remember an event known as episodic memory along with the memory related to remembering facts and figures known as declarative memory (Glick 2011). In terms of the left temporal lobe, the lobe typically facilitates verbalization and comprehension of language (Glick 2011). In terms of the right temporal lobe, the lobe facilitates the understanding of vocal tone and overall vocal subtleties (Glick 2011).

11.2.8 Pituitary Gland

Located immediately beneath the hypothalamus, the pituitary gland controls other endocrine organs (Schoenberg and Scott 2011). Notably, neurons from the hypothalamus extend directly into the pituitary gland, and the hypophyseal nerve tract forms from 100,000 axons (Nussey and Whitehead 2001). The pituitary gland works with the hypothalamus and the adrenal gland to control the body's response to stress (Adinoff et al. 1998). Specifically, the pituitary gland represents the location where CRH (corticotropin-releasing hormone) and arginine vasopressin (AVP) work together to achieve the release of adrenocorticotropic hormone (ACTH), which upon reaching the adrenal cortex works to achieve cortisol secretion (Adinoff et al. 1998).

11.2.9 HPA Axis

In response to stress, both the sympathetic nervous system and the HPA (hypothalamo-pituitary-adrenal) axis become active (Stratakis and Chrousos 2006). Along with the adrenal and hypothalamus, research indicates that the hypothalamus–pituitary–adrenal (HPA) axis plays a role in mood disorders (Schoenberg and Scott 2011). Upon activation of the HPA axis, the neurons located in the parvocellular division of the paraventricular nucleus (PVN) regulate the secretion of peptide corticotropin-releasing hormone (CRH) (Al'Absi 2007). Once the CRH peptide releases into the adenohypophyseal portal circulation, CRH interacts with receptors on the anterior pituitary corticotropin, thereby synthesizing proopiomelanocortin (POMC), a forerunner protein, broken down using water into smaller active peptides including beta endorphin and adrenocorticotropin (ACTH)

(Al'Absi 2007). Once the ACTH reaches the adrenal gland through circulation, adrenocorticosteroids such as glucocorticoids and cortisol, which are secreted from the adrenal cortex, are formed (Al'Absi 2007). Notably, the HPA axis works differently in adults who were the victims of trauma in childhood (Heim et al. 2008).

11.2.10 Cellular Brain

Understanding how the brain functions requires an understanding of the cells within the brain or cellular brain, which represent all the cells of the brain (Glick 2011). Different kinds of brain cells exist including neurons and glia (Glick 2011). While neurons interact via electrochemical means, the glia supports the structure and care for the neurons (Glick 2011). Moreover, neurons represent self-contained units, of which 100 billion exist within the brain, comprised of the cell body, dendrites, and the axon (Glick 2011). In terms of functionality, the dendrites accept incoming information, and the axon transmits information outward (Glick 2011). To facilitate enhanced processing speed, myelin wraps around the axon and also serves as a means of insulation (Glick 2011).

11.2.11 The Limbic System: Structures and Functions

Although there is no agreement in terms of the structures that are parts of the limbic system, certain regions of the brain are considered part of the limbic system including the limbic cortex, the hippocampal formation, amygdala, septal area, and hypothalamus (RajMohan and Mohandes 2007). Together, these regions work to control emotion (RajMohan and Mohandes 2007).

11.2.12 Limbic Cortex

In terms of location, the limbic cortex is located by the margin also known as the limbus of the cortical mantle (Ravoso and Dagosto 2007). The limbic cortex consists of the cingulate gyrus and the parahippocampal gyrus (RajMohan and Mohandes 2007). Research has shown that the cingulate cortex controls the processing of emotional and social information (Hadland et al. 2003).

11.2.13 Cingulate Gyrus

In terms of location, the cingulate gyrus is located immediately above the middle of the surface of the ventromedial prefrontal cortex (Lovallo 2005). The cingulate gyrus works with the prefrontal cortex in formulating thoughts necessary for the individual to select a particular choice (Lovallo 2005).

11.2.14 Hippocampus

In terms of location, the hippocampus is located within the medial temporal lobe (Purves et al. 2001). The hippocampus consists of two parts the Ammon's horn and the dentate gyrus (Andersen et al. 2007). Scientists typically define the hippocampus in terms of the CA fields (CA3, CA2, and CA1) identified by Lorente de No (Andersen et al. 2007). Interestingly, stimulating the hippocampus results in changes both respiratory and cardiovascular in nature, as well as changes in movement (Jacobson and Marcus 2008).

11.2.15 Hippocampal Formation

Located within the temporal lobe, the hippocampal formation functions primarily to facilitate the creation of memories and spatial navigation (Schoenberg and Scott 2011). Included in the structures of the hippocampal formation are the dentate gyrus, subiculum (e.g., subicular complex), presubiculum, parasubiculum, and entorhinal cortex (Andersen et al. 2007; RajMohan and Mohandes 2007).

11.2.16 Subicular Complex

Notably, the subicular complex is composed of the presubiculum, the parasubiculum, and the subiculum (RajMohan and Mohandes 2007). The subiculum represents the transitional area located between the entorhinal cortex and the hippocampus (RajMohan and Mohandes 2007). Research shows that the subicular complex has a role in a variety of neurological diseases including Alzheimer and epilepsy (Ding 2013). Moreover, it also serves as an important function within the memory system that occurs in the medial temporal area (Ding 2013).

11.2.17 Dentate Gyrus

The dentate gyrus functions as part of the hippocampus and represents one of the few areas for the formation of new neurons (e.g., neurogenesis), which results in improvements to spatial memory attributable to this structure (Nakashiba et al. 2012). In terms of functionality, the dentate gyrus has several functions including helping to form new memories and making decisions based on spatial perception (Kee et al. 2007). Researchers have hypothesized that the process of neurogenesis may be part of the processes aimed at controlling the symptoms related to stress as well as depression (Surget et al. 2011). Surget et al. noted that their research found that mild stress both reduced neurogenesis and decreased the body's ability to respond to the primary stress hormone (2011).

11.2.18 Amygdala

Within the anterior lobe of each hemisphere of the brain, the amygdala constitutes a collection of nuclei that projects to those areas located in the brain stem as well as in regions of the hypothalamus responsible for controlling how the body reacts in response to emotion (Koenigs and Grafman 2009). On the medial side of the temporal lobe in front of the hippocampi, somatosensory areas send the amygdala information for processing emotional conditions and references memory with emotional functioning (Schoenberg and Scott 2011). In essence, the amygdala serves as part of a responsive system to emotions providing a means for avoiding danger by communicating information between the prefrontal cortex and the hypothalamus (Pastorino and Doyle-Portillo 2013; RajMohan and Mohandes 2007).

11.2.19 Hypothalamus

Located near the base of the brain, the hypothalamus serves as the means of maintaining homeostasis or the equilibrium of the systems of the body (Schoenberg and Scott 2011). In order to achieve homeostasis, the hypothalamus links the nervous and endocrine systems, both of which regulate proper body functioning (Schoenberg and Scott 2011). In addition to controlling the autonomic nervous system to regulate breathing, blood pressure, and heart rate along with temperature, the hypothalamus also uses chemicals sent to the pituitary gland, which is located immediately below the hypothalamus, to increase or decrease hormone release (Schoenberg and Scott 2011). Among the processes regulated by the hypothalamus are blood sugar levels, sleep cycles, thirst, hunger, 24-h rhythms, energy levels, and emotions (Schoenberg and Scott 2011). Moreover, the hypothalamus regulates fear and rage reaction (Schoenberg and Scott 2011).

11.2.20 Septal Area

Behind the commissure, the septal area is comprised of gray matter and connects to the hippocampus by way of the fornix (RajMohan and Mohandes 2007). Despite the existence of research focused on the area, the function of this area remains unclear (Melillo and Leisman 2004). However, the term septal syndrome describes a condition related to the damage of the area, which manifests in terms of overreaction to a majority of stimuli of the environment resulting in rage, as well as changes in sexual, eating, and hydration patterns of behavior (Melillo and Leisman 2004). Other studies indicate that septal lesions impair allocentric functionality (Noonan, Penque, Axelrod 1996). Hence, in the Noon, Penque & Axelrod study, the rats being tested who developed septal lesions lacked the ability to use their internal mapping system as a means of location (1996). Generally, most of the functions of the septum appear to be connected to the hippocampus and secondarily to the amygdala and hypothalamus (Melillo and Leisman 2004).

11.3 Nervous System: Structures and Functions

To place the nervous system into context, one needs to understand that the nervous system represents the broad network of the neurons within the human body (Pastorino and Doyle-Portillo 2013). Within the overall nervous system, two subsystems exist: the central nervous system (CNS) and the peripheral nervous system (PNS) (Brodal 2010). In terms of function, the nervous system controls how our body functions and how we behave (Pastorino and Doyle-Portillo 2013). Understanding its function facilitates ultimately understanding its role in PTSD.

11.3.1 Central Nervous System

The brain and the spinal cord comprise the components of the central nervous system (Brodal 2010). The central nervous system uses neurons to quickly transmit electrical charges known as nerve impulses within the system or between itself and the organs in other body systems (Brodal 2010). To convey chemical signals from one neuron to another, neurotransmitters serve as the facilitating substance released when a neuron reaches a synapse or the area in between the axon and the next neuron (Brodal 2010). Hence, the central nervous system receives information from the sense organs and transmits it throughout the body and represents the first step in the process of responding to stress (Van Der Kolk and Saporta 1991).

11.3.2 Peripheral Nervous System

11.3.2.1 Parts of the Peripheral Nervous System

Meanwhile, the peripheral nervous system represents all the parts of the nervous system that do not constitute a part of the central nervous system (Pastorino and Doyle-Portillo 2013). Notably, two distinct systems comprise the peripheral nervous system: the somatic nervous system and the autonomic nervous system (Pastorino and Doyle-Portillo 2013). The somatic nervous system controls actions that are voluntary, which require conscious thought in order to occur, along with controlling how the senses function (Pastorino and Doyle-Portillo 2013). Meanwhile, the autonomic nervous system controls organ functions, which represent an involuntary process (Pastorino and Doyle-Portillo 2013). Notably, the autonomic nervous system divides into two additional systems known as parasympathetic nervous system and the sympathetic nervous system (Pastorino and Doyle-Portillo 2013). In terms of function, the parasympathetic nervous system manages how organs function when an individual is calm (Pastorino and Doyle-Portillo 2013). Conversely, the sympathetic nervous system controls how organs function during times of stress by controlling the mechanism that determines if an individual reacts to stress by fighting or fleeing (flight) (Pastorino and Doyle-Portillo 2013).

11.3.2.2 Purpose of the Peripheral Nervous System

In terms of overall purpose, the peripheral nervous system performs two functions: transmitting updates to the central nervous system and following directions provided to it by the central nervous system (Pastorino and Doyle-Portillo 2013). Within the peripheral nervous system, the sensory neurons focus on carrying visual information to the brain and provide the central nervous system with updates on ongoing occurrences external to the body, including what is seen and heard as well as relaying the status of internal functions such as pain (Pastorino and Doyle-Portillo 2013). After the central nervous system receives the updates, the information moves across interneurons or nerve cells focused on only participating in processing information on a local basis (Purves et al. 2001). Subsequently, the interneurons focus on examining information from the brain for processing and allow the brain to process the information (Pastorino and Doyle-Portillo 2013). In terms of the second function, the motor neurons, which represent a component of the motor pathways aimed at transporting information out from the brain within the peripheral nervous system, direct information to muscles from the central nervous system (Pastorino and Doyle-Portillo 2013).

11.3.3 The Nervous System Response to Stress

Activation of the sympathetic nervous system occurs after an individual recognizes a stressor (Pastorino and Doyle-Portillo 2013). Notably, what constitutes stress includes not only events that threaten but also those events that an individual perceives as threatening (Weinstock 2005). Once the level of stress that the organism can adapt to is exceeded, the stress response occurs (Cohen et al. 2002). To respond to the stressor and allow the individual to cope properly with stress in order to reestablish homeostasis, several systems in the body work together (Al'Absi 2007). To facilitate the coping mechanisms responsive to stress, the sympathetic nervous system functions with the endocrine system that regulates hormones (Pastorino and Doyle-Portillo 2013). As a result of this collaboration, the adrenal gland releases adrenalin and corticosteroids (Pastorino and Doyle-Portillo 2013).

Once released, the body transports both the adrenalin and corticosteroid hormones to each of the major organs (e.g., liver, heart, and lungs) in order to prepare the body to either fight or flee (i.e., flight) (Pastorino and Doyle-Portillo 2013). As a result of the influx of these two hormones, individuals experience increased heart rate, blood pressure, lung expansion, and the release of sugar by the liver to provide the energy needed for a response (Pastorino and Doyle-Portillo 2013). Meanwhile, the brain engages the pituitary gland in order to have it release endorphins to naturally diminish pain (Pastorino and Doyle-Portillo 2013). Notably, the use of resources continues for the duration of the individual's exposure to the stressor, and upon elimination of the stressor, the sympathetic nervous system discontinues release of stress hormones thereby allowing the body to return to parasympathetic mode, which involves calm as well as relaxation (Pastorino and Doyle-Portillo 2013).

11.3.3.1 General Adaptation Syndrome

In terms of how stress occurs, Hans Selye defined the three phases known as general adaptation syndrome that occur following the identification of a stressor (Selye 1950). The first phase known as the alarm reaction represents what transpires as the body reacts by releasing chemicals in response to a stressor (Selye 1950). However, as a result of the release of chemicals, the resistance of the body decreases (Selye 1950). Subsequently, the resistance stage occurs and enables the body to reinforce added resources to enable the individual to cope with the stressor (Selye 1950). Finally, upon persistence of stressors, the exhaustion stage occurs, and as a result of diminished resources, functioning occurs at less than optimal levels (Selye 1950).

11.3.4 Cortisol Level Changes Attributable to Stress

As a whole, stress disorders represent conditions that create pressure within the system of a person that exceed their coping abilities, including individual abilities, family support, group support, or community support (Lazarus and Folkman 1984). Such pressures result from stress that represents a risk to the body's ability to remain at homeostasis or in balance (De Kloet et al. 2006). Stress changes the level of many hormones including glucocorticoids, catecholamines, growth hormone, and prolactin (Ranabir and Reetu 2011).

Cortisol represents the primary glucocorticoid involved in the stress response (Ranabir and Reetu 2011). Some studies indicate that increased levels of cortisol result from stress disorders, while others indicate decreased levels of cortisol (Miller et al. 2007). This variation in cortisol levels appears to be attributable to a depletion in cortisol levels in those suffering from long-term stress (Miller et al. 2007)

PTSD represents a condition that remains unique among psychologically based conditions given that the trauma necessitates a stressor to manifest itself (Copeland et al. 2007). Because the response to fear is altered in those with PTSD, researchers initially examined the role played by the brain, chemicals, and hormones relative to fear in order to identify the systems within the individual that function in PTSD development (Yehuda et al. 2011). As a result of these studies, researchers ultimately determined that PTSD did not elicit the elevated levels of cortisol typically associated with stress disorders (Yehuda et al. 2000). After considering the low cortisol levels exhibited in those with PTSD in conjunction with the fact that PTSD only occurs in some of those exposed to trauma, research began to transition toward examining the connection between PTSD and genetics as a means of explaining this variation in susceptibility (Yehuda et al. 2011). Given that long-term exposure to stress results in lower cortisol levels, the levels of cortisol measured in the studies aligns with that which would be expected based on the chronic nature of PTSD.

11.3.5 Allostatic Load

Stress can occur for varying periods of time: short and long periods. Since homeostasis occurs within a very narrow range of levels, changes that exceed those levels require allostatic regulation in order to restore homeostasis (McEwen and Stellar 1993). Notably, heart damage and decreased functioning of the immune system occur as a result of over exposure to high levels of adrenalin and corticosteroid across a protracted period of time due to chronic stress (Pastorino and Doyle-Portillo 2013). Such damage represents the allostatic load that occurs due to exposure and fluctuation over long periods (McEwen and Seeman 1999). Though initially the release of mediators (e.g., cortisol, catecholamines, and DHEA-S) results in positive outcomes, long-term exposure to these same mediators can result in disease (McEwen and Seeman 1999; McEwen and Stellar 1993). Hence, research shows that protracted periods of stress weaken the immune system thereby placing the body at an increased risk of falling prey to disease.

11.3.6 Toxic Stress and Children

Events involving stress can be tolerated or even positive in nature (National Scientific Council on the Developing Child 2014). What constitutes trauma varies across different groups (Nelson 1996). However, a child's health development depends on the natural occurrence of stress (National Scientific Council on the Developing Child 2014).

Different levels of stress exist including positive, tolerable, and toxic stress (National Scientific Council on the Developing Child 2014). Positive stress results when stress occurs in moderation and represents a normal component of everyday life (National Scientific Council on the Developing Child 2014). During positive stress, brief changes deviating away from homeostasis occur such as a change in the stress hormone (National Scientific Council on the Developing Child 2014). Meanwhile, the National Scientific Council on the Developing Child defines tolerable stress as stress that could negatively affect brain architecture but due to the fact that it occurs over short periods of time fails to rise to the level necessary to create long-term negative effects (2014). According to the American Academy of Pediatrics, toxic stress in children occurs when stress is severe and occurs frequently thereby extending over the course of long periods during which the stress response is activated without adequate support from an adult (Garner et al. 2011). Typical scenarios for the occurrence of toxic stress include physical or emotional abuse, neglect that occurs chronically, being exposed to violence, substance or mental abuse perpetrated by their caregiver, or economic hardship (Shonkoff et al. 2009). As a result of toxic stress, how the brain and other organ systems develop may change thereby increasing the potential for stress-related disease and cognitive impairment that lasts in adulthood (Johnson and Thompson 2007). Moreover, those children who lack parental support either by way of being a ward of the state or harmful treatment at the hand of their families become at risk for disorders relative to emotional along with attention regulation (Loman et al. 2010). Research shows that chronic abuse results in changes in the brain, especially early during sensitive development periods (National Scientific Council on the Developing Child 2014). Such changes exponentially increase the potential for the development of physical and mental illnesses attributable to stress (Shonkoff et al. 2009). A discussion of what research indicates in terms of the extent of such changes is examined within the section labeled neurobiology and PTSD.

11.4 Overall Prevalence of Posttraumatic Stress Disorder

As a greater awareness of those afflicted with the condition occurred, researchers began examining demographics regarding the prevalence of the condition (Jayawickreme et al. 2012). Within a context of many studies trying to identify the root cause of PTSD, researchers examined large populations to see if a pattern of prevalence served as a means of explaining why the condition occurs (Jayawickreme et al. 2012). As a result, references to prevalence exist within many studies (Jayawickreme et al. 2012) As one such study noted, 75 % of the American population experienced a traumatic situation in their lifetime (Jayawickreme et al. 2012). Yet, the same study noted that only 6.7 % of the 75 % developed posttraumatic stress disorder (Jayawickreme et al. 2012).

Significant information about prevalence stemmed from the US National Comorbidity Survey that occurred from February 2001 to April 2003 (Kessler et al. 2004). According to data from the US National Comorbidity Survey, an estimated 7.7 million Americans aged 18 and above suffer from PTSD (Kessler et al. 2005b). Moreover, the National Comorbidity Survey also indicated an average onset of PTSD in those aged 23 (Kessler et al. 2005a). Overall, 1 in 12 individuals suffer from PTSD within their lifetime (Kessler et al. 1995). Hence, in excess of one-third of those suffering from PTSD do not recover despite the passage of several years (Kessler et al. 1995).

11.4.1 Comparing Prevalence Between the United States and Germany

When the National Institute of Mental Health conducted the last National Comorbidity Survey Replication (NCS-R) in 2005, researchers determined that PTSD occurred to 3.5 % of the US population (Kessler et al. 2005a). Moreover, those researchers estimated lifetime PTSD prevalence at 7 % (Kessler et al. 2005b). However, a similar study of data from Germany indicated a much lower prevalence of 2.3 % (Maercker et al. 2008). Rather than trying to explain this significant difference in data relative to the same condition resulting from the study population being from different countries, researchers instead attributed the variation to other factors (Schmidt et al. 2013). One study contended that the prevalence in PTSD represents the difference within the groups of persons being studied (Schmidt et al. 2013). Meanwhile, other researchers identified specific factors affecting those suffering from PTSD as contributors to increased prevalence (Mollica et al. 1998). The number of traumatic events experienced represents one such event contributing to an increased prevalence of PTSD (Mollica et al. 1998). Similarly, the intensity of the stress involved in the traumatic incidence contributes to an increased prevalence of PTSD occurring (Mollica et al. 1998).

11.4.2 Prevalence by Group

With regard to the prevalence within specific groups, several studies occurred aiming to identify the degree to which certain groups developed PTSD. Research examined a variety of groups seeking to determine if any one group had a higher prevalence overall. Moreover, researchers sought to identify the exact prevalence that applied to each group as delineated more specifically in the sections that follow.

11.4.3 Prevalence in Children and Adolescents

Research conducted relative to children suffering from PTSD indicated that approximately 5 million children in the United States face exposure to a traumatic stressor and of those in excess of 30% develop PTSD, which manifests across demographic lines (Perry and Azad 1999). Moreover, responses to the National Comorbidity Survey-Adolescent Supplement NC-A findings indicate a 22.2 % prevalence of a mental disorder (Merikangas et al. 2010). However, a study of prevalence in adolescents compared to children (i.e. those under age 13) found that adolescents experienced a higher prevalence of PTSD than they experienced major depression following a trauma when compared to the children in the study (Cuffe et al. 1998).

11.4.4 Prevalence in Veterans

Research on veterans varied in prevalence. A study inclusive of veterans from Korea and Vietnam resulted in a determination of prevalence of PTSD at a rate of 24 % overall (Blake et al. 2006). A study of Gulf War veteran resulted in a finding of 3.3 % prevalence of minimal PTSD as opposed to 6.1 % moderate PTSD (Kang et al. 2003). Moreover, Kang et al. also found increasing levels of high level PTSD along with symptoms similar to chronic fatigue in those experiencing multiple military service tours involving combat with prevalence within those groups ranging from 7 % for those with no combat exposure to 22.6 % for those who saw someone die during their tour (Kang et al. 2003). Although this is less than those veterans in the prior study, such a small decrease may be the result of the addition of chronic fatigue as a compound condition.

11.4.5 Prevalence Based on Gender

In terms of gender, researchers have examined the degree to which PTSD prevalence differs based on the gender of the individual (Pastorino and Doyle-Portillo 2013). As a result of such research, Tolin and Foa reviewed 25 years of research and identified a pattern indicative of a higher level of females developing PTSD following trauma (2006). In theorizing on the explanation behind these phenomena, Tolin and Foa observed that this may be the result of the types of trauma women experience, which may be more severe in nature (2006). Moreover, they noted that prior research had shown an increased prevalence of PTSD if rape was involved in the traumatic experience (Kessler et al. 1995). Similarly, higher rates of PTSD occurred when combat was involved in the traumatic situation (Weiss et al. 1992). Based on these increased incidence of PTSD in incidents involving both rape and combat, Tolin and Foa proposed that in both situations a higher number of female participants might be involved (2006).

11.4.5.1 Increased Prevalence in Women

To understand this increased prevalence of women developing PTSD despite men being exposed to trauma to a higher degree, researchers examined a variety of potential factors (Tolin and Foa 2006). One study examined sexual abuse either in childhood or as an adult (Tolin and Foa 2006). However, data indicated that men remained more likely to experience an event of a traumatic nature (Tolin and Foa 2006). When comparing both genders relative to the same types of trauma, women continued to experience higher rates of PTSD (Tolin and Foa 2006).

Research performed also found retention of more vivid memories in women if the event was a negative emotional event (Seidlitz and Diener 1998). Moreover, research also indicated that men do not experience the same degree of emotion as women in the face of PTSD (Tolin and Foa 2006). Hence, there do appear to be biological factors in play based on how the feminine gender reacts that increases their potential for developing PTSD after a traumatic event.

Yet, to date, no definite explanation of what results in higher incidence of PTSD in women have been identified (Tolin and Foa 2006). Rather, a variety of potential explanations exist (Tolin and Foa 2006). Included among the explanation are gender-based predisposition to PTSD and research indicates an overall susceptibility of women to specific mental disorders (Kessler et al. 1993). Clearly, there is more to learn in terms of what factors result in an increased propensity for women to develop PTSD to a larger degree than men even when faced with similar types of traumatic events.

11.4.6 Prevalence of Childhood Trauma

As the data indicates, exposure of children to the stressors typically associated with development of trauma represents a common occurrence (Kaffman 2009). According to the data from the United Nations, sexual victimization occurs to one in four girls and one in five boys (Tolin and Foa 2006). More recent data regarding victimization indicates that girls were more likely to be victimized at a rate of 9.5 per 1000 girls as compared to 8.7 per 1000 for boys (U.S. Department of Health and Human Services Children's Bureau -Administration on Children, Youth and Families, Administration for Children and Families 2012).

Moreover, ethnicity also appears to factor into prevalence. Data regarding ethnicity indicates that African American children have a greater chance of being victimized at a rate of 14.2 per 1000 children (U.S. Department of Health and Human Services Children's Bureau -Administration on Children, Youth and Families, Administration for Children and Families 2012). On the other end of the spectrum, Asian children had a 1.7 rate of victimization. Meanwhile, American Indian/Alaskan Natives exhibited a 12.4 rate of victimization, but Pacific Islanders experienced a rate of 8.7. Finally, Hispanics exhibited a rate of 8.4 as compared to non-Hispanic Whites who had a rate of 8. Comparitively, multiracial children exhibited a higher victimization rate of 10.3 (U.S. Department of Health and Human Services Children's Bureau -Administration on Children, Youth and Families, Administration for Children and Families 2012).

According to the U.S. Department of Health and Human Services 2012 report, child protective agencies across the United States receive approximately 3.4 million referrals related to childhood abuse or neglect. In terms of breaking down the incidence into subgroups, the data indicate that 686,000 children experienced maltreatment and 78 % suffered from neglect (U.S. Department of Health and Human Services Children's Bureau -Administration on Children, Youth and Families, Administration for Children and Families 2012). Of those suffering from a type of abuse, 18 % suffered from physical abuse and 9 % suffered from sexual abuse. Meanwhile, the remaining 11 % suffered from other forms of maltreatment such as emotional abuse or threatened abuse, consequences related to a parent's drug abuse, or overall lack of supervision (U.S. Department of Health and Human Services Children's Bureau -Administration on Children, Youth and Families, Administration for Children and Families 2012). Further illustrating the incidence of child abuse, researchers determined that twice as many women suffered child abuse than were raped (Fikelhor et al. 1990).

Although these represent significantly large figures, some researchers contend that they underrepresent the extent of child maltreatment occurring (Finkelhor et al. 2009). Such a contention results from other research including a study that was not conducted by child protection services; in that study, the data revealed a lifetime prevalence rate of one in four children experiencing one of the various types of maltreatment (Finkelhor et al. 2009).

11.4.7 Prevalence Rates of Comorbid Conditions

Some conditions occur to a higher degree in those with PTSD. Comorbidity represents separate clinical conditions that coexist or occur while a patient is being treated for another condition (Feinstein 1970). If an individual develops PTSD, the likelihood of developing an addiction or vice versa appears to be between 35 and 50 % (Brown et al. 1995; Mills et al. 2005). Moreover, those with PTSD typically also have at least one other mental disorder (Brady et al. 2000). In fact, a substantial number of PTSD sufferers have been diagnosed with three or more psychiatric diagnoses (Brady et al. 2000). Typically, the most common types of comorbid conditions include anxiety, substance abuse, and other anxiety-related disorders (Brady et al. 2000).

11.4.8 Diminished Quality of Life and Disability

Researchers documented a correlation between PTSD diminishing quality of life and a disability that lasted long term (Sareen et al. 2007). Another study demonstrated a higher prevalence of disability at a rate that exceeded the rate of PTSD (Marshall et al. 2001). Moreover, a 1997 study examined the relationship between PTSD, functioning, and quality of life in Vietnam veterans and found that those with PTSD suffered from significantly worst quality of life outcomes (Zatrick et al. 1997). Similarly, a 2009 study examined quality of life data on veterans of the Iraq wars and determined that though the PTSD they experienced lasted for a shorter length than Vietnam veterans the diminished functioning and decreased quality of life remained just as likely to occur to them (Schnurr et al. 2009).

11.4.9 Suicidal Behavior

Research also consistently demonstrated a higher incidence of suicidal behavior associated with PTSD (Selaman et al. 2013). One study found that PTSD increased suicidal ideations (Krysinska and Lester 2010). However, the same study noted that their review of 50 research articles failed to prove that research to date indicated a connection between PTSD and completed suicide (Krysinska and Lester 2010). Notably, the researchers highlighted several other factors that may impact the relationship between PTSD and suicidal behavior including the existence of a psychiatric condition prior to the trauma as well as a coexisting depression (Krysinska and Lester 2010).

11.4.10 Psychiatric Conditions

Researchers recognize that comorbidity exists between PTSD and other psychiatric conditions (Vaccarino et al. 2014). In fact, research indicates that those with PTSD typically suffer from one of the following psychiatric disorders: depression, substance abuse, and anxiety (Brady et al. 2000). In fact, a significant percent suffer from three or more diagnoses related to psychiatry (Brady et al. 2000). Meanwhile, other research has focused on determining the degree to which PTSD increases the prevalence of other types of conditions.

11.4.11 Drug Use

Various studies identified a connection between PTSD and drug use (Stine and Kosten 1995; Brown and Wolfe 1994). One study indicated that PTSD increases the prevalence of cannabis use disorder in adolescents: those with PTSD exhibited three times the prevalence of cannabis use disorder (CUD) (Cornelius et al. 2010a). In fact, a study of teens diagnosed with CUD also indicated a correlation between PTSD and CUD (Cornelius et al. 2010b). Moreover, various studies indicated increased prevalence of PTSD among those who suffer from substance abuse (Brown et al. 1995; Stewart 1996). Hence, substance abuse represents a comorbid condition to PTSD.

11.4.12 Insulin Resistance

Other researchers continue to focus on research related to a potential connection between PTSD and insulin resistance Type-2 diabetes (Vaccarino et al. 2014; Boyko et al. 2010). One study focused on determining if a connection exists between those suffering from self-reported diabetes and their development of PTSD (Goodwin and

Davidson 2005). However, the research did not indicate what caused the link between diabetes and PTSD (Goodwin and Davidson 2005).

In addition to studies focused on diabetes, researchers also focused on other conditions. Another study found a correlation between PTSD as a manifestation of trauma and arthritis; yet, it did not find a similar causal relationship between PTSD and diabetes (Norman et al. 2006). Meanwhile, some studies focused on the correlation between obesity and PTSD (Farr et al. 2014). Moreover, other studies indicate a causal connection between PTSD and chronic pain (Defrin et al. 2008). As the variety of conditions that arise following PTSD development indicate, PTSD appears to indicate a decrease in an individual's ability to remain resilient in the face of other conditions. Therefore, those with PTSD appear to be susceptible to a variety of comorbid conditions that further diminish their overall quality of life. Based on the findings already made, additional research needs to be conducted in order to further understand the types of conditions that correlate to PTSD as well as if any other factors determine the degree of severity of the correlated condition.

11.4.13 Impact of PTSD

PTSD directly impacts individuals as well as posing an impact to society as a whole (Iribarren et al. 2005). For instance, PTSD symptoms interfere with daily functioning in terms of social interactions, educational aspirations, and functioning within an occupation (National Institute for Health and Clinical Excellence 2005). Notably, some researchers have determined that PTSD qualifies as a public health concern that in some scenarios may pose a threat to the individual suffering from the (Iribarren et al. 2005). Hence, the level of stress that occurs during a traumatic event traumatizes the psyche resulting in PTSD, while also adversely impacts the individual's overall health in a variety of ways (Norman et al. 2006). Such ways include the development of other chronic conditions as a result of allostatic changes that cause long-term damage (McFarlane 2010). Without examining each of these aspects, one cannot understand the impact of PTSD or the compelling nature of the need behind identifying the cause of this life altering condition becomes clear.

11.4.14 Costs of Care

As a result of the detrimental effects of PTSD to the health of those suffering from the condition, PTSD results in higher use of medical care and costs associated with obtaining such care (Seng et al. 2005). A 1996 study assessing the costs in the United States of anxiety disorders as a whole estimated those costs at $46.6 billion (DuPont et al. 1996). A subsequent 1999 study estimated overall anxiety treatment costs in the United States at $42.3 billion, which represents $23 billion in nonpsychiatric related medical treatment, $13.3 billion related to treatment for psychiatric treatment, $4.1 billion in indirect costs related to work, $1.2 billion of costs related to costs, and $0.8 billion in prescription related costs (Greenberg et al. 1999).

11.4.15 Limited Value of Representative Cost Analysis

However, those costs do not reflect the true costs because the costs only included "direct psychiatric treatment costs, unnecessary medical treatment costs, work performance costs" due to sickness and shorter workdays as well as costs related to death (Kessler and Greenberg 2002). As the overall data indicates, PTSD results in an economic impact relative to costs directly expended in obtaining mental health services along with indirect costs related to public programs to assist with supporting those with disabilities as well as lost wages for those unable to work as a result of their conditions (National Institute of Mental Health 2002).

11.5 Etiology of PTSD

As the prevalence analysis indicates, a disparity exists in terms of the exposure to trauma and the incidence of PTSD. As a result, some researchers focused on identifying the etiology of PTSD from various perspectives including genetic and environmental factors as catalyst for PTSD development. A 1999 study aimed to assess if previous traumatic incidences served as a precursor to subsequent development of PTSD (Breslau et al. 1999). Researchers performed the 1999 study on a sample of 2181 individuals between age 18 and 45 in the Detroit area (Breslau et al. 1999). Analysis of data derived from the 1999 study demonstrated that if even one violent incident of trauma occurred, a higher risk of PTSD was present (Breslau et al. 1999). Moreover, if two or more violent traumatic events occurred, the chance of developing PTSD increased by five times as much as those who only experienced only one incident of trauma (Breslau et al. 1999). Additionally, the data indicated that if an individual experienced a violent trauma, the risk of them developing PTSD from a subsequent trauma was higher regardless of how many years had passed from the original violent trauma (Breslau et al. 1999). Ultimately, these studies all have the common goal of determining why some individuals exhibit resilience to developing PTSD in the face of trauma whereas others appear to lack such resilience, and the data indicates a correlation in terms of preceeding trauma resulting in higher prevalence of PTSD development.

11.5.1 Overview of Inheritance

Inheritance explains the conditions or features that manifest in an individual based on the genetic code of their blood relatives. Understanding inheritance requires an examination of the initial research on inheritance performed by Mendel. Such research highlights how genes function to facilitate the inheritance of certain characteristics. Moreover, understanding how inheritance functions provides a means for putting into context the research on the development of PTSD through inheritance.

Research relative to inheritance traces back to Mendel whose research on large numbers of plants allowed him to identify a connection in terms of factors that impacted inheritance (Sadava et al. 2014). As a result of the observations made following crossbreeding of pea plants, Mendel developed hypotheses relative to what caused traits to be inherited, and then he performed further experiments as a means of testing those hypotheses (Sadava et al. 2014).

11.5.1.1 Blending Theory of Inheritance

Prior to Mendel's discovery, Francis Galton recognized that there were two types of characteristics passed by way of inheritance (Castle 1933). Galton classified those two types of characteristics as those that alternated and those that blended (Castle 1933). The blending theory held that if parents with two different traits were bred, the resulting child exhibited blended characteristics inherited from the parents (Sadava et al. 2014). Alternation occurred in situations such as when each child had different eye colors, each of which represented the eye color of one of the parents (Castle 1933). Children who exhibited height in between the height of each of their parents represent an example of blending (Castle 1933). However, Mendel's discovery of the role of genes in inheritance disproved the blending theory (Castle 1933; Sadava et al. 2014).

11.5.1.2 Dominant and Recessive Genes

Moreover, although one trait might appear in the first child or filial generation, the trait from the parent that did not manifest itself in the first generation could ultimately arise in subsequent generations (Sadava et al. 2014). Based on these findings, Mendel categorized the trait that arose in the first generation as the dominant trait, while the one that did not manifest itself in the first generation he named the recessive trait (Sadava et al. 2014).

11.5.1.3 Law of Segregation

Moreover, he developed the law of segregation to explain how parental traits were passed onto children with alternate gene versions or alleles separating from one another prior to the formation of the gamete then randomly combining during fertilization (Sadava et al. 2014). Today, the scientific community understands that DNA sequences on chromosomes are responsible for facilitating the random combination that Mendel explained as segregation (Sadava et al. 2014).

11.5.1.4 Law of Independent Assortment

Further experimentation led Mendel to develop his second principle known as the law of independent assortment (Sadava et al. 2014). The law of independent assortment explains why alleles of distinct genes group independently of the other alleles during the formation of gametes (Sadava et al. 2014). Notably, the law of independent assortment does not apply if genes are located in close proximity within the same chromosome (Sadava et al. 2014).

11.5.1.5 The Role of Heredity and PTSD

In seeking to explain individual susceptibility to PTSD, some researchers sought to establish that a hereditary connection existed without specifically identifying the specific genetic catalyst responsible (Sack et al. 1995). Instead, these researchers focused on PTSD resulting based on parents of an individual diagnosed with PTSD having PTSD themselves (Sack et al. 1995). Sack and Clarke (1995) conducted a study of Khmer adolescents and one of their parents. Within the Sack and Clarke study, a significant number of those whose parents had PTSD also developed PTSD (1995). Similarly, a subsequent study of children of holocaust survivors also supported the overall conclusion that those whose parents had PTSD experienced a higher incidence of PTSD (Yehuda et al. 1998). Ultimately, these studies only indicate that if a parent has PTSD their offspring are likely to develop the condition. However, none of these studies to date have definitively identified any specific genes common to the parent and child that could be causative agents in the development of PTSD (Cornelis et al. 2010).

11.5.1.6 Using Heredity as a Means of Explaining PTSD

Later studies attempted to explain exactly to what degree heritability explained PTSD. One study of 2,224 identical twins and 1,818 nonidentical twins who were Vietnam vets indicated that 30 % of PTSD could be attributed to genetics (Trevor and Way 2001; True et al. 1993). However, a more recent study demonstrated heritability of 70 % (Sartor et al. 2011). Whether the degree of PTSD that is attributable to hereditary is 30 % or 70 %, there exists another percentage of those who develop PTSD unrelated to hereditary. A true understanding of how PTSD manifests itself in individuals remains impossible without first understanding the mechanism that allows PTSD to develop in individuals who do not inherit the condition. Moreover, given the fact that no specific genes could be identified as causative of PTSD another mechanism may be the causative agent even in those instances where an individual's parent suffered from PTSD.

11.5.2 Genetic Variants and PTSD

11.5.2.1 DNA

Serving as a blueprint, the genome represents all the DNA within an individual that codes for the appearance and functionality of an organism (Sadava et al. 2014). From a genetic standpoint, DNA or deoxyribonucleic acid contains the codes that provide a means for cell differentiation to occur during the first stages of cellular development (Sadava et al. 2014). Each DNA strand contains a combination of the four nucleotide bases known as adenine (A), cytosine (C), guanine (G), and thymine (T) (Sinden 2012). Of the four nucleotide bases, A and T pair with each other (A–T), while G and C pair only with each other (Sinden 2012) These bases represent the rungs of the double helix shaped DNA with the sides of the helix composed of sugar and phosphate molecules (Sinden 2012). DNA molecules packed tightly around histone proteins are known as chromosomes, which are present in 23 pairs in the nucleus of each cell (Kornberg and Baker 2005).

11.5.2.2 Transcription

Genes represent the exact portions of DNA encoding the information used by the cell for the creation of amino acids in the translation phase (Sadava et al. 2014). However, such translation only occurs after DNA unwinds to become accessible within the transcription process, thereby allowing for the creation of mRNA known as messenger RNA, which contains the instructions for making amino acids (Sadava et al. 2014). When three codons known as a triplet form a codon, they function as a group to create a specific amino acid (Nelson and Cox 2013). After the final amino acid is coded, termination of the translation process occurs (Nelson and Cox 2013). One of the three codons for termination within mRNA (UAA, UAG, UGA) signals the termination, which has resulted in these codons being named termination codons (Nelson and Cox 2013).

11.5.2.3 Translation

Ultimately, the amino acids become proteins as part of the translation process (Sadava et al. 2014). These amino acids manage the cell's chemical reactions as well as forming a significant percent of the structure of an organism (Sadava et al. 2014). However, this represents the ideal process and does not explain when a mutation occurs thereby resulting in an unexpected outcome (Malcolm 2001).

11.5.2.4 Gene Mutations

Gene mutations represent scenarios where an unexpected phenotypic result occurs (Malcolm 2001). A variety of mutations can occur (Malcolm 2001). Genes can be deleted from a sequence as a result of a small deletion or through an insertion of other gene(s) that results in a shift that deletes a gene previously in the sequence (Malcolm 2001). Moreover, an amino acid can mutate into a termination codon (known as a nonsense mutation), and an amino acid can be altered to either decrease or increase the activity of a gene (Malcolm 2001).

11.5.3 Studies Regarding Genetics and Heredity as Causative of PTSD

Scientists attributed differences in development to the order in which the four bases within each chromosome were organized (Sinden 2012). Hence, some studies related to genetics focused on identifying a specific gene responsible for the development of a condition (Sinden 2012). Other studies focused on the existence of an overall hereditary connection as a means of explaining why PTSD developed in some but not all exposed to traumatic events (Sinden 2012). By reviewing studies and examining genetics and heredity, one gains an understanding of any connections identified in terms of either as a means of explaining the development of PTSD.

One approach to identifying specific genes to which PTSD could be attributed involved examining genes to find gene variants (alleles) potentially associated with PTSD (Yehuda et al. 2011). To conduct this review, specific locations or loci on genes are examined to determine the frequency of such genes occurring in a group of people who experience trauma to compare the incidence of such genes in those suffering from PTSD as opposed to those not suffering from PTSD (Yehuda et al. 2011). Typically, alleles are either single nucleotide polymorphisms (SNPs) or alleles in which the variation in the gene appears within sequences that repeat known as variable number tandem repeats (VNTRs) (Yehuda et al. 2011). Despite studies aimed at identifying alleles responsible for PTSD, few alleles have been linked with PTSD (Yehuda et al. 2011). As a result, some researchers recommended focus on more than just a genetic variant to which PTSD can be attributed (Schmidt et al. 2011). Moreover, these researchers recommend changing the focus toward the investigation of specific alleles including epigenetic profiles (Schmidt et al. 2011).

11.6 Childhood Trauma

11.6.1 Overview

A range of traumatic conditions plague children in addition to physical and sexual abuse (D'Andrea et al. 2012). The common thread among the scenarios in which children suffer trauma is their roles as victims within such situations (D'Andrea et al. 2012). Hence, researchers have begun classifying childhood trauma under the category of victimization (D'Andrea et al. 2012). Within the context of victimization, a victim suffers harm because other humans act in ways that go against social norms (D'Andrea et al. 2012). However, some situations involving childhood trauma are more egregious in nature and involve a breach of trust relative to an interpersonal relationship. The term interpersonal victimization adds to the definition of victimization malevolence, betrayal, injustice, and immorality (D'Andrea et al. 2012).

11.6.2 Relationship Between Childhood Trauma and PTSD

In seeking another means of explaining what could be causing PTSD unrelated to heredity, other researchers looked at the environment as a means of identifying a predisposition to PTSD (Voisey et al. 2014). Some researchers also studied trauma and posttraumatic stress manifesting in childhood (Copeland et al. 2007). Other researchers began to focus on the specific environmental factors such as child abuse in early childhood as a trauma that ultimately created a predisposition to PTSD based on the environmental effects of the trauma from an epigenetic standpoint (Mehta et al. 2013).

11.6.3 Deaths Attributable to Childhood Trauma

The 2012 figures indicate 1,640 children died from maltreatment, which translates into a rate 2.2 per 100,000 children (U.S. Department of Health and Human Services Children's Bureau -Administration on Children, Youth and Families, Administration for Children and Families 2012). Of those dying as a result of maltreatment, 70 % suffered from neglect while 44 % experienced abuse that was physical in nature with their abuse specifically physical or combined with other maltreatment. Seventy percent of the fatalities in 2012 involved children who were less than 3 years of age. Moreover, African American and Pacific Islander children experienced the highest rates of death of 4.7 in comparison to the following other groups: American Indian/ Alaskan Natives 2.2, Hispanics 1.7, Non-Hispanic Whites 1.6, and Asians 0.6 (U.S. Department of Health and Human Services Children's Bureau -Administration on Children, Youth and Families, Administration for Children and Families 2012)

11.6.4 Costs of Childhood Trauma

11.6.4.1 Overall Costs

As of 2012, researchers estimate the total costs of childhood trauma at $124 billion with such costs inclusive of both new fatal and nonfatal cases as of 2008 (Fang et al. 2012). Notably, a 2007 study set the costs at $103.8 billion divided into $33.1

billion in direct costs and $70.7 billion in indirect costs (Wang and Holton 2007). For 2010, costs were assessed as $32,648 for healthcare during childhood, $10,530 in medical care for adults, $144,360 due to lost productivity, $7,728 relative to child welfare, $6,747 costs related to criminal justice, and $7,999 in special education costs (Fang et al. 2012).

11.6.4.2 Direct and Indirect Costs

Direct costs include costs that arise directly from the trauma such as hospitalization, obtaining mental health care, support of child welfare systems, and law enforcement (Wang and Holton 2007). Additionally, some costs constitute indirect costs such as obtaining special education, support of the juvenile delinquent system, obtaining long-term mental health care, obtaining long-term healthcare, costs for those who enter the criminal justice system as adults, and the overall costs in lost productivity of these individuals relative to the time taken off from work to deal with their condition (Wang and Holton 2007).

11.6.5 Overall Long-Term Effects of Childhood Trauma

According to the data from the Adverse Childhood Experience (ACE) Study that explores the long-lasting effect of childhood trauma into adulthood, over 60 % of adults who experienced childhood trauma experienced physical and mental health related outcomes as adults (Felitti et al. 1998). One study found long-term effects that extended into adulthood and adversely impacted productivity over the course of the lifetime of individuals who had previously suffered abuse (Daro 1988).

11.6.6 Childhood Trauma and PTSD

Within those studying childhood trauma as a whole, one study examined 1420 children who ranged in age from 9 to 13 at the beginning of the study and reached age 16 at the end of the study (Copeland et al. 2007). Researchers noted that the study confirmed the resiliency of children in the face of trauma (Copeland et al. 2007). As a result of this resiliency, PTSD did not manifest in most children following trauma (Copeland et al. 2007). As a result of this resiliency, most children who suffered only one instance of trauma that rose to the level of potentially causing PTSD did not develop the condition (Copeland et al. 2007). However, children participating in the study who suffered multiple traumatic episodes, already suffering from anxiety, or suffered from family adversity or disadvantage developed PTSD (Copeland et al. 2007).

To assess if any particular type of premilitary traumatic event appeared to be a catalyst for subsequent development of PTSD, researchers studied a group of 66 Vietnam Veterans all of whom had experienced combat in order. Within the group, 38 were diagnosed with PTSD (Bremner et al. 1993). At the conclusion of the study, analyzed data indicated that study participants with PTSD reported higher rates of childhood physical abuse as compared with those participants who did not have PTSD (Bremner et al. 1993). However, as Bremner noted himself in the conclusion of his study, a definitive correlation between childhood abuse and PTSD could not be drawn because of the small study population size thereby preventing further generalizations (1993). Moreover, Bremner noted that the findings of his study appeared to align with the findings of neurobiological studies, which indicated that the neurobiology systems associated with stress response experienced changes as a result of early life exposure to stress (1993). Nevertheless, Bremner noted that the study suggested that the previous concept of early life stress serving as a means of protecting a person from subsequent trauma (through a concept known as the hypothesis of trauma related to stress inoculation) appeared incorrect based on his research (1993). To place such a finding in context, one needs to understand that stress inoculation represents a cognitive behavioral therapy developed by Meichenbaum in order to treat anxiety in part through prolonged exposure (1972). Notably, subsequent studies also found as Bremner did that absent other treatment mechanisms prolonged exposure alone failed to improve PTSD symptoms (Grunert et al. 2007).

Subsequently, Zaidi and Foy's 1994 preliminary study also examined within a pool of 22 subjects whether or not those who experienced childhood trauma in terms of physical punishment and subsequently were exposed to combat during their military service developed PTSD. At the conclusion of the study, Zaidi and Foy found a statistically significant correlation in terms of those who experienced physical punishment and development of PTSD (1994). Moreover, the researchers noted that the data in their study illustrated that of the combat veterans admitted for inpatient PTSD treatment approximately 45 % experienced physical abuse during childhood (Zaidi and Foy 1994). In explaining the high percentage, the researchers explained that many fail to label themselves as having been abused but when posed with questions focused on their experiences, rather than the label itself, approximately twice as many veterans report experiencing treatment that would qualify as abuse (Zaidi and Foy 1994).

In 1997, Cloitre, Scarvalone, and Difede conducted a study focused on determining if women who were assaulted as children and adults exhibited higher levels of social impairment. The women studied belonged to one of three groups: those who had suffered sexual assault in both childhood and as adults, which they classified as retraumatized; those assaulted only as adults; and those never having been assaulted (Cloitre et al. 1997). After applying several instruments to assess variances between the groups, the researchers determined that retraumatized study participants experienced PTSD along with alexithymia or difficulty in accurately assessing the emotions of others, which might predicate their susceptibility to subsequent assault (Cloitre et al. 1997).

11.6.7 Impact of Stress Following Child Abuse on Brain Development

Several studies examined biological abnormalities relative to children suffering from maltreatment (Bevans et al. 2008; DeBellis et al. 2002). One study identified decreases in the volume of the corpus callosum, prefrontal cortices, as well as the temporal lobe along with increases in the volume of the superior temporal gyrus (DeBellis et al. 2002). A study of adult women who had suffered abuse as children also indicated decreased brain volumes (Kitayama et al. 2007). One study found decreased volumes in different brain structures depending on when the abuse occurred (Andersen et al. 2008). Notably, early childhood abuse resulted in decreases to the hippocampus (Andersen et al. 2008). Meanwhile, abuse suffered in middle school reduced the size of the corpus callosum, and abuse suffered during adolescence resulted in decreased prefrontal cortex volumes (Andersen et al. 2008). As a result of these findings, the researchers opined that stress alters the central nervous system due to childhood abuse and such changes may extend into adulthood though potentially in a less severe fashion as a result of adaptation (D'Andrea et al. 2012).

11.6.8 Neurobiological Effects Resulting from PTSD

Neurobiological investigation of PTSD focuses on several areas of the brain including the hippocampus, the amygdala, and the prefrontal or anterior region (Bremner 2006). Several studies have been conducted to determine the role the brain plays in the development of PTSD. Overall, these studies focused on identifying an explanation as to why PTSD occurs based on the results obtained by examining the brains of affected individuals (Karl et al. 2006).

Many studies used neuroimaging as a means of studying PTSD and identifying the circuit within the brain that plays a role in the disorder (Hughes and Shin 2011). Using a model of the neurocircuitry of the brain, improper functioning of the amygdala, medial prefrontal cortex, and the hippocampus is implicated as occurring in those with PTSD (Rauch et al. 2000).

Within the context of those examinations, the studies can be further divided into studies related focusing on the amygdala, studies focusing on the hippocampus, or studies focused on the hormones released during long-term hyperarousal, characteristic of PTSD. One study identified increased reactivity as a result of change in the amygdala in PTSD patients (Rauch et al. 2000). In Rauch et al.'s study, the researchers recommended further studies to determine if the increased amygdala responses indicate an underlying susceptibility for PTSD development in existence prior to trauma development (2000). Moreover, the researchers noted that their results could not be generalized without conducting research involving subjects other than combat veterans, which represented the population that they studied (Rauch et al. 2000).

Most importantly, the researchers note that their study represents the first study aimed at assessing the effects of general negative stimuli rather than traumatic stimuli (Rauch et al. 2000). Another study exposed both veterans and the control group to stimuli to determine to what degree the limbic regions of the brain, including the amygdala, hippocampal region, and the limbic cortex reacted to exposure to combat sounds (Lizberzon et al. 1999). Data elicited from the study performed by Lizberzon et al. revealed an increased response in the amygdala region for the study subjects diagnosed with PTSD at higher levels than non-PTSD study subjects when exposed to combat sounds, which the researchers noted aligned with the generally understood role of this area in the processing of emotion (1999).

While the previous two studies identified amygdala changes in terms of degree of reactivity, other studies identified the influence of other segments of the limbic system such as the hippocampus (Shnuff et al. 1997). Interestingly, these differences in hippocampal volume presented whether the individual suffered combat-related trauma or trauma precipitated by an event that did not involve trauma (Bremner et al. 1997). However, a subsequent study evaluating the relationship between volume reduction in the hippocampus and PTSD in those who specifically had no history of alcohol abuse did not indicate a difference in the volume of the hippocampus and the entorhinal cortex of those subjects with PTSD and the control subjects (Schuff et al. 2001). Notably, the researchers observed a 23 % reduction of *N*-acetylaspartate within the hippocampus along with a 26 % reduction in compounds containing creatine within the right segment of the hippocampus (Schuff et al. 2001). In summarizing their findings, the researchers opined that these substances (i.e., *N*-acetylaspartate and creatine) appear to be more accurate indicators of PTSD (Schuff et al. 2001). Other studies have also identified creatine as a substance that plays a role in PTSD (Koga et al. 2005). In the study performed by Koga et al., chicks exposed to stress exhibited increased creatine from which the researchers concluded that creatine's role in the brain is to protect the brain from stressful conditions (Koga et al. 2005). In creatine as a whole, Koga et al. 2005 noted that creatine results from the methylation of GAA, which is methylated to create creatine (Neu et al. 2002). Moreover, Neu et al. noted that lower concentrations of creatine activated GABA-A receptors (2002). To further explain this phenomenon, Koga et al. referenced research by Sieghart (1995). Sieghart explained that the two types of GABA receptors (i.e., GABA-A and GABA-B) facilitate the use of GABA as the neurotransmitter (e.g., message carrier) that estimates indicate between 20 % and 50 % of the central synapses use (1995). Based on these findings, Koga et al. opined that creatine functions only with the GABA-A receptor and when it does it protects the brain from stress escalating to the level of PTSD (2002). In so doing, the researchers also observed that GABA-A receptors contain binding sites for Benzodiazepines (BZ) (Study and Barker 1981). BZ increases the effect of GABA at the GABA-A receptors by opening up the chloride channel more often, thereby enhancing GABA's effect (Study and Barker 1981). Given as Trevor and Way (2001) noted BZ represented the most effective agent for sedation-hypnosis, Koga et al. opined that this presented itself in decreased vocalization and decreased mobility in the chicks who were exposed to stress in their study following creatine administration (2005).

Other studies have focused more specifically on the hormones created during the state of long-term hyperarousal (Kendall-Tackett 2000). In one study, the researchers hypothesized that the success of SGB can be attributed to a reduction in nerve growth factor, which functions to maintain a state of hyperarousal characteristic of PTSD, as well as noradrenalin, which reduces the level of central noradrenalin (Takitori et al. 2006).

11.7 Epigenetics

11.7.1 Overview

Genetics answers some questions relative to why conditions develop (Omoto and Lurquin 2004). However, it clearly does not explain all human development (Omoto and Lurquin 2004). Sometimes, genes exist but do not express themselves (Omoto and Lurquin 2004). However, epigenetics appears to provide a means of explaining changes not attributable to the changes to DNA that occur during human development (Jaenisch and Bird 2003).

Approximately half of the 25,000 genes responsible for coding proteins express within any type of cell (Romanoski et al. 2015). Researchers within the ENCODE project identified hundreds of regions that function to enhance gene expression in mammals from a distance (Romanoski et al. 2015). The researchers noted that approximately between 20,000 and 40,000 enhancers regulate genes thereby determining how the gene expresses itself (Romanoski et al. 2015). Activation of these enhancers occurs by way of their interaction with factors involved in the transcription process that recognize and attach to particular sequences of DNA within a region responsible for enhancement (Romanoski et al. 2015). How genomic information is interpreted depends on how its DNA elements are regulated (Romanoski et al. 2015). When a developmental transition occurs, the epigenomic signatures either activate or suppress the transcription factors (Romanoski et al. 2015). Notably, the researchers highlighted the fact that understanding health as well as disease requires an understanding of the differences between the genome and epigenome of cells thereby allowing the scientific community to gain an understanding as to what is driving diseases (Romanoski et al. 2015).

By definition, epigenetics represents the changes that occur to cells above or outside the entire set of human genes (Bender 2004; Jaenisch and Bird 2003). Such cellular changes occur without changing the DNA, yet impact traits that are observable in individuals (Kiefer 2007; Diwakar 2014). Notably, epigenetic changes either randomly or as a result of the environment and not only within the course of development, cell production, and growth (Jaenisch and Bird 2003). Based on the research performed, scientists identified specific epigenetic mechanisms such as DNA methylation, histone modifications, "ATP-dependent chromatin remodeling," and "non-coding RNA-mediated gene silencing" (Issa and Just 2011). Each of these epigenetic mechanisms presents a means of controlling DNA transcription, yet in so doing does not change the structure of the DNA (Zhang and Meaney 2010). Rather, the epigenetic mechanisms modify the DNA biochemically, as well as the histone proteins, which represent a significant portion of what comprises chromatin (Graff and Mansuy 2008). When an epigenetic modification occurs, the rate at which chromatin condenses changes (Graff and Mansuy 2008). Changing how the condensation occurs affects the ability of transcription (Graff and Mansuy 2008). The factors that regulate transcription cannot bind to the DNA, thereby preventing the affected genes from being expressed (Klose and Bird 2006).

Among these epigenetic mechanisms, DNA methylation has been confirmed as the most defined mechanism and also represents the mechanism thought to play a significant role in a variety of diseases including relative to PTSD (Alvarado et al. 2014). Using the cytosine and adenine bases, DNA methylation occurs (Klose and Bird 2006). Meanwhile, methylation occurs when a methyl group (CH_3) is added to DNA nucleotides (Medina 2008). Some instances of DNA methylation have been shown to result in the activation of a gene (Chahrour et al. 2008). However, typically following methylation, the methylated gene becomes inactive (Medina 2008). Two distinct mechanisms result in the inhibition of gene expression (Klose and Bird 2006). The first mechanism involves changing the cytosine bases to prevent binding of the factors at the recognized points on the DNA (Watt and Molloy 1988). The second mechanism involves methyl-CpG co-repressor molecules stopping transcription and changing the chromatin surrounding the DNA (Boyes and Bird 1991).

11.7.2 Epigenetics and PTSD

As it became clear that not all traumas resulted in PTSD, researchers began examining catalyst for the exacerbation of trauma to the level necessary to precipitate PTSD (Yehuda et al. 2011). In explaining the direction such research took, Yehuda et al. (2011) reference the neurological studies suggestive of an alteration within the hormones in existence prior to the trauma that placed certain individuals at higher risk of developing PTSD (Yehuda et al. 2011). Notably, Yehuda et al. highlight the need for further investigation regarding how the environment through epigenetics and other such molecular level basis alter the way that genes express themselves (2011). Rather than highlighting one or the other, Yehuda et al. recommend reviewing epigenetics in conjunction with gene expression to understand the role the environment plays in PTSD (2011). Given that epigenetic modifications, including DNA methylation, result in response to the influence of the environment and alter how genes function long term, epigenetics presents a means of explaining the individual variation along with the long-term exposure effects of trauma that cannot otherwise be (Yehuda and Bierer 2009). Hence, Yehuda and Bierer note that a variety of findings relative to how genes express themselves as a result of stress appear to be explainable through epigenetics (2009).

A variety of examples of research findings relative to how stress manifests itself appear to be best explained by epigenetic changes. For instance, a predisposition to mental illness based on a specific genotype/epigenotype combination may explain how certain conditions are triggered in response to certain environmental exposure (Uddin et al. 2011). Moreover, epigenetic changes involved in PTSD determine how chromatin is regulated in the brain cells that do not divide (Zovik and Sweatt 2013). Additionally, researchers hypothesize that the neuronal brain cells are marked as a result of epigenetics as triggered by an environmental catalyst (Zovik and Sweatt 2013).

In the case of PTSD, the environmental exposure that triggers methylation appears to be the traumatic event, which triggers a change in how an individual processes stress (Uddin et al. 2011). The occurrence of such an event appears to trigger the epigenetic change that culminates in predisposing an individual to later development of PTSD following a subsequent trauma (Uddin et al. 2011). To understand how such a predisposition works, a review of the research focused on the neurobiology of PTSD needs to be examined as a means of explaining the impact of PTSD neurologically. However, a neurobiological examination cannot occur until after a review of both the nervous system and the brain occurs in particular to further understand how PTSD manifests itself.

11.7.3 New Treatments for PTSD Support an Underlying Epigenetic Basis

Another factor to consider in evaluating the viability of epigenetics as the basis for the decreased resiliency that occurs in some but not all exposed to trauma is performing an examination of the treatments found to be effective. One such treatment involves stellate ganglion block (SGB) (Navaje et al. 2014). In recent research, SGB requires injection of an anesthetic to the bundle of nerves known as the stellate ganglion (Navaje et al. 2014). Notably, some who were administered SGB reported immediate relief (Navaje et al. 2014). In terms of explaining why the SGB functions, the researchers referenced studies related to how parts of the insular cortex and other structures within the cerebrum function relative to the levels of hormones involved in the hyperarousal state characteristic of PTSD (Osuch et al. 2001; Shin et al. 2004; Takitori et al. 2006).

11.8 Barriers to Obtaining Reliable Results in Future Research

Several barriers exist that make performing research related to PTSD challenging. Included among the barriers that need to be addressed prior to designing a study are statistical significance, sensitive topics, anonymity of reported data, reporting bias, as well as ethical considerations. Understanding these limitations proves necessary to allow for definitive findings to be made upon which an actionable plan can be developed to reduce the incidence of PTSD. Moreover, those reviewing a study need to keep these considerations in mind when evaluating the ability to generalize the findings. Hence, the degree to which such barriers were addressed needs to be assessed given that failing to address these barriers in the design of the research may result in unreliable or ungeneralizable data.

11.8.1 Statistical Significance

A contributing factor to the success of Mendel's research stemmed from his use of large sample sizes that allowed him to identify distinct patterns upon which he could reliably formulate theories (Sadava et al. 2014). As other scientists became aware of Mendel's research, they too incorporated probability into their research as a means of predicting results (Sadava et al. 2014). Using statistics, these scientists tested the predicted results against the actual results (Sadava et al. 2014). To date, these same principles of probability represent a crucial component of how results are determined to be statistically significant with only those results where probability is above 5 % falling into the significant category, which allows researchers to determine if a real difference exists within the results being examined (Banerjee et al. 2009; Guyatt et al. 1995). Such a difference leads to the identification of statistically significant results (Guyatt et al. 1995). However, such statistical significance cannot be evaluated without a sample size large enough to discount the observation as occurring by chance (Guyatt et al. 1995).

11.8.2 Randomization

Although in the nineteenth century entire populations were surveyed to perform research on that data, today researchers work with a small group or a sample of the larger whole. Using the data elicited from the sample, researchers infer about the larger population (Banerjee and Chadhury 2010). A significant number of studies focus on veterans, but such a focus results in the use of a nonrandom study to then make inferences about the general population (Johnson and Thompson 2007). Use of a nonrandom population for a study limits the ability to generalize the findings to the overall population, thereby diluting the value of such results (Johnson and Thompson 2007). As one researcher observed, relying only on data relative to veterans may not accurately depict the prevalence of the condition because what constitutes a trauma is lesser in severity in those who are not in the military (Meichenbaum 1994). In fact, failing to randomize essentially excludes individuals, qualifies as selection bias, and selection bias results in the data signaling an altered interpretation (Ards et al. 1998).

11.8.3 Sensitive Topic

The term sensitive subject covers issues that may adversely affect the reputation of the individual due to personal information involved (Bankert and Amdur 2006). PTSD qualifies as a sensitive subject (Bankert and Amdur 2006). One set of researchers noted that such sensitive topics require adoption of approaches aimed at protecting privacy as well as confidentiality (Mealer and Jones 2014).

11.8.4 Anonymity of Reported Data

Many studies involving PTSD allow the study participant to anonymously self-report using a questionnaire (Lelkes et al. 2012). Yet, self-reporting typically results in inaccuracy that makes the data potentially unreliable because there is no accountability encouraging honesty given that no one knows who answered (Lelkes et al. 2012). One study addressed the reliability and accountability issues incidental to using online surveys as a means of anonymity by creating a study in which some of the participants were surveyed online and some received the questionnaire by way of an in-person meeting with an interviewer (Campos et al. 2011). Such a safeguard presents a viable means of avoiding the creation of inaccurate data by adding accountability through interaction with the research team.

11.8.5 Reporting Bias

In the realm of epidemiology, reporting bias represents selective disclosure of results (Porta et al. 2008). As Last elaborated, bias results from an error in how data are collected, analyzed, interpreted, published, or reviewed (Last 2001). As a result of such errors, the results may be overestimated or underestimated relative to how frequently a pattern occurs within data or the extent of any effect (Van den Broeck and Brestoff 2013).

From a research standpoint, there are many biases including those related to the nonpublication of results as well as publication of only favorable results known as outcome reporting (McGauran et al. 2010). Publication bias occurs when some research fails to be published (McGauran et al. 2010). Similarly, outcome bias represents a scenario where only some or the selected outcomes are the only ones published (McGauran et al. 2010). Given that the scientific community relies on the research to make generalizations in terms of patterns that further the understanding on a topic as well as how any condition is addressed in terms of treatment, any type of reporting bias, including those listed, represent a risk likely to undermine the direction of any initiative crafted based on research. A number of studies have found widespread evidence of bias that results in treatments that caused harm to patients, and they recommend mandatory registration of trials as a means of ensuring that all results of research are disclosed (Dwan et al. 2008; McGauran et al. 2010; Harrison and Mayo-Wilson 2014).

11.8.6 Ethics

Ethics or the rules that represent the standards of conduct represent an essential consideration in research (Resnik 2011). Certain general principles apply one of which is the principle of doing no harm, which represents a component of many codes of ethics (Resnik 2011). Following the principle of first do not harm, it would not be ethical to create the level of stress necessary to create PTSD due to the known impact of the condition as well as the identified comorbidities. Therefore, when trying to measure what actually occurs, researchers must use lab subjects such as rats (Goswami et al. 2012). However, such tests represent the only means available that allows researchers to infer causality between the trauma and the development of PTSD, which represents a shortcoming of cross-sectional studies given that the variables are measured at the same time (Jaranson et al. 2004).

11.8.7 Failure to Apply Evidence-Based Research Approach

Some researchers have reviewed the PTSD related research to date to identify the shortcomings in how it was conducted (Iribarren et al. 2005). In order to avoid a reliance on faulty data, they have recommended a new approach centered on doing more than relying on the literature at face value (Iribarren et al. 2005). Instead, they recommend moving to an evidence-based approach to research that requires validation of the existing research by communicating with the researchers as well as the authors (Iribarren et al. 2005). Such an approach allows for those reviewing the literature to conduct a thorough investigation aimed at evaluating both the strength and the shortcoming of existing data (Iribarren et al. 2005).

11.9 Conclusion

Homeostasis represents a means of achieving health. Stress represents a challenge to the natural coping mechanism that allows us to remain resilient in the face of stress. In some individuals, the levels of resiliency that prevent PTSD from occurring are absent. Specifically relative to those who suffered child abuse, the data indicate that adverse functional changes potentially precipitated by epigenetics have altered how the natural response to stress occurs, thereby decreasing the level of stress necessary for allostatic load to be reached.

When the body responds to stress chronically, an individual may develop chronic conditions as a result of over exposures to the hormones that regulate stress, which represents the allostatic load. When an individual's allostatic load is exceeded, an individual develops one of many comorbid conditions based on the body having to chronically be in a state of responding to stress. Debilitating conditions that pose a risk to the lives as well as the quality of life of individuals occur as a result of protracted periods of stress that rises to the level of trauma incidental to the development of PTSD.

Though significant amounts of research have been undertaken to understand what causes PTSD, researchers have not been able to attribute PTSD to any specific alleles or gene variants. Along the path of conducting this research, scientists learn more and more about PTSD thereby resulting in the evolving definition. To date, research has revealed a higher susceptibility in those who suffered prior trauma to develop PTSD. Moreover, women appear to be more susceptible to PTSD. Additionally, those who suffer another mental condition also appear to be at increased levels of susceptibility. However, given that no gene variant appears to be the cause of PTSD manifestation, researchers have begun exploring epigenetics as a means of explaining why PTSD manifests in one individual as opposed to another. Given that epigenetic manifestations in individuals are known to occur as a result of the environment, it appears likely that epigenetics facilitates the environmental influence on individuals that function as a catalyst to the development of PTSD.

Establishing a definitive connection between epigenetics and PTSD requires further research. Moreover, this research needs to include larger, random samples, rather than specific groups such as veterans, to provide for statistically significant results that can be generalized across the greater population. To ensure the validity of any findings, all research needs to be published whether the findings support the generally accepted position in order to avoid outcome bias; and even what might seem inconsequential research needs to be published to eliminate the potential for publication bias. To ensure that neither of the named types of reporting bias occurs, a registration system needs to be implemented to track all research conducted across disciplines thereby avoiding elimination of data that might provide crucial insight to our understanding of PTSD as a whole and PTSD resulting from childhood trauma specifically. Moreover, future research also needs to be designed to avoid any ethical constraints relative to imperative to not do any harm and look toward avoid self-reporting due for the potential for bias.

Achieving a cross-population sample of adequate size upon which to make generalization requires a very aggressive approach. Rather than having researchers perform research individually, a national program aimed at fostering wide-scale research among the scientific community including research based on long-term follow-up appears to be likely to achieve the type of definitive findings and allow for the type of quality assurance necessary to avoid inaccurate conclusions. Such a program is predicated on also increasing the overall awareness of PTSD to facilitate further diagnosis of individuals who suffer from the condition without benefit of diagnosis; through a public awareness campaign that destigmatizes PTSD, the potential exists to capture the number of participants necessary to make generalizable findings.

Once such a wide-spread effort occurs, additional data will be available for analysis. Such data will increase the understanding of the role of epigenetics as a contributing factor in the development of PTSD.

If the data obtained from large-scale studies continue to support epidemiological changes decreasing the ability of an individual to remain resilient in the face of protracted trauma, the potential exists to examine gene therapy as a means of reversing the epidemiological changes. Alternatively, awareness of an already occurring epigenetically driven susceptibility to PTSD on an individual basis as a result of childhood trauma provides a means of designing cognitive behavioral therapy prior to the actual development of PTSD with the aim being to enhance resiliency in order to avoid the consequences of allostatic load. Only through such wide-scale research and proactive measures does the potential exist to decrease the significant impact of PTSD on all levels thereby offsetting the epigenetic prediposition resulting from childhood abuse.

Abbreviations

PTSD	Posttraumatic stress disorder
CNS	Central nervous system
PNS	Peripheral nervous system
DNA	Deoxyribonucleic acid

References

Adinoff, B., Iranmanesh, A., Veldhuis, J., & Fisher, L. (1998). Disturbances of the stress response: The role of the HPA axis during alcohol withdrawal and abstinence. *Alcohol Health and Research World, 22*(1), 67–72.

Al'Absi, M. (Ed.). (2007). The hypothalamic-pituitary-adrenal axis and addiction. In Al'Absi, M. (Ed.), *Stress and addiction: Biological and psychological mechanisms* (pp. 21–40). Boston: Academic Press.

Alexander, D. A., & Wells, A. (1991). Reactions of police officers to body handling after a major disaster. A before and after comparison. *British Journal of Psychiatry, 159*, 547–555.

Alvarado, S., Fernald, R. D., Storey, K. B., & Szyf, M. (2014). The dynamic nature of DNA methylation: A role in response to social and seasonal variation. *Integrative and Comparative Biology, 54*, 68–76. doi:10.1093/ich.icu034.

American Psychiatric Association. (1987). *Diagnostic and statistical manual of mental disorders* (3rd ed.). Washington, DC: American Psychiatric Association.

Andersen, P., Morris, R., Amaral, D., Bliss, T., & O'Keefe, J. (2007). *The hippocampus book.* New York, NY: Oxford University Press.

Andersen, S. L., Tomada, A., Vincow, E. S., Valente, E., Polcari, A., & Teicher, M. H. (2008). Preliminary evidence for sensitive periods in the effect of childhood sexual abuse on regional brain development. *Journal of Neuropsychiatry and Clinical Neurosciences, 20*, 292–301.

Ards, S., Chung, C., & Myers, S. L. (1998). The effects of sample selection bias on racial differences in child abuse reporting. *Child Abuse and Neglect, 22*(2), 103–115.

Banerjee, A., & Chadhury, S. (2010). Statistics without tears: Populations and samples. *Industrial Psychiatry Journal, 19*, 60–65.

Banerjee, A., Jadhav, S. L., & Bhawalkar, J. S. (2009). Probability, clinician decision making and hypothesis testing. *Industrial Psychiatry Journal, 18*, 64.

Bankert, F. A., & Amdur, R. J. (2006). *Institutional review board: Management and function*. Boston, MA: Jones & Bartlett.

Bender, J. (2004). DNA methylation and epigenetics. *Annual Review of Plant Biology, 55*, 41–68. doi:10.1146/annurev.arplant.55.031903.141641.

Bennet, G. (2011). Shakespeare and post-traumatic stress disorder. *The British Journal of Psychiatry, 198*(4), 255.

Bevans, K., Cerbone, A., & Overstreet, S. (2008). Relations between recurrent trauma exposure and recent life stress and salivary cortisol among children. *Development and Psychopathology, 20*(257-272).

Blake, D. D., Keane, T. M., Wine, P. R., Mora, C., Taylor, K. L., & Lyons, J. A. (2006). Prevalence of PTSD symptoms in combat veterans seeking medical treatment. *Journal of Traumatic Stress, 3*(1), 15–27. doi:10.1002/jts.2490030103.

Boyes, J., & Bird, A. (1991). DNA methylation inhibits transcription indirectly via a methyl-CpG binding protein. *Cell, 64*, 1123–1134.

Boyko, E. J., Jacobson, I. G., Smith, B., Ryan, M. K., Hooper, T. I., & Amoroso, P. J. (2010). Risk of diabetes in U.S. military service members in relation to combat deployment and mental health. *Diabetes Care, 33*(8), 1771–1777.

Brady, K. T., Killeen, T. K., Brewerton, T., & Lucerini, S. (2000). Comorbidity of psychiatric disorders and posttraumatic stress disorder. *Journal of Clinical Psychiatry, 61*(Suppl 7), 22–32.

Bremner, J. D. (2006). Traumatic stress: Effects on the brain. *Dialogues Clinical Neuroscience, 8*(4), 445–461.

Bremner, J. D., Randall, P., Vermetten, E., Staib, L., Bronen, R. A., & Mazure, C. (1997). Magnetic resonance imaging-based measurement of hippocampal volume in posttraumatic stress disorder related to childhood physical and sexual abuse--A preliminary report. *Biological Psychiatry, 41*(1), 23–32.

Bremner, J. D., Southwick, S. M., Johnson, D. R., Yehuda, R., & Charney, D. S. (1993). Childhood physical abuse and combat-related posttraumatic stress disorder. *American Journal of Psychiatry, 150*, 234–239.

Breslau, N., Chilcoat, H. D., Kessler, R. C., & Davis, G. C. (1999). Previous exposure to trauma and PTSD effects of subsequent trauma: Results from the Detroit area survey of trauma. *The American Journal of Psychiatry, 156*(6), 902–907.

Brodal, P. (2010). *The central nervous system: Structure and function* (4th ed.). New York, NY: Oxford University Press.

Brown, P. J., Recupero, P. R., & Stout, R. I. (1995). PTSD substance abuse comorbidity and treatment utilization. *Addictive Behavior, 20*, 251–254.

Brown, P. J., & Wolfe, J. (1994). Substance abuse and post-traumatic stress disorder comorbidity. *Drug and Alcohol Dependence, 35*, 51–59.

Campos, J. D., Zucoloto, M. L., Bonafé, F. S., Jordani, P. C., & Maroco, J. (2011). Reliability and validity of self-reported burnout in college students: A cross randomized comparison of paper-and-pencil vs. online administration. *Computer in Human Behavior, 5*, 1875–1883. doi:10.1016/j.chb.2011.04.011.

Castle, W. E. (1933). The Gene theory in relation to blending inheritance. *Proceedings of the National Academy of Sciences, 19*(2), 1011–1015.

Chahrour, M., Jung, S. Y., Shaw, C., Zhou, X., Wong, S. T., Qin, J., & Zoghbi, H. Y. (2008). MeCP2, a key contributor to neurological disease, activates and represses transcription. *Science, 320*, 1224–1229.

Cloitre, M., Scarvalone, P., & Difede, J. (1997). Posttraumatic stress disorder, self- and interpersonal dysfunction among sexually retraumatized women. *Journal of Traumatic Stress, 10*, 437–452.

Cohen, J., Perel, J., DeBellis, M., Friedman, M., & Putnam, F. (2002). Treating traumatized children: Clinical implications of the psychobiology of posttraumatic stress disorder. *Trauma, Violence & Abuse, 3*(2), 91–108. doi:10.1177/15248380020032001.

Copeland, W. E., Keeler, G., Angold, A., & Costello, E. J. (2007). Traumatic events and posttraumatic stress in childhood. *Archives of General Psychiatry, 64*(5), 577–584.

Cornelis, M. C., Nugent, N. R., Amstadter, A. B., & Koenen, K. C. (2010). Genetics of post-traumatic stress disorder: Review and recommendations for genome-wide association studies. *Current Psychiatry Reports, 12*(4), 313–326. doi:10.1007/s119420-010-0126-6.

Cornelius, J. R., Kirisci, L., Reynolds, M., Clark, D. B., Hayes, J., & Tarter, R. (2010a). Genetics of post-traumatic stress disorder: Review and recommendations for genome-wide association studies. *Current Psychiatry Reports, 12*(4), 313–326. doi:10.1007/s119420-010-0126-6.

Cornelius, J. R., Kirisci, L., Reynolds, M., Clark, D. B., Hayes, J., & Tarter, R. (2010b). PTSD contributes to teen and young adult cannabis use disorder. *Addictive Behavior, 35*(2), 91–94. doi:10.1016/j.addbeh.2009.09.007.

Cuffe, S. P., Addy, C. L., Garrison, C. Z., Waller, J. L., Jackson, K. L., McKeown, R. E., & Chilappagari, S. (1998). Prevalence of PTSD in a community sample of older adolescents. *Journal of the American Academy of Child and Adolescent Psychiatry, 37*(2), 147–154.

D'Andrea, W., Ford, J., Stolbach, B., Spinazzola, J., & Van der Kolk, B. A. (2012). Understanding interpersonal trauma in children: Why we need a developmentally appropriate trauma diagnosis. *American Journal of Orthopsychiatry, 82*(2), 187–200. doi:10.1111/j.q939.0025-201201.154x.

DaCosta, J. M. (1871). On irritable heart: A clinical study of a form of functional cardiac disorder and its consequences. *American Journal Medical Science, 61*, 17–52.

Daro, D. (1988). *Confronting child abuse: Research for effective program design*. New York, NY: The Free Press.

De Kloet, C. S., Vermetten, E., Geuze, E., Kavelaars, A., Heijnen, C. J., & Westenberg, H. G. (2006). Assessment of HPA-axis function in posttraumatic stress disorder: Pharmacological and non-pharmacological challenge tests, a review. *Journal Psychiatric Research, 40*(6), 550–567.

DeBellis, M. D., Keshavan, M. S., Shifflett, H., Iyengar, S., Beers, S. R., Hall, J., & Moritz, G. (2002). Brain structures in pediatric maltreatment-related posttraumatic stress disorder: A sociodemographically matched study. *Biological Psychiatry, 52*, 1066–1078.

Defrin, R., Ginzburg, K., Solomon, Z., Polad, E., Bloch, M., Govezensky, M., & Schreiber, S. (2008). Quantitative testing of pain perception in subjects with PTSD--implications for the mechanism of the coexistence between PTSD and chronic pain. *Pain, 138*(2), 450–459. doi:10.1016/j.pain.2008.05.006.

Ding, S. L. (2013). Comparative anatomy of the prosubiculum, subiculum, presubiculum, postsubiculum, and parasubiculum in human, monkey, and rodent. *Journal of Comparative Neurology, 521*(18), 4145–4162. doi:10.1002/cne.23416.

Diwakar, D. (2014). Epigenetics. *Journal of Pharmaceutical Sciences and Research, 6*(3), 161–163.

Domschke, K. (2012). Patho-genetics of posttraumatic stress disorder. *Psychiatria Danubina, 24*(3), 264–273.

DuPont, R. I., Rice, D. P., Miller, L. S., Shiraki, S. S., Rowland, C. R., & Harwood, H. J. (1996). Economic costs of anxiety. *Anxiety, 2*, 167–172.

Dwan, K., Altman, D. G., Arnaiz, J. A., Bloom, J., Chan, A. W., & Cronin, E. (2008). Systematic review of the empirical evidence of study publication bias and outcome reporting bias. *PLoS One, 3*(8), e3081. doi:10.1371/journal.pone.0003081.

Everly, G. S., & Lating, J. M. (2013). *A clinical guide to the treatment of the human stress response*. London: Springer.

Fang, X., Brown, D. S., Florence, C. S., & Mercy, J. A. (2012). The economic burden of child maltreatment in the United States and implications for prevention. *Child Abuse and Neglect, 36*(2), 156–165. doi:10.1016/j.chiabu.2011.10.006.

Farr, O. M., Sloan, D. M., Keane, T. M., & Mantzoros, C. S. (2014). Stress and PTSD associated obesity and metabolic dysfunction: A growing problem requiring further research and novel treatments. *Metabolism, 63*(12), 1463–1468.

Feinstein, A. R. (1970) The pre-therapeutic classification of co-morbidity in chronic disease. *Journal of Chronic Disease, 23*, 455–468.
Felitti, V. J., Anda, R. F., Nordenberg, D., Williamson, D. F., Spitz, A. M., & Edwards, V. (1998). Relationship of childhood abuse and household dysfunction to many of the leading causes of death in adults. The Adverse Childhood Experiences (ACE) Study. *American Journal of Preventative Medicine, 14*(4), 245–258.
Fikelhor, D., Hotalin, G., Lewis, I. A., & Smith, C. (1990). Sexual abuse in a national survey of adult men and women: Prevalence, characteristics, and risk factors. *Child Abuse and Neglect, 14*(1), 19–28.
Finkelhor, D., Turner, H., Ormrod, S., & Hamby, S. L. (2009). Violence, abuse, and crime exposure in a national sample of children and youth. *Pediatrics, 124*(5), 1411–1423.
Fuster, J. (2015). *The prefrontal cortex* (5th ed.). Oxford, England: Elsevier.
Garner, A. S., Shonkoff, J. P., & The Committee on Psychosocial Aspects of Child and Family Health, Committee on Early Childhood, Adoption, and Dependent Care and Section on Developmental and Behavioral Pediatrics. (2011). The lifelong effects of early childhood adversity and toxic stress. *American Academy of Pediatrics, 129*, e232. doi:10.1542/peds.2011-2663.
Glick, M. (2011). A brain primer-Major structures and their functions. In *The instructional leader and the brain: Using neuroscience to inform practice.* (pp. 1–20). Thousand Oaks, CA: Corwin Press. doi:10.4135/9781483387437.n1
Goodwin, R. D., & Davidson, J. R. (2005). Self-reported diabetes and posttraumatic stress disorder among adults in the community. *Preventive Medicine, 40*, 570–574.
Goswami, S., Samuel, S., Sierra, O. R., Cascardi, M., & Paré, D. (2012). A rat model of post-traumatic stress disorder reproduces the hippocampal deficits seen in the human syndrome. *Frontiers in Behavioral Neuroscience, 6*, 26. doi:10.3389/fnbeh.2012.00026.
Graff, J., & Mansuy, I. M. (2008). Epigenetic codes in cognition and behavior. *Behavioral Brain Research, 192*, 70–87.
Greenberg, P. E., Sisitsky, T., Kessler, R. C., Finkelstein, S. N., Berndt, E. R., & Davidson, J. R. (1999). The economic burden of anxiety disorders in the 1990s. *Journal of Clinical Psychiatry, 60*(7), 427–435.
Greenstein, B., & Greenstein, A. (2007). *Concise clinical pharmacology*. London: Pharmaceutical Press.
Grunert, B. K., Weis, J. M., Smucker, M. R., & Christianson, H. F. (2007). Imagery rescripting and reprocessing therapy after failed prolonged exposure for post-traumatic stress disorder following industrial injury. *Journal of Behavior Therapy and Experimental Psychiatry, 38*, 317–328.
Guyatt, G., Jaeschke, R., Heddle, N., Cook, D., Shannon, H., & Walter, S. (1995). Basic statistics for clinicians: 1. Hypothesis testing. *Canadian Medical Association Journal, 152*(1), 27–32.
Hadland, K. A., Rushworth, M. S., Gaffan, D., & Passingham, R. E. (2003). The effect of cingulate lesions on social behavior and emotion. *Neuropsychologia, 41*(8), 919–931.
Hale, J. B., & Fiorello, C. A. (2004). *School neurology a practitioner's handbook*. New York, NY: Guilford Press.
Harrison, B. A., & Mayo-Wilson, E. (2014). Trial registration: Understanding and preventing reporting bias in social work research. *Research on Social Work Practice, 24*(3), 372–376.
Heim, C., Mletzko, D. J., Miller, T., & Nemeroff, A. H. (2008). The link between childhood trauma and depression: Insights from HPA axis studies in humans. *Psychneuroendocrinology, 33*, 693–710.
Hughes, K. C., & Shin, L. M. (2011). Functional neuroimaging studies of post-traumatic stress disorder. *Expert Review of Neurotherapeutics, 11*(2), 275–285. doi:10.1586/ern.10.198.
Iribarren, J., Prolo, P., Neagos, N., & Chiappelli, F. (2005). Post-traumatic stress disorder: Evidence-based research for the Third Millennium. *Evidence-Based Complementary and Alternative Medicine, 2*(4), 503–512. doi:10.1093/ecam/neh127.
Issa, J. P., & Just, W. (2011). Epigenetics. *FEBS Letters, 585*(13), 7. doi:10.1016/j.febslet2011.06.007.
Jacobson, S., & Marcus, E. M. (2008). *Neuroanatomy for the neuroscientist*. New York: Springer.

Jaenisch, R., & Bird, A. (2003). Epigenetic regulation of gene expression: How the genome integrates intrinsic and environmental signals. *Nature Genetics, 33*, 245–254. doi:10.1038/ng1089.

Jaranson, J. M., Butcher, J., Halcon, L., Johnson, D. R., Robertson, C., & Savik, K. (2004). Somali and Oromo refugees: Correlates of torture and trauma history. *American Journal of Public Health, 94*(4), 591–598.

Jayawickreme, N., Yasinski, C., Williams, M., & Foa, E. B. (2012). Gender-specific associations between trauma cognitions, alcohol cravings, and alcohol-related consequences in individuals with comorbid PTSD and alcohol dependence. *Psychology of Addictive Behaviors, 26*(1), 13–19. doi:10.1037/a0023363.

Joëls, M. (2007). Functional actions of corticosteriods in the hippocampus. *European Journal of Pharmacology, 583*(2–3), 312–321. doi:10.1016/j.ephar.2007.11.064.

Johnson, H., & Thompson, A. (2007). The development and maintenance of post-traumatic stress disorder (PTSD) in civilian adult survivors of war trauma and torture: A review. *Clinical Psychology Review, 28*(1), 36–47.

Kaffman, A. (2009). The silent epidemic of neurodevelopmental injuries. *Biological Psychology, 66*(7), 624–626. doi:10.1016/j.biopsych.2009.08.002.

Kane, M. J., & Engle, R. W. (2002). The role of prefrontal cortex in working-memory capacity, execution attention, and general fluid intelligence: An individual-differences perspective. *Psychonomic Bulletin and Review, 9*(4), 637–671.

Kang, H. K., Natelson, N. H., Mahan, C. M., Lee, K. Y., & Murphy, F. M. (2003). Post-traumatic stress disorder and chronic fatigue syndrome-like illness among Gulf War veterans: A population-based survey of 30,000 veterans. *American Journal of Epidemiology, 157*(2), 141–148.

Karl, A., Schaefer, M., Malta, L. S., Dörfel, D., Rohleder, N., & Werner, A. (2006). A meta-analysis of structural brain abnormalities in PTSD. *Neuroscience and Biobehavioral Reviews, 30*, 1007–1031.

Kee, N., Teixeira, C. M., Wang, A. H., & Frankland, P. W. (2007). Preferential incorporation of adult-generated granule cells into spatial memory networks in the dentate gyrus. *Nature Neuroscience, 10*(3), 355–362.

Kendall-Tackett, K. A. (2000). Physiological correlates of childhood abuse: Chronic hyperarousal in PTSD, depression and irritable bowel syndrome. *Child Abuse and Neglect, 24*(6), 799–810.

Kessler, R. C., Berglund, P., Chiu, W. T., Demler, O., Heeringa, S., & Hiripi, E. (2004). The US National Comorbidity Survey Replication (NCS-R) design and field procedures. *International Journal of Methods of Psychiatric Research, 13*(2), 69–92.

Kessler, R. C., Berglund, P. A., Demler, O., Jin, R., & Walters, E. E. (2005a). Lifetime prevalence of age-of-onset distributions on DSM-IV disorders in the National Comorbidyt Survey Repolication (NCS-R). *Archives of General Psychiatry, 62*(6), 593–602.

Kessler, R. C., Chiu, W. T., Demler, O., & Walters, E. E. (2005b). Prevalence, severity, and comorbidity of twelve-month DSM-IV disorders in the National Comorbidity Survey Replication (NCS-R). *Archives of General Psychiatry, 62*(6), 617–627.

Kessler, R. C., & Greenberg, P. E. (2002). The economic burden of anxiety and stress disorders. In K. Davis (Ed.), *Neuropsychopharmacology: The Fifth generation of progress*. Baltimore, MD: American College of Neuropsychopharmacology and Lippincott Williams & Wilkins.

Kessler, R. C., McGonagle, K. A., Swartz, M., Blazer, D. G., & Nelson, C. B. (1993). Sex and depression in the National Comorbidity Survey. I: Lifetime prevalence chronicity and recurrence. *Journal of Affective Disorders, 29*(2–3), 85–96.

Kessler, R. C., Sonnega, A., Bromet, E., Hughes, M., & Nelson, C. B. (1995). Posttraumatic stress disorder in the National Comorbidity Survey. *Archives of General Psychiatry, 52*, 1048–1060.

Kiefer, J. C. (2007). Epigenetics in development. *Developmental Dynamics, 236*(4), 1144–1156.

Kitayama, N., Brummer, M., Hertz, L., Quinn, S., Kim, Y., & Bremner, J. D. (2007). Morphologic alterations in the corpus callosum in abuse-related posttraumatic stress disorder: A preliminary study. *Journal of Nervous and Mental Disease, 195*, 1027–1029.

Klose, R. J., & Bird, A. P. (2006). Genomic DNA methylation: The mark and its mediators. *Trends in Biochemical Sciences, 31*(2), 89–97.

Koenigs, M., & Grafman, J. (2009). Posttraumatic stress disorder: The role of medial prefrontal cortex and amygdala. *Neuroscientist, 15*(5), 540–548. doi:10.1177/1073858409333072.

Koga, Y., Takahashi, H., Oikawa, D., Tachibana, T., Denbow, D. M., & Furuse, M. (2005). Brain creatine functions to attenuate acute stress responses through GABAnergic system in chicks. *Neuroscience, 132*(1), 65–71.

Kornberg, A., & Baker, T. A. (2005). *DNA replication* (2nd ed.). Mill Valley, CA: University Science.

Krysinska, K., & Lester, D. (2010). Post-traumatic stress disorder and suicide risk: A systematic review. *Archives of Suicide Research, 14*(1), 1–23. doi:10.1080/13811110903478997.

Last, J. M. (2001). *A dictionary of epidemiology* (4th ed.). London: Oxford University Press.

Lazarus, R. S., & Folkman, S. (1984). *Stress appraisal and coping*. New York: Springer Publishing Company. Retrieved from https://books.google.com/books?id=i-ySQQuUpr8C&printsec=frontcover&source=gbs_ge_summary_r&cad=0#v=onepage&q&f=false.

Lelkes, Y., Krosnick, J. A., Marx, D. M., Judd, C. M., & Park, B. (2012). Complete anonymity compromises the accuracy of self-reports. *Journal of Experimental Social Psychology, 48*, 1291–1299.

Lizberzon, I., Taylor, S. F., Amdur, R., Jung, T. D., Chamberlain, K. R., & Minoshima, S. (1999). Brain activation in PTSD in response to trauma-related stimuli. *Biological Psychiatry, 45*(7), 817–826.

Loman, M. M., Gunnar, M. R., & Early Experience, Stress and Neurobehavioral Development Center. (2010). Early experience and the development of stress reactivity and regulation in children. *Neuroscience and Biobehavioral Reviews, 34*(6), 867–876. doi:10.1016/j.neurobiorev.2009.05.007.

Lovallo, W. R. (2005). *Stress and health: Biological and psychological interactions* (2nd ed.). Thousand Oaks, CA: Sage Publications.

Maercker, A., Frostmeier, S., Wagner, B., Glaesmer, H., & Brähler, E. (2008). Post-traumatic stress disorder in Germany. Results of a nationwide epidemiological study. *Nervenarzt, 79*(5), 577–586. doi:10.1007/ss00115-008-2467-5.

Malcolm, S. (2001). Genotype to phenotype: Interpretations of the human genome project. In S. Malcolm & J. Goodship (Eds.), *Genotype to phenotype* (pp. 1–10). Oxford: BIOS Scientific Publishers.

Marshall, R. D., Olfson, M., Hellman, F., Blanco, C., Guardino, M., & Struening, E. I. (2001). Comorbidity, impairment and suicidality in subthreshold PTSD. *American Journal of Psychiatry, 158*, 1467–1473.

McEwen, B. S., & Seeman, T. (1999). Protecting and damaging effects of mediators of stress. Elaborating and testing the concepts of allostasis and allostatic load. *Annals of the New York Academy of Sciences, 896*, 30–47.

McEwen, B. S., & Stellar, E. (1993). Stress and the individual. Mechanisms leading to disease. *Archives of Internal Medicine, 153*(18), 2093–2101.

McFarlane, A. C. (2010). The long-term costs of traumatic stress: Intertwined physical and psychological consequences. *World Psychiatry, 9*(1), 3–10.

McGauran, N., Wieseler, B., Kreis, J., Schüler, Y. B., Kölsch, H., & Kaiser, T. (2010). Reporting bias in medical research-a narrative review. *Trials, 11*, 37.

Mealer, M., & Jones, J. (2014). Methodological and ethical issues related to qualitative telephone interviews on sensitive topics. *Nurse Researcher, 21*(4), 32–37.

Medina, J. J. (2008). Neurobiology of PTSD. *Psychiatric Times, 21*(4), 32–37.

Mehta, D., Klengel, T., Conneely, K. N., Smith, A. K., Altmann, A., & Pace, T. W. (2013). Childhood maltreatment is associated with distinct genomic and epigenetic profiles in posttraumatic stress disorder. *Proceedings of the National Academy of Sciences of the United States of America, 110*(20), 8302–8307. doi:10.1073/pnas.1217750110.

Meichenbaum, D. H. (1972). Cognitive modification of test anxious college students. *Journal of Consulting and Clinical Psychology, 39*(3), 370–380. doi:10.1037/h0033896.

Meichenbaum, D. (1994). *Treating adults with PTSD*. Clearwater, FL: Institute Press.

Melillo, R., & Leisman, G. (2004). *Neurobehavioral disorders of childhood: An evolutionary perspective*. New York, NY: Springer Science+Business Media.

Merikangas, K. R., He, J. P., Burstein, M., Swanson, S. A., Avenevoli, S., & Cui, L. (2010). Lifetime prevalence of mental disorders in the U.S. adolescents: results from the National Comorbidity Survey Replication-Adolescent Supplement (NCS-A). *Journal American Academy of Child & Adolescent Psychiatry, 49*(10), 980–989. doi:10.1016/j.jaac.2010.05.017.

Milad, M. R., & Quirk, G. J. (2002). Neurons in medial prefrontal cortex signal memory for fear extinction. *Nature, 420*(6911), 70–74.

Miller, G. E., Chen, E., & Zhou, E. S. (2007). If it goes up, must it come down? Chronic stress and the hypothalamic-pituitary-adrenocortical axis in humans. *Psychological Bulletin, 133*(1), 25–45.

Miller, B. L., & Cummings, J. L. (Eds.). (2007). Social cognition in frontal injury. *The human frontal lobes: Functions and disorders*. (2nd ed.). New York, NY: Guilford Press.

Mills, K., Lynskey, M., Teeson, M., Ross, J., & Darke, S. (2005). Post-traumatic stress disorder among people with heroin dependence in the Australian treatment outcome study (ATOS): Prevalence and correlates. *Drug and Alcohol Dependence, 77*(3), 243–249.

Mollica, R. F., McInnes, K., Poole, C., & Tor, S. (1998). Dose-effect relationships of trauma to symptoms of depression and post-traumatic stress disorder among Cambodian survivors of mass violence. *British Journal of Psychiatry, 173*, 484–488.

Morgan, M. A., Romanski, L. M., & LeDoux, J. E. (1993). Extinction of emotional learning: Contribution of medial prefrontal cortex. *Neuroscience Letters, 163*(1), 109–113.

Nakashiba, T., Cushman, J. D., Pelkey, K. A., Renaudineau, S., Buhl, D. L., & McHugh, T. J. (2012). Young dentate granule cells mediate pattern separation, whereas old granule cells facilitate pattern completion. *Cell, 149*(1), 188–201. doi:10.1016/j.cell.2012.01.046.

National Institute for Health & Clinical Excellence. (2005). *Post-traumatic stress disorder: the management of PTSD in adults and children in primary and secondary care*. London: Gaskell and the British Psychological Society.

National Institute of Mental Health. (2002). Annual total direct and indirect costs of serious mental illness. Retrieved from http://www.nimh.nih.gov/health/statistics/cost/index.shtml

National Scientific Council on the Developing Child. (2014). Excessive stress disrupts the architecture of the developing brain: Working paper 3. Retrieved from http://developingchild.harvard.edu

Navaje, M., Keefe, M. S., Hickey, A. H., McLay, R. N., Ritchie, E. C., & Abdi, S. (2014). Use of stellate ganglion block for refractory post-traumatic stress disorder: A review of published cases. *Anesthesia and Clinical Research, 5*(4), 1–9. doi:10.4172/2215-6148.1000403.

Nelson, C. B. (1996). Post-traumatic stress disorder in the National Comorbidity Survey. *Archives of General Psychiatry, 52*, 1049–1060.

Nelson, D., & Cox, M. M. (2013). *Principles of biochemistry*. New York, NY: W.H. Freeman and Company.

Neu, A., Neuhoff, H., Trube, G., Fehr, S., Ullrich, K., Roeper, J., & Isbrandt, D. (2002). Activation of GABA(A) receptors by guanidinoacetate: A novel pathophysiological mechanism. *Neurobiology of Disease, 11*(2), 298–307.

Noonan, M., Penque ,M., Axelrod ,S. (1996). Septal lesions impair rats' Morris test performance but facilitate left-right response differentiation. *Physiology & Behavior, 60*(3): 895–900.

Norman, S. B., Means-Christensen, A. J., Craske, M. G., Sherbourne, C. D., Roy-Bryne, P. P., & Stein, M. B. (2006). Associations between psychological trauma and physical illness in primary care. *Journal of Traumatic Stress, 19*(4), 461–470.

Nussey, S., & Whitehead, S. (2001). *Endocrinology: An integrated approach*. Retrieved from http://www.ncbi.nlm.nih.gov/books/NBK22/

Omoto, C. K., & Lurquin, P. F. (2004). *Genes and DNA, A beginner's guide to genetics and its applications*. New York, NY: Columbia University Press.

Osuch, E. A., Benson, B., Geraci, M., Podell, D., Hersovitch, P., McCann, U. D., & Post, R. M. (2001). Regional cerebral blood flow correlated with flashback intensity in patients with post-traumatic stress disorder. *Biological Psychiatry, 50*(4), 246–253.

Ouimette, P. C., Brown, P. J., & Najavits, L. M. (1998). Course and treatment of patients with both substance abuse and post-traumatic stress disorders. *Addictive Behaviors, 23*(6), 785–795.

Pastorino, E., & Doyle-Portillo, S. (2013). *What is psychology?* (2nd ed.). Belmont, CA: Cengage Learning.
Perry, B. D., & Azad, I. (1999). Posttraumatic stress disorders in children and adolescents. *Current Opinion in Pediatrics, 11*, 310–316.
Porta, M., Greenland, S., & Last, J. M. (2008). *A dictionary of epidemiology. A handbook sponsored by the I.E.A.* (5th ed.). New York, NY: Oxford University Press.
Purves, D., Augustine, G. J., Fitzpatrick, D., Katz, L. C., LaMantia, A. S., McNamara, J. O., & Williams, S. M. (Eds.). (2001). *Neuroscience* (2nd ed.). Sunderland, MA: Sinauer Associates.
RajMohan, V., & Mohandes, E. (2007). The limbic system. *Indian Journal of Psychiatry, 49*(2), 132–139.
Ranabir, S., & Reetu, K. (2011). Stress and hormones. *Indian Journal Endocrinology and Metabolism, 15*(1), 18–22. doi:10.4103/2230-8210.77573.
Rauch, S. L., Whalen, P. J., Shin, L. M., McInerney, S. C., Macklin, M. L., & Lasko, N. B. (2000). Exaggerated amygdala response to masked facial stimuli in posttraumatic stress disorder: A functional MRI study. *Biological Psychiatry, 47*, 769–776.
Ravoso, M. J., & Dagosto, M. (2007). *Primate origins: Adaptations and evolution*. Chicago, IL: Springer.
Resnik, D. B. (2011). What is ethics in research and why is it important? Retrieved from http://www.niehs.nih.gov/research/resources/bioethics/whatis/
Romanoski, C. E., Glass, C. K., Hendrick, G., Stunnenberg, L. W., & Almouzni, G. (2015). Epigenomics: Roadmap to regulation. *Nature, 518*, 314–316. doi:10.1038/518314a.
Sack W, Clarke G, Seeley J. Posttraumatic stress disorders across two generations of Cambodian refugees. *Journal of the American Academy of Child and Adolescent Psychiatry*. 1995;34: 1160–1166.
Sack, W. H., Clarke, G. N., & Seeley, J. (1995). Posttraumatic stress disorder across two generations of Cambodian refugees. *Journal American Academy Child and Adolescent Psychiatry, 34*(9), 1160–1166.
Sadava, D., Heller, H. C., Orians, G. H., Purves, W. K., & Hillis, D. M. (2014). *Life: The science of biology* (10th ed.). Sunderland, MA: Sinauer Associates.
Sareen, J., Cox, B. J., Stein, M. B., Afifi, T. O., Fleet, C., & Asmundson, G. J. (2007). Physical and mental comorbidity, disability and suicidal behavior associated with posttraumatic stress disorder in a large community sample. *Psychosomatic Medicine, 69*(3), 242–248.
Sartor, C. E., Grant, J. D., Lynskey, M. T., McCutcheon, V. W., Waldron, M., & Statham, D. J. (2011). Common heritable contributions to low-risk trauma, high-risk trauma, posttraumatic stress disorder, and major depression. *Archives of General Psychiatry, 69*, 293–299.
Schmidt, U., Holsboer, F., & Rein, T. (2011). Epigenetic aspects of posttraumatic stress disorder. *Disease Markers, 30*(2-3), 77–87. doi:10.3233/DMA-2011-0749.
Schmidt, U., Kaltwasser, S. F., & Wotjak, C. T. (2013). Biomarkers in posttraumatic stress disorder: Overview and implications for future research. *Disease Markers, 35*(1), 43–54. doi:10.1155/2013/835876.
Schnurr, P. P., Lunney, C. A., Bovin, M. J., & Marx, B. P. (2009). Posttraumatic stress disorder and quality of life: Extension of findings to veterans of the wars in Iraq and Afghanistan. *Clinical Psychology Review, 29*(8), 727–735. doi:10.1016/j.cpr.2009.08.006.
Schoenberg, M. R., & Scott, J. G. (Eds.). (2011). Neuroanatomy primer: Structure and function of the human nervous system. *The little black book of neuropsychology a syndrome based approach* (pp. 59–126). New York: Springer. Retrieved from https://www.google.com/url?sa=t&rct=j&q=&esrc=s&source=web&cd=1&ved=0CB8QFjAAahUKEwiftZOjhubGAhWGnIgKHXKrAl8&url=http%3A%2F%2Fm.friendfeed-media.com%2F5b30e067d11b854fd98b7235e7e76acee6f27345&ei=JP6qVZ-TEIa5ogTy1or4BQ&usg=AFQjCNEtD9MGxgNBKK4R4jhDL5MTZxfCtA&sig2=7KwDqw6HeeP4VyyZ0V9yVA
Schuff, N., Neylan, T. C., Lenoci, M. A., Du, A. T., Weiss, D. S., Marmar, C. R., & Weiner, M. W. (2001). Decreased hippocampal N-acetylaspartate in the absence of atrophy in posttraumatic stress disorder. *Biological Psychiatry, 50*(12), 952–959.
Seidlitz, L., & Diener, E. (1998). Sex differences in the recall of affective experiences. *Journal of Personality and Social Psychology, 74*(1), 262–271.

Selaman, Z. M., Cartrand, H. K., Bolton, J. M., & Sareen, J. (2013). Which symptoms of post-traumatic stress disorder are associated with sucide attempts? *Journal of Anxiety Disorders, 28*, 246–251.

Selye, H. (1950). Stress and the general adaptation syndrome. *British Medical Journal, 1*(4667), 1383–1392.

Seng, J. S., Graham-Bermann, S. A., Clark, M. K., McCarthy, A. M., & Ronis, D. L. (2005). Posttraumatic stress disorder and physical comorbidity among female children and adolescents: Results from service-use data. *Pediatrics, 116*(6), e767–e776.

Shin, L. M., Orr, S. P., Carson, M. A., Rauch, S. L., Macklin, M. L., Lasko, N. B., & Pitman, R. K. (2004). Regional cerebral blood flow in the amygdala and medial prefrontal cortex during traumatic imagery in male and female Vietnam veterans with PTSD. *Archives of General Psychiatry, 61*(2), 168–176.

Shnuff, N., Marmar, C. R., Weiss, D. S., Neyland, T. C., Schoenfeld, F., Fein, G., & Weiner, M. W. (1997). Reduced hippocampal volume and n-acetylaspartate in post-traumatic stress disorder. *Annals of the New York Academy of Science, 821*, 516–520.

Shonkoff, J. P., Boyce, W. T., & McEwen, B. S. (2009). Neuroscience, molecular biology and the childhood roots of health disparities: Building a new framework for health promotion and disease prevention. *JAMA, 301*(21), 2252–2259. Retrieved from http://jama.ama-assn.org/cgi/content/full/301/21/2252.

Sieghart, W. (1995). Structure and pharmacology of g-amino butyric acidA receptor subtypes. *Pharmacology Review, 47*, 181–234.

Sinden, R. R. (2012). *DNA structure and function*. St. Louis, MO: Elsevier.

Stewart, S. H. (1996). Alcohol abuse in individuals exposed to trauma: A critical review. *Psychological Bulletin, 120*, 83–112.

Stine, S. M., & Kosten, T. R. (1995). Complications of chemical abuse and dependency. In M. J. Friedman, D. S. Charney, & A. Y. Deutch (Eds.), *Neuroendrocrinology and pathophysiology of the stress system* (pp. 447–464). Philadelphia, PA: Lippincott-Raven.

Stratakis, C. A., & Chrousos, G. P. (2006). Neuroendocrinology and pathophysiology of the stress system. *Annals of the New York Academy of Sciences, 771*, 1–18. doi:10.1111/j.1749-6632.1995.tb44666.x.

Study, R. E., & Barker, J. L. (1981). Diazepam and --pentobarbital: Fluctuation analysis reveals different mechanisms for potentiation of gamma-aminobutyric acid responses in cultured central neurons. *Proceedings of the National Academy of Science, 771*, 1–18.

Surget, A., Tanti, A., Leonardo, E. D., Laugeray, A., Rainer, Q., & Touma, C. (2011). Antidepressants recruit new neurons to improve stress response regulation. *Molecular Psychiatry, 16*(12), 117–1188. doi:10.1038/mp.2011.48.

Takitori, M., Kuroda, Y., & Hirose, M. (2006). Local anesthetics suppress nerve growth factor-mediated neurite outgrowth by inhibition of tyrosine kinase activity of TrkA. *Anesthesia and Analgesia, 102*(2), 462–467.

Talilaz, D., Loya, A., Gersner, R., Haramati, S., Chen, A., & Zangen, A. (2011). Resilience to chronic stress is mediated by hippocampal brain-derived neurotropic factor. *The Journal of Neuroscience, 31*(12), 4475–4483. doi:10.1523/JNEUROSCI.5725-10-2011.

Tolin, D. F., & Foa, E. B. (2006). Sex differences in trauma and posttraumatic stress disorder: A quantitative review of 25 years of research. *Psychology Bulletin, 132*(6), 959–992.

Trevor, A. J., & Way, W. L. (2001). Sedative-hypnotic drugs. In B. G. Katzung (Ed.), *Basic and clinical pharmacology* (8th ed., pp. 364–381). New York, NY: Lange Medical Books/McGraw-Hill.

True W, Rice J, Eisen S, Heath A, Goldberg J, Lyons M, et al. Twin Study of Genetic and Environmental Contributions to Liability for symptoms of posttraumatic stress symptoms. *Archives of General Psychiatry*. 1993;50:257–264.

Turnbull, G. J. (1998). A review of post-traumatic stress disorder. Part I: Historical development and classification. *Injury, 29*(2), 87–91.

U.S. Department of Health and Human Services Children's Bureau (Administration on Children, Youth and Families, Administration for Children and Families). (2012). Child maltreatment. Retrieved from http://www.acf.hhs.gov/sites/default/files/cb/cm2012.pdf

Uddin, M., Galea, S., Chang, S. C., Aiello, A. E., Wildman, D. E., De los Santos, R., & Koenen, K. C. (2011). Gene expression and methylation signatures of MACN2CI are associated with PTSD. *Disease Markers, 30*, 111–121. doi:10.3233/MDA-2011-0750.

Vaccarino, V., Goldberg, J., Magruder, K. M., Forsberg, C. W., Friedman, M. J., & Litz, B. T. (2014). Posttraumatic stress disorder and incidence of type-2 diabetes: A prospective twin study. *Journal of Psychiatric Research, 56*, 158–164.

Van den Broeck, J., & Brestoff, J. R. (Eds.). (2013). Basic concepts in epidemiology. *Epidemiology: Principles and practical guidelines* (pp. 19–36). London: Springer.

Van Der Kolk, B. A., & Saporta, J. (1991). The biological response to psychic trauma: Mechanisms and treatment of intrusion and numbing. *Anxiety Research, 4*, 199–212.

Voisey, J., Young, R. M., Lawford, B. R., & Morris, C. P. (2014). Progress towards understanding the genetics of posttraumatic stress disorder. *Journal of Anxiety Disorders, 28*(8), 873–883. doi:10.1016/j.janxdis.2014.09.014.

Wang, C. T., & Holton, J. (2007). *Total estimated costs of child abuse and neglect in the United States*. Chicago, IL: Prevent Child Abuse America.

Watt, F., & Molloy, P. L. (1988). Cytosine methylation prevents binding to DNA of a HeLa cell transcription factor required for optimal expression of the adenovirus major late promoter. *Genes and Development, 2*, 1136–1143.

Weinstock, M. (2005). The potential influence of maternal stress hormones on development and mental health of the offspring. *Brain, Behavior, and Immunity, 19*(4), 296–308.

Weiss, D. S., Marmar, C. R., Schlenger, W. E., Fairbanks, J. A., Jordan, B. K., Hough, R. L., & Kulka, R. A. (1992). The prevalence of lifetime and partial post-traumatic stress disorder in Vietnam theater veterans. *Journal of Traumatic Stress, 5*(3), 365–376.

Yehuda, R., & Bierer, L. M. (2009). The relevance of epigenetics to PTSD: Implications for DSM-V. *Journal of Traumatic Stress, 22*(5), 427–434.

Yehuda, R., Bierer, L., Schmeidler, J., Aferiat, D., Breslau, I., & Dolan, S. (2000). Low cortisol and risk for PTSD in adult offspring of holocaust survivors. *American Journal of Psychiatry, 157*(8), 1252–1259.

Yehuda, R., Koenen, K. C., Galea, S., & Flory, J. D. (2011). The role of genes in defining a molecular biology of PTSD. *Disease Markers, 30*(2–3), 67–76. doi:10.3233/dma-2011-0794.

Yehuda, R., Schmeidler, J., Wainberg, M., Binder-Brynes, K., & Duvdevani, T. (1998). Vulnerability to posttraumatic stress disorder in adult offspring of Holocaust survivors. *American Journal of Psychiatry, 155*(9), 1163–1171.

Zaidi, L. Y., & Foy, D. W. (1994). Childhood abuse experiences and combat-related PTSD. *Journal of Traumatic Stress, 7*(1), 33–42.

Zatrick, D. F., Marmar, C. R., Weiss, D. S., Browner, W. S., Metzler, T. J., & Golding, J. M. (1997). Posttraumatic stress disorder and functioning and quality of life outcomes in a nationally representative sample of male Vietnam Veterans. *American Journal of Psychiatry, 154*(12), 1690–1695.

Zhang, T. Y., & Meaney, M. J. (2010). Epigenetics and the environmental regulation of the genome and its function. *Annual Review of Psychology, 6*, 439–466. doi:10.1146/annurev.psych.60.110707.163625.

Zovik, I., & Sweatt, J. D. (2013). Epigenetic mechanisms in learned fear: Implications for PTSD. *Neuropsychopharmacology, 38*, 77–93.

Chapter 12
Ethical Implications of Epigenetics

Steven G. Gilbert

12.1 Introduction

Periodically a new discovery provides insights that are so fundamental as to alter the ethical basis of our decision making. The discovery of epigenetics, the phenomenon where the expression of DNA can be altered without changing the DNA sequence, is such an event. Epigenetic changes can potentially occur throughout the lifespan and even be passed on to future generations. The epigenome is particularly vulnerable during development to epigenetic changes that can manifest as disease later in life and become sources of the developmental origins of health and disease (DOHaD). Epigenetic changes occur in response to exposure to a range of environmental contaminants such as heavy metals, pesticides, organic pollutants, and air pollutants, as well as in response to nutritional factors and stress. These subtle changes to the epigenome can result in increased susceptibility to cancers, immunological disorders, neurodegenerative diseases, developmental disorders, and other conditions.

Recognition of the adverse effects of epigenetic changes coincides with a growing appreciation that even very low levels of exposure to toxic agents can cause adverse effects, especially in the developing organism. A classic example of this phenomenon is lead exposure. After decades of often repetitive research, most public health professionals and government agencies recognize that there is no safe level of exposure during development. Many researchers agree that endocrine disrupting chemicals (EDCs), which come in a variety of kinds, act at unexpectedly low doses (Vaiserman 2014). EDCs also illustrate another important issue, which is the challenge of establishing a good risk-assessment paradigm that addresses

S.G. Gilbert, Ph.D., D.A.B.T. (✉)
Institute of Neurotoxicology and Neurological Disorders (INND),
3711 47th Place NE, Seattle, WA 98105, USA

Interdisciplinary Arts and Sciences, University of Washington, Bothell, WA, USA
e-mail: sgilbert@innd.org; sgg@u.washington.edu

D. Hollar (ed.), *Epigenetics, the Environment, and Children's Health Across Lifespans*, DOI 10.1007/978-3-319-25325-1_12

exposures to not just one chemical but to multiple chemicals that have similar mechanisms of action.

Application of the science of epigenetics requires reconsideration of an ethical framework for decision making. While there are a number of ethical frameworks that could be considered, the precautionary principle captures a number of important elements. The precautionary principle is commonly stated as such: "When an activity raises threats of harm to human health or the environment, precautionary measures should be taken even if some cause and effect relationships are not fully established scientifically (Wingspread Conference, 1998)" (Raffensperger and Tickner 1999). The primary elements of the precautionary principle can be summarized as "do no harm" and "take action even if there is uncertainty." When this principle is applied along with the standard risk assessment approach, the outcome is to reduce or minimize the chemical exposure. However, epigenetic changes move beyond chemicals to include conditions caused by nutrition or stress. A new ethical framework for decision making must move beyond the precautionary principle to one of "epiprecaution." Epiprecaution requires us to include the entire environment of the developing and mature individual into consideration. In other words, "do no harm" is not good enough, we must "do good" and create a nurturing environment. It is not sufficient to simply reduce chemical exposure but we must take the next step to create a nurturing environment.

12.2 Epigenetics: The Genes but More

The concept of epigenetics, meaning something acting above or in addition to genes, is both old and new. One of the greatest scientific questions has been how life replicates, adapts, changes, and reproduces. Gregor Johann Mendel (July 20, 1822–January 6, 1884), an Austrian scientist, founded the new science of genetics by demonstrating that the inheritance of certain traits in pea plants follows particular patterns. Mendel did not know it but he was describing the functionality of DNA. About 100 years later, James Watson and Francis Crick discovered the double helical structure of the DNA molecule (Watson and Crick 1953). Decoding the genome and looking for variations in an individual's DNA was a focus of research to help explain individual variability and susceptibility to disease or toxic effects. We have been accustomed to blaming a lot on "our genes." But then the story got both more complicated and intriguing with the recognition of the subtle power of epigenetics.

Conrad Hal Waddington (November 8, 1905–September 26, 1975) postulated that it was not just the genes that shaped development but also the environment that shaped the genes. In the late 1930s, he wrote a textbook describing developmental epigenetics, a term that then meant how the environment shaped genetic activity. Understanding a possible mechanism for this had to wait until a far deeper understanding of DNA and its role in development was realized. Epigenetics is now defined as changes in gene expression caused by factors other than changes in the

DNA sequence (epi meaning "above" or "in addition to"). When epigenetic changes occur, gene expression typically is silenced or suppressed by DNA methylation or histone deacetylation. These ideas were first articulated in 1975. The power and subtlety of epigenetics are that these changes can be passed to the next generation. The ability of maternal conditions to influence fetal development was also articulated by David J.P. Barker in what became known as the Barker Hypothesis (Barker 1997). The Barker Hypothesis suggested that reduced fetal development due to poor maternal nutrition could result in cardiovascular diseases or diabetes later in life. A possible benefit is that the maternal conditions are preparing the next generation for these adverse conditions.

While we are well aware of the damage thalidomide, alcohol, lead, mercury, or PCBs can cause to the developing organism, more subtle epigenetic changes can also result from environmental exposures. A dramatic example of the power of epigenetic information transfer was demonstrated by research on the endocrine-disruptive fungicide vinclozolin, which is used on several agricultural crops such as raspberries and lettuce (Anway et al. 2006). Male offspring of the exposed mother rat transferred disease susceptibility to the next generation (Skinner and Anway 2007). Further research has indicated that epigenetic changes can result from exposure to environmental hazards such as cigarette smoke, arsenic, alcohol, phthalates, and BPA. Furthermore, epigenetic changes can occur through nutritional factors, methyl content of the diet, intake of folic acid and vitamins, or even social and maternal behavior toward the offspring. Epigenetic changes can cause cancer, neurodevelopment disorders, autoimmune disease, diabetes, asthma, behavioral disorders, and other changes.

The "environment" may be taken broadly to include exposure to toxic chemicals, nutritional factors, maternal and paternal stress, and behavioral well-being: in other words, anything that can influence fetal genomic development. Epigenetic changes offer an explanation of how an organism can rapidly adapt to the changing nature of the environment (Jirtle and Skinner 2007).

12.3 Ethical Consideration and Responsibility

The challenge is what to do with the tremendous amount of knowledge available not only to scientists but also to the business community, government, and general public. In many ways, epigenetics is a record of environmental exposures and organisms' efforts to adapt to environmental changes and even pass along these experiences to future generations. What are our responsibilities stemming from our understanding of epigenetics?

Humans now have incredible knowledge to reshape the environment and affect human health, but we have yet to fully acknowledge the responsibility that this implies. We need to take more responsibility around the manufacture, use, exposure, and disposal of chemicals.

It is estimated that there are more than 80,000 chemicals in commerce and 2000 new chemicals are added each year. Trillions of pounds of these chemicals are manufactured or imported every year just in the US. Unfortunately, we know very little about the specific health effects of these chemicals because industry has not generated or made available the data. We do know, however, that children are more vulnerable to the effects of these chemicals and that annual costs of childhood disease due to environmental contaminants is in the range of $55 billion in 2002 (Landrigan et al. 2002) and certainly more now. Children and adults are exposed to a wide range of chemicals in homes, schools, workplaces, and from products used. Exposure to some of these chemicals can cause significant adverse health effects such as cancer, Parkinson's disease, immunological disorders, and neurobehavioral deficits, resulting in a needless loss of potential for both the individual and society. Our growing understanding of epigenetics is an important link to a mechanism by which various agents affect gene expression, particularly at low doses.

12.4 Responsibility: An Overview

Humans have amassed an enormous amount of power to change the physical environment as well as affect human and environmental health. Aldo Leopold, America's first bioethicist, summarized our ethical responsibilities in a simple statement in 1949: "A thing is right when it tends to preserve the integrity, stability, and beauty of the biotic community. It is wrong when it tends otherwise" (Leopold 1949). When we expose our children to lead, mercury, or alcohol, we are robbing them of their integrity, stability, and beauty. In essence, we are robbing them of their potential, reducing their ability to do well in school and contribute to society (Gilbert 2005). We have the knowledge to inform our decisions and must accept the responsibility to preserve the biotic community, which will preserve ourselves, as well as future generations. Key institutions in our society, as well as individuals, must address different aspects of a shared responsibility to ensure a sustainable biotic community. This means we must have a better understanding of the effects of epigenetic changes over the lifetime of an individual.

12.5 Corporate Responsibility

Under current corporate rules and regulations, the primary responsibility of a corporation is to make money for its shareholders. Corporate management's primary responsibility is to increase the value of the corporation for its shareholders, which is accomplished by increasing revenue or product sales and by reducing or externalizing costs. In 1994, an array of tobacco executives stood before the U.S. Congress Subcommittee on Health and the Environment and swore that nicotine was not addictive (Henningfield et al. 2006). This was clearly false, but they were protecting the interest of their corporations and shareholders to profit at the expense of people's health. The health effects of tobacco are borne by the individual and collectively through taxes and healthcare costs. The tobacco companies have a long history of externalizing the health costs of their product onto taxpayers while reaping profits for the executives and shareholders, a point of criticism that is debatable and that could be applied to many corporations.

Other corporations have also externalized, or not accounted for, the costs of dumping chemicals into the air, water, or land, which result in disease and environmental damage (Flores et al. 2011). Some corporations contaminate the environment because it is cost effective and our laws shield executives from personal responsibility. In other words, they operate this way because they can make larger profits by not investing in pollution control or adapting sustainable practices and they can get away with it. Of course not all corporations operate irresponsibly, but enough do that their behavior merits serious scrutiny. The challenge is to discuss the ethical responsibilities of corporations to protect and enhance human and environmental health. What is the corporate responsibility to control the use of brominated flame retardants, BPA, and other EDCs found in products, which may have epigenetic effects that increase the probability of diabetes or obesity?

12.6 Government Responsibility

The primary responsibility of the government is to protect and preserve common wealth for the greater good of the people. Government has a duty and responsibility to ensure the "integrity, stability, and beauty of the biotic community." In essence, government must ensure that future generations have an environment in which they can reach and maintain their full genetic potential. The U.S. Government has made various attempts to control chemicals, while the governments of many developing countries such as China are just beginning to consider the problems of uncontrolled corporate exploitation of the environment and people. A failed effort by the U.S. Congress was the passage of Toxic Substances Control Act of 1976 (TSCA). This law was meant to empower the Environmental Protection Agency (EPA) to control the introduction of new chemicals into the environment. Unfortunately, corporations are not required to generate or make available health effects data, which impedes the government and the public from making informed decisions on the safety of products. Our representatives in government must take seriously their responsibility to protect common wealth for the greater good of all. A first step would be to fix TSCA by requiring greater chemical testing and disclosure of this information. But even more challenging is how to incorporate new information such as on epigenetic changes into laws that protect human and environmental health. Should we require some form of epigenetic assessment that can be incorporated into risk assessment?

12.7 Media Responsibility

The primary responsibility of the media is to create an informed and engaged public, not just to inform the educated public. The media has an obligation to produce socially responsible material that is fair, objective, and balanced. This does not mean giving equal time to minority views, as was the case with global warming. Most importantly, the media has a responsibility to be open and transparent about sources of information and to acknowledge any potential conflicts of interest. The burden and obligations of the media to be responsible must also be shared with the listeners, viewers, and readers. The media has great power to inform and influence people and with that comes a grave responsibility. It is essential that the media have people that understand and can explain epigenetics to not only the public but also to the political decision makers.

12.8 Academic Responsibility

The academic community, particularly those engaged in issues related to public health, has a responsibility to be thoughtful public health advocates and to share their knowledge beyond narrow academic journals and conferences. Being a scientist includes the obligation to seek the truth and question the facts, and there is also an obligation and responsibility to speak out on public health issues. Scientists and educators have tremendous amounts of knowledge that can be shared with K-12 students, media, legislators, and the general public. Educators and researchers have a responsibility to help create an informed public by sharing their knowledge and being thoughtful public health advocates.

12.9 Individual Responsibility

Individuals have the greatest burden of responsibility because we must take into account not only the above responsibilities of our professional lives, but we must also address the responsibilities of our personal lives. We must confront individually and collectively that we have the power, and the means, to reshape or even destroy the world. Individually, it may seem as if we have little control over global warming, nuclear weapons, or food imported from other countries. But we have a responsibility to consider how our individual actions combine to collectively shape the world and society around us. This extends from whom we elect for office to what we buy in the store, to the temperature in our homes, and the pesticides on our farms and lawns. We also have a responsibility to stay informed and demand that the media inform us. Democracy is a participatory sport and we must be well informed to participate. Our corporations run on, and will respond to, what we purchase. Our government and corporations will respond to our opinions and demands for a fair, just, and sustainable society. We must translate responsibility into action to create a just and sustainable world.

12.10 Expanding the Ethical Framework

Our expanding appreciation of the influence of epigenetic changes on development will have profound effects on risk assessment and strongly argues for a precautionary approach to managing chemicals. Much of our basic ethical construct has been based on "do no harm." But is this simple approach good enough now that we are aware of epigenetic changes? In rodents, maternal grooming or lack of grooming results in significant epigenetic changes. This means that it is not just enough to have a developmental environment free of chemical contaminants but that there must be a loving and supportive environment during development. We must move beyond just "doing no harm" to "doing good." The concept of epigenetics provides the scientific and biological foundation for the necessity of "doing good." This concept is called "epiprecaution" to signify the need to move above and beyond preventing exposures to harmful material to creating an environment that is nurturing and supportive. Our expanding appreciation of the influence of epigenetics on development will have profound effects on our ethical thinking. We have an ethical responsibility to ensure that our children have an environment in which they can reach and maintain their full potential, not just free of exposure to chemicals but an environment that is supportive and nurturing. Epiprecaution moves beyond just doing no harm to one of creating a positive and supportive environment for our children.

The power of epigenetics is that it describes a plausible mechanism of action for the low-dose effects of chemicals and the onset of disease later in life. With this knowledge come other challenges, such as the potential for misuse of information. Many of these issues are similar to ethical situations addressed in genomics and gene therapy. These include issues of privacy and confidentiality of epigenetic information. But all these issues must be considered within the framework of taking responsibility for individual and community health. Epigenetics provides yet another layer of knowledge emphasizing individual vulnerability to disease. The high levels of uncertainty make cause and effect difficult to determine, but the critical aspect is understanding that stress and chemical exposure can undermine health and well-being. This means that some form of an epiprecautionary approach is necessary to ensure an individual can reach and maintain their full potential. Our ethical challenge or responsibility is to create an environment that is free of toxic agents and nurtures physical and emotional health, allowing all individuals to reach and maintain their full potential.

Abbreviations

DOHaD Developmental Origins of Health and Disease
EDCs Endocrine disrupting chemicals
EPA Environmental Protection Agency
TSCA Toxic Substances Control Act of 1976

References

Anway, M. D., Leathers, C., & Skinner, M. K. (2006). Endocrine disruptor vinclozolin induced epigenetic transgenerational adult-onset disease. *Endocrinology, 147*(12), 5515–5523. doi:10.1210/en.2006-0640. en.2006-0640 [pii].

Barker, D. J. (1997). Maternal nutrition, fetal nutrition, and disease in later life. *Nutrition, 13*(9), 807–813.

Flores, M. I. Grineski, S. E. & Collins, T. W. (2011). Using visual images to achieve environmental justice: a case study of the Asarco Copper Smelter in El Paso, Texas. Environmental Justice, 4(1), 45–53.

Gilbert, S. G. (2005). Ethical, legal, and social issues: Our children's future. *Neurotoxicology, 26*(4), 521–530. doi:10.1016/j.neuro.2004.12.006. S0161-813X(04)00168-8 [pii].

Henningfield, J. E. Rose, C. A. & Zeller, M. (2006). Tobacco industry litigation position on addiction: continued dependence on past views. Tobacco Control, 15 (suppl. 4), iv27-iv36. doi:10.1136/tc.2005.013789.

Jirtle, R. L., & Skinner, M. K. (2007). Environmental epigenomics and disease susceptibility. *Nature Reviews Genetics, 8*(4), 253–262. doi:10.1038/nrg2045. nrg2045 [pii].

Landrigan, P. J., Schechter, C. B., Lipton, J. M., Fahs, M. C., & Schwartz, J. (2002). Environmental pollutants and disease in American children: Estimates of morbidity, mortality, and costs for lead poisoning, asthma, cancer, and developmental disabilities. *Environmental Health Perspectives, 110*(7), 721–728.

Leopold, A. (1949). *A Sand County Almanac*. New York: Oxford University Press.

Raffensperger, C., & Tickner, J. (Eds.). (1999). *Protecting public health and the environment: Implementing the precautionary principle*. Washington, DC: Island Press.

Skinner, M. K., & Anway, M. D. (2007). Epigenetic transgenerational actions of vinclozolin on the development of disease and cancer. *Critical Reviews in Oncogenesis, 13*(1), 75–82. 03e4ce9b1 9ac1381,05554a2461c72327 [pii].

Vaiserman, A. (2014). Early-life exposure to endocrine disrupting chemicals and later-life health outcomes: An epigenetic bridge? *Aging and Disease, 5*(6), 419–429. doi:10.14336/ AD.2014.0500419.

Watson, J. D., & Crick, F. H. (1953). The structure of DNA. *Cold Spring Harbor Symposia on Quantitative Biology, 18*, 123–131.

Chapter 13
Child Obesity and Epigenetics

Caroline Hohensee, Tricia Varela, and Dustin Harris

13.1 Introduction

Childhood obesity is a preventable disease affecting millions of youth in the United States. Its prevalence is spreading throughout the world and has been declared one of the most serious health challenges of the twenty-first century by the World Health Organization (2015).

Obesity in children is defined as body mass index (BMI) equal to or greater than the 95th percentile, while overweight is defined as the 85th to less than or equal to the 95th percentile (Kuczmarski et al. 2002). Approximately 17 % of children aged 2–19 are obese in the United States. When overweight is included, an estimated 32 % of children, or almost one-third, are affected by higher than normal BMI (Ogden et al. 2014). While the proportion of young children 2–5 years has decreased slightly from 13.9 % in 2004 to 8.4 % in 2012, the number of overweight and obese among 6–19-year olds remains high (Ogden et al. 2014). When stratified by ethnicity, obesity prevalence was higher among Hispanics (22.4 %) and non-Hispanic black children (20.2 %) than non-Hispanic white (14.1 %) and non-Hispanic Asian children (8.6 %) (Ogden et al. 2014).

The obesity epidemic is associated with major health consequences. The most common condition is cardiovascular disease, which involves high blood pressure,

C. Hohensee, Ph.D., M.H.A. (✉)
Department of Health Administration, Pfeiffer University, 4701 Park Road, Charlotte, NC 28209, USA
e-mail: Caroline.Hohensee@Pfeiffer.edu

T. Varela, M.H.A.
Novant Health Group, 9431 Bayview Pkwy, Charlotte, NC 28216, USA
e-mail: Tajames333@aol.com

D. Harris, B.S.
HireRight, Charlotte, NC, USA
e-mail: dustin.harris@hireright.com; DSHarris191@my.pfeiffer.edu

D. Hollar (ed.), *Epigenetics, the Environment, and Children's Health Across Lifespans*, DOI 10.1007/978-3-319-25325-1_13

elevated cholesterol, and abnormal glucose tolerance. One study of 5- to 17-year-olds demonstrated that 60 % of overweight children had at least one of the risk factors for cardiovascular disease (Freedman, Dietz, Srinivasan, and Berenson 1999). Other health conditions related to overweight that are equally serious include asthma, diabetes, cancers, orthopedic conditions, and cerebrovascular disease (Dietz 1998; Witmer 2007).

Social consequences of obesity are also problematic. Overweight children are "targets of early and systematic social discrimination" (Schwartz and Puhl 2003). The effects of psychosocial stress can decrease self-esteem leading to lower academic performance and decreased social functioning (Reilly 2007). These effects can persist into adulthood and have economic repercussions.

Economic consequences of obesity have the potential to be devastating to the nation.

Hammond and Levine (2010) found the total annual economic costs associated with obesity to be greater than $215 billion. The cost areas reviewed included (1) direct medical spending, (2) productivity costs such as absenteeism, presenteeism, or decreased productivity due to illness, disability, and premature mortality, (3) transportation costs, e.g., excess fuel costs of commercial jets due to excess weight of obese passengers, and (4) human capital accumulation, which include quantity and quality of educational attainment. The economic consequences of obesity are produced mostly by adults. However, because most obese children become obese adults (Freedman et al. 2005), it is important to keep in mind the long-term economic consequences of child obesity.

13.2 Epigenetics and Child Obesity: An Overview

Energy imbalance is known to be the underlying cause of obesity. That is, obesity results when more energy, or calories, are consumed versus energy expended (Hall et al. 2011). However, science has shown the equation to be more complex. Excess adiposity gain for an individual involves a complex interaction of biochemical mechanisms, hormones, and most importantly genes (Choquet and Meyre 2011). Without a genetic predisposition to obesity, one is at low risk for excess weight gain.

Genes, or deoxyribonucleic acid (DNA), are the blueprint of our biological makeup. However, not all genes are fully expressed all of the time. When genes are not expressed they lie dormant and have *no* impact on our health (Campbell and Campbell 2006). Some genes remain dormant our entire life. This phenomenon of why genes are "turned on" is epigenetics. The role of epigenetics in the development of poor health outcomes such as obesity is gaining increased attention.

In the China Study (Campbell and Campbell 2006), researchers found that people of roughly the same ethnic background who likely had similar genes had large differences in disease rates, depending on their environment. Other research studies have similarly demonstrated that when people migrate, they most often assume the disease risk of the country to which they move (Koya and Egede 2007). Such studies alongside the relatively short timeframe of our childhood obesity epidemic indicate that the source of the problem is one of gene expression via environmental change and not one of a change in the genetic makeup of our society. Geneticists postulate that genetic shifts occur over a period of hundreds to thousands of years. So, what then causes child obesity related genes to turn on?

Scientists are finding that epigenetic markers on the DNA sequence trigger overweight and that the environment is causing these markers to activate. According to the Center for Disease Control and Prevention (2013), child obesity varies by environmental and behavioral factors such as racial/ethnic group, level of education for head of household, and household income for preschool children. These research studies suggest that behavior and environment play a crucial role in turning "on" obesity genes in children.

13.3 Prenatal Care and Epigenetics

The number of studies conducted to date attempting to understand childhood obesity is extensive. Many of these studies have provided irrefutable facts that parental obesity, specifically maternal obesity, has profound effects on a child's body mass index. Recognizing that the mother is often the primary caregiver, as well as the primary grocery shopper and meal preparer in the home, the influence a mother has on a child's exercise and nutrition is evident. A sedentary parent may encourage a sedentary lifestyle either consciously or indirectly through role modeling inactive choices. A mother who struggles with purchasing and preparing nutritious foods balanced with vitamins and minerals either due to budget constraints or lack of education will develop a child's palate for the foods often linked to childhood obesity. If these external influences were the sole sources for childhood obesity, then the myriad of community programs focused on getting children moving at least 60 min a day, or eating healthier surely should have an impact on childhood obesity, agree?

Another explanation for childhood obesity may be an epigenetic modification in utero to the gene determined to influence obesity. Most people assume their DNA predetermines obesity. If this were the only influence then how does one explain an obese child when neither parent is obese? Studies suggest that mothers who limit their carbohydrate intake during their pregnancy, otherwise dieting to limit weight gain or through malnutrition, create an environment for the fetus in utero that will otherwise increase a child's body fat in the first 9 years of life (Godfrey et al. 2011a, b). Simply put, carbohydrates are fuel for the body. A portion of the carbohydrates consumed is used by the body immediately to create instant energy. The balance of the carbohydrates consumed becomes stored as glycogen in the body's muscles and liver. Eliminating carbohydrates as a source of fuel forces the body into a fasting or starvation mode. This starvation mode becomes a learned behavior for the fetus. During pregnancy, researchers collected DNA samples from fetus' umbilical cords. Nine years later researchers weighed and measured the children's body fat. Researchers conducted this study with two separate groups of participants. In both groups, there was a distinct corollary observed among mothers with decreased carbohydrate intake during pregnancy and their children who at age of nine shared a change, anomaly, or affect to the retinoid X receptor-α (RXR-α) gene that influence's a child's body fat.

Godfrey et al.'s (2011a, b) study suggested a cause and effect. What if the epigenetic change is in the translation of the message sent, then received? Gluckman and Hanson (2008) suggested that childhood obesity is the result of developmental plasticity resulting from epigenetics versus the effect of ketosis the child's body learned in utero. The study suggested that children whose mothers experience a low nutrient prenatal diet, thus being born with a lower birth weight, have a greater risk at or predisposition for being obese children. There is no distinction as to whether the low birth weight was the result of socioeconomic factors, mothers who self-imposed prenatal diets with low nutrition in the effort to reduce weight gain during pregnancy or mothers who experienced severe morning sickness during pregnancy. By eliminating the varying hypotheses for the low birth weights, clinical evidence suggests that these children, instead of maintaining a natural appetite control, develop a preference for higher fat foods.

The studies available to date focus primarily on the possible physical effects that may manifest in childhood from documented low nutrient prenatal diets. No articles were found that tackle the emotional or psychological impact, or the reactance theory, of withholding higher fat foods from children from birth to age nine and how this would influence a child's food preference when offered a choice. However, an alternate perspective that is valid and remains to be explored is the epigenetic effect of a high-fat diet during pregnancy on childhood obesity. If a fetus is exposed to hyper nutrition, or overconsumption, in utero with no other mitigating factors such as gestational diabetes or maternal obesity present, would this translate into a higher satiety threshold during childhood resulting in obesity (Gluckman et al. 2008)?

There is little research available linking a high-fat prenatal diet to childhood obesity in children born to otherwise normal weight mothers. Most studies focus on the effects of gestational diabetes during pregnancy or mothers whose body mass index already places her in the category of obese. These common themes present in the prenatal environment have been linked to higher birth weights that later translated into childhood obesity. Shapira (2008) reported that hypernutrition, or overconsumption, does not necessarily ensure increased or even adequate nutrient intake—only higher fat or caloric intake. In fact, Shapira (2008) determined that the level of folic acid, iron, and vitamin D present in the women who reported overconsumption during the prenatal period was comparable as low as the levels present in women who reported a poor quality prenatal diet, or malnutrition.

Based on the work of Gluckman et al. (2008) and Shapira (2008), the logical deduction is that both hypernutrition and malnutrition have similar epigenetic influences on the RXR-a gene. Creating these conditions during pregnancy create an environment for the fetus in utero that translates into obesity during childhood. Therefore, the key to reducing childhood obesity, or uncrossing the wires, is to focus on the quality of the nutrients consumed during pregnancy and not the quantity. Yajnik (2010) coined the term thin-fat. During his research, Yajnik (2010) recognized that infants born to mothers who had low nutrient intake during pregnancy had babies with less lean mass and more fat mass. Low nutrient intake was determined by the low levels of vitamin B_{12} and folate present.

Unfortunately, during pregnancy, and when the socioeconomic level of the mother is not a factor, food cravings often determine many of the calories consumed during the prenatal diet. Sweet, sour, salty, or spicy are the primary sensations of taste. During pregnancy, many women report overconsumption of foods such as chocolate, pickles, and ice cream. Although all provide high fat and high calories, none of these selections is a nutrient rich food source. Women who experience severe or long-term morning sickness during pregnancy often retreat to only consuming popsicles or crackers. Although these foods are low fat and low calorie, they also do not provide the fetus with a complex, rich nutrient diet in utero.

Identifying women during pregnancy who are experiencing the adverse nutrient effects of either malnutrition or overconsumption may be the key to unlocking the epigenetic predisposition to childhood obesity. Identifying fetuses while still in utero who are at risk of being obese children allows the medical community, nutritionists, and parents a key indicator to determine which children require greater intervention and monitoring to ensure they maintain a healthy body mass index to puberty.

13.4 Postnatal Environment and Epigenetics

Monozygotic twins will have the same genes; however, as they age they will behave differently, make distinct decisions, adapt their personalities, be affected by variations in health, and illustrate changes in appearance (Powledge 2011). These variations are called epigenetics, which is the study of changes in gene expression in organisms without affecting the genetic coding (Bollati and Baccarelli 2010). Postnatal epigenetic occurrences tend to cause variation amongst appearance, physiology, cognition, and behavior; these occurrences are described as phenotypes (Powledge 2011). The epigenetic occurrences are dependent on environmental factors including social experience, nutrition, hormones, and postnatal and adult exposure to toxins (Powledge 2011). These factors can affect the expression of genes that code for certain diseases, creation of proteins needed for proper bodily function, cellular function, and the levels of DNA methylation and histone modifications needed for learning and remembering (Powledge 2011). One example of this was shown in a study looking at the influences of social interactions with child rats with their mothers. It was shown that more nurturing mothers resulted in offspring that were less anxious than those with lethargic mothers (Powledge 2011). Another example shows where a human study illustrated higher levels of morbidity and mortality from cardiorespiratory disease when there was exposure to particulate matter air pollution (Bollati and Baccarelli 2010). The study showed that when levels of particulate matter increased there was a decrease in the methylation of the iNOS gene, therefore resulting in inflammation and oxidative stress, which correlate to the mechanisms that link particulate matter inhalation to the aforementioned health effects (Bollati and Baccarelli 2010). Lastly, one of the most imperative postnatal epigenetic factors is nutrition.

Early diets play a vital role in child development and correlate with the future health of all organ systems (Amarasekera et al. 2013). This can be easily illustrated by the drastic rises in infant immune diseases, obesity, and allergens, which denote the fragility of the early postnatal immune system to environmental changes (Amarasekera et al. 2013). Epigenetic mechanisms are the likely pathway in explaining how early diets and nutritional metabolic introductions can affect fetal gene expression and the correlating disease risks (Amarasekera et al. 2013). One of the likely culprits in the rise of immune diseases and allergens amongst infants is the introduction of the "modern diet" (Amarasekera et al. 2013, p. 175). The "modern diet" contains more processed and artificial foods that are high in fats and refined carbohydrates (Amarasekera et al. 2013). Also, the "modern diet" contains less of the beneficial nutrients that are found in fiber, fresh fruit, fish, and vegetables (Amarasekera et al. 2013). The modern diet is correlated with changes in the gut microbiome and the immune functions of infants (Amarasekera et al. 2013). Dietary exposures have been shown to lead to variations in epigenetic regulation of immune gene expression (Amarasekera et al. 2013). This can result in severe effects on the function of the immune system and the likelihood of disease and illness (Amarasekera et al. 2013). The gut microbe plays a vital part in the overall health of infants; however, the overly hygienic environments are contributing to alterations in the gut microbe that is increasing the probability of immune disease, obesity, and allergens (Amarasekera et al. 2013). In addition, there is fundamental research that now demonstrates that high-fat, low-fiber diets modify the gut microbiome (Amarasekera et al. 2013). These modifications have been shown to increase the likelihood of the development of allergic diseases in infants due to the altered pattern of gut microbiota (Amarasekera et al. 2013). This indicates the pivotal role that the gut microbiome plays in immune programming (Amarasekera et al. 2013). The gut microbiota has been shown to interact with immune pathways through epigenetic mechanisms—one example is shown by the variations in the enzymes involved in the translation of histone proteins and gene expression by the byproducts produced by dietary fiber (Amarasekera et al. 2013).

The "modern diet" has also resulted in the rise in obesity and metabolic disease (Amarasekera et al. 2013). The "modern diet" affects the gut microbiome by influencing epigenetic variations in gut permeability, increase in systemic antigen load, and increases in serum levels of inflammatory cytokines (Amarasekera et al. 2013). Human studies have shown that individuals that have been diagnosed as obese have detailed variations in gut microbiota compared to their lean, fit counterparts (Amarasekera et al. 2013). This research highlights the complex interactions between the metabolic and immune programming in the gut, and how alterations to the gut microbiome can have systematic effects that result in obesity (Amarasekera et al. 2013). With obesity rates measuring at 35 % of the United States' population (Amarasekera et al. 2013), we are left wondering what role pediatric physicians play in the control of epigenetics.

Pediatricians have been well educated in the environmental factors that affect a child's social development; however, there has been less emphasis placed on the importance of an infant's diet and the need for documenting family medical histories (Hall 2014). Most troubling, there is research that states the environmental epigenetic changes can actually be passed generation to generation, thus resulting in generational shifts in unbeneficial gene phenotype expression (Hall 2014). One of the most significant policy changes needed by pediatricians is to study and record infant patient's family medical records because this information may help future epigenetic therapies if and when they become available (Hall 2014). Other edifying information to note would be documented diet, assisted reproductive technologies, intrauterine growth, birth weight, stress and illness during pregnancy, and the presence of illness in early stages of infant development (Hall 2014). Future pediatricians should look at the diet of their infant patients, while looking into the possible benefits of probiotics during infant development. Pediatricians should also design their practice to ensure they have the proper time needed to accurately and thoroughly document all of the aforesaid information that may provide epigenetic solutions later in the patient's life.

To illustrate the complexity epigenetics has on research we examined a study performed by Buchbinder et al. (2011) on the role epigenetics played in phenotype variations on a mutated gene sequence shared by a monozygotic twin pair (2011). The twin brothers were found to have missense mutations of the WAS gene, which equated to a reduced expression of the WAS protein (Buchbinder et al. 2011). This mutation causes two phenotypes: the less sever immunodeficiency X-linked thrombocytopenia (XLT) and the most troublesome immunodeficiency Wiskott–Aldrich syndrome (WAS) (Buchbinder et al. 2011). Both brothers were originally diagnosed with the less severe phenotype XLT and with the high possibly of being asymptomatic; however, a disputatious phenotype evolved in one of the twin brothers that resulted in the expression of the more severe phenotype WAS.

The authors believed that epigenetic alterations were conceivable, unrecognized mechanisms that could affect the variability of immunodeficiency phenotypes (Buchbinder et al. 2011). The authors based this prediction on the fact that epigenetics' role in alterations in transcriptional activity and gene expression (Buchbinder et al. 2011). The authors reviewed the medical histories of the twins. They found that through 8 years of age, Twin B remained asymptomatic XLT with no evidence of complications related to immunodeficiency; however, Twin A developed a fever and severe respiratory distress at the age of seven. Twin B then developed arthritis and began corticosteroids at the age of 8 (Buchbinder et al. 2011).

The authors measured the levels of transcription and protein expression in the twin brothers. Twin B with XLT had 160 % more protein expression than Twin A with WAS (Buchbinder et al. 2011). The authors also found another epigenetic change in the levels of methylation, resulting in Twin A with WAS having higher levels of methylation than Twin B with XLT (Buchbinder et al. 2011). Despite the twins being raised in similar environments, there are vast differences in the WAS mutation in the monozygotic twin brothers, which lead the authors to suggest that one plausible explanation for these results are due to epigenetic changes resulting in alteration in gene expression and variations in cellular differentiations and functions (Buchbinder et al. 2011). The authors conclude by stating that the "epigenetic phenomenon may be able to fill the gap in our understanding of the phenotype variations" of certain diseases and gene expressions (Buchbinder et al. 2011, p. 777).

13.5 Challenges and Future Directions

The role of epigenetics in child obesity remains to be determined. While it is known that the environment affects the expression of certain genes associated with excess adiposity, the nature of the cause–effect relationship is not yet clear. In other words, it is not known whether obesity causes the epigenetic modification or whether epigenetics causes obesity. Some epigenetic changes have been shown to occur in utero, yet research has demonstrated the ability of such changes to the epigenome to be modified after birth.

The American government and the Department of Health and Human Services have carried out many interventions aimed at reducing child obesity with limited success (Hafekost et al. 2013). It is thus more effective to educate parents and mothers on obesity prevention. The study of epigenetics sheds new light on prevention. Unfavorable biomarkers in the prenatal and postnatal phase can be reduced through optimum behaviors such as improved prenatal and infant nutrition and reduced exposure to environmental toxins such as smoking. Nevertheless, research points out that there is still potential after birth for unfavorable epigenetic markers to undergo modification to more favorable epigenetic status through behavior-based interventions. Studies show that interventions during the first year of life are most effective (Huh et al. 2011).

References

Amarasekera, M., Prescott, S. L., & Palmer, D. J. (2013). Nutrition in early life, immune-programming and allergies: The role of epigenetics. *Asian Pacific Journal of Allergy and Immunology, 31*(3), 175–182.

Bollati, V., & Baccarelli, A. (2010). Environmental epigenetics. *Heredity, 105*(1), 105–112. http://dx.doi.org/10.1038/hdy.2010.2.

Buchbinder, D., Nadeau, K., & Nugent, D. (2011). Monozygotic twin pair showing discordant phenotype for X-linked thrombocytopenia and Wiskott–Aldrich syndrome: A role for epigenetics? *Journal of Clinical Immunology, 31*(5), 773–777.

Campbell, T. C., & Campbell, T. M. (2006). The China Study: the most comprehensive study of nutrition ever conducted and the startling implications for diet, weight loss and long-term health. Dallas, Texas: BenBella Books.

CDC. Obesity – United States 1999-2010. MMWR 2013: 62(03);120–128 (Suppl; November 22, 2013).

Choquet, H., & Meyre, D. (2011). Genetics of obesity: What have we learned? *Current Genomics, 12*(3), 169–179. doi:10.2174/138920211795677895.

Dietz, W. H. (1998). Health consequences of obesity in youth: Childhood predictors of adult disease. Pediatrics, 101(3), 518–525. Retrieved from http://search.proquest.com/docview/228391080?accountid=13172.

Freedman, D. S., Dietz, W. H., Srinivasan, S. R., & Berenson, G. S. (1999). The relation of overweight to cardiovascular risk factors among children and adolescents: The bogalusa heart study. Pediatrics, 103(6), 1175–1182. Retrieved from http://search.proquest.com/docview/228372557?accountid=13172.

Freedman, D. S., Kettel, L., Serdula, M. K., Dietz, W. H., Srinivasan, S. R., & Berenson, G. S. (2005). The relation of childhood BMI to adult adiposity: The Bogalusa Heart Study. *Pediatrics, 115*, 22–27.

Gluckman, P. D., & Hanson, M. A. (2008). Developmental and epigenetic pathways to obesity: An evolutionary-developmental perspective. *International Journal of Obesity, 32*, S62–71. doi: http://dx.doi.org/10.1038/ijo.2008.240.

Gluckman, P. D., Hanson, M. A. Cooper, C., & Thornburg, K. L. (2008). Mechanisms of disease: Effect of in utero and early-life conditions on adult health and disease. *The New England Journal of Medicine, 359*(1), 61–73. doi: http://dx.doi.org/10.1056/NEJMra0708473.

Godfrey, K. M., Sheppard, A., Gluckman, P. D., Lillycrop, K. A., Burdge, G. C., McLean, C., Rodford, J., Slater-Jeffries, J. L., Garratt, E., Crozier, S. R., Emerald, B. S., Gale, C. R., Inskip, H. M., Cooper, C., & Hanson, M. A. (2011a, June 20). Mom's pregnancy diet linked to DNA changes, child's obesity. Retrieved May 18, 2015, from http://www.environmentalhealthnews.org/ehs/newscience/moms-pregnancy-diet-linked-to-dna-changes-childs-obesity/

Godfrey, K. M., Sheppard, A., Gluckman, P. D., Lillycrop, K. A., Burdge, G. C., & McLean, C., et al. (2011b). Epigenetic gene promoter methylation at birth is associated with child's later adiposity. *Diabetes, 60*(5), 1528–1534.

Hafekost, K., Lawrence, D., Mitrou, F., O'Sullivan, T. A., & Zubrick, S. R. (2013). Tackling overweight and obesity: Does the public health message match the science? *BMC Medicine, 11*, 41.

Hall, J. G. (2014). Epigenetics: What does it mean for paediatric practice? *Paediatrics and Child Health, 19*(1), 27–30.

Hall, K. D., Sacks, G., Chandramohan, D., Chow, C. C., Wang, Y. C., Gortmaker, S. L., & Swinburn, B. A. (2011). Quantification of the effect of energy imbalance on bodyweight. *Lancet, 378*(9793), 826–837. PMID:21872751.

Hammond, R. A., & Levine, R. (2010). The economic impact of obesity in the United States. *Diabetes, Metabolic Syndrome and Obesity: Targets and Therapy, 3*, 285–295. doi:10.2147/DMSOTT.S7384.

Huh, S. Y., Rifas-Shiman, S. L., Taveras, E. M., Oken, E., & Gillman, M. W. (2011). Timing of solid food introduction and risk of obesity in preschool-aged children. *Pediatrics, 127*(3), e544–e551.

Koya, D. L., & Egede, L. E. (2007). Association between length of residence and cardiovascular disease risk factors among an ethnically diverse group of United States immigrants. *Journal of General Internal Medicine, 22*(6), 841–846. doi:10.1007/s11606-007-0163-y.

Kuczmarski, R. J., et al. (2002). 2000 CDC growth charts for the United States: Methods and development. *Vital and Health Statistics, 11*(246), 1–190.

Ogden, C. L., Carroll, M. D., Kit, B. K., & Flegal, K. M. (2014). Prevalence of childhood and adult obesity in the United States, 2011-2012. *JAMA, 311*(8), 806–814. doi:10.1001/jama.2014.732.

Powledge, T. M. (2011). Behavioral epigenetics: How nurture shapes nature. *BioScience, 61*(8), 588–592.

Schwartz, M. B., & Puhl, R. (2003). Childhood obesity: A societal problem to solve. Obesity Reviews, 4(1), 57–71. doi: http://dx.doi.org/10.1046/j.1467-789X.2003.00093.x.

Shapira, N. (2008). Prenatal nutrition: A critical window of opportunity for mother and child. *Women's Health, 4*(6), 639–656. http://dx.doi.org/10.2217/17455057.4.6.639.

World Health Organization. (2015). *Global strategy on diet, physical activity and health*. Geneva, Switzerland: World Health Organization.

Yajnik, C. S. (2010). Fetal programming of diabetes: Still so much to learn! *Diabetes Care, 33*(5), 1146–1148.

Chapter 14
Children's Exposure to Alcohol, Tobacco, and Drugs: Long-Term Outcomes

F. Elisa Melvin

14.1 Introduction

Epidemiological data increasingly provide evidence that suggests that respective individuals' environmental exposures early in development have a role in susceptibility to disease in later life (Jirtle and Skinner 2007, p. 253). Additionally, some of these environmental effects seem to be passed on to future generations (Jirtle and Skinner 2007, p. 253). According to scholarly literature, modifications in epigenetic adaptations provide a credible link between the alterations in gene expression and the environmental entities that might lead to disease phenotypes. A cumulative body of evidence from animal studies supports the role of environmental epigenetics in disease susceptibility and methods are now becoming accessible to examine the relevance of these genetic phenomena to human disease (Jirtle and Skinner 2007, p. 253).

Prior to the recognition of epigenetic mechanisms, traditional research examined the combined cascade of genetics and respective environments of individuals in addition to investigating disease, health, and the correlation between one's susceptibility to disease and respective environmental exposures (Dolinoy et al. 2006, p. 298). Emerging evidence depicts a paradigmatic shift from the aforementioned concepts to the actual study of epigenetics. Epigenetics is defined as "the study of the heritable changes in gene expression that occur without a change in DNA sequence, including epigenetic mechanisms of gene regulation such as the modification of DNA methylation and chromatin remodeling," which

F.E. Melvin, Ph.D., M.Ed. (✉)
Department of Health Administration, Pfeiffer University, 4701 Park Road, Charlotte, NC 28209, USA
e-mail: Lisa.melvin@pfeiffer.edu

D. Hollar (ed.), *Epigenetics, the Environment, and Children's Health Across Lifespans*, DOI 10.1007/978-3-319-25325-1_14

are also influenced by the environment and play a very important role in the fetal basis (early origins) of adult disease susceptibility (Dolinoy et al. 2006, p. 297). According to scholarly literature, the original concept of a fetal basis of adult disease has more recently evolved to that of Developmental Origins of Health and Disease (DOHaD), as adverse effects may already develop in childhood and adolescence and disease states are an integral part of a continuum of outcomes in response to risk factors (Barouki et al. 2012, p. 4). The aforementioned mechanisms are sensitive to environmental stimuli other than physiological factors such as drugs, industrial chemicals, tobacco smoke, and other environmental exposures that ultimately lead to long-term adverse consequences such as increased disease risks later in life (Barouki et al. 2012, p. 1). Experts continue to postulate that environmental factors during prenatal and early postnatal development influence developmental plasticity, thereby altering susceptibility to adult chronic diseases (Dolinoy et al. 2006, p. 298).

14.2 Epigenetic Programming

According to Rothstein et al., epigenetics is "one of the most scientifically important…cutting edge subjects of scientific discovery (Rothstein et al. 2009, p. 1)." Conceptually, epigenetics link genetic and environmental influences that can inevitably affect the future development and health of an individual and their respective offspring based on the fact that the epigenome is highly receptive to toxic and environmental exposures (Rothstein et al. 2009, p. 1). Weaver et al.'s findings provided the first evidence that maternal behavior yields stable modifications of DNA methylation and chromatin structure, which ultimately provides a mechanism for the long-term effects of maternal care on gene expression in their respective offspring (Weaver et al. 2004, p. 847). Epigenetic changes can be transmitted from one generation to the next (Rothstein et al. 2009, p. 1). These studies offer an opportunity to clearly define the nature of gene–environment interactions during development and how such effects result in the sustained mechanism of environmental programming of gene expression and function over the lifespan (Weaver et al. 2004, p. 852). Conceptually, it is important to note that maternal effects on the expression of defensive responses, such as increased HPA activity, are a common theme in biology among mammals, and natural selection may have shaped offspring to respond to subtle variations in parental behavior as a prognosis of the environmental conditions they will inevitably face once they become independent of the parent (Weaver et al. 2004, p. 852). Epigenetic modifications of targeted regulatory sequences in response to even reasonably subtle variations in environmental conditions might then serve as a major source of epigenetic variation in gene expression and function and ultimately as a process mediating such maternal effects (Weaver et al. 2004, p. 852).

According to *Impact on Genetic and Epigenetic Factors*, there is sufficient evidence that long-term prenatal and infant cardiovascular development can be predicted by events that transpire during an individual's early-life experiences especially as it pertains to children categorized as lower socioeconomic status delivered by mothers who smoke throughout pregnancy (Tremblay and Hamet 2008, p. 27). These children, who have physiologically acquired toxins derived maternally, are more susceptible to hypertension in the long term. This "mismatch" that occurs during early exposure to toxic environments and adult environments increases the risk of the long-term consequences of adult diseases (Tremblay and Hamet 2008, p. 28).

14.3 The Effects of Maternal Smoking

According to Knopik (2009), mothers' smoking during pregnancy continues to be a significant public health concern with at least half of all females who present as smokers continuing to smoke throughout their respective pregnancies. According to Rothstein et al. (2009), exposure to toxic pollutants in tobacco smoke has been found to influence the development and health of subsequent generations. Moreover, another study found that grandchildren of grandmothers who smoked during the fetal period had an increased risk of developing asthma (Rothstein et al. 2009, p. 7). Additionally, tobacco use during pregnancy adversely affects not only prenatal but also postnatal growth and significantly increases the risk of behavioral and developmental deficits in children and adolescents (Vaglenova et al. 2004, p. 167). Additionally, previous studies have concluded a direct correlation between the following: prenatal smoke exposure with reduced birth weight, increased risk for diseases and behavioral disorders later in life, and poor developmental and psychological outcomes (Knopik et al. 2012, p. 1). Prenatal nicotine exposure throughout pregnancy produced offspring that had a marked impact on their cognitive abilities, physical maturation, kinetic movement, and anxiety levels which extended from birth through early adulthood (Vaglenova et al. 2004, p. 165). Also noted within the literature, there is a direct correlation between male nicotine-exposed offspring and later growth retardation, with most of the significant differences seen during the adolescent period (Vaglenova et al. 2004, p. 165). Scholarly literature also states that prenatal nicotine exposure would be expected to act primarily through its direct actions at neuronal nicotinic acetylcholine receptors. In addition, alterations in the functional activity of nicotinic receptors could have unique long-term consequences because of their potential role in brain development and as presynaptic modulators of numerous neurotransmitter systems (Vaglenova et al. 2004, p. 165). Additionally, Knopik et al. found in regard to brain and neurodevelopmental outcomes that prenatal exposure to maternal cigarette smoking is directly associated with a higher degree of DNA methylation in the BDNF exon 6 in adolescents whose mothers smoked during pregnancy, which further suggests that exposure to cigarette smoke while in utero may have long-term consequences that are still measurable in the adolescent stage of life (Knopik et al. 2012, p. 9).

Furthermore, scholarly literature states that there is a direct link between developmental nicotine exposure and the clear exertion of a variety of effects on the developing nervous system. Experts also found that these modifications, which encompass changes at all levels of analysis, from individual molecules to neuronal structure, can clearly continue into adulthood and may underlie the behavioral, cognitive, and attentional variances observed in humans and animals that are developmentally exposed to the aforementioned toxins (Heath and Picciotto 2009, p. 10).

14.4 Epigenetics and the Effects of Alcohol

According to Perkins et al. (2013), although treatment regimens for fetal alcohol spectrum disorder remain in the formative stage, research shows that there are definite changes in the epigenome, and in the brain following developmental alcohol exposure these epigenetic changes are long-lasting. Although experts have studied four decades of clinical and preclinical research documenting the teratogenicity of ethanol, knowledge of the epigenetic changes induced by prenatal ethanol exposure is only recently developing (Krishnan et al. 2014, p. 2). Scholars also state that research into long-term epigenetic changes caused by developmental alcohol exposure is paramount and very much needed (Perkins et al. 2013, p. 10).

14.5 Additional External Factors and Effects on the Epigenome

According to the literature, external factors such as chemical, social, and nutritional exposures are noted as predictors of disease risk in later life (Roberts et al. 2014, p. 7). The literature also states that the developmental origins of health and disease hypothesis (postulates) that gene–environment interactions during early life result in long-lasting effects and points to epigenetic inheritance as a prime underlying mechanism (Roberts et al. 2014, p. 7). Rothstein et al. (2009) state it is possible that prenatal and early-life exposures to environmental toxins with epigenetic impacts could manifest in behavioral modifications later in life. Moreover, literature states that the concept of "lifestyle" includes not only factors such as behavior, stress, and physical activity but also smoking and alcohol consumption (Alegria-Torres et al. 2011, p. 1). In addition, studies show that lifestyle and environmental factors may have a direct link that influences epigenetic mechanisms which include DNA methylation, microRNA expression, and histone acetylation. In other words, the lifestyle factors that might modify epigenetic patterns include tobacco smoking and alcohol consumption (Alegria-Torres et al. 2011, p. 2).

14.6 Implications and Challenges

The developing systematic approach to understanding various long-term effects of developmental exposures adds a new dimension to the importance of preventing the deleterious effects of environmental chemicals (Barouki et al. 2012, p. 8). This new research paradigm is challenging as it requires a multidisciplinary team of collaborative scientists that includes toxicologists, clinicians, and epidemiologists, etc., to scrutinize viable external contributing factors in the past, as far back as each individual's intrauterine life; the life of their parents and/or grandparents; prospective studies to define and identify early-life exposures to chemicals and nutrients and provide a new emphasis on multigenerational studies (Barouki et al. 2012, p. 8). This ideal multidisciplinary team can identify applicable epigenetic predictive changes in order to help mitigate long-term health consequences and promote disease prevention. Unfortunately, because all human development "involves a plastic element," the mere presence of epigenetic modification may not predict a significant effect on health in later life (Barouki et al. 2012, p. 8).

Additionally, conceptual shifts are needed to understand the dynamic interactions of genes, environment, epigenetics, social processes, and behavioral choices (Jackson et al. 2013, p. 40). Moreover, health can no longer be discussed solely in the simple terms of individual responsibility alone; the causative factors of many chronic diseases occur earlier in life, and even in previous generations. Additionally, the implications that gene–environment interactions have upon our health cannot be truly understood or actually acted upon, unless key decision makers consider a holistic approach that involves environmental, genetic, and epigenetic factors that together drive our phenotypes (Jackson et al. 2013, p. 40). This paradigmatic shift should lead to a more improved approach—the continuum of the implementation of a shift involving an epigenetic holistic approach in regard to the development of healthcare policies.

14.7 Conclusion

Epigenetic modifications are dynamic and unlike genetic changes potentially reversible, they hold promise for public health as targets for preventive and therapeutic interventions. Experts believe that persistent epigenetic adaptations occur very early in human development in response to maternal nutritional factors and are directly associated with increased disease susceptibility later in life among genetically identical mice (Roberts et al. 2014, p. 7). A new approach toward disease prevention is needed with a new emphasis on early development. A rational approach is to improve maternal nutrition and reduce environmental chemical exposures pre-pregnancy, during pregnancy, and during the first few years of life (Barouki et al. 2012, p. 9). This change is likely to have a very large impact on reducing disease incidence and the cost of health care, while at the same time increasing the quality of life at a global level (Barouki et al. 2012, p. 9). Future studies will also provide a more all-inclusive picture of epigenetic mechanisms regulated by alcohol in order to increase our understanding of overall genomic function in alcoholism. These early research findings clearly provide evidence that compounds that inhibit histone deacetylase (HDACs) inhibitors "may be promising future therapeutic agents in the treatment of alcoholism" (Starkman et al. 2011, p. 4).

Abbreviations

DOHaD Developmental Origins of Health and Disease
HDACs Histone deacetylase

References

Alegria-Torres, J. A., Baccarelli, A., & Bollati, V. (2011). Epigenetics and lifestyle. *Epigenomics, 3*(3), 267–277.

Barouki, R., Gluckman, P., Grandjean, P., Hanson, M., & Heindel, J. (2012). Developmental origins of non-communicable disease: Implications for research and public health. *Environmental Health, 11*, 42.

Dolinoy, D. C., Weidman, J. R., & Jirtle, R. L. (2006). Epigenetic gene regulation: Linking early developmental environment to adult disease. *Reproductive Toxicology, 23*(3), 297–307.

Heath, C. J., & Picciotto, M. R. (2009). Nicotine-induced plasticity during development: Modulation of the cholinergic system and long-term consequences for circuits involved in attention and sensory processing. *Neuropharmacology, 56*(Suppl 1), 254–262.

Jackson, L. C., Niculescu, M. D., & Jackson, R. T. (2013). Conceptual shifts needed to understand the dynamic interactions of genes, environment, epigenetics, social processes and behavioral choices. *American Journal of Public Health, 103*(S1), S33–S42.

Jirtle, R. L., & Skinner, M. K. (2007). Environmental epigenomics and disease susceptibility. *Nature Reviews Genetics, 8*, 253–262.

Knopik, V. (2009). Maternal smoking during pregnancy and child outcomes: Real or spurious effect? *Developmental Neuropsychology, 34*(1), 1–36.

Knopik, V. S., Maccani, M. A., Francazio, S., & McGeary, J. E. (2012). The epigenetics of maternal cigarette smoking during pregnancy and effects on child development. *Development and Psychopathology, 24*(4), 1377–1390.

Krishnan, H. R., Sakharkar, A. J., Teppen, T. L., Berkel, T. D., & Pandey, S. C. (2014). The epigenetic landscape of alcoholism. *International Review of Neurobiology, 115*, 75–116.

Perkins, A., Lehmann, C., Lawrence, R. C., & Kelly, S. J. (2013). Alcohol exposure during development: Impact on the epigenome. *International Journal of Developmental Neuroscience, 31*(6), 391–397.

Roberts, J. S., Dolinoy, D., & Tarini, B. (2014). Emerging issues in public health genomics. *Annual Review Genomics and Human Genetics, 15*, 461–480.

Rothstein, M. A., Cai, Y., & Marchant, G. E. (2009). The ghost in our genes: Legal and ethical implications of epigenetics. *Health Matrix Cleveland, 19*(1), 1–62.

Starkman, B. G., Sakharkar, A. J., & Pandey, S. C. (2011). Epigenetics—Beyond the genome in alcoholism. *Alcohol Research: Current Reviews, 34*(3), 293–305.

Tremblay, J., & Hamet, P. (2008). Impact of genetic and epigenetic factors from early life to later disease. *Metabolism Clinical and Experimental, 57*(Suppl 2), S27–S31.

Vaglenova, J., Birru, S., Pandiella, N. M., & Breese, C. R. (2004). An assessment of the long-term developmental and behavioral teratogenicity of prenatal nicotine exposure. *Behavioural Brain Research, 150*, 159–170.

Weaver, I. C., Cervoni, N., Champagne, F. A., D'Alessio, A. C., Sharma, S., Seckl, J. R., Dymov, S., Szyf, M., & Meaney, M. J. (2004). Epigenetic programming by maternal behavior. *Nature Neuroscience, 7*, 847–854.

Chapter 15
Epigenetics and Development

Ankita Das

15.1 History of the Term Epigenetics and How It Relates to Development

15.1.1 Coining of the Term and History of the Term

Epigenetics stands for information that can be inherited through cell divisions beyond what is coded in genetic information or DNA itself. It was coined by Conrad Hal Waddington in 1942 as an attempt to define epigenesis (Van Speybroeck 2002), which involves understanding of the basis of how phenotypes arise by studying genotypes. In summary, at that time, the term epigenetics defined what we understand today as embryology or development. In his efforts to explain processes for development in an embryo, he emphasized on two key concepts in trying to understand epigenetics or embryo development: one being "canalization" and the other "epigenetic landscape" (Van Speybroeck 2002; Hemberger et al. 2009). While cells in an embryo have the option to adopt fates based on their genetics, there exists a range of influences from the environment, known as the epigenetic landscape that influences the outcomes. These options can be thought of canals, but what drives canalization is the need for the embryo to adapt to their environment.

A. Das, Ph.D. (✉)
Department of Embryology, Carnegie Institution for Science, 3520 San Martin Drive, Baltimore, MD 21218, USA
e-mail: das@ciwemb.edu

D. Hollar (ed.), *Epigenetics, the Environment, and Children's Health Across Lifespans*, DOI 10.1007/978-3-319-25325-1_15

15.1.2 Relevance of Epigenetic Analysis in Understanding Embryonic Development

It is well documented that developmental outcomes in cells in an embryo are not just an effect of the genetic factors but also a result of external environmental influences. Therefore, understanding the epigenetic mechanisms that underlie developmental decisions will aid in gaining a complete understanding of the process of development (Burdge and Lillycrop 2010). Furthermore, it is well accepted that the fine-tuning of the signals in the environment really influences and gives rise to very different outcomes in terms of cell lineages. More specifically, in the paragraphs below, I have discussed the two key properties of developing embryos that are attributed by epigenetic mechanisms. These properties are robustness and plasticity (Bateson and Gluckman 2012).

The robustness of developmental outcomes is determined by epigenetic factors during development. In order to ensure proper embryonic development, it is often seen that there are multiple layers of epigenetic regulation, and upon perturbation of any of the processes, there are severe consequences in embryonic development.

The developmental plasticity of cell types is ensured by epigenetic mechanisms: In a developing embryo, whereas intrinsic genetic mechanisms and cross talk between tissues determine the fate choices that early cell types make, it has been found that there are various early cell progenitor cells that maintain the ability to switch lineages upon being given proper instructions (Le Douarin et al. 2004). It is rather amazing how rapidly these cells are able to respond to the signals in the environment, which is attributed to the epigenetic makeup of the cells (Kim et al. 2010). In summary, various epigenetic mechanisms in the embryo impart crucial developmental plasticity to cells (Boyer et al. 2006), thereby making it critical to understand these mechanisms to fully understand the developmental potential of the cells.

15.2 Key Cellular Concepts in Development and the Role of Epigenetics

Specification or Formation of Cell Types (Epigenetic States of Cells) In a developing embryo, cells of various lineages have distinct epigenetic marks on their chromatin, which define their epigenetic states or developmental potential. In an embryo, epigenetic states have to be first established and then maintained throughout development to ensure maintenance and carry-over of the developmental identity of cells as the embryo develops. Whereas in most cases the epigenetic states are stable and maintained, they are also reversible, which then imparts developmental plasticity to the cells that were discussed earlier in the chapter.

15.3 Key Determinants in Establishment and Maintenance of Epigenetic States

15.3.1 Cis-Regulatory Elements and Trans-Acting Factors

The key determinants for regulators of epigenetic states can be separated into two categories, namely, the cis-regulatory elements and the trans-acting factors. Whereas the cis-regulatory elements provide cell-type-specific signatures during development and can be used to predict cell states, the trans-acting factors are critical in establishing those states and in some cases even maintaining them. For example, specific covalent modifications on the histone tails not only dictate the accessibility of the DNA for further modifications, it regulates where cell-type-specific factors can bind and regulate lineage decisions. In parallel to that, specific modifications of the histone tails dictate the states of the promoters and enhancers, which then dictate the genes that are expressed, and the cell identities.

Taken together, in order to drive proper development, proper establishment of epigenetic states by modifications of histones and binding of transcription factors to the promoters, also signals and modifiers to the enhancers, is critical for normal development to proceed. These determinants comprise specific DNA sequences, the so-called cis-regulatory factors, and the trans-acting factors comprising of histone proteins, modifying enzymes, and cell-type-specific transcription factors. The cis-regulatory elements can be further classified broadly into two categories, the *promoter elements* of genes and the distal-regulatory elements, the *enhancers*. While the promoter elements govern the gene expression profiles of specific cell types, allowing for transcription of lineage-specific genes, the enhancers further fine-tune the establishment of identity by facilitating the cross talk between the transcriptional machinery and the upstream regulators like signals from the environment. Recent studies have highlighted the importance of enhancer sequences in regulating the plasticity and adaptability of cells. These studies highlight the role of enhancer sequences as platforms for signal integration from the environment, pinpointing again to the importance of epigenetic regulation in determining cell fates (Buecker and Wysocka 2012). As a consequence of the dynamics of regulation of the epigenetic states of the chromatin in various cell types, there lies a possibility of regulation of the states, thereby leading to a completely different outcome (Buecker et al. 2014).

Whereas changes in the regulation can lead to various developmental disorders, these inherent mechanisms are also used for the evolution of natural variations in morphology and forms during development. A great example of that is what is thought to underlie the variation and formation of distinct craniofacial morphologies in humans (Rada-Iglesias et al. 2011). It has now been shown that alterations in the enhancer sequences of the DNA can lead to very distinct developmental outcomes.

In addition to the observed and well-documented consequences of epigenetic deregulation during embryo development, numerous studies in the field of stem cell biology have suggested that the so-called epigenetic signatures of various stem cells, both embryonic and adult derived, can be used to predict their developmental potential. These studies have documented specific epigenetic states and have been able to define "signatures" for each cell type. Not only have these studies clearly demonstrated that embryonic and adult stem cell types have distinct epigenetic states and that again specific promoter and enhancer regions in the cell types can be used to define cellular identities, they have also led to the idea that control of epigenetic mechanisms by chemical and genetic modifications can lead to "reprogramming" of the cell fates. Therefore, *simply by leveraging the known mechanisms of cellular and developmental plasticity regulated via epigenetic mechanisms and by utilizing those mechanisms, one can precisely direct cell fates.*

15.3.2 DNA Methylation as a Mechanism for Epigenetic Memory and Plasticity

In addition to the abovementioned modifications of DNA–histone complexes, the DNA sequences themselves can be modified by methylation, which then regulates the accessibility of the DNA to various enzymes and the state of the chromatin. DNA methylation can further be classified into two types, which encompasses methylation of the promoters and also the gene bodies themselves (Adrian 2002; Suzuki and Bird 2008; Jones 2012). The prevalence of DNA methylation at the promoters is more commonly studied in the context of understanding the role of mechanisms of regulation of gene transcription. In recent times, DNA methylation at gene bodies is also being investigated in the context of genome instability and cancer, but for the purpose of this review, I will discuss the processes of DNA methylation at the promoter of genes. DNA methylation has been shown to be important both for the establishment and maintenance of methylation states of specific cell types during development. Whereas the previously discussed mechanisms of epigenetic regulation are thought to be primary events in regulation of gene expression, DNA methylation is historically thought to be a secondary event regulating transcription (Suzuki and Bird 2008). In most cases, it has been suggested that DNA methylation occurs on already repressed genes to further ensure irreversible repression. *Although most of the stretches of CG nucleotides called CpG islands typically remain unmethylated at the promoter regions of genes*, there are specific regions in the genome that maintain methylated CpG islands. This phenomenon is thought to be important both in X-chromosome inactivation and also genome imprinting (Adrian 2002). The mechanism by which DNA methylation is maintained on genomes is by the activity of enzymes encoded by the Polycomb family of proteins that specifically maintain methylated histones at these regions (Ringrose and Paro 2007). Interestingly enough, it has also been observed that there are mechanisms of active demethylation occurring in the embryo. It has long been speculated that this "latter" mechanism might be important for tissue-specific activation of genes, although the evidence for that was not very conclusive. Recent loss of function studies of DNA methylation enzyme Dnmt1 has shown reduced methylation at tissue-specific promoters and aberrant activation of genes in other domains (Suzuki and Bird 2008; Jones 2012). These studies provide strong evidence for DNA methylation being more than a secondary event in gene regulation and also are deterministic of cell lineages in an early embryo.

15.3.3 Noncoding RNAs as a Mechanism for Epigenetic Regulation

The other commonly studied mechanism of epigenetic regulation involves noncoding RNA-mediated regulation. Noncoding RNAs can be divided into two categories, the long and short noncoding RNAs. For this review, I will focus on the functions of small noncoding RNAs to discuss their functions in establishing robustness and developmental plasticity in early embryogenesis. Two major classes of small noncoding RNAs have been described to date, namely, the microRNAs (miRNAs) and the piwi-RNAs (pi-RNAs) (Stefani and Slack 2008). Extensive gene perturbation analyses of the microRNA pathways in various model organisms have revealed that in the majority of cases, these molecules function to modulate and fine-tune signaling activity in the embryos. In other cases, they regulate robustness by dampening active signaling by balancing expression levels of upstream regulators of the signaling pathways. At the molecular level, miRNAs function by either *blocking the transcripts of various genes to degradation* or by blocking protein synthesis. It is to be noted that these mechanisms have been shown to be very important for the establishment of gradients of signaling and also for the formation of axis during embryo development. In summary, this mode of regulation is thought to be important for morphogenesis of very early embryos. In addition to the function of this pathway in early embryogenesis, there are also several known functions of these pathways in tissue-specific developmental processes, namely, neuronal, muscle, cardiovascular, and lymphatic system development. In addition to the widespread activity and functions of the miRNAs, the pi-RNAs have been implicated in germline development.

15.4 Developmental Disorders Due to Perturbations in Epigenetic Mechanisms

Epigenetic diseases can arise from defects or alterations in various aspects of the abovementioned epigenctic mechanisms. Defects in epigenetic processing can lead to defects both in embryonic development and in development of cancers. For the purpose of this chapter, I will focus the discussion on developmental disorders. These can arise from alterations in a single gene expression, as well as by alterations of multiple genes at once due to changes in the global epigenetic landscape (Feinberg 2007). With regard to the types of disorders that can occur, one can think of alterations in the epigenetic regulation of genes, such as changes in imprinting, or changes in the epigenetic machinery itself that can give rise to developmental disorders. One feature that has been seen in epigenetic disorders is that the cell types in the embryo that are highly plastic early in development are preferentially affected by alterations in epigenetic machinery. There are several instances of changes in expression of chromatin-modifying enzymes like histone methyltransferases (Feinberg 2007), as well as chromatin-remodeling enzymes, which when mutated lead to severe neurocristopathies and craniofacial disorders. Another interesting feature of these diseases is that they present a very broad spectrum of phenotypes, even though the cells arise from similar kinds of cells in the embryo. This observation again emphasizes the function of epigenetic mechanisms in establishing plasticity as well as maintaining robustness, such that when these mechanisms are perturbed, you get an imbalance or loss of plasticity or adaptability of the cells and manifestation of a phenotype. Some of the well-documented cases of these kinds of disorders include CHARGE syndrome and Prader–Willi and Angelman syndromes amongst many others. In addition, recent studies have also revealed that factors like environmental toxins can also alter the DNA and chromatin composition.

References

Adrian, B. (2002). DNA methylation patterns and epigenetic memory. *Genes and Development, 16*, 6–21.

Bateson, P., & Gluckman, P. (2012). Plasticity and robustness in development and evolution. *International Journal of Epidemiology, 41*, 219–223.

Boyer, L. A., Plath, K., Zeitlinger, J., Brambrink, T., Medeiros, L. A., Lee, T. I., Levine, S. S., Wernig, M., Tajonar, A., Ray, M. K., et al. (2006). Polycomb complexes repress developmental regulators in murine embryonic stem cells. *Nature, 441*, 349–353.

Buecker, C., Srinivasan, R., Wu, Z., Calo, E., Acampora, D., Faial, T., Simeone, A., Tan, M., Swigut, T., & Wysocka, J. (2014). Reorganization of enhancer patterns in transition from naive to primed pluripotency. *Cell Stem Cell, 14*, 838–853.

Buecker, C., & Wysocka, J. (2012). Enhancers as information integration hubs in development: Lessons from genomics. *Trends in Genetics, 28*, 276–284.

Burdge, G. C., & Lillycrop, K. A. (2010). Nutrition, epigenetics, and developmental plasticity: Implications for understanding human disease. *Annual Review of Nutrition, 30*, 315–339.

Feinberg, A. P. (2007). Phenotypic plasticity and the epigenetics of human disease. *Nature, 447*, 433–440.

Hemberger, M., Dean, W., & Reik, W. (2009). Epigenetic dynamics of stem cells and cell lineage commitment: Digging Waddington's canal. *Nature Reviews. Molecular Cell Biology, 10*, 526–537.

Jones, P. A. (2012). Functions of DNA methylation: Islands, start sites, gene bodies and beyond. *Nature Reviews. Genetics, 13*, 484–492.

Kim, K., Doi, A., Wen, B., Ng, K., Zhao, R., Cahan, P., Kim, J., Aryee, M. J., Ji, H., Ehrlich, L. I., et al. (2010). Epigenetic memory in induced pluripotent stem cells. *Nature, 467*, 285–290.

Le Douarin, N. M., Creuzet, S., Couly, G., & Dupin, E. (2004). Neural crest cell plasticity and its limits. *Development, 131*, 4637–4650.

Rada-Iglesias, A., Bajpai, R., Swigut, T., Brugmann, S. A., Flynn, R. A., & Wysocka, J. (2011). A unique chromatin signature uncovers early developmental enhancers in humans. *Nature, 470*, 279–283.
Ringrose, L., & Paro, R. (2007). Polycomb/Trithorax response elements and epigenetic memory of cell identity. *Development, 134*, 223–232.
Stefani, G., & Slack, F. J. (2008). Small non-coding RNAs in animal development. *Nature Reviews. Molecular Cell Biology, 9*, 219–230.
Suzuki, M. M., & Bird, A. (2008). DNA methylation landscapes: Provocative insights from epigenomics. *Nature Reviews. Genetics, 9*, 465–476.
Van Speybroeck, L. (2002). From epigenesis to epigenetics: The case of C. H. Waddington. *Annals of the New York Academy of Sciences, 981*, 61–81.

Chapter 16
Lifespan Development, Instability, and Waddington's Epigenetic Landscape

David W. Hollar Jr.

16.1 Introduction

The rapid industrial and technological advances from the nineteenth century to the present greatly improved human health and welfare, with the exception of systems applied by tyrannies. Biologically, the advances outpaced our understanding of their ecological genetic and epigenetic effects, which continue to remain incalculable due to the lack of controlled experimental studies to establish necessary and sufficient, cause-and-effect relationships (Freedman et al. 2015; Ioannidis 2005; Rothman and Greenland 1998). Nevertheless, the accumulating evidence that environmental toxins have epigenetic effects during the life span indicates the need for research, health policy, research-based interventions, and both patient and clinician education to improve transgenerational health (Bollati and Baccarelli 2010; Skinner et al. 2011). This observation is strongly supported by the prepubescent and perinatal nutritional stressors (Pembrey et al. 2006; Heijmans et al. 2008), dramatic cellular epigenetic controls over tissue differentiation throughout normal and aberrant development (Bonnin et al. 2011; Dixon et al. 2015; Maruyama et al. 2011; Parle-McDermott and Ozaki 2011; Waterland et al. 2009; Wilkins 2010), and tentative maternal stress (Yehuda and Bierer 2009).

16.2 Germline Versus Soma Epigenetics

The prepubescent nutritional stress effects of grandparents on same-gender grandchildren (Pembrey et al. 2006) implicate epigenetic reprogramming of sperm and egg, reproductive cells with high chromatin methylation protection compared to

D.W. Hollar Jr., Ph.D., M.S. (✉)
Department of Health Administration, Pfeiffer University, Misenheimer, NC, USA
e-mail: David.Hollar@pfeiffer.edu

D. Hollar (ed.), *Epigenetics, the Environment, and Children's Health Across Lifespans*, DOI 10.1007/978-3-319-25325-1_16

somatic tissues. Singh et al. (2013) described two major phases of epigenetic remodeling events in mammals, the first occurring in the early embryo to establish the soma/germline separation and the second phase occurring mid-gestation in prospermatogonia and following birth for oogonia with large-scale demethylation and differential prospermatogonia/oogonia methylation patterns. Loukinov et al. (2002) discovered that prospermatogonial demethylation corresponds with upregulation of the gene BORIS and downregulation of the gene encoding the CTCF zinc-finger protein. Manikkam et al. (2012) documented three-generation epigenetic effects and gene networks involved in the altered methylation of prospermatogonia and oogonia due to gestational endocrine disruptor exposure. Consequently, sperm and egg will be vulnerable to environmental insults at different times for the gestational/perinatal period through early adolescence, as Pembrey et al. (2006) observed.

Furthermore, a third event, somatic aging, begins just before puberty and appears to be an epigenetic phenomenon (Georgieva et al. 2015; Hayflick 1994; Pogribny and Vanyushin 2010). Aging is thermodynamic in nature (Hayflick 2007a, b; Skulachev 2001, 2002; Sedivy et al. 2008) and will be discussed below.

Both gametic development and somatic aging involve dynamic spatiotemporal changes in chromatin methylation/demethylation. In prospermatogonia, histone H3K9ac, H3K4me, H3K27me3, H3K79me2, H3K79me3, and HK9me3 levels elevate in tandem with methylation (Abe et al. 2011; Singh et al. 2013). Changes in chromatin methylation and especially global demethylation are similar between aging individuals and individuals with Hutchinson–Gilford progeria syndrome, the latter involving a homozygous splicing defect in Lamin A gene exon 11 (http://omim.org/entry/176670; Sedivy et al. 2008). Lee et al. (2009) demonstrated a clear relationship between the one-carbon metabolism pathway and DNA methylation. The one-carbon pathway includes methionine, choline, and pyrimidine synthesis, disruption of which leads to metabolic disorders that are detected in newborn screening (Hollar 2012).

16.3 Waddington's Model and Two Distinctive Epigenetic Pathways

Waddington's (1957) epigenetic landscape model metaphorically described a given cell's state as a ball rolling down hills and valleys, with its trajectory depending upon bumps (i.e., disturbances in epigenetic programming/reprogramming) along its journey (Fig. 16.1). Thom (1972) attempted to map cellular differentiation and tissue formation using catastrophe theory and following Waddington's (1957) model. Whereas such attempts have incompletely explained tissue structures, they provide a theoretical approach to studying and interpreting epigenetic changes in children's health and lifespan health.

The epigenetic state/phase of a cell is very relevant for the gene methylation/histone chromatin acetylation events that occur with cell fate commitment/

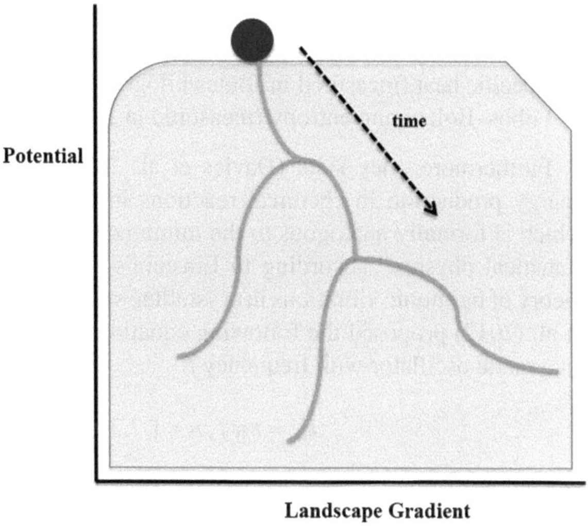

Fig. 16.1 Waddington's (1957) epigenetic landscape model

differentiation, including the early embryonic gametogenesis/somatogenesis division. The gametogenesis commitment adds genetic control protections that insure faithful transmission of genetic information and genetic controls to subsequent generations via sexual reproduction. In contrast, the disposable soma (i.e., nonreproductive body) (Kirkwood and Rose 1991; Tabatabaie et al. 2011) accumulates epigenetic dysregulation and chaos exponentially as aging. This point is not to say that gametes are impervious to epigenetic changes, just that such changes are more difficult, and they do occur (Pembrey et al. 2006; Heijmans et al. 2008; Skinner et al. 2011). Thus, epigenetics follows two tracks: (1) lifespan degradation from entropy/aging (Hayflick 2007a, b) and (2) transgenerational gene reprogramming affecting morbidity/mortality and behavior, via gametes (Pembrey et al. 2006; Yehuda and Bierer 2009). Waddington's (1957) model of epigenetic reprogramming is limited for gametes but global for the soma.

16.4 "Quantum" Models and Molecular Instability

Davies et al. (2012) followed Hayflick's (2007a, b) aging entropy model to implicate cancer, a partial genetic/epigenetic cell transformation/dedifferentiation event, as a quantum metabolic phenomenon. They use the term "quantum" in the sense that biological structures such as mitochondria oscillate with the following bioenergetics properties:

1. Metabolic cycle time (measured in seconds)
2. Metabolic rate (measures in Joules/kilogram second)

3. Entropy production rate (measured in Joules/Kelvin seconds) corresponding to the thermodynamic properties of temperature
4. Specific heat (measured in Joules/kilogram Kelvin)
5. Gibbs–Boltzmann entropy (measured in Joules/Kelvin)

Furthermore, they state (Davies et al. 2012, pp. 011101–5) that "the rate of energy production in chemical reactions in cells satisfies a variational principle which is formally analogous to the minimization of the free energy in equilibrium statistical physics," according to Einstein's (1907) and Debye's (1912) quantum theory of harmonic vibrations in crystalline solids (and their heat capacities). Davies et al. (2012) proposed the following equation of metabolic energy E_n stored by an enzymatic oscillator with frequency f:

$$E_n = n\kappa f, n = 1, 2, 3, \ldots \text{multiples.}$$

This equation borrows from Einstein's (1907) $E_n = nhf$, where h is Planck's constant (6.63×10^{-34} J s). Estimating 3×10^4 ATP phosphorylations per mitochondrion per second, $f = 1000$ Hz, and a characteristic enzyme kinetic time scale of 10^{-3} s, Davies et al. (2012) estimated the biological equivalent of Planck's constant h to be $\kappa = 10^{-24}$ J s. This parameter is 1.8×10^{11} times larger than h, which is the upper energy limit for quantum events, although Davies et al. (2012) argued that 1.9×10^{14} atoms/cell and approximately 1×10^3 mitochondria per cell yields 1.9×10^{11} atoms energized per mitochondria, assuming equal spacing but not considering optimum mitochondrial fission/fusion dynamics (Picard et al. 2013). Therefore, they are arguing that bioenergetic events within cells represent quantum-like phenomena and that these events are prone to atomic/molecular instability, as per Hayflick's (2007a, b) thermodynamic aging model and Waddington's (1957) perturbation epigenetic landscape.

Numerical estimations notwithstanding, what has all of this to do with epigenetics? Davies et al. (2012) presented their quantum model of bioenergetics as a thermodynamic instability hypothesis for cancers, where cells bioenergetically shift from mitochondrial oxidative phosphorylation to anaerobic glycolysis. Additionally, we know that cancer occurrence goes beyond just genetic mutations in some instances to epigenetic reprogramming by methylation/demethylation, chromatin remodeling, lysogenic viruses, and transposable genetic elements. Returning to Waddington's (1957) epigenetic landscape (Fig. 16.1) and Thom's (1972) potential well catastrophe modeling of the landscape, environmental disturbances (e.g., stress, toxins, nutrition) thermodynamically alter gene regulation, in the process creating cellular molecular instability and/or gametic imprinted instability that further can lead to transgenerational effects. These events translate throughout cellular phenotypes or throughout the entire organism via gametes. This includes the energy-converting mitochondria, which function by generating a catastrophic potential well of $\Delta p \sim 220$ mV (i.e., the proton-motive force; Nicholls and Budd 2000) as a transmembrane potential across the inner mitochondrial membrane. Only recently have researchers turned their attention to the central role of mitochondria in cellular and organismal development as well as mitochondrial epigenetics, given that approximately 1500 mitochondrial genes positioned in the mitochondrial DNA and in the nuclear genome regulate many critical biochemical pathways (Minocherhomji et al. 2012; Shaughnessy et al. 2014; Wallace and Fan 2010). Within quantum models of epigenetic instability related to aging and Waddington's (1957) landscape, the proton-motive force figures prominently given that each of the human body's 5×10^{13} cells contains between 200 and 3000 mitochondria, cardiac myocytes contain as many as 7000, sperm have 16, and oogonia can contain up to 100,000 (Brunk and Terman 2002; Gray 1989; Kurz et al. 2010; Nicholls and Budd 2000; Pieczenik and Neustadt 2007; Szewczyk and Wojtczak 2002; Veltri et al. 1990).

16.5 Genomic and Epigenomic Stability/Instability

Hayflick (2007a, b) similarly discussed genomic instability and the entropic thermodynamics of aging, although not with the general mathematical formalism of Davies et al. (2012). Genomic instability illustrates the thermodynamic pulsatility of all matter, living and nonliving, that Hayflick (2007b) emphasizes. From another quantum perspective, Levy-Leblond and Balibar (1990, p. 103) described the well-known iterative stationary phenomenon, ad infinitum: "A harmonic time dependence, given by exp $i(\omega t + \varphi)$ corresponds to a phenomenon which is 'always the same,' or in other words, remains identical to itself in the course of time." In this harmonic equation, similar to Davies et al. (2012), $\omega = 2\pi f$ is the pulsation or cyclic frequency, t is time, ωt is the phase or argument of the harmonic function, and φ is the original phase.

Epigenetic reprogramming is locked into genetic, maternal, egg, and mitochondrial controls and is reliably passed with generally minimal changes ("always the same") from generation to generation. This gametic, early in utero development/differentiation set of instructions maintains strong relative stability for maintenance of the species while allowing room for small changes (i.e., genetic mutation and epigenetic reprogramming) that create variation across individuals, variation that is selected for, against, or neutrally by the given environment in which individuals reside (i.e., Waddington's epigenetic landscape). Post-reproduction, the somatic cells (i.e., the individual organism) experience logistic and then exponentially increasing epigenetic/genomic instability (Davies et al., 2012; Hayflick 2007a, b) in cellular/tissue functioning, mitochondrial bioenergetics, chromatin remodeling and methylation, telomere shortening, etc., eventually resulting in systemic catastrophic failure (Thom 1972) that is irreversible by the second law of thermodynamics (Prigogine 2002).

Thus, environmental exposures to stress, toxins, etc., impact transgenerational development of individuals, for good or for bad, and lifespan functioning, increasingly bad over time. For both situations, early epigenetic reprogramming research offers the potential to moderate negative transgenerational epigenetic changes as well as possible slowing of lifespan aging, using epigenetic control reprogramming to approximately earlier life phenotypes. This last possibility would be daunting, as every change could not be corrected or reversed, necessary gene functional cycles must not be interrupted, and the reprogramming will be different from cell to cell, tissue to tissue. This could be a next step for aging research as we gain improved understanding of the epigenome, a system vastly more complex than the Human Genome Project. Nevertheless, the epigenetics of aging would be realistic to explore in light of grounded biogerontological research, in contrast to the mostly fraudulent antiaging industry (Binstock 2003; Olshansky et al. 2002).

16.6 Trajectory Analysis and Epigenetics

Returning to Waddington's (1957) epigenetic landscape and Davies et al.'s (2012) quantum model, the epigenetic state of all controls for a given cell's nuclear DNA plus mitochondrial DNA can be represented by a spatiotemporal manifold or surface $M(fx_i, t_i)$ at a static, stable time point. Unfortunately, each cell of an organism is not isolated, as it interacts with neighboring cells, substances delivered in the bloodstream, heat transfer and electromagnetic radiation, waste products of cellular metabolism, reactive oxygen species (ROS) generated by cellular metabolism, global physiological changes, viruses, transposable genetic elements, billions of mutualistic bacteria, occasional bacterial pathogens, etc.

Consequently, the epigenetic phase state of the cell changes from an initial $M(fx_i, t_i)$ to some future or temporally subsequent $M'(fx_i', t+1)$ (Fig. 16.2). We can mathematically map this dynamical phase transition for the epigenetic cell landscape. For Fig. 16.2, the expansion of the phase state due to epigenetic changes in the phase state genes (x_i, y_i two-dimensional coordinates) can be evaluated via the Jacobian matrix:

$$\mathbf{J} = \delta(x,y)/\delta(x_0,y_0) = \begin{matrix}[\delta x/\delta x_0 & \delta x/\delta y_0] \\ [\delta y/\delta x_0 & \delta y/\delta y_0]\end{matrix}$$

This represents a simplified equation for one gene. Each element (e.g., $\delta x/\delta x_0$) represents the spatiotemporal effects of a previous variable onto a current variable. The determinant of the Jacobian matrix yields the eigenvalues, each of which measures the stretching or contracting of the phase space along a given dimension or direction (e.g., the x-axis). The real parts of the eigenvalues of the Jacobian matrix are termed Lyapunov exponents, again one per dimension of stretching/contracting. The Lyapunov exponents represent the shear or contraction/expansion of the phase space (Cvitanovic et al. 2004; Glass and Mackey 1988; Ruelle 1989; Tufillaro et al. 1992). Positive Lyapunov exponents represent expansion, or instability, whereas negative Lyapunov exponents and exponents near zero represent contraction, or stability. Therefore, the Jacobian matrix and its Lyapunov exponents represent useful

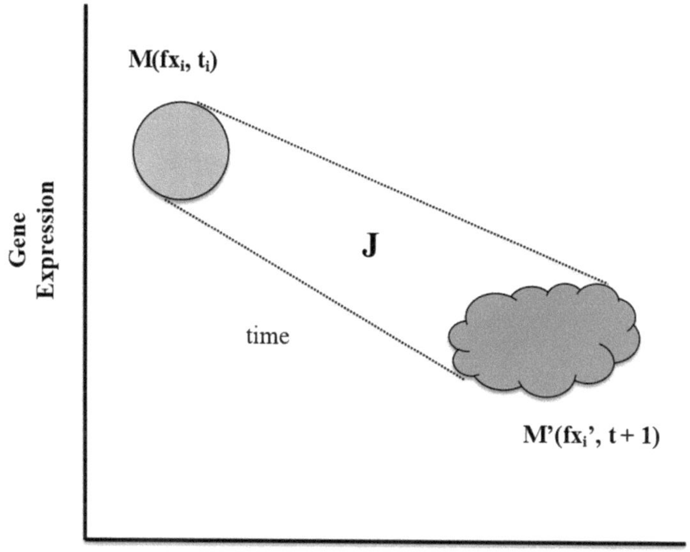

Fig. 16.2 Phase-state change in gene expression due to methylation or chromatin remodeling can be mapped for a temporally evolving genomic manifold M using the Jacobian matrix **J** Lyapunov exponents

mathematical tools for mapping the dynamical change of any information system, including the epigenetic gene regulatory networks.

To illustrate, consider a single gene under epigenetic control. If the gene is methylated to be inactive (or demethylated to be active) during "normal" development, then partial or complete reversal of this condition would represent a disturbance to the phase state of the gene, such that its messenger RNA transcription and protein translation would be upregulated (or downregulated for a previously demethylated gene), barring other posttranscriptional or post-translation epigenetic controls. Such gene activity easily is evaluated using DNA methylation profiling and microarray analysis (Eckhardt et al. 2006; He et al. 2008; Lister et al. 2009; Meissner et al. 2008). The degree change in gene expression as a function of time can be quantified using the Jacobian matrix of partial differential equation change from the resulting Lyapunov exponents.

If genes are epigenetically stable over time, there will be no change (Lyapunov exponent $\lambda = 0$). If the initial methylation or demethylation gene programming state is tightened, then the Lyapunov exponents would become negative. If their states are relaxed, the exponents become more positive. If the gene is central to a much larger network of gene activities and cell/tissue physiologies, then the sequential effects could be disruptive, and the overall Lyapunov exponents for system stability could diverge significantly more positively such that beyond $\lambda \geq +3$, the system could become chaotic (Ruelle 1989; Tufillaro et al. 1992). One famous chaotic model is

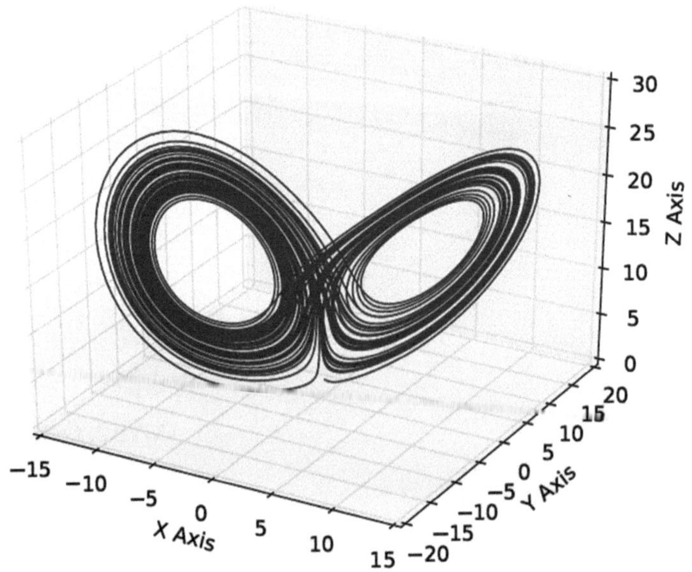

Fig. 16.3 Lorenz attractor showing two periodic cycle attractors with even disparate distributions, $F(x, y, z)$, parameters $s = 10$, $r = 15$, $b = 2.667$ computed using Enthought Canopy Python

the often cited meteorological Lorenz attractor (Fig. 16.3; Lorenz 1963), where $\lambda \geq +1$ as the system oscillates between two attractors/phases. As weather can be dynamic and chaotic between extremes of stability and instability, so can the intricate molecular networks of the cell. The redundancy of these networks likely promotes stability, but increasing epigenomic instability with aging can push the Lyapunov threshold to $\lambda \geq +3$ and chaotic dynamics (e.g., morbidity and mortality). This mathematical formalism relates directly to the Hayflick (2007b) and Davies et al. (2012) models. It further illustrates the critical importance of epigenetic stability in utero, in the soma for transgenerational inheritance, and in healthy versus unhealthy aging/physical functioning with continuous bodily exposures to stress, radiation, and environmental toxins.

16.7 Epigenetic Reprogramming, Mitochondria, and the Waddington Model

Thom (1972) mathematically modeled Waddington's (1957) epigenetic landscape to model developmental change. Thom's (1972) seven catastrophe models described variations in parameters around critical points that could maintain stability or that could collapse the system to lower energy potentials. At the cellular level, the interactions between gene products and other genes/gene products are staggering for dozens of interacting biochemical pathways. Any change in epigenetic programming can lead to a cascade of additional effects across multiple biochemical systems, although redundancies can buffer such changes to a certain extent. Whereas such systems tend to be very resilient, especially for the germline cells, they become increasingly chaotic with aging and infant metabolic disorders. The same situation holds with large blocks of genes that can be activated or deactivated by histone methylation/acetylation. For such complexity, the Jacobian matrix approach described above represents a robust measure to assess temporal changes in epigenetic programming activity.

With Hayflick's (2007a, b) and Davies et al.'s (2012) molecular instability models of aging and cancer, the potential epigenetic reprogramming effects implicate mitochondrial bioenergetics, as described above (Brunk and Terman 2002; Kurz et al. 2010; Pieczenik and Neustadt 2007; Szewczyk and Wojtczak 2002). Pieczenik and Neustadt (2007) identified numerous vitamins and nutrients that are essential for electron transport/oxidative phosphorylation on the inner mitochondrial membrane: vitamins B_1, B_3, B_5, B_6, and C and L-carnitine, ubiquinone, lipoic acid, iron, sulfur, copper, zinc, magnesium, and manganese. Nutrient deficiencies can impact pathologies across multiple body systems, including the central nervous system, heart, skeletal muscles, eyes, liver, and kidneys. Pathologies include mitochondrial diseases and disorders such as Kearns–Sayre syndrome, mitochondrial encephalomyopathy, Leber hereditary optic neuropathy, myoclonic epilepsy and ragged red fibers, neuropathies, and seizures (Pieczenik and Neustadt 2007).

Minocherhomji et al. (2012) noted that there are approximately 1500 mitochondrial genes scattered between the mitochondrial and nuclear genomes. They cited several research studies demonstrating mitochondrial methyltransferase activity and nuclear chromatin remodeling that impacts mitochondrial genes. Several breast, renal, and hematopoietic cancers are implicated, and they argued that epigenetic changes lead to molecular instabilities and mitochondrial diseases, including cancer (Minocherhomji et al. 2012).

Wallace and Fan (2010) similarly found that *S*-adenosylmethionine (SAM) and calorie-limited chromatin dephosphorylation and deacetylation are involved in environmental epigenetic reprogramming. They noted that mitochondrial diseases are similar to various epigenetic diseases such as Angelman, Rett, and fragile X syndromes as well as several cancers. They proposed a bioenergetics–epigenomic hypothesis for gene regulation. These findings directly impact research on mitochondrial diseases and metabolic conditions that are diagnosed via newborn screening/tandem mass spectrometry (Hollar 2012).

Pogribny and Vanyushin (2010) related aging to DNA hypomethylation and the one-carbon pathway that is connected by amino acid and mitochondrial bioenergetics. Their work supports the Sedivy et al. (2008) epigenetic model of aging, a follow-up to Skulachev's (2001, 2002) model. Shaughnessy et al. (2014) provided a comprehensive model linking environmental stressors to cellular biochemical pathways, nuclear and mitochondrial DNA methylation/reprogramming, calcium ion fluxes across the inner mitochondrial membrane, synthesis of lipids, pyrimidines, and heme compounds, various metabolites, and cytokines/growth factors. Furthermore, mitochondria are structurally dynamic in relation to their function under these stressors, with fusion or fission into larger or smaller units, representing a dynamic chondriome (Brunk and Terman 2002; Huang et al. 2011; Oka et al. 2012; Passos et al. 2007; Picard et al. 2013).

The complexities of cellular and mitochondrial signals impacted by epigenetic programming are further magnified by the estimated 5×10^{13} cells in the adult human body and the intricate connections between many of these cells and tissues. The human cerebral cortex contains approximately 3×10^{10} neurons and the cerebellum 1×10^{11} neurons, with approximately 5×10^{10} glial cells (Herculano-Houzel 2009). Most importantly, each neuron averages 7000 synaptic connections, resulting in a staggering 1×10^{15} connections for a single brain. Just from a neural perspective, epigenetic changes in single genes not only impact biochemical events within the cell but potentially thousands of directly or indirectly connected cells. The cascade of effects could transmit to body regions controlled by specific brain regions.

The situation is even more pronounced if the stressors or environmental toxins alter epigenetic reprogramming in utero during critical periods in brain development. For example, neural tube folding, cell division, and folding of the encephalons occur during the early weeks of gestation. Environmental exposures would affect more brain regions and body functions the earlier that the exposure occurred. This holds for mercury, lead, alcohol and other drug use, phthalates, etc. Folate intake has a profound impact on neural tube development and for other body functions, hence its fortification in many foods. Consequently, early exposures have the strongest reprogramming effects that can impact the individual for their entire life and that can be transmitted to future generations via the germline. In utero epigenetic reprogramming affects the individual for the entire life span and, via the germline, affects future generations. Neurologically, endocrine disruptors can permanently alter the hypothalamic–pituitary–adrenal (HPA) axis for physiological stress and allostatic load (Hollar 2013; Seeman et al. 2001, 2002; Yehuda and Bierer 2009). HPA axis effects impact cellular cytokines whose molecular effects include Th1/Th2 immune responses as well as mitochondrial functioning (Hollar 2009; Salminen et al. 2012; Sternberg 1997).

16.8 Conclusions

Epigenetic programming and reprogramming enable proper sequential development and differentiation of individuals from conception onward. Stress, radiation, environmental stressors (e.g., endocrine disruptor), and other agents can alter/disturb normal development by reprogramming cells and tissues into unpredictable directions. Cellular metabolic pathways are intricate and resilient against many such changes, but many agents (e.g., alcohol, lack of important nutrients, organic solvents) acting on genes at critical developmental periods create systemic instabilities that affect the individual via morbidity and increased risk for mortality across the life span or that can be transmitted via the germline to future generations. The most susceptible critical periods are gestation for the unborn child, especially the first trimester when most organs form and brain development is established, the germline/soma separation, and prepuberty, the latter two of which promote transgenerational epigenetic inheritance.

Instability reprograms the epigenome, cumulatively so in aging (Hayflick 2007a, b). Waddington's (1957) model and Thom's (1972) mathematical treatment illustrate how environmentally induced "change of pathways" drives cells to altered functioning and/or collapse. Epigenetic changes can be further mathematically described with specific measures, including Lyapunov exponents (Ruelle 1989) from Jacobian matrices, to inform research into epigenomic stability. From these approaches, researchers can examine possibilities for carefully reverse programming the epigenome (Cakir et al. 2009; Morgan et al. 2005) to improve health and functioning, treat disease, and modulate aging processes.

Abbreviations

ATP	Adenosine triphosphate
ROS	Reactive oxygen species
DNA	Deoxyribonucleic acid
SAM	*S*-adenosylmethionine
HPA	Hypothalamic–pituitary–adrenal axis

References

Abe, M., Tsai, S. Y., Jin, S. G., Pfeifer, G. P., & Szabo, P. E. (2011). Sex-specific dynamics of global chromatin changes in fetal mouse germ cells. *PLoS One, 6*, e23848. http://dx.doi.org/10.1371/journal.pone.0023848.

Binstock, R. H. (2003). The war on "anti-aging medicine". *The Gerontologist, 43*(1), 4–14.

Bollati, V., & Baccarelli, A. (2010). Environmental epigenetics. *Heredity, 105*, 105–112.

Bonnin, A., Goeden, N., Chen, K., Wilson, M. L., King, J., Shih, J. C., Blakely, R. D., Deneris, E. S., & Levitt, P. (2011). A transient placental source of serotonin for the fetal forebrain. *Nature, 472*, 347–350.

Brunk, U. T., & Terman, A. (2002). The mitochondrial-lysosomal axis theory of aging: Accumulation of damaged mitochondria as a result of imperfect autophagocytosis. *European Journal of Biochemistry, 269*, 1996–2002.

Cakir, T., Hendriks, M. M. W. B., Westerhuis, J. A., & Smilde, A. K. (2009). Metabolic network discovery through reverse engineering of metabolome data. *Metabolomics, 5*, 318–329.

Cvitanovic, P., Artuso, R., Dahlqvist, P., Mainieri, R., Tanner, G., Vattay, G., Whelan, N., & Wirzba, A. (2004). *Chaos: classical and quantum,* version 14.4.1 (April 21, 2013). Accessed February 1, 2015 at ChaosBook.org.

Davies, P., Demetrius, L. A., & Tuszynski, J. A. (2012). Implications of quantum metabolism and natural selection for the origin of cancer cells and tumor progression. *AIP Advances, 2*, 011101. http://dx.doi.org/10.1063/1.3697850.

Debye, P. (1912). Zur Theorie der spezifischen Wärmen. *Annalen der Physik, 344*(14), 789–839.

Dixon, J. R., Jung, I., Selvaraj, S., Shen, Y., Antosiewicz-Bourget, J. E., Lee, A. Y., Ye, Z., Kim, A., Rajagopal, N., Xie, W., Diao, Y., Liang, J., Zhao, H., Lobanenkov, V. V., Ecker, J. R., Thomson, J. A., & Ren, B. (2015). Chromatin architecture reorganization during stem cell differentiation. *Nature, 518*, 331–336.

Eckhardt, F., Lewin, J., Cortese, R., Rakyan, V. K., Attwood, J., Burger, M., Burton, J., Cox, T. V., Davies, R., Down, T. A., Haefliger, C., Horton, R., Howe, K., Jackson, D. K., Kunde, J., Koenig, C., Liddle, J., Niblett, D., Otto, T., Pettett, R., Seemann, S., Thompson, C., West, T., Rogers, J., Olek, A., Berlin, K., & Beck, S. (2006). DNA methylation profiling of human chromosomes 6, 20, and 22. *Nature Genetics, 38*(12), 1378–1385.

Einstein, A. (1907). Die Plancksche Theorie der Strahlung und die Theorie der spezifischen Wärme. *Annalen der Physik, 327*(1), 180–190.

Freedman, L. P., Cockburn, I. M., & Simcoe, T. S. (2015). The economics of reproducibility in preclinical research. *PLoS Biology, 13*(6), e1002165. http://dx.doi.org/10.1371/journal.pbio.1002165.

Georgieva, M., Staneva, D., & Miloshev, G. (2015). Epigenetic significance of chromatin organization during cellular ageing and organismal lifespan. In D. Hollor & D. Vasudevan (Eds.), *Epigenetics, the environment and children's health across lifespans*. New York: Springer.

Glass, L., & Mackey, M. C. (1988). *From clocks to chaos: The rhythms of life*. Princeton, NJ: Princeton University Press.

Gray, M. W. (1989). Origin and evolution of mitochondrial DNA. *Annual Review of Cell Biology, 5*, 25–50.

Hayflick, L. (1994). *How and why we age*. New York: Ballantine.

Hayflick, L. (2007a). Biological aging is no longer an unsolved problem. *Annals of the New York Academy of Sciences, 1100*, 1–13.

Hayflick, L. (2007b). Entropy explains aging, genetic determinism explains longevity, and undefined terminology explains misunderstanding both. *PLoS Genetics, 3*(12), 2351–2354.

He, X., Chang, S., Zhang, J., Zhao, Q., Xiang, H., Kusonmano, K., Yang, L., Sun, Z. S., Yang, H., & Wang, J. (2008). MethyCancer: The database of human DNA methylation and cancer. *Nucleic Acids Research, 36*, D836–D841.

Heijmans, B. T., Tobi, E. W., Stein, A. D., Putter, H., Blauw, G. J., Susser, E. S., Slagboom, P. E., & Luney, L. H. (2008). Persistent epigenetic differences associated with prenatal exposures to famine in humans. *Proceedings of the National Academy of Sciences of the United States of America, 105*(44), 17046–17049.

Herculano-Houzel, S. (2009). The human brain in numbers: A linearly scaled-up primate brain. *Frontiers in Human Neuroscience, 3*, 31. http://dx.doi.org/10.3389/neuro.09.031.2009.

Hollar, D. W. (2009). Risk for intentional violent death associated with HLA genotypes: A preliminary survey of deceased American organ donors. *Genetica, 137*(3), 253–264.

Hollar, D. (2012). Genetic and metabolic conditions for children with special health care needs. In D. Hollar (Ed.), *Handbook of children with special health care needs* (pp. 271–288). New York: Springer.

Hollar, D. (2013). Cross-sectional patterns of allostatic load among persons with varying disabilities, NHANES: 2001-2010. *Disability and Health Journal, 6*, 177–187.

Huang, P., Galloway, C. A., & Yoon, Y. (2011). Control of mitochondrial morphology through differential interactions of mitochondrial fusion and fission proteins. *PLoS One, 6*(5), e20655. http://dx.doi.org/10.1371/journal.pone.0020655.

Ioannidis, J. P. A. (2005). Why most published research findings are false. *PLoS Medicine, 2*(8), e124. http://dx.doi.org/10.1371/journal.pmed.0020124.

Kirkwood, T. B., & Rose, M. R. (1991). Evolution of senescence: Later survival sacrificed for reproduction. *Philosophical Transactions of the Royal Society of London, Series B: Biological Sciences, 332*, 15–24.

Kurz, F. T., Aon, M. A., O'Rourke, B., & Armoundas, A. A. (2010). Spatio-temporal oscillations of individual mitochondria in cardiac myocytes reveal modulation of synchronized mitochondrial clusters. *Proceedings of the National Academy of Sciences of the United States of America, 107*(32), 14315–14320.

Lee, Y. L., Xu, X., Wallenstein, S., & Chen, J. (2009). Gene expression profiles of the one-carbon metabolism pathway. *Journal of Genetics and Genomics, 36*(5), 277–282.

Levy-Leblond, J.-M., & Balibar, F. (1990). *Quantics: Rudiments of quantum physics*. New York: North-Holland.

Lister, R., Pelizzola, M., Dowen, R. H., Hawkins, R. D., Hon, G., Tonti-Filippini, J., Nery, J. R., Lee, L., Ye, Z., Ngo, Q.-M., Edsall, L., Antosiewicz-Bourget, J., Stewart, R., Ruotti, V., Millar, A. H., Thomson, J. A., Ren, B., & Ecker, J. R. (2009). Human DNA methylomes at base resolution show widespread epigenomic differences. *Nature, 462*, 315–322.

Lorenz, E. N. (1963). Deterministic nonperiodic flow. *Journal of the Atmospheric Sciences, 20*, 130–141.

Loukinov, D. I., Pugacheva, E., Vatolin, S., Pack, S. D., Moon, H., Chernukhin, I., Mannan, P., Larsson, E., Kanduri, C., Vostrov, A. A., Cui, H., Niemitz, E. L., Rasko, J. E. J., Docquier, F. M., Kistler, M., Breen, J. J., Zhuang, Z., Quitschke, W. W., Renkawitz, R., Klenova, E. M., Feinberg, A. P., Ohlsson, R., Morse, H. C., III, & Lobanenkov, V. V. (2002). BORIS, a novel germ-line-specific protein associated with epigenetic reprogramming events, shares the same 11-zinc-finger domain with CTCF, the insulator protein involved in reading imprinting marks in the soma. *Proceedings of the National Academy of Sciences of the United States of America, 99*(10), 6806–6811.

Manikkam, M., Guerrero-Bosagna, C., Tracey, R., Haque, M. M., & Skinner, M. K. (2012). Transgenerational actions of environmental compounds on reproductive disease and identification of epigenetic biomarkers of ancestral exposures. *PLoS One, 7*(2), e31901. http://dx.doi.org/10.1371/journal.pone.0031901.

Maruyama, R., Choudhury, S., Kowalczyk, A., Bessarabova, M., Beresford-Smith, B., Conway, T., Kaspi, A., Wu, Z., Nikolskaya, T., Merino, V. F., Lo, P.-K., Liu, X. S., Nikolsky, Y., Sukumar, S., Haviv, I., & Polyak, K. (2011). Epigenetic regulation of cell type-specific expression patterns in the human mammary epithelium. *PLoS Genetics, 7*(4), e1001369. http://dx.doi.org/10.1371/journal.pgen.1001369.

Meissner, A., Mikkelsen, T. S., Gu, H., Wernig, M., Hanna, J., Sivachenko, A., Zhang, X., Bernstein, B. E., Nusbaum, C., Jaffe, D. B., Gnirke, A., Jaenisch, R., & Lander, E. S. (2008). Genome-scale DNA methylation maps of pluripotent and differentiated cells. *Nature, 454*, 766–770.

Minocherhomji, S., Tollefsbol, T. O., & Singh, K. K. (2012). Mitochondrial regulation of epigenetics and its role in human diseases. *Epigenetics, 7*(4), 326–334.

Morgan, H. D., Santos, F., Green, K., Dean, W., & Reik, W. (2005). Epigenetic reprogramming in mammals. *Human Molecular Genetics, 14*, R47–R58.

Nicholls, D. G., & Budd, S. L. (2000). Mitochondrial and neuronal survival. *Physiological Reviews, 80*(1), 315–360.

Oka, T., Hikoso, S., Yamaguchi, O., Taneike, M., Takeda, T., Tamai, T., Oyabu, J., Murakawa, T., Nakayama, H., Nishida, K., Akira, S., Yamamoto, A., Komuro, I., & Otsu, K. (2012). Mitochondrial DNA that escapes from autophagy causes inflammation and heart failure. *Nature, 485*, 251–255.

Olshansky, S. J., Hayflick, L., & Carnes, B. A. (2002). Position statement on human aging. *Journal of Gerontology, Biological Sciences, 57A*(8), B292–B297.

Parle-McDermott, A., & Ozaki, M. (2011). The impact of nutrition on differential methylated regions of the genome. *Advances in Nutrition, 2*, 463–471.

Passos, J. F., Saretzki, G., Ahmed, S., Nelson, G., Richter, T., Peters, H., Wappler, I., Birket, M. J., Harold, G., Schaeuble, K., Birch-Machin, M. A., Kirkwood, T. B. L., & von Zglinicki, T. (2007). Mitochondrial dysfunction accounts for the stochastic heterogeneity in telomere-dependent senescence. *PLoS Biology, 5*(5), e110. http://dx.doi.org/10.1371/journal.pbio.0050110.

Pembrey, M. E., Bygren, L. O., Kaati, G., Edvinsson, S., Northstone, K., Sjostrom, M., Golding, J., & the ALSPAC Study Team. (2006). Sex-specific, male-line transgenerational responses in humans. *European Journal of Human Genetics, 14*, 159–166.

Picard, M., Shirihai, O. S., Gentil, B. J., & Burelle, Y. (2013). Mitochondrial morphology transitions and functions: Implications for retrograde signaling? *American Journal of Physiology. Regulatory, Integrative and Comparative Physiology, 304*, R393–R406.

Pieczenik, S. R., & Neustadt, J. (2007). Mitochondrial dysfunction and molecular pathways of disease. *Experimental and Molecular Pathology, 83*, 84–92.

Pogribny, I. P., & Vanyushin, B. F. (2010). Age-related genomic hypomethylation. In T. O. Tollefsbol (Ed.), *Epigenetics of aging* (pp. 11–28). New York: Springer.

Prigogine, I. (2002). Dynamical roots of time symmetry breaking. *Philosophical Transactions of the Royal Society of London A, 360*, 299–301.

Rothman, K. J., & Greenland, S. (1998). *Modern epidemiology* (2nd ed.). Philadelphia: Lippincott-Raven.

Ruelle, D. (1989). *Chaotic evolution and strange attractors*. New York: Cambridge University Press.

Salminen, A., Kaarniranta, K., & Kauppinen, A. (2012). Inflammaging: Disturbed interplay between autophagy and inflammasomes. *Aging, 4*(3), 166–175.

Sedivy, J. M., Banumathy, G., & Adams, P. D. (2008). Aging by epigenetics – A consequence of chromatin damage? *Experimental Cell Research, 314*(9), 1909–1917.

Seeman, T. E., McEwen, B. S., Rowe, J. W., & Singer, B. H. (2001). Allostatic load as a marker of cumulative biological risk: MacArthur studies of successful aging. *Proceedings of the National Academy of Sciences of the United States of America, 98*(8), 4770–4775.

Seeman, T. E., Singer, B. H., Ryff, C. D., Love, G. D., & Levy-Storms, L. (2002). Social relationships, gender, and allostatic load across two age cohorts. *Psychosomatic Medicine, 64*, 395–406.

Shaughnessy, D. T., McAllister, K., Worth, L., Haugen, A. C., Meyer, J. N., Domann, F. E., Van Houten, B., Mostoslavsky, R., Bultman, S. J., Baccarelli, A. A., Begley, T. J., Sobol, R. W., Hirschey, M. D., Ideker, T., Santos, J. H., Copeland, W. C., Tice, R. R., Balshaw, D. M., & Tyson, F. L. (2014). Mitochondria, energetics, epigenetics, and cellular responses to stress. *Environmental Health Perspectives, 122*, 1271–1278.

Singh, P., Li, A. X., Tran, D. A., Oates, N., Kang, E.-R., Wu, X., & Szabo, P. E. (2013). De novo DNA methylation in the male germ line occurs by default but is excluded at sites of H3K4 methylation. *Cell Reports, 4*(1), 205–219.

Skinner, M. K., Manikkam, M., & Guerrero-Bosagna, C. (2011). Epigenetic transgenerational actions of endocrine disruptors. *Reproductive Toxicology, 31*(3), 337–343.

Skulachev, V. P. (2001). The programmed death phenomena, aging and the Samurai law of biology. *Experimental Gerontology, 36*(7), 995–1024.

Skulachev, V. P. (2002). Programmed death phenomena: from organelle to organism. *Annals of the New York Academy of Sciences, 959*, 214–237.

Sternberg, E. M. (1997). Neural-immune interactions in health and disease. *The Journal of Clinical Investigation, 100*(11), 2641–2647.

Szewczyk, A., & Wojtczak, L. (2002). Mitochondria as a pharmacological target. *Pharmacological Reviews, 54*(1), 101–127.

Tabatabaie, V., Atzmon, G., Rajpathak, S. N., Freeman, R., Barzilai, N., & Crandall, J. (2011). Exceptional longevity is associated with decreased reproduction. *Aging, 3*(12), 1202–1205.

Thom, R. (1972). *Structural stability and morphogenesis*. New York: W.A. Benjamin.

Tufillaro, N. B., Abbott, T., & Reilly, J. (1992). *An experimental approach to nonlinear dynamics and chaos*. New York: Addison-Wesley.

Veltri, K. L., Espiritu, M., & Singh, G. (1990). Distinct genomic copy number in mitochondria of different mammalian organs. *Journal of Cell Physiology, 143*(1), 160–164.

Waddington, C. H. (1957). *The strategy of the genes: A discussion of some aspects of theoretical biology*. London: Allen & Unwin.

Wallace, D. C., & Fan, W. (2010). Energetics, epigenetics, mitochondrial genetics. *Mitochondrion, 10*(1), 12–31.

Waterland, R. A., Kellermayer, R., Rached, M.-T., Taterian, N., Gomes, M. V., Zhang, J., Zhang, L., Chakravarty, A., Zhu, W., Laritsky, E., Zhang, W., Wang, X., & Shen, L. (2009). Epigenomic profiling indicates a role for DNA methylation in early postnatal liver development. *Human Molecular Genetics, 18*(16), 3026–3028.

Wilkins, A. S. (2010). The enemy within: An epigenetic role of retrotransposons in cancer initiation. *Bioessays, 32*, 856–865.

Yehuda, R., & Bierer, L. (2009). The relevance of epigenetics to PTSD: Implications for the DSM-IV. *Journal of Trauma and Stress, 22*(5), 427–434.

Index

D. Hollar (ed.), *Epigenetics, the Environment, and Children's Health Across Lifespans*, DOI 10.1007/978-3-319-25325-1

C

D

E

MIX
Papier aus verantwortungsvollen Quellen
Paper from responsible sources
FSC® C105338
FSC
www.fsc.org

If you have any concerns about our products,
you can contact us on
ProductSafety@springernature.com

In case Publisher is established outside the EU,
the EU authorized representative is:
Springer Nature Customer Service Center GmbH
Europaplatz 3, 69115 Heidelberg, Germany

Printed by Libri Plureos GmbH
in Hamburg, Germany